D0114622

Nuclear Fear

SPENCER R. WEART

Nuclear Fear

A HISTORY

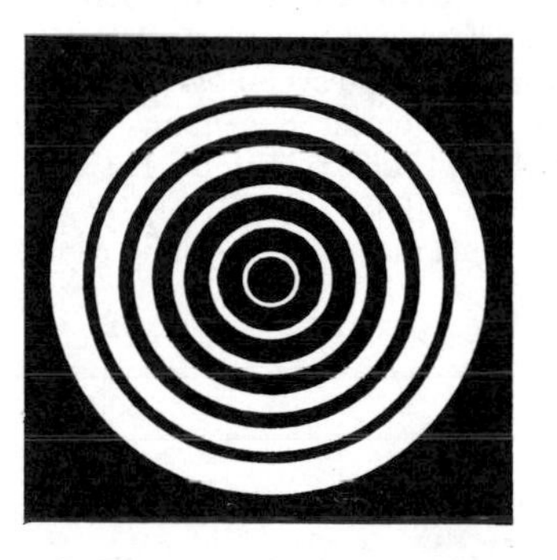

OF IMAGES

HARVARD UNIVERSITY PRESS
CAMBRIDGE, MASSACHUSETTS
LONDON, ENGLAND

Printed in the United States of America
10 9 8 7 6 5 4 3 2

This book is printed on acid-free paper, and its binding
materials have been chosen for strength and durability.

Library of Congress Cataloging-in-Publication Data

Weart, Spencer R., 1942–
Nuclear fear: A history of images

Bibliography: p.
Includes index.
1. Nuclear energy—History.
2. Nuclear energy—Psychological aspects—History.
3. Nuclear warfare—Psychological aspects—History.
4. Antinuclear movement—History.
5. Radiation—Public opinion—History. I. Title.
QC773.W43 1988 327.1'74 87-25995
ISBN 0-674-62835-7 (alk. paper)

Designed by Gwen Frankfeldt

For my parents

Spencer A. Weart

Janet Streng Weart

Contents

Preface

Fifteen years ago when I began studying the history of nuclear energy I did not think that images were important in themselves. I was wrong. Radioactive monsters, utopian atom-powered cities, exploding planets, weird ray devices, and many other images have crept into the way everyone thinks about nuclear energy, whether that energy is used in weapons or in civilian reactors. The images, by connecting up with major social and psychological forces, have exerted a strange and powerful pressure within history.

Early in my research, as I riffled through yellowing magazines in a library or watched a fifty-year-old movie on television long past midnight, I would snort with laughter at the incredible fantasies. Later I would toss in my bed with nightmares, for the images were at work in my own head. Hence a warning. This is no story of things locked away safely in the past: the images are more powerful today than ever.

The study of images is a relatively new field of history, built up chiefly by French scholars working on the "mentalities," the widely expressed and static beliefs, of times long past. I have gone beyond that to study also inarticulate pictures and even emotional patterns, all changing rapidly in our times. Further, I have looked into how these things have been spread in specific historical circumstances—sometimes by deliberate decision—and have then worked upon specific minds to influence subsequent crucial decisions. Nobody before has written a history of images in this broad a sense, and I have had to work out my own methodology.

Preface

How can an investigator connect the colorful images in people's heads with the attitudes that affect historical events? For an answer I rely on commonsense ideas that have been validated by rigorous experiment. Minds form associations between separate things; the sound of a can opener brings my cat to the kitchen and may even make him salivate. This mechanism is fundamental to all thought, including raw perception. Experiments show that we must have a rough idea of what to look for, some previously learned mental picture, or we even have trouble recognizing an object in a drawing. In short, as a result of experience every simple image, from direct perception to elaborate mental representation, becomes connected with various other things in a web of associations.

Thinking is swayed not only by a person's direct experience with objects but also by associations that are taught, connections developed over the course of history. Certain things become associated at second hand with all sorts of other things, gathering into one or another cluster: a public image in a general sense. Such a cluster may include various kinds of simple images from mental pictures to complex stereotypes. It may also include carefully thought out ideas, unconsidered beliefs, inarticulate feelings, and visceral emotions. All these are internal experiences that people may project back onto the external world by associating the experiences with particular things. The result of all this is the attitudes that finally determine action.

A few clusters develop into tightly structured symbolic representations standing for crucial matters. For example, with two zigzag strokes of a pencil I can draw a design that most people will call a swastika. To a Hindu gentleman in 1925 the design would recall symbols that he had seen carved on white temples, evoking feelings of religious devotion and ideas about withdrawal from the world. To a Nazi youth in 1935 the associated material would include flag parades and national pride. To a Jew in 1945 the same simple design would evoke nightmares of mass death. As this example shows, such associations may come through age-old tradition (as with the Hindu), through conscious manipulation by propagandists (as with the Nazi), through historical events (as with the Jew), or from all of these together.

Thus the historian of images can take up a straightforward task: to look through materials of every description and find for a given group of people what pictures, symbols, beliefs, rational concepts, feelings, and emotions have become strongly associated with one another in a cluster that includes a particular subject, such as nuclear energy. The historian should also investigate how the associations were forged—deliberately or by chance.

An especially revealing source of information on the existence of associations is the millions of hours of experimental and clinical evidence accumulated by psychologists and psychiatrists. Possibly they were only observing matters peculiar to the culture of their patients, educated members of Western civilization since about 1900, but these groups happen to be the main actors in my history. The patterns of their imagery and emotions can be studied, not as "psychohistory" interpreted through some disputed theory, but simply as events in history. Beyond this, biographical study can seek out similar patterns within the minds of specific individuals who were important in purveying images.

Further help comes from the disciplines of anthropology and sociology, which may turn up images arranged in structures that resemble the patterns found in psychology and biography. These disciplines, with their field observations and polls, can also suggest how a particular social system would sustain a particular image structure. But the main sources of social information are still the historian's, the records of the past, and particularly the written and visual images seen by many people. For experiments have proved what anyone would suspect: that such images do have lasting impact on the beliefs and emotions of their audience.

This book will mention only in passing the historical and social forces concerned with nuclear energy. These forces are important, but they have been thoroughly analyzed in many other books. I will focus more on personality and imagery because they are less well understood. Indeed, nuclear imagery stands before us as a screen that prevents us from truly understanding how the other forces work.

The main premise of this book is also its main conclusion: that the images we cherish have a greater role in history than has commonly been thought. I do not mean that our minds are helpless before inexorable pressures, for images do not just happen among people; they are built. In pointing out the tremendous dangers we face from the power of imagery, whether this power is used consciously by propagandists or whether we impose it unwittingly upon one another, I am not preaching pessimism. On the contrary, it is only when we see clearly all the strengths of a danger that we can feel capable of opposing it.

At this point an author is supposed to thank all who have helped him. But this work has dug so deep into me that my thanks should embrace people I have met over my entire lifetime, not least the physicists and the professional and lay psychologists who instructed me long before I planned this book. I also should thank people I have never met at all, such as the graduate students whose little-known doctoral dissertations have been a special help. So I will leave my acknowledgments to the specific ones in the notes. I must say a word of praise, however, for the resources of New York City—the Niels Bohr Library of the Center for History of Physics at the American Institute of Physics, where much of this work was done; the New York Public Library, and especially its theater collection at Lincoln Center; the Museum of Broadcasting; and the many other sources, from collections of old newsreels to editorial offices of science fiction magazines, that exist nowhere else. Scarcely less helpful were the public institutions and agencies of Washington, D.C., including the National Archives, the Department of Energy, the Nuclear Regulatory Commission, the Library of Congress, and particularly the National Science Foundation, whose Program in History and Philosophy of Science aided me with a grant at an early stage. For reading and advising me on various drafts I owe particular thanks to Lawrence Badash, Paul Boyer, Mary Ellen Geer, Jack Holl, George Mazuzan, Pat McCullough, Dorothy Nelkin, Angela von der Lippe, J. Samuel Walker, and Joan Warnow.

Hastings-on-Hudson, New York
August 1987

Nuclear Fear

part one

YEARS OF FANTASY

1902–1938

1

Radioactive Hopes

Once there was a man who sought after hidden knowledge. The story says that he hoped to make human civilization more noble, and if there was an ugly, mad streak in him, as in all of us, he controlled it strictly. This man arduously studied not only modern science but also alchemy, and it was after pondering the arcane philosophers' stone that he discovered the most prodigious secret of physics: the release of vast energy from within atoms. He knew at once that this energy would change the world. He feared vast explosions, but at the same time he hoped that atomic energy would save civilization, which he believed was otherwise destined to collapse when its fossil fuels ran out. A vision came to him of white towers rising from gardens, a peaceful and prosperous future city centered upon gleaming atomic power plants.

Up to here the story is historical fact, but the rest becomes increasingly like a dream. The man built a shining cylindrical device that could project atomic rays. He was pleased but not surprised to find that, among various remarkable effects, the radiation could cure cancer and other ills. However, in his experiments the rays sometimes did not cure people but gave them cancer, or horribly deformed their flesh, or changed their very genes so that their children were monsters.

The destructive power of atomic rays might be useful, the man thought, for his nation was under deadly threat from enemies. If he could make an all-powerful weapon, surely nobody would dare to start a war. He went to a laboratory hidden down a shaft deep in the earth, and there he used his rays to construct a weird creature, a sort of living robot armed with irresistible energies.

In this story there was also a woman who might have been the scientist's lover. He had found little time to court her when all his efforts were going after knowledge and power, but she nevertheless visited his workroom. Just then he had been thinking of a ray that might possibly render living creatures immortal. As the woman approached he aimed a ray device toward her and proposed an experiment; she fled in horror. Rage exploded in the man's overtaxed brain, and he screamed that everyone had abandoned him, leaving him alone in the world. Climbing into a recess in his robotic creature, he rode it to the surface of the earth. But when he emerged his enemies were frightened and attacked him, which automatically activated his weapons. Enormous clouds mushroomed into the sky; radioactive poisons swept the planet. In the ashen landscape lay the robot, blackened and deformed.

From the underground room where she had taken shelter, the woman emerged. When she tried to lift the ruined creature it cracked apart like a shell and the man crawled out, his madness purged away. The pair joined hands. A new world would rise on the ashes of the old, a purified and wiser race, perhaps with a white city after all . . .

There are some curious things about this legendary tale, which I have constructed as a composite of numberless stories familiar to every citizen of the twentieth century. A close look will show that such tales included divergent and even contradictory ideas. Yet in some odd fashion the ideas fitted neatly together. Still more remarkable, the images were plausible. Atom-powered city, potent ray, strange creature, blasted plain—each could happen. Images so plausible, and also so impressive, might have been expected to exert some kind of influence on the people who made the political, economic, and military decisions that determined the history of nuclear energy.

The most curious and unsettling thing is that every theme in such tales was already at hand early in the twentieth century, decades before the discovery of nuclear fission showed how to actually release the energy within atoms. The imagery, then, did not come from experience with real bombs and power plants. It came from somewhere else.

Legends conceal grave truths, but not truths about nuclear physics. Such tales are really about more important matters: the forces of human history, social structure, and psychology. In this book I will explore these matters as seen in the interaction of imagery with the history of nuclear energy. In the first four chapters I will deal with the years prior to the discovery of nuclear fission, focusing in turn on the white city, the destroyed planet, the transforming ray, and the monstrous creature—images that have more to do with the history of our times, and indeed more to do with one another, than appears on the surface.

WHITE CITIES OF THE FUTURE

There really was a man who studied both science and alchemy, found the secret of atomic energy, and exclaimed that it would lead humanity to paradise or doomsday. His name was Frederick Soddy. As a youth at the turn of the century he had been bright, ambitious, quarrelsome, and lonely; photographs show his features set firmly and scornfully, like a gentleman boxer about to enter the ring. His chance to become famous came while he was teaching chemistry in Montreal and fell in with Ernest Rutherford. At thirty Rutherford was a hearty, well-liked, and well-established professor, all the things that Soddy was not. But the two men shared a gift for science, and they also shared an ambition: to crack the puzzle of radioactivity.

Radioactivity, discovered in 1896, had at first attracted scant attention. It seemed only a curiosity that a few minerals such as thorium and uranium emitted feeble rays resembling a sort of invisible light. Then Marie Curie discovered the new metal radium, whose rays, compared with the whisper from uranium, were like a piercing shout. When the cream of the world's physicists gathered in Paris for a congress in 1900, Marie and her husband, Pierre, proudly displayed little vials of radium compounds so active that they glowed with a pearly light. The newspapers began to pay attention to radioactivity, and so did Soddy and Rutherford.

Late in 1901 the pair discovered that radioactivity is a sign of fundamental changes within matter. A pulse of radiation signals that an atom is changing into a different kind of atom, a different element with its own chemical properties. Rutherford and Soddy found, for example, that radioactive thorium, atom by atom, was gradually turning itself into radium. At the moment he realized this, Soddy recalled, "I was overwhelmed with something greater than joy—I cannot very well express it—a kind of exaltation." He blurted out, "Rutherford, this is transmutation!"

"For Mike's sake, Soddy," his companion shot back, "don't call it *transmutation*. They'll have our heads off as alchemists." But the next moment Rutherford was waltzing around the laboratory, booming out "Onward Christian Soldiers." Already at the instant the new science was born, it could stir strong emotions.[1]

What did it mean, this word *transmutation*, which elated Soddy and gave Rutherford pause? To most people the word was associated only with gold-making charlatans and crackpots, relics of the Middle Ages, still active here and there at the turn of the century. But in fact the concept of transmutation had once been the central strand of a far-reaching and ancient web of thought. It was a clue that could help to explain almost every strange image that would later appear in nuclear energy tales.[2]

At first Soddy and Rutherford could not quite say why transmutation seemed to be special, but during 1902 they found one of the reasons: energy. Any radioactive substance was, of course, radiating energy. But now the pair, and meanwhile Pierre Curie in Paris, showed that radioactivity released vastly more energy, atom for atom, than any other process known. Scientists had been speculating for decades that the universe might contain a reserve of unseen power, but few had imagined that there was so much of it.

Soddy explained the discovery to the public promptly in May 1903 in a British magazine read by cultivated ladies and gentlemen. Radioactivity, he said, pointed to "inexhaustible" power; henceforth matter must be considered not just as inert stuff but as a storehouse of energy. A year later, while taking a long journey by steamship to Australia where he was to lecture on radium, he made a more specific calculation, the kind of tangible example that an audience would not forget: a pint bottle of uranium contained enough energy to drive an ocean liner from London to Sydney and back!

Soddy had more in mind than such mundane tricks. Writing in an American magazine, he said that if we could manage to tap the energy within atoms, "the future would bear as little relation to the past as the life of a dragonfly does to that of its aquatic prototype." He summed up his ideas in 1908 in a widely read book, *The Interpretation of Radium*. "A race which could transmute matter would have little need to earn its bread by the sweat of its brow," he declared. "Such a race could transform a desert continent, thaw the frozen poles, and make the whole world one smiling Garden of Eden."[3]

Eden restored, the dragonfly springing from its larva—where did Soddy find these extraordinary images? The few facts then known about radioactivity gave no support for such language. Rather, these were images that the solitary chemist had already cherished before nuclear energy was discovered.

It is not enough to look at the facts of physics in order to understand the fateful enthusiasm for nuclear energy that would sweep the world during the first half of the twentieth century. It is equally important to look at certain peculiar images, which in Soddy's day lay scattered about in odd corners of Western culture. While taking this look I will refer to related social factors and historical events, for certainly they were as important as imagery. But they are relatively well understood, and I will chiefly address what has been less often discussed. To see how the world could be altered by fantastic hopes for nuclear energy (and fantastic fears as well), it is necessary to dissect the associated images. First I will consider the shining future city, the goal toward which my legendary scientist, and Soddy, labored.

At the start of the twentieth century many felt that science would lead humanity to an abundance not only of material goods but of brotherhood and wisdom. From Moscow to Mexico City, progressive elites believed that even politics would soon be made "scientific." The greatest enthusiasts were scientists themselves, such as the eminent French chemist Marcellin Berthelot, who fascinated people with his books and with his conversation at fashionable Parisian soirées. By the year 2000, he declared, the earth would be a garden where a kinder and happier humanity would live amid the abundance of a Golden Age. He explained that it was the discoveries of scientists like himself that would bring this to pass, for example by providing a limitless source of energy. Many of Soddy's declarations could have been taken straight from Berthelot's discourses.[4]

In 1893 a perfected city had actually been constructed, if only for a summer. The fairgrounds of the Chicago International Exposition, dubbed the "White City," were a fairyland of broad avenues and sparkling fountains, incandescent at night under the new electric lamps, with steel dynamos gleaming alongside alabaster sculptures of virgins, a picture of the future harmony between technology and art. Such a creation lent to utopian vision a conviction of reality, which would be inherited by generations not yet born when Chicago's White City was dismantled.[5]

Even brighter evidence for enthusiasm in the scientific future was discovered in the planet Mars. A few astronomers, peering for long hours through their telescopes, convinced themselves that they saw spidery lines across the blurred disc. Camille Flammarion, a French astronomy popularizer, said in 1892 that the lines were probably canals, enormous engineering projects of an elder race. Percival Lowell of Boston explained that Mars must represent a later stage of evolution than the Earth, a planet grown dry over the ages. Thus the future of our planet was written upon our neighbor. The colossal scale of the Martian aqueducts showed how far technical prowess could take a civilization, and the fact that the canals spanned the entire planet proved that the Martians had outgrown war. Although most professional astronomers looked askance at such ideas, sensational stories on Mars crowded the press for over a decade. The public was half convinced that the Martian engineers existed, and almost entirely convinced that our own planet's future would fulfill such technological visions.[6]

Of course Flammarion and Lowell had simply projected onto the ambiguous, blurred disk an image of civilization that existed only in their heads. I say this to emphasize a mechanism that is central to any history of images: associations already in the mind can creep into the picture that people think they perceive. In this case a history of the associations might note that water systems were a main factor in the growth of ancient civilizations, and would go on to study how the link remained famous into the nineteenth century with its land reclamation projects and far-flung canal transportation networks. Such widely shared ideas could help explain Lowell's beliefs and even his perceptions at the telescope. Much the same kind of study can be made for nuclear energy.

By the start of the twentieth century the image of a White City, expanded to planetary scale and projected onto the blurred screen of the future, was at a peak of popularity; it was well positioned to become the first symbol associated with the energy of atoms. The connection came through another widely held idea: that modern civilization was founded, no longer on canals, but on energy. For coal and electricity were visibly transforming nations. It was a change more rapid than has happened to any generation before or since. As a child Soddy had seen rooms lit with smoking oil lamps and people traveling behind plodding horses, scarcely different from Roman times; by the time he grew up, electric trams were humming down the avenues of brilliantly illuminated cities.

That seemed only a beginning, for much remained to be done. Industry was carried on the stooped backs of coal miners, brutes working in the dark until their health broke. Coal smoke choked the cities, yet millions preferred the soot and tuberculosis of industrial slums to traditional rural life, which was still ruled by superstitious ignorance, exhausting labor, and malnutrition (farm laborers had an even higher mortality rate than the miners). Progressive thinkers had good reason to say that civilization could lift itself far higher—provided it continued to replace human muscle with inanimate supplies of energy.

The energy of atoms might be crucial here, for some doubted whether coal would suffice. Back in 1865 the problem had been popularized by an economist, W. S. Jevons. He pointed out to the British that if they continued doubling their use of energy every decade or so, then eventually they would have to dig so deep for the remaining coal that it would be too costly to sustain their industrial leadership. Renowned scientists repeated the warning, stirring governments to make geological surveys. The results were reassuring, for geologists said Britain's supply of cheap coal would last at least until the mid-twentieth century (which was quite accurate). The mid-twentieth century was generations away, too far in the future to worry about.[7]

Scientists promised that before the coal gave out they would find other sources of energy. There was energy in rivers, in the tides and the wind, in the internal heat of the earth, and especially in sunlight. Businessmen put hard cash into solar steam boilers, although they never quite recovered their investments. Meanwhile visionaries wrote of a sun-powered civilization where people could abandon slums for rural life, or where perhaps the slums themselves would be cleansed of coal dust to become true white cities. Imagining the future age of solar energy, one scientist wrote in 1896, "The rivers are clean again, the harbor shows only white sails, and England's 'black country' is green once more!" Others predicted that civilization would migrate toward the tropical deserts where sunlight abounded, effecting a Martian solution with solar-powered irrigation projects.[8]

Soddy was only repeating familiar ideas, then, when he wrote that the mainstay of civilization was energy, and that "the world's demand for energy is ever increasing and will continue to increase, while the available supply of fuel is ever diminishing." He saw poor prospects for solar power or the like, and warned that the "inevitable coming struggle" over dwindling fuel supplies might be nearer than people guessed. He insisted that there would be only one way to move toward a finer future, or even to prevent a collapse into poverty: exploit the energy within atoms.[9]

MISSIONARIES FOR SCIENCE

It was not enough that the image of the White City existed and, through concern over energy supply, was in a position to link up with enthusiasm for radioactivity. There would have to be people with reasons to teach this linkage to the public. I now turn to these people, some of the most important but least known actors in the history of nuclear energy. They came to their work through an uneasy alliance between science and the press.

There was ample reason for scientists to tell their dreams for the future to newspaper reporters, or indeed to go directly to the public, as Soddy did in numerous lectures and writings. One motive was the scientists' pride in their profession and in their personal discoveries. Beyond that, they could not hope to find money for their laboratories and students for their classes unless the public joined in their belief in the value of science. What is more interesting is how avidly journalists took up the same belief.

The press would eventually purvey every part of the legendary tale of nuclear energy, world doom and all, but at first the White City stood in the foreground. Journalists zealously preached the praises of science in general, and of atomic science in particular. As soon as the Curies had exhibited radium glowing in their little vials, newspapers began to entice readers by declaring, for example, that radium "rivals the sun in giving out everlasting heat and brilliant light." When Rutherford and Soddy pointed out that radioactive forces might be the long-sought source of the sun's own energy, the press took up the idea with relish. Instead of sustaining future civilization with solar steam boilers, perhaps scientists would create solar energy itself in a bottle! After Rutherford and Soddy's discovery of what even scientists now called transmutation, every medium from newspapers to public lecturers exclaimed that it gave scientists a tool to revolutionize civilization. Radium might be harnessed to illuminate cities, propel vehicles, create new metals, and do almost anything else imaginable. The most memorable statements came from Gustave Le Bon, a brash French writer who called himself an important scientist, although to real scientists like the Curies he was only a pest. In best-selling books Le Bon expounded the grand meaning of radioactivity. Once people learned to use its energy, "The poor will be equal to the rich and there will be no more social problems."[10]

Every literate person heard the news. When the Museum of Natural History in New York City put a speck of radium on display in 1903, one of the largest crowds the museum had ever admitted swarmed in, squeezing and elbowing, to stare at the dull pinch of powder. By the 1920s, when a professor or a reporter remarked that a bottle of uranium might propel a steamship across an ocean, he was repeating a tired cliché known to schoolchildren around the world.[11]

One obvious reason that journalists and popularizers of science said so much about radium was that it was a fine source for sensational claims, tales that could attract a paying audience. But this was only one of their motives. As much as any citizen if not more, these writers were concerned with the future of civilization, and that meant giving the closest attention to scientific discoveries. Besides, as the twentieth century wore on, not only Sunday supplement curiosities but more and more of the front page had to do with technology and science. That posed a problem for the press.

Educated people had customarily learned about science directly from the books and lectures of scientists themselves, but the literate public was now growing to include far more than the small educated elite. Unfortunately, the new mass press snipped only the most striking remarks from eminent scientists' pronouncements, or ignored them entirely in favor of sensational claims by pseudo-scientists like Le Bon. Both scientists and publishers began to see that specially trained journalists were needed to separate the wheat from the chaff.

Science journalists, a new breed of writer, rose to the challenge. They were young men with a solid background of scientific training. In their work as reporters they spent many hours in the company of leading scientists, remarkable men and women whom they admired, and upon whom they relied for their information. These journalists had come to their specialty in the first place because they deeply appreciated science. They saw themselves as missionaries for the scientific viewpoint, believing that, for the sake of progress, it was their duty to make everyone see things as scientists did. Public admiration of science seemed not only proper but necessary, for in the end it was the support of society that sustained science (and therefore, of course, science journalism).

The dean of these writers was Waldemar Kaempffert. In 1927, when he became science editor of the *New York Times*, Kaempffert had already been reporting on subjects like radium for a quarter century. He had been a student of science and then of law, but had found that his real talent lay in turning out clear, enticing prose. Over the years he came to resemble the famous professors he reported on, robust and impressive, by turns charming and pompous. Agreeing with his scientist friends, Kaempffert wrote that they had already done more than all the emperors of history to transform society. We could look forward to still greater marvels: rockets to the moon, precooked meals, electric power projects creating "thousands of small towns with plenty of garden space, low rents, breathing space . . . health, and a finer outlook on life." As energy sources for the future civilization he mentioned winds and tides, the sun, the internal heat of the earth and the oceans, but above all, atoms.[12]

The supply of energy was no mere passing interest for Kaempffert and his fellow science journalists; it was the prime example of a technological subject that had become front-page news. During the First World War coal had been so necessary that mines had been worked within gunshot of the Front while Americans were advised to keep their homes chilly; after the war, strikes in the coal fields brought a temporary crisis of scarcity. Even more worrisome was the supply of oil, which was beginning to replace coal as the world's chief fuel. Fierce arguments broke out over regulation of the oil industry, while in the early 1920s magazines warned that future oil shortages might lead to international conflict. Nations threw themselves into hydroelectric projects on a scale to rival the engineers of Mars, changing landscapes from the Colorado River to the Dnieper. By the 1930s electricity seemed so important that many Americans expected that all national development would be controlled by either government-owned utilities or private power companies. Battle was joined as propagandists exclaimed that "Red socialists" or, according to taste, "selfish monopolies" were literally seizing the levers of power.

Most scientists and science journalists remained convinced that such problems would be reconciled in the future by the advance of science. Experts continued to predict that new energy sources—solar, hydroelectric, atomic, or whatever—would resolve every difficulty. In place of bulky fuels there would be clean and efficient electricity, giving citizens all the power they could desire. Industry would disperse into the countryside; people would become more free in every way. The public (these pronouncements implied) need only wait, passively and patiently, for authorities and experts to achieve the inevitable progress.

Anyone with an interest in legitimizing the existing social order would be glad to spread such beliefs. This was all the more true where existing authorities planned for changes. Lenin declared that true Communism meant Soviet power plus the electrification of the whole country; C. P. Steinmetz of the General Electric Company suggested that electricity would become "so cheap that it is not going to pay to meter it."[13]

As for Kaempffert, he told readers of the *New York Times* that someday scientists would release the energy within atoms so that we might (the inevitable example) propel a ship across the ocean on a glassful of fuel. But that was only a beginning. In 1934 he wrote, "Probably one building no larger than a small-town postoffice of our time will contain all the apparatus required to obtain enough atomic energy for the entire United States." With transmutation under control, gold might be a waste by-product of the new industry, used for roofing material. In the hands of a writer like Kaempffert, radioactivity had become not just a sensational promise in itself, but a guarantee of all the wonders that modern civilization would inevitably bring.[14]

THE MEANING OF TRANSMUTATION

Buildings roofed with gold? Kaempffert's casual remark showed that radioactivity could sound far more exciting than, for example, a hydroelectric dam. The social and historical factors that brought atomic physics into association with the utopian scientific White City may seem plain enough, but these moved alongside other forces, deeper and harder to explain, that would also have their say. Something pushed the imagery farther than anyone intended, and in a different direction. For the concept of "transmutation" was preternaturally fascinating—a doorway through which many kinds of archaic images would make their way into nuclear energy tales. To see what drove this process I must back off a long way from the modern world, into areas that in Soddy's day seemed almost forgotten.

A clue to what was happening lay in the fact that physicists and the press loved to compare the atomic scientists who altered elements with the medieval alchemists who had sought to turn lead into gold. By the 1930s even level-headed Rutherford titled his popularizing book on atomic physics *The Newer Alchemy*. He and others had made the technical concept of atomic transmutation into a useful paradigm for research, an approach that could explain almost every radioactive process in terms of an atom changing into a kindred atom. It became a cliché that "the famous problem of the alchemists has been solved." But in reality the problem of the alchemists remained unsolved. It was a problem with a very different kind of transmutation, although still one that could be associated only too easily with radioactivity.[15]

For some two thousand years men and women in Egypt, the Middle East, India, China, and Europe had devoted lifetimes of intense effort to blending chemicals and metals in their furnaces, producing kaleidoscopic displays, grotesque lumps and stenches, glittering crystal formations. Arduously studying these substances day and night, the alchemists were seized by a conviction that the chemical changes pointed to something tremendously significant. In fact they were projecting their most deeply held feelings and cultural symbols onto the ambiguous transformations in their crucibles. Thus alchemy opened a window onto a secret inner landscape of images.[16]

By late medieval times the wiser adepts knew that they were working less with matter than with their own minds. Masters warned their followers not to aim merely at physically changing lead into gold, for such work should be only an aid and a symbol for something infinitely more important. They explained that the true philosophers' stone—the central goal and secret of transmutation—was a spiritual matter. To achieve transmutation meant the perfection of the soul, its transformation from a dull leaden state into the golden state of mystical illumination, divine grace, joy and peace.

Some writers went still farther, hinting that the true alchemical gold would be found in perfecting not just the spirit of one individual, but of everyone. This meant achieving a world of justice, peace, and plenty, a world described in the legends of many peoples and known in Western culture, through no coincidence, as the Golden Age. That realm might be located in the past, like the Garden of Eden, but in Middle Eastern and European tradition the Golden Age might also lie in the future. Sometimes whole populations eagerly awaited the Jewish Messiah or the Mahdi of Islam with their kingdoms of righteousness, and many times in the history of Christendom there were mass movements proclaiming that the millennium was at hand. When Soddy and others exclaimed that transmutation of the elements could mean a new Eden, a society emerging like a dragonfly from its larva, they were constructing a nuclear energy legend out of ideas that had already been associated with one another centuries earlier.

Less obvious at first to Soddy, although familiar to alchemists, was the fact that "transmutation" also involved death. The adepts explained that in their crucibles matter must literally die before it could be refashioned into a golden state. In chemical practice this meant turning substances into a black mass, inflicting a fall into corruption and putrefaction. That was a deliberate metaphor of a process that religious mystics insisted upon: humans must make an agonizing descent into darkness and chaos, purging themselves in the "divine furnace," painfully surrendering attachments before they could achieve spiritual renewal. After all, since antiquity it had been common knowledge that people might need to break down ingrained patterns of thinking before they could find inner peace. In the end transmutation became a symbol for still more, the greatest of all human themes—the passage into death and beyond. This was the theme symbolized by the man in my nuclear energy legend when he crawled from his blackened shell after the catastrophe, remade in body and soul. I will show him again in many other guises.[17]

More obviously related to atoms was a parallel concept: passage of all the world through fire. The alchemists themselves said that creating gold in the furnace was symbolically connected with the springtime resurrection of the sun (the sun being a standard alchemical symbol for gold). This solar cycle was linked in turn with another archaic theme, the Wheel of Time. Many peoples expected that the world would eventually wear out and society would dissolve into a cosmic winter of war, strife, and chaos, after which everything could be remade. The nadir might be the world conflagration that ethnographers have recorded in native tales from every continent. In Christian culture during every century including the nineteenth and twentieth, a great many preachers have insisted that someday, and perhaps soon, before humanity could enjoy the Golden Age, the fire and bloodshed of Armageddon must come.[18]

Alchemists themselves were convinced that the secret of transmutation held boundless perils, both spiritual and mundane; that was why it was a secret. The adepts hid their chemical formulas amid a confusion of arcane symbols, suns and green lions and so forth, to ward off misuse by unworthy people. In short, long before Soddy's day the idea of transmutation was tangled up with ideas about vast hidden forces, cosmic transformations, and apocalyptic perils.

The last point in time at which the historian can see all these ideas consciously connected together was in the seventeenth century as the scientific revolution began. The central thinker of that revolution, Isaac Newton, reflected his times when, persuading himself that the apocalypse might be imminent, he devoted more of his life to religious studies than to physics. He devoted himself most indefatigably of all to the study of alchemy. Newton suspected that he was on the track of a great and ancient secret and that (as he wrote in private) the transmutation of gold might be a key to "something more noble, not to be communicated without immense danger to the world." At the same time he hoped that scientific knowledge would restore something of the lost Golden Age. Thus all the concepts I have mentioned fitted neatly together, along with modern physics, in the great scientist's mind.[19]

Soon afterward the ideas unwound from one another, each to go its own way. For example, when Renaissance visionaries described utopian societies they had been steeped in transmutation symbolism, but by the nineteenth century the image of a White City, even if its streets were paved with gold, would not remind most people of the alchemist's laboratory. Likewise the concepts of the transformation of the soul and the end of the world, as well as the stereotyped alchemist himself with his peculiar secrets, each began to follow separate historical paths.

Yet the original connections could be recalled. They were known, for example, to the chemist Berthelot, who took many hours away from his Paris laboratory to peruse antique manuscripts and write a multi-volume history of alchemy. Soddy read it around 1900. As a university lecturer, he had chosen to teach the history of chemistry, so he read up on alchemy and was fascinated by the imagery. From the moment in 1901 when he told Rutherford that what they had discovered was "transmutation," Soddy deliberately dealt in potent old symbols. In his popular writings throughout the following decade he hinted at tremendous long-hidden secrets, primordial powers of evolution, and cosmic cycles of creation and decay, which he now linked with the modern question of energy supply and with radium.

The meager facts known about atomic energy in the early twentieth century were only an ambiguous screen upon which scientists and their admirers could project more urgent ideas. They were building up a tale in which the nuclear physicist was like the legendary alchemist, a man who could work mighty transformations. Atomic energy was coming to stand for things more important than atomic energy itself. First of all it was coming to stand for all the powers of science, powers for the better—or perhaps for the worse.

2

Radioactive Fears

Would the future hold not a White City but a desert of ash? Put another way, was the man in the nuclear energy legend good or evil? This matter of imagery stood for a question that sophisticated people would debate with increasing passion as the twentieth century advanced: could science be trusted to lead civilization in the right direction? That debate, as expressed in pictures of the future world and stereotypes of the scientist who might create it, would do much to influence images of nuclear energy, and would be influenced by the images in return. The debate is the subject of this chapter and somewhat of the entire book.

At the base of the arguments lay divergent beliefs about the role of all expert authority in society. But people only slowly came to see expertise as a problem, and at first the critics of science were more concerned with immediate practical matters. In particular they worried about scientific weapons, which turned out to be easily associated with the energy in atoms. Right from the start this new idea of atomic weapons was linked with an even more impressive idea: the end of the world.

When Soddy first told the public about atomic energy, in May 1903, he said that our planet is "a storehouse stuffed with explosives, inconceivably more powerful than any we know of, and possibly only awaiting a suitable detonator to cause the earth to revert to chaos." This was an entirely new idea: that it might be technically possible for someone to destroy the world deliberately. Yet the idea slipped into the public mind with suspicious ease. In the tales that people were beginning to hear about atomic devices, the vision of a utopian city had to make room from the outset for the reckless man who would doom the earth.[1]

Atomic doom had concrete details. For example, in 1903 the irrepressible Gustave Le Bon got into newspaper Sunday supplements in various countries by imagining a radioactive device that could "blow up the whole earth" at the touch of a button. Rutherford explained the danger almost plausibly to another physicist, who published it in a magazine: "It is conceivable that some means may one day be found for inducing radio-active change in elements which are not normally subject to it. Professor Rutherford has playfully suggested to the writer the disquieting idea that, could a proper detonator be discovered, an explosive wave of atomic disintegration might be started through all matter which would transmute the whole mass of the globe into helium or similar gases."[2]

Serious scientific thinking lay behind Rutherford's little joke, as his friend called it, that "some fool in a laboratory might blow up the universe unawares." Rutherford and Soddy had found that everything in their laboratory tended to become mildly radioactive. Therefore they suspected that radioactivity was contagious from one atom to another. Soon they realized this was a mistake; their laboratory had simply been contaminated by radioactive gas. But the idea remained that a radioactive atom might somehow make neighboring atoms radioactive in turn, starting a chain reaction that could spread without limit. Fortunately, as Rutherford later pointed out, this was impossible. For billions of years radioactive atoms had been randomly trying every conceivable reaction, so that if the planet were unstable it would long since have disintegrated. Yet the hobgoblin of a global chain reaction made an indelible impression on the public. One reason was that the image could serve as a symbol, reflecting rising anxieties about the entire technological future.[3]

SCIENTIFIC DOOMSDAYS

By the late twentieth century the idea of doomsday would be inseparable from ideas about nuclear energy, but this came only as a late stage in a long process. It was in the early nineteenth century that ideas about the end of the world first began to separate from their original mythical and religious contexts, joining up instead with science. A first tentative step came in 1805 when a fantastic novel, *The Last Man*, was published in France. The author, Cousin de Grainville, still referred to the Judgment Day of Christian tradition, but the bulk of his story was unlike anything published before; it showed the human race dwindling through natural processes to a lonely end. Probably the author was contributing something personal, for he was himself a desperately lonely man—late one winter night after finishing his book, he drowned himself. His unhappy image of a last man survived him, inspiring dozens of poems, paintings, and stories during the next decades.[4]

The most substantial of these was a *Last Man* novel that Mary Wollstonecraft Shelley published in 1826. Her long book was filled with melancholy passages about how war and plague extinguished humanity in ways that had nothing to do with religion. According to her biographers, this author too had reasons for writing about desolation, for she had grown up motherless, and before writing this novel she had lost her husband and other good friends through tragic early deaths. Perhaps such authors used tales of world devastation to stand for more personal problems? That is a question I will defer until the point in history where talk of atomic doom began to be plausible.[5]

Whatever its emotional meanings, destruction of the planet could also be a plain scientific idea. In the late nineteenth century physicists suggested that over millions of years the sun would cool off and the earth would freeze. The public learned these ideas through various channels, such as a popular novel by Flammarion, whose Last Man huddled in a greenhouse as the earth's air dwindled away. In both popular and intellectual thought, the scientific idea that the world's energy must slowly fail was a severe blow to optimism.[6]

Many people kept up their faith in progress by suggesting that decline would lead to renewal; the Wheel of Time would revolve again. For example, Flammarion's novel ended with a cataclysm giving birth to a new universe. This idea of cosmic cycles remained potent even in the theories of eminent scientists. From time to time astronomers sighted the flaring of novas, explosions that convulsed entire stars; some hypothesized that the earth had been born in just such a convulsion and that the process could repeat.

The idea was presented most vividly in a 1903 magazine story by the noted American astronomer Simon Newcomb. After sketching a utopian future city, he had a wandering star smash into the sun to make a nova: "Even the stones of the buildings, cracked and pulverized by the heat, were now blown through the air like dust, and, churned with the rain, buried the land under a torrent of mud," and so forth. A handful of survivors remained in a buried shelter, doomed wretches who (as Newcomb put it) could only envy the dead. Their one solace was that under the new and brighter sun, a superior race of beings might evolve. Later in the century images like these would appear mainly in nuclear energy tales, but Newcomb's story showed that atoms were not needed to make doomsday a scientifically plausible idea.[7]

Meanwhile another traditional element in doomsday, the idea of future social breakdown, also became separated from the religious context and forged ahead on its own toward a linkup with science. If technology was driving back natural perils such as plague and famine, it was also overturning accustomed social relationships; some people began to feel that the chief danger to society came from social change itself. Opposing the general self-confidence of the late nineteenth century, some prominent authors predicted that war and insurrection would smash civilization to ruins. The most outspoken writers were those who felt personally alienated from society. One example was the bitter and lonely Henry Adams, who felt it was a demonstrable fact that industrial civilization would collapse of its own weight within a few decades. Another example was Jack London, whose novel *The Iron Heel*, about a brutal future oligarchy degenerating into savagery, reflected his personal disgust with modern social conditions. For whatever reason, these writers kept vividly before the public the image of a doom brought by the advance of civilization itself.[8]

By the beginning of the twentieth century almost every conceivable world disaster had appeared in widely read books and stories. A review of these tales by the scholar Warren Wagar showed that two-thirds of them featured natural dooms such as a nova. The remaining, man-made apocalypses were divided between social revolutions and wars, made more inevitable and awful by technology.[9]

Soddy's atomic nightmare differed from all these catastrophes in one essential way. An astronomical cataclysm or a rising tide of insurrection each involved forces beyond the control of the individual, but atomic energy might bring disaster through the foolishness or malice of a single man. That thought would give rise to the most central and most troubling feature of the nuclear energy legend: the scientist himself.

THE DANGEROUS SCIENTIST

The stereotyped scientist is another nuclear energy image that is best understood by looking into its ancestry. Before modern times people believed that only divine forces could wholly extinguish the world, yet they still feared the sorcerer. Here was a stereotype reaching back through medieval wizard legends and the biblical Simon Magus to prehistoric shamans. Such an individual might release pestilence and other evils, as Pandora did, or might unleash demons, as witches did, or might simply propagate heretical ideas, as some proto-scientists in fact did. In short, people have always feared that an overweening individual, through evil thoughts or magical acts, might contaminate his neighbors or even the entire world.[10]

This idea seemed to find confirmation in the lives of real people. Most notorious was Dr. Faust, a sixteenth-century itinerant rakehell who boasted that he controlled demonic powers. Pious churchmen mixed Faust's actual career together with legends of black magicians; their aim was to issue an impressive warning against the rising breed of humanist skeptics and scientists who were not guided by established Christian belief. Many other quasi-scientists stirred up the same feelings. Of special note was the charismatic Dr. Franz Mesmer, who in the late eighteenth century attracted tens of thousands of disciples, including leaders of Parisian society, with his ability to work cures. He claimed that his powers were based on a new "magnetic" science that he had discovered, but medical authorities denounced him as a charlatan and worried that his following was a cult that threatened the social order.

Through the nineteenth century both Faustian tales and mesmeric cults grew in popularity. A stereotype was being refashioned, one that would eventually bear heavily on the careers of real nuclear physicists. Popular authors from hack newspaper writers up to Nathaniel Hawthorne modified the old tales of witch and sorcerer to create a new fictional figure, the mesmeric "scientist" who endangered himself and those near him with a mixture of demonic and scientific powers.[11]

A few writers (for example the Marquis de Sade) suggested that such a figure might desire to personally end the world, but at first authors did not suggest that such a deed could really be possible. As technology altered entire landscapes, and sometimes for the worse, opinions began to shift. The change from a sorcerer contaminating his neighborhood to a technical expert who threatened all the world could be seen as early as 1817 in a few key sentences of Mary Shelley's *Frankenstein*. The scientist was forced by his monstrous creation to prepare an artificial bride, but at the last moment he hesitated. Would the two fiends mate, the scientist wondered, and produce "a race of devils," polluting the earth and perhaps exterminating humanity? Magical release of demons, a witchcraft theme that had become commonplace in the gothic thriller novels that preceded Shelley's, was turning into a threat of scientific annihilation.[12]

A new stereotype was emerging, the dangerous scientist-inventor. He was associated with surface characteristics of the legendary alchemist and magus, such as laboratory glassware, and also with more essential features. Exactly like the traditional magus, he was remote from everyday life; he avoided women's love; he was obsessively devoted to the search for dangerous secrets; he seized powers over life and death; he boasted an impious pride; and he tended to die an uncanny and violent death. I will come later to the meanings underlying each of these standard features.[13]

Jules Verne did more than anyone to develop the new stereotype. His first widely read story, *Five Weeks in a Balloon*, written in 1862, contained the earliest remark I have seen on an accidental technological doomsday. The great new feat of those years was steam power, but boiler explosions occasionally killed scores of people at a stroke (with the usual results of technological accidents: public outcry, committee investigations, and official regulation). Verne had a character remark, "I sometimes think that the end of the world will come when some immense boiler, heated to three thousand atmospheres, blows up the earth." A more typical Verne novel told of an insane chemist who had invented a mighty explosive. On an island hideout he used the invention to destroy an attacking fleet, and then to blow up, if not the planet, at least his island and himself. The premises seemed plausible; Verne had apparently based his fictional character on a living inventor of explosives, who promptly sued for libel.[14]

In 1901 an author published the first extended treatment of an individual destroying the entire world. *The Purple Cloud*, an avant-garde novel by Matthew Shiel, began as the story of an expedition to discover the North Pole. Although warned by a preacher that God had forbidden men to see that mighty place, the protagonist, a sort of scientist, contrived to be first to discover its secret. When he returned he found that clouds of poison gas, released from the earth when he approached the pole, had spread southward to kill everyone else.[15]

As if to validate the new fantasy of world doom brought on by overweening curiosity, two years later Soddy and Rutherford's concept of a planetary chain reaction provided the first superficially rational description of how a person might in fact destroy the world. Leading scientists passed along the idea. For example, a famous German chemist said in 1921 that we are like a primitive race living on an island of gunpowder, a race to whom no Prometheus had yet brought the perilous gift of fire. A well-known Dutch textbook on atoms carried the thought a step farther, wondering whether the novas seen in the sky might be outbursts of atomic energy "brought about perhaps by the 'super-wisdom' of the unlucky inhabitants themselves." Newspapers were happy to repeat such startling thoughts. A writer in the *New York Times* in 1929 stretched the idea to its limit, claiming that eminent scientists suspected that not just one planet, but the whole universe, could be accidentally "fired like a train of gunpowder." By the 1930s even schoolchildren had heard about the risk of a runaway atomic experiment.[16]

Still more chilling was the thought that someone might bring world cataclysm on purpose. Bomb-throwing anarchists had become notorious in the late nineteenth century, and the public connected their powerful new explosives with science. Thus scientists were coming to be personally linked with the old idea that civilization would perish in social upheaval. More than one novel featured the cold-blooded Professor, a chemical expert who, in the name of freedom, plotted to blast society apart, to make "a world like a shambles" with his infernal devices. In Anatole France's *Penguin Island* of 1908, the evil chemists were replaced by physics-minded terrorists who destroyed entire cities with pocket-sized atomic explosives.[17]

The dangerous scientist was usually shown as more or less mad. For example, a 1938 adventure novel by J. B. Priestley, *The Doomsday Men*, featured a fanatic scientist who declared that since life is only suffering, it would be best to end the world forthwith. He prepared to set off an atomic reaction that would peel the skin off the earth "like an orange, only faster."[18]

The man in the nuclear energy legend was most convincingly portrayed in *Wings over Europe*, a play that enjoyed success with the critics, a modest run on the stage in London and New York beginning in 1928, and frequent revivals among college theater groups during the 1930s. Center stage was held by Francis Lightfoot, a youthful scientist, brilliant but reclusive and a bit unbalanced. Lightfoot announced to the British Cabinet that he had found the secret of releasing atomic energy. Staring into space, he exclaimed hoarsely that he could give them the power to make a Golden Age, "the power of . . . a god, to slay and to make alive." The ministers in the play dithered about what transmutation might do to the gold standard and took an unwholesome interest in the prospects for weaponry. With his naive idealism trampled, Lightfoot lost his bearings, decided the world was wicked, and made ready to blow it into dust.[19]

Of course this was not the only way scientists were pictured; the majority of writings always showed scientists in a less lurid light, as people working for the good of humanity. Throughout the nineteenth and the early twentieth century the most common stereotype showed the scientist as downright harmless—well-meaning, unworldly, and absent-minded. For example, in 1908 a Sunday newspaper comic strip showed a "Doctor Radium" who was a skinny, nearsighted old pedant, serving as the butt of small boys' pranks while he mumbled scientific gibberish to himself.

Over the years the scientist began to seem less of a joke. A 1941 *Batman* comic book featured a very different Professor Radium, a robust genius who meant to work wonders with radioactivity, but accidentally turned himself into a murderously insane "human radium ray." This was no isolated case. Even in the new pulp science fiction magazines, which seemed firmly committed to the idea of scientific wonders, the heroic atomic inventor sometimes had to share space with an evil one. Pressure from somewhere was gradually altering the public image of scientists, and by implication the image of the radioactivity they worked with.[20]

This pressure had nothing to do with any new discoveries about radioactivity itself. *Wings over Europe,* for example, was a literary vehicle for high-flown speeches about humanity, and *The Doomsday Men* was a popular thriller with a bit of preaching on the side; neither showed the least interest in real nuclear physics. The authors were using commonly understood images of mad scientists and doomsday as tools for social commentary, within a growing controversy over the direction modern civilization was taking.

SCIENTISTS AND WEAPONS

The mad scientist and his doomsday were images too crude to sustain serious social commentary. Debates over science would have to deal with more realistic problems, and the most acutely felt of these was the advance of weaponry. During the early decades of the twentieth century it was worries about bombs in particular that would forge links between sober debate over modern civilization and legendary tales of nuclear perils.

The first well-publicized hint of a link between atomic energy and weapons came from Sir William Crookes, a scientist with a spectacular white goatee and mustachios, well known to the British public. Even before radium was discovered, newspapermen had found the genial Crookes a fine source for quotable speculations on such questions as the exhaustion of fuel resources and the prospects of transmuting elements. In early 1903 he set forth his ideas on radioactivity, and in passing devised a startling image of atomic energy. Any physics teacher, asked to illustrate a quantity of energy, would be inclined to talk about lifting a weight. Crookes chose to lift the British Navy. The energy locked within one gram of radium, he calculated, could hoist the entire fleet several thousand feet into the air.[21]

The idea developed in a revealing way. From Crookes's scientific discourse the newspapers picked up the naval example as the stuff of headlines, adding helpful engravings of a cluster of battleships suspended in midair. The public quickly caught the image's latent meaning. Soddy, passing through Boston, wrote to Rutherford that people were repeating Crookes's remark that a gram of radium could "blow the British Navy sky high." Scientist, press, and public had together crafted a new thought.[22]

It was not news that scientists could invent weapons. After all, Crookes himself was on the British War Office's Explosives Committee. Typical of public thinking was an 1892 rumor that Thomas Edison was building an electrical device that could annihilate a city from a distance, followed by a newspaper satire about the great inventor destroying England with a pushbutton "doomsday machine." In 1903 Soddy, speaking to the Royal Corps of Engineers, became the first scientist to explicitly add atomic energy to the roster of possible weapons. The idea promptly became a cliché, with radium missiles adding spice to newspaper Sunday supplements and science fiction books. All this was more exciting than frightening at a time when the existing weapons, let alone imaginary ones, seemed a matter for colonial battlefields rather than a threat to the ordinary citizen.[23]

In 1913 a more disturbing atomic weapons tale came from the pen of H. G. Wells—aside from Jules Verne, perhaps the most influential author of the era, at least among science-minded young men. *The World Set Free,* one of the worst written yet best considered of Wells's novels, was dedicated to Soddy and was directly inspired by his writings on radium. Wells took up the idea of a radioactive chain reaction to show how it could lead to (coining a phrase) "atomic bombs." He described a world war of the 1950s with aviators biting the fuse on round black devices and leaning out of their cockpits to drop them, erasing an entire city with each bomb. Colossal pillars of fire raged while "puffs of luminous, radio-active vapor" drifted downwind, "killing and scorching all they overtook." The near extinction of civilization taught the survivors a lesson, and they created a world government that nurtured a brilliant new society. At the story's end citizens could travel where they chose in atom-powered aircars, building atom-powered garden cities in deserts and arctic wastes, enjoying liberty and free love. Wells had neatly fitted together fragmentary notions about science and atomic energy to craft the first full-scale scientific legend of atomic Armageddon and millennium.[24]

It was a legend with special meaning for Wells himself. From a wretched beginning as a subservient clerk, where thoughts of destroying society had seemed almost attractive, he had saved himself through an education in science. Wells thought that society should be rearranged so that science-minded men like himself would rise higher still. The science fiction writer thus resembled the missionary science journalists, with his own reasons to preach the power of science.[25]

Wells's book was barely in print when real warfare made the power of science an urgent problem. Soon everyone knew what could come from aeronautical engineering, poison gas research, and the like. After the war's end scientists worked to drive home the lesson, advising nations to spend more money on their laboratories or be defeated in the next war. The thought was taken to an extreme in many novels and stories. Whereas before 1914 two-thirds of fictional apocalypses had been due to natural causes, after 1914 two-thirds were caused by humans, and of these, three-quarters of the doomsdays came in world wars with scientific weapons.[26]

Talk of such things became common among the most serious thinkers. Winston Churchill remarked that humanity already had the tools to "pulverize" civilization; Sigmund Freud wrote that science had given us the power to kill one another "to the last man." Leading scientists admitted that another war might turn entire nations into deserts.[27]

What would pulverize civilization in the next war? The most commonly mentioned threat was attack from the skies—usually not with atomic bombs, for experts expected those only in the distant future if at all, but with weapons already at hand. For example, the liberal commentator Stuart Chase warned that fleets of airplanes could attack cities with poison gas and chemical explosives so that within hours not even a roach would be left alive. Most widely influential was an Italian military theorist, General Giulio Douhet, who claimed that victory would come within weeks to whichever side struck with its bombers quickest and hardest.[28]

"The bomber will always get through," Prime Minister Stanley Baldwin told a dismayed House of Commons in 1932, reaffirming the new supremacy of technology. "The only defense is in offense, which means that you have to kill more women and children more quickly than the enemy, if you wish to save yourselves." The European public grew increasingly frightened. During the previous war German bombs had touched off hysteria and riots at public air raid shelters in Britain, and during the 1930s gruesome attacks by Japanese bombers in China, by Italian bombers in Ethiopia, and by German bombers in Spain drove home the impression that bombs could wreck a society.[29]

Many ideas that became central to debates over nuclear weapons in later decades got a trial run during the 1930s. For example, the "deterrent": a nation should develop bomber fleets and poison gas so that its enemies would fear to start a war. Or the pre-emptive "knockout blow": one side might annihilate the other by surprise the day the war started. Therefore owning a large bomber force might be a dangerous provocation, and disarmament negotiations were pursued, aiming at an international inspection system for bombers. The loudest public debate was over civil defense. Some said that gas masks and bomb shelters would dissuade an enemy from launching an attack; others declared that defense against air raids was hopeless, a ruse of the ruling classes to lull the public and keep militarism alive.[30]

Especially cherished was the idea that war had become too horrible to persist for long. The playwright George Bernard Shaw wrote that white flags would go up all over a city, fluttering frantically, the moment citizens saw enemy bombers draw near. General Douhet, scarcely less hopeful, argued that citizens would not suffer bombs for long before they would force their leaders to end the war; a few days of bombing and the white flags would spring up. Stuart Chase wrote that two warring nations might exterminate each other, but the rest of the world could learn a lesson and break out the white flags of universal peace.[31]

It was no new idea that terror of modern weapons would bring an age of peace. Back in 1892 the inventor of dynamite, Alfred Nobel, had told a promoter of pacifist meetings, "My factories may make an end to war sooner than your congresses." Total war already seemed unthinkable, and if a world war nevertheless began, many agreed with H. G. Wells when (coining another phrase) he called it a "war that will end wars." A few learned the First World War's great lesson: once modern nations start fighting they find it terribly hard to stop. But others saw only proof that weapons had become ghastly enough to compel peace the next time around.[32]

The idea of atomic weapons in particular inspired as much hope as fear. "When we have discovered the secret of the atom," a journalist wrote, "it is likely that all nations will be ready and willing to lay down their arms and abolish their armies and navies." The scientist's dream of inventing weapons to deter war, a persistent part of the nuclear legends, would eventually exert a mighty influence on armament programs. Belief in the virtues of science and technology could be so strong that even a threat of destruction might sound like a promise of peace.[33]

DEBATING THE SCIENTIST'S ROLE

If some felt new misgivings about the role of science in war, many others insisted more than ever that science was not the problem but the solution. Moral preaching and diplomacy had failed to bring peace; science would succeed, and not only through the terror of new weapons. After all, scientists themselves had already formed their own international community—peaceful, cooperative, and dedicated to reason. "Religion may preach the brotherhood of man," said Kaempffert, but "Science practices it." Many began to insist that progress depended not just on technical projects such as atomic energy, but on the specific role that scientists could play in society. These beliefs promptly came under attack. As the debate raged, public thinking about atomic energy began to link up with a key social question: could scientific authorities, the would-be liberators of that energy, be trusted?[34]

The view of some scientists, reduced to a primitive cartoon, was revealed in a novel that the astronomer Simon Newcomb published around the turn of the century. His hero was a physics professor who, having discovered a source of ethereal energy, raised an army of idealistic youths and armed them with flying machines. Using skinny metal limbs like the legs of a daddy longlegs spider, the flying machines, called "daddies," hovered over armies and snatched rifles out of the hands of soldiers. At the end the hero took the modest title "His

Wisdom, Defender of the Peace of the World" and ushered in the standard Golden Age. Newcomb had exposed a secret wish: to be a benevolent scientific daddy setting right the naughty world.[35]

The book was not exceptional, for since the 1890s juvenile fiction had swarmed with technological heroes like Frank Reade, Jr., who rode in on his electric tricycle to break up the Nigerian slave trade. The play *Wings over Europe* went farther; at the end a group of scientists who had learned the secret of atomic energy came soaring over the world's cities in huge green airplanes, bearing atomic bombs, to enforce the will of a "League of United Scientists of the World." As usual, the most thoughtful development of such ideas came from H. G. Wells. In the script of a 1936 film, *Things to Come*, he described an Everytown shattered by bombs in 1939. The city fell into a new Dark Age, with ragged survivors herding sheep through the ruins. Fortunately a group of airplane pilots had survived, and they came to the rescue in a fleet of shining aircraft—the virtuous Airmen. Wells saw the Airmen as a caste of austere technocratic guardians, devoting their lives to the service of humanity, casting out superstition and transforming Everytown into a great white city. It was much the same tale that Wells had set forth decades earlier in *The World Set Free*.[36]

Soddy, who had inspired Wells's 1913 novel, was in turn attracted to the writer's ideas about the social role of science. In 1915 when his fellow Oxford graduate Henry Moseley, a fine young physicist, died in battle, Soddy had been outraged. "Something snapped in my brain," he recalled. "I felt that governments and politicians, or man in general, was not yet fitted to use science." A month after Moseley fell, Soddy became the first prominent scientist to give a vehement public warning about atomic weapons. "Imagine, if you can," he cried, "what the present war would be like if such an explosive had actually been discovered." He said the world would be doomed unless it was reformed by the time the discovery came.[37]

It seemed urgent to rethink all of society from the start, and Soddy set himself this task. He took up economics and spent years working out the nature of money, beginning with the idea that all wealth is ultimately based on energy. The move from chemistry into monetary reform separated Soddy from most of his university colleagues, and he further isolated himself by launching biting tirades. But he found an audience among certain scientists who were coming to feel that it was not enough simply to make discoveries. Science would do more harm than good, they decided, unless society was reshaped to use the discoveries wisely. "The solution," Soddy wrote, "is for the public to acknowledge its real master," namely science, and to choose as rulers "those who are concerned with the creation of its wealth," namely scientists.

A number of young scientists and science journalists, mostly on the political left, agreed that the proper way to reshape society was to give a greater role to scientifically trained people, that is, to people like themselves. Among the most widely heard of such thinkers were James Crowther, an Englishman trained in science and one of the first journalists to devote himself to writing about it almost exclusively, and J.B.S. Haldane, a fiery Marxist biologist who likewise threw himself into writing for the public. They insisted that science was languishing under the thumb of capitalists who used technology only to enrich themselves. Let affairs instead be efficiently organized by science-minded people, they said, and progress would be amazingly swift; atomic energy was only one of many wonders they promised. Their program was already being attempted in the Soviet Union, where the government was lavishing funds and prestige on scientists, as Haldane pointed out with frank envy.[38]

Applying science through rational planning implied a particular type of Golden Age. The gleaming cities of the planners, ranging from Plato's authoritarian Republic to the industrial society projected by Marxists, would fit individuals into their roles as harmoniously as the components of a machine. But many visionaries over the centuries had imagined an opposite sort of paradise—a rural Arcadia where couples would lie like flowers in the gentle breeze, blending into nature, ignorant of technology and social authority. Some writers, like Kaempffert, hoped for a compromise, a middle state between untouched nature and orderly civilization—a world of electrically powered villages, a landscape gardened with the help of machinery. But if one had to choose between the two extremes, people like Kaempffert would choose the benefits of order and technology. "We may cry 'back to nature' as loudly as we please," he wrote, "but the scientist within us answers, 'forward to the laboratory and the machine.'"[39]

Such ideas were made visible in the 1933 Chicago World's Fair. In place of the intricate white palaces constructed on the fairgrounds forty years earlier, there now stood buildings resembling giant machine parts; in place of the alabaster virgins, angular statues of irresistible robots. Aided by hundreds of leading scientists, the fair's corporate sponsors spread the message that science, as applied by the great industrial corporations, would solve society's problems. The 1933 official guidebook frankly offered the slogan, "Science Finds—Industry Applies—Man Conforms."[40]

Not everyone agreed that science should set the pace. In the 1920s and early thirties, popular German writers argued that heartless scientific analysis was separating people from the intuitively grasped wholeness of life. Essays in American literary magazines declared that because science by definition had nothing to do with such ideas as beauty and morality and holiness, it was helpless to deal with human problems and was destroying traditional values. Humanists recalled the spiritual horrors of the First World War, where bombardment had left soldiers helpless in the mud, the old knightly virtues like bravery rendered useless. A leader of the French Senate complained that technological breakthroughs had upset industry and brought on the Great Depression. A prominent British bishop suggested that there should be a ten-year moratorium on research to give society time to adjust.

The debate was not new, for scientific discoveries had often collided with particular social customs and traditional features of religion. It did not necessarily follow that rational skepticism weakened genuine civic virtue and religious morality. Nevertheless, claims that it did so were as old as the trial of Socrates. Now world war and depression had raised the debate to a fever pitch.[41]

This debate had an influence on images of nuclear energy. The most obvious connection came when the new concern helped motivate authors to write works like *Wings over Europe* and *The Doomsday Men*. Something more subtle than direct social criticism was at work here; otherwise the authors would have done better to portray realistic military or industrial scientists, not the stereotypical evil scientist descended from tales of witches and impious magicians. What was the connection between this stereotype and the debate over the actual position of science in society?

The problem with real scientists was that their role in society was anomalous. Most of them were either underlings in a corner of some industrial organization or poorly paid teachers, yet their discoveries were radically transforming civilization. Thus they had an incongruous sort of power over the future, nothing like the power normally granted to a businessman or government official. A few outspoken men like Haldane proclaimed that this situation must be rectified by honoring scientists and engineers above others, reorganizing society to conform to their ideals. The people who were used to shaping society's ideals—clergymen, humanists, popular writers, and the like—could well feel that their world was threatened by something from outside the established order of things. These were the groups who in fact helped to spread talk about blasphemous scientists.

The mechanism resembled one that was later studied by anthropologists. In many societies, anyone who upset others might be accused of attacking people with secret magic. These accused witches or sorcerers or heretics were not the decrepit hags of modern children's books, but sharp-tongued men and women, often precisely the ones whose ambiguous social or moral role disturbed established patterns. It was symbolically fitting to accuse incongruous "moral entrepreneurs" by saying that they wielded incongruous powers. And in practical terms, the attack could put them back in their place. I believe that stories of wicked scientists similarly reflected social tensions and uneasiness about anomalous, disquieting roles.[42]

An evil-scientist story could scarcely revolve around the design of a hydroelectric power distribution system; atomic energy proved useful for the tales because it already had a reputation as the grandest and weirdest of all powers. Therefore atomic energy became particularly closely associated with all the uneasiness that people felt over matters of science and technology. For similar reasons, ideas about atomic energy were sometimes dragged directly into sophisticated social criticism.

One example of the critics who invoked the atom was Raymond B. Fosdick, an idealistic American lawyer. Deeply impressed by the destructive forces revealed in the First World War, Fosdick went to London in 1919 to help the League of Nations. There he happened to meet Soddy, who warned him about the future of science in general and atomic weapons in particular. In 1928 Fosdick repeated the warnings in a book, *The Old Savage in the New Civilization*. He explained that savage humanity, like a child playing with matches, could scarcely cope with the powers of technology. We needed to give less heed to scientists and more to humanists who would boldly question society, that is, people like himself.[43]

Fosdick's challenge was answered by Robert Millikan, the leading spokesman of American science. A physicist famous for his work on atomic particles, Millikan might seem to have had an inflated confidence in himself, except that so many people shared that confidence. Newspapers photographed him alongside his friend Herbert Hoover, and it was Millikan, with his handsome features and silver hair, who looked more like a president. In fact he was the head of the California Institute of Technology, and when he gave a speech to one or another group of industrialists, Millikan wanted not only the audience's abstract approval of science but their cash as donations for his laboratories. Although his political views were at the opposite pole from Haldane's Marxism, Millikan was no less insistent that pure science was the key to progress. He spoke convincingly about this, pointing out to his audiences that all around them, in electric lights and appliances, fertilizers and medicines, lay the proof of how science had made daily life comfortable and secure, far beyond what they had known in their childhoods. As for atomic transmutation itself, he said it would be worth a billion dollars to study whether it could be controlled. Even if it could not, the by-products of the research would be worth the money—"And if it succeeded, a new world for man!"

When he took up the cudgels against Fosdick and like-minded critics of science in a 1930 magazine article, Millikan scoffed at the "advocates of a return to the 'glories' of a pre-scientific age," the people who talked about diabolical scientists tinkering, "like a bad small boy," with atomic energy and blowing the earth to bits. Atomic apocalypse was a mere hobgoblin for the ignorant, he insisted. God would not have made the world such an unstable place. Millikan said it was not scientists but their opponents, those who would hold back the increase of human knowledge, who threatened our spiritual life and our very survival.[44]

If sensational claims for atomic energy made people nervous, Millikan was more than willing to tone them down. In fact, in the decades since radium was discovered, physicists had turned up no hints about how radioactivity might be released wholesale. Millikan deflated the exaggerated public hopes along with the fears, announcing that solar energy would be a better bet than atoms. Faced with such authoritative skepticism, one magazine sighed about "the necessity of surrendering this great expectation of future wealth of energy."[45]

Still more widely heard in the 1930s were Rutherford's denials. He was more famous than ever, for he was the one who had proved that the source of all radioactivity is not the atom as a whole but its nucleus, that tiny kernel buried among the cloud of electrons that fills most of an atom's volume. (Thus radioactivity did not really use "atomic" energy but "nuclear" energy, although most people used the terms interchangeably.) When a popular science book mentioned the usual ideas about the revolution to come if atomic energy were liberated, it was natural for the author to add that Rutherford was "working on this subject," and newspapers paid close attention to his pronouncements. What he said was that the world would not blow up, and that no process yet discovered could produce atomic energy for practical use. As he grew older he grew more blunt: "Anyone who says that . . . with our present knowledge we can utilize atomic energy is talking moonshine."[46]

This did not stop the colorful newspaper stories. Kaempffert pointed out that Rutherford and Millikan were speaking only of the known techniques of physics, and that other methods might be invented one day. Rutherford himself, when he addressed not the public but his colleagues, hinted that the release of nuclear energy was conceivable. As for Millikan, his most widely read statement on atoms—probably the most widely read of all statements on the subject—was one he let stand in the many editions of his textbook on physics, used in the majority of American high schools at a time when most high school students took a course in physics. Highlighting the question of whether atomic energy could ever be put to work, Millikan told generations of students that "such a result does not now seem likely or even possible; and yet the transformations which the study of physics has wrought in the world within a hundred years were once just as incredible as this." Other scientists, almost as famous as Millikan and Rutherford, were more openly optimistic. Prominent physicists in Italy, France, Germany, the United States, the Soviet Union, and in Rutherford's own Cambridge all said publicly that an astonishing atomic revolution might begin at any time.[47]

Some younger scientists also thought about dangers. In France, a leader of the new generation was Madame Curie's son-in-law, Frederic Joliot-Curie, a dashing and eloquent physicist. In 1934 he and his wife, Irène Curie, discovered artificial radioactivity: not only could they provoke a transmutation at will in certain elements, but the newly transmuted atoms were radioactive. It looked like a step toward contagious radioactivity, the fateful chain reaction that Soddy and Rutherford had wondered about decades earlier. A Nobel Prize made the discovery famous, and Joliot-Curie used his acceptance speech to warn scientists to take care lest they make the planet explode like a nova. Yet Joliot-Curie also told the public of the utopian wonders that atomic energy might bring, and worked to raise funds for his laboratory so that nuclear research could push ahead full speed. After all, he would never have chosen a career in science had he not believed that over the long run knowledge would do more good than harm.[48]

Not even humanist critics denied the value of science altogether. For example, by 1939 Raymond Fosdick was a leader of the Rockefeller Foundation, and when the physicist Ernest Lawrence requested money for a nuclear research apparatus, a giant cyclotron, Fosdick enthusiastically voted in favor of the proposal. He saw the device as a tool of pure, disinterested science, an emblem of "the noblest expression of the human spirit." He little imagined that parts of the cyclotron would soon be converted into a pilot plant to make material for atomic bombs.[49]

The press and with it the public also paid less attention to possible dangers from science than to expected benefits. Even articles that reported cautious predictions ran under titles that made grandiose promises, and in between the articles were pages of advertisements for "scientifically" improved products. Indeed, advertising (including its more subtle forms, from planted newspaper stories to fairground exhibits) never ceased to present an image of a future world that delighted in advanced technology. The only other route for the public to learn about science was in school, where physics teachers seldom neglected to impress students with the old tale of a bit of uranium driving a ship across an ocean.[50]

These forces were not confined to the capitalist democracies. The whole set of futuristic images was accepted by the educated elites in less developed countries, and perhaps most of all in the Soviet Union. For a central premise of Marxism was that improved technology inevitably drove history toward a Wellsian future of abundance and amity. H. G. Wells's writings were in fact highly popular in Russia. Soddy's writings too were translated into Russian and exerted a powerful influence, as one of many channels that brought atomic energy into the Soviet dream of a perfected future. Occasionally a Russian author or city planner would explicitly describe the White City of the future, its towers rising from gardens into the clouds. More common were short-term claims about increased steel production and the like, but the glow of utopian myth showed clearly through the threadbare screen of statistics. Communist thinkers made no bones about the role of such propaganda: it mobilized the populace to sacrifice themselves in fulfilling the leadership's plans for industrial development.[51]

Ceaseless technological advance was structurally embedded in the cultural, economic, and political institutions of modern society. If people had not expected progress, many institutions would have tottered. The promise that technology would improve their followers' lives was indispensable to liberal politicians, Fascist and Communist dictators, labor union leaders, advertisers, and the writers of corporate reports to stockholders. The critics of technology, a half-hearted minority of humanities professors, clerics, and authors, swayed no other social groups to their concerns. By the late 1930s the ripple of criticism had subsided. The evil scientist lived on only in lurid fiction, not in sober factual writings. From the capitalists with their Millikans to the Marxists with their Haldanes there was only one ideology, Progress through Science.

3

Radium: Elixir or Poison?

If there had been no visions of atom-powered cities or of planetary explosions, if there had been no thought of ever releasing energy wholesale from radioactive atoms, radioactivity would still have impressed the public as a symbol of the tremendous possibilities of science. There was another theme in the nuclear energy legend: the rays that could transform living flesh, healing ills or creating monsters. From the outset newspapers hinted that radium might "solve the problem of life." The idea was readily accepted. For there was already a connection between radiation and life, a connection found everywhere from ancient transmutation myths to modern science news.[1]

It was clear almost from the start that radioactivity had powerful effects on living creatures. Pierre Curie and other radium researchers found that if they carried the substance around in a pocket, it would burn their skin. Curie killed a mouse with a dab of radium, and impressed reporters by remarking that he would not care to share a room with a kilogram of the element. In the hands of criminals, he announced, it could be a great danger. However, Curie and other scientists followed up such warnings with reassurances that the problem was not radioactivity itself but only the chance of misuse. In the hands of experts who took proper precautions, the prodigious power of radioactivity would be all to the good. The patches of sore, red skin caused by radium were akin to the wounds caused by concentrated X-rays, and physicians were finding X-rays extraordinarily helpful; atomic rays seemed likely to be better still.[2]

Already in 1903 Soddy suggested that people with tuberculosis, the most dreaded disease of the time, might benefit from inhaling radioactive gas. Physicians raced to try out such ideas. Radium proved a help in reducing certain skin cancers and tumors, for if the rays were harmful to ordinary flesh, they were far more harmful to some types of cancer cells. The press quoted doctors who speculated that radium would cure every type of cancer and many other diseases. Meanwhile Pierre Curie and other scientists, studying the waters in health spas, found that the springs carried mild radioactivity up from the deep rocks, and speculated that the waters were beneficial because of their dose of radioactivity. Did the rays somehow stimulate the body toward health?[3]

In the absence of solid knowledge, fantasy had free play. Newspapers proclaimed, "Old Age May be Stayed by Radium." When a doctor held radium next to the optic nerve of a blind girl, she claimed to have sensations of light, which suggested that "Radium Makes Blind Girl See." A small-town newspaper editor drew a logical conclusion, saying that if radium could revive a dead nerve, why not other parts of the body too? Might it not raise the dead? A young science graduate at Cambridge University mixed sterile bouillon with radium in a test tube and created curious shapes—had radioactivity created altogether new life? The chemist Sir William Ramsay pointed out that the shapes were merely coagulated bubbles, but he went on to declare that the idea of radium vitalizing inert matter was indeed attractive.[4]

Radioactivity somehow reminded people irresistibly of life. In their first paper on atomic energy, Rutherford and Soddy speculated about the "evolution" of atoms, and soon they were calling transmuted atoms the "daughters" of radioactive "parent" atoms. Gustave Le Bon made the idea explicit in a popular book titled *The Evolution of Matter*, exclaiming that the transmutations of radioactive matter were precisely comparable to changes in living creatures. "All Nature Now Alive!" cried a London review in 1903. Radioactivity might not only stimulate health but reflect a vitality stirring within all matter, the "Secret of Life."[5]

Some guessed that the secret extended to things beyond life. Le Bon said that radioactivity showed matter continually transforming into a more ethereal, not to say spiritual, form of existence. Preachers compared the eerie glow emitted by radium with mystic illumination; like radium itself, "Radium Christians are constantly giving off energy." After Ramsay did some research with Soddy on transmutation he received a number of letters from spiritualists, who declared that radioactivity was a supernatural agent and solemnly warned scientists to keep their hands off this mystic force.[6]

After a few years the newspapers turned to other sensations, but the idea that radioactivity was associated with the secret of life remained in the back of people's minds. The idea revived in the 1920s when Robert Millikan announced that he had solved the problem of the radiation that he and others had detected streaming down from outer space. He believed that these "cosmic rays" were a sign of the creation of atoms out between the stars—the "birth-cries" of atoms. To Millikan, atomic rays suggested that all matter was in a sense alive, continually at birth or in decay. Newspapers went further, with headlines declaring that Millikan had found the "Secret of Eternal Life."[7]

In plain fact, of course, a rock is no more alive than ever if it is radioactive; a person exposed at random to radiation will not necessarily become healthier; cosmic rays are simple particles that can yield no secret of immortality. Then where did these extraordinary ideas come from? Soddy gave a clue that can help lead through the maze. The philosophers' stone, he pointed out, "was accredited the power not only of transmuting the metals, but of acting *as the elixir of life.*"[8]

THE ELIXIR OF LIFE

If the forces within atoms suggested much more than industrial power and explosives, this was because ideas about physical forces had always been inseparable from ideas about life. The tangle of associations could be found everywhere from primitive folklore to child psychology. To see the imagery in its most elaborate and revealing development, I turn again to the medieval alchemists.

Chinese, Middle Eastern, and European alchemists shared theories about a fundamental life-force, an occult power perhaps like a fire or fluid, a quicksilver elixir underlying every process from the growth of plants to chemical change. They believed, for example, that this force makes gold grow in the earth much as a plant grows, if more slowly. By manipulating the life-force in their furnaces, the adepts hoped to master such processes. They would not only make gold but transmute human flesh itself into incorruptible perfection; they would achieve not only spiritual rebirth but also bodily immortality.

The alchemists reinforced these ideas with striking imagery. When they said that gold would grow in their laboratories, they meant growth just like that of a baby within its mother; they deliberately shaped their sealed vessels in the form of an egg or a womb. Some of them explained that their transmutational work was like a manipulation within a mother's body. Furthermore, just as human life begins with a union of man and woman, so to move matter toward rebirth the adept aimed to bring about a marriage of "male" and "female" chemical substances. Manuscripts illustrated this concept with frank drawings of a naked couple in coitus. The life-force, then, could be a specifically sexual energy.

Wiser adepts used the marriage of chemicals as another symbol of processes leading to spiritual transformation. Perhaps they sensed what modern psychologists have insisted: to shake free from infantile problems, a person has to reconcile male and female influences and come to terms with sexual drives. However, the alchemists' explicit thinking followed a more religious direction. From early times into the twentieth century, in religions from Catholicism to Tantric Buddhism, the "mystical marriage" was a powerful symbol for the way to spiritual renewal.[9]

The mystical marriage carried even grander associations, extending to cycles of cosmic creation. Long before Soddy and Millikan connected radioactive transmutation with the primeval evolution of matter, some alchemists had imagined that in their retorts they were re-enacting not just any birth, but the destruction and rebirth of the world. Thus transmutational marriage was associated with the Wheel of Time.

Ethnographers have found among peoples on every continent the myth of a couple who had survived world catastrophe to engender a new cycle of humanity. In modern Europe the theme remained visible in such places as the *Last Man* novel of 1805, where the lonely protagonist fruitlessly sought a mate, and the 1901 novel *The Purple Cloud*, where after years of wandering his poisoned world the protagonist did find a last woman, and the couple became the Adam and Eve of a new and better race. However, this image would not become a ubiquitous feature of modern legends until after the first atomic bombs exploded.[10]

A main reason for the persistence of all this imagery since early times was the way it fitted with other common patterns of thought. The most basic idea, the life-force, was not just a Eurasian historical tradition but a remarkably widespread feature of human culture and perhaps even human psychology. Anthropologists found that most native peoples tended toward animism, seeing signs of life in everything that moved or changed. Some native thinkers abstracted the idea of a living energy, perhaps called "heat" and connected with fire or fluids. Following this hint, Jean Piaget discovered that in modern Europe too, young children spontaneously tended to treat all things as somehow alive.

To children, the life-force and related bodily changes were not abstractions. Questioned with care, children said that they felt surrounded by a world of mysterious powers, literally the powers of life and death. They were especially interested in the mysteries of birth, yearning to understand the mother's manner of creating life. A still more fascinating problem was that focus of energy and genuine source of life, the parents' sexuality. Curiosity about such matters was a childish anticipation of more mature aspirations to understand the mysteries of life, but the associations forged in childhood tended to persist. A wide-ranging symbolism of sexuality and procreative life-force was unequivocally observed in many cultures, shaping images that people projected onto all sorts of things—not least, as I will presently show, onto radioactivity.[11]

Ideas about using the life-force for marvelous transformations, perhaps through some sort of marriage, were particularly close to infantile thinking. Facing the overwhelming forces of life and death, children commonly had fantasies of wielding mysterious magic powers, setting everything right like a fairy-tale hero. And the hero would not act entirely alone. "They married and lived happily ever after": the philosophers' stone in its most primitive form.

In summary, a number of themes were permanently associated with the concept of life-force. That potent force entered into speculations about the cosmic order, questions of marriage and procreation, and fantasies of magical powers capable of anything up to bodily immortality and world rebirth. These associations were evidently important, for they not only arose spontaneously in modern children but persisted among adults in traditions around the world. The whole tangle of themes was woven together most explicitly in medieval alchemical thought, but it could show up in modern stories as well.

How did all this become associated with nuclear energy? I cannot pursue every strand in the tangle. To show how these themes came into the twentieth century I will take one case—a crucial one for the history of nuclear energy—the theme of radiation.

RAYS OF LIFE

An association between life-force and rays, which became a key element in the public image of radioactivity, was inherited from the distant past. Primitive peoples naturally connected sunlight with the growth of crops, and hence with procreation. Sometimes light rays took an explicitly sexual form, as in folk tales where sunlight brought conception (the many variants of the Danae legend that persisted in Europe into modern times, along with tales found in Siberia, Mexico, Fiji, and so forth). In Western culture, from ancient Egypt onward, people saw light as a general symbol of life and energy. Respected medieval philosophers said that sunbeams and the invisible rays that carried down astrological influences from the stars conveyed occult forces that might be the cause of every kind of change. Radiation in the form of light also seemed close to the divine illumination of ecstatic mystics, and even to the light that the Bible connected with God's original Creation (Genesis 1:3; John 1:7–9). As alchemists and churchmen contemplated radiant light, cosmic energies and cycles opened before their dazzled eyes. All this and more would eventually find a home within the gleaming ray-devices of the nuclear energy tales.[12]

Up to the time of the scientific revolution, the tangle of images associated with transmutation and life-force explicitly included rays; it all formed a coherent pattern, which leading thinkers like Newton swallowed whole. Later scientists struggled to sort out the true agents of physical change, for example electricity. But the old associations were hard to shed. Into the eighteenth century people saw electricity in particular as a fiery fluid which could stream across space in rays and which was intimately involved with life processes. Scientists speculated that electrical forces could energize mineral slime into primitive life; poets compared electricity with the divine spark that animated souls; sick people flocked to quacks like the Englishman James Graham, in whose electrified bed couples were guaranteed to beget children.[13]

In the nineteenth century physicists made sprays of electricity visible in evacuated tubes, enchanting the public with displays of colorful rays in a thousand darkened lecture halls. These electrical rays turned out to be simply streams of tiny particles, electrons, with no more life than a shower of sand. At the end of the century came the discovery that when such a beam of electrons struck something and made it glow, the radiation might include a new form of energy, X-rays. Interest in these invisible rays led directly to the discovery of radioactivity—only the latest step in an ancient tradition of inquiry into radiation. But by then all the rich symbolism of life-force had finally been pared away . . . or had it?

Well into the twentieth century most people spoke of "energy" without distinguishing the physicists' term, which was now precisely defined, from spiritual energy, sexual energy, and so forth. The symbolism of light rays remained part of the package, whether in children's stories, in futurist paintings, or in nationalistic festivals that deliberately used beams of light as archaic symbols of vitality and rebirth. It was above all electricity that carried forward the old associations. Reporters likened Edison to a wizard on the track of an electrical philosophers' stone. The ancient and sacred symbol of thunderbolts now streamed out, ray-like, in advertisements for electrical appliances, while quacks sold tens of thousands of battery-powered "electrical belts" guaranteed to cure everything from impotence to old age. In the 1914 movie serial *The Exploits of Elaine*, a heroine was brought back from the dead with electricity; that had long since become a hackneyed plot device.[14]

Not even scientists could disentangle themselves from these pervasive ideas. Around the time radioactivity was discovered, widely accepted physics theories suggested that rays and electricity were both manifestations of an invisible "ether," which some linked with immortal spiritual forces. Medical authorities still believed that vitality had to do with a mysterious force that was much like electricity and probably close to sexuality. When Freud first began to treat neurotics he followed a method recommended by prominent physicians, sending electric currents through his patients.[15]

Radiation imagery remained particularly fascinating, as the case of René Blondlot showed. Blondlot, a respected French physics professor, announced in 1903 the discovery of a new sort of radiation, "N-rays." Although the rays were almost too dim to see, dozens of scientists around France managed to observe them. A physiologist found that N-rays signaled the presence of life and even of mental activity, for a working brain emitted more rays than a quiescent one. The radiation in return seemed likely to stimulate nerves and benefit health. Some people claimed that the rays revealed psychic energy, and might explain the séances where spiritualists brought dead souls back to earth.[16]

Alas, N-rays never existed. The international community of scientists, after a year or two of careful study, dismissed Blondlot's claims. He and his followers, straining their eyes for hours in darkened laboratories, had projected onto the shadowy apparatus their hopes for a special light. The most noteworthy projection had been an association between radiation and life-force. Scientists might keep patiently working to unwind the strands of imagery, but each strand survived and kept twining back around the others.

By 1910 the concept of radioactive transmutation had knitted the strands firmly together. The ancient transmutational imagery had spoken of heat, primeval light, and the creation of the world; as if in confirmation, excited physicists pointed out that for countless ages radium had been emitting heat and light while transmuting, as though it contained the energies of the original creation. When doctors reported that atomic rays could heal, they seemed only to confirm that radioactivity meant life-force. Some people even suspected that, as a headline proclaimed, "Secret of Sex Found in Radium." If electricity was becoming a humdrum household matter, then the mysteries once associated with it could be transferred to radioactivity, where they seemed to fit even better. A 1935 movie serial, updating *The Exploits of Elaine*, brought cowboy actor Gene Autry back from the dead in a "radium reviving room," although the apparatus also crackled with traditional electric sparks.[17]

DEATH RAYS

The imagery of rays and life-force carried not only hopes but fears: a power great enough to transfigure flesh could also destroy it. The idea of a malevolent radiated force may seem straightforward, but a review of its history will show that, even more than other nuclear themes, this one had long been encrusted with complex and emotive meanings. These meanings, which would one day appear in sober political debate over bomb tests and nuclear reactors, first became visible in the monstrous death rays of legendary tales.

In their earliest form, malevolent energies were not embodied in visible rays but were simply thrown from a distance. Native peoples around the world believed firmly in witches who caused disease by projecting some sort of magical contamination into their enemies. The more specific idea of harmful radiation occurred to various peoples, for example the Maya of Central America, who believed that the god-star

Venus sometimes shot down rays like spears to cause pestilence and other harms. Evil stars were still more familiar in Europe, where into the eighteenth century physicians believed that people could literally be struck ill by astrological forces, depicted sometimes as vague influences and sometimes as explicit rays.[18]

Ideas of projected life-force, whether harmful or beneficial, were carried into modern culture particularly by doctors, and not least by the notorious Dr. Mesmer. He claimed that during his astrological studies he had discovered a universal "magnetic" fluid which could stream forth much like electricity. In the minds of his followers the new fluid quickly picked up the usual traits of life-force, from claims of remarkable medical effects to reports of Mesmerists raising the dead. Rival physicians whispered that Mesmer was a menace who dominated impressionable women with a more familiar kind of life-force, his male sexual energy.[19]

Many people associated the dangerous projected forces, as much as the life-giving ones, with sex. For example, seventeenth-century Europeans and Americans accused witches particularly of misusing procreative powers, causing impotence and the death of infants. Perhaps the most obvious symbolism was found in an Australian tribe where a male witch would hold a bone under his penis and point it toward his intended victim; the modern American jabbing a middle finger at an enemy was scarcely less explicit.[20]

Life-force was still more easily associated with mental powers. There were self-styled witches who sometimes caused real harm, and Mesmerists who worked real cures, through their unwitting use of hypnotism. Going far beyond that, stories throughout the nineteenth century told of fiery-eyed Mesmerists putting victims into trances, using a sort of magic willpower to enslave them. In 1895, the year X-rays were discovered, America's best-selling book and most popular touring stage show featured the scientific hypnotist Svengali, who controlled his female victim with a combination of sexual and mental energies. The scientist of nuclear energy tales who pointed a weird ray device at a woman was Svengali's heir.[21]

The psychological meaning of commanding rays can be seen most clearly in the image of willpower radiated out through the eyes. Pictures of a powerful, radiating eye can be found everywhere from ancient Egyptian papyri to the drawings of modern schizophrenics, and in Irish and Persian myths a divine eye could even hurl thunderbolts. The ancient concept that eyes can emit a sort of luminous beam survived in modern figures of speech, as when an 1890s political leader was said to have a gaze as "penetrating" as X-rays. In all these cases the radiated power was associated with a strong authority. This idea was not unreasonable, for animals and people do tend to submit when they face a dominating stare.[22]

Stern eyes could also seem threatening when they saw "into things," for example into a child's misdeeds or forbidden sexual matters. The discovery of X-rays called forth nervous jokes about people using gadgets to peep through closed doors, and a brash London firm offered X-ray-proof underclothes for sale. Decades later, a psychologist reported the case of a male X-ray technician who had chosen the profession because, as he learned under analysis, he had a secret wish to see within his mother's body. A symmetrically opposite case was a female psychology student who fainted when being X-rayed, fearful of being "seen through." Thus baneful penetration and discovery of secrets were added to the many powers latent in the symbol of rays.[23]

The whole cluster of symbols was most openly displayed in minds not restrained by logic. Clinical studies of schizophrenic people found that they tended to believe passionately in magical forces of life and power. The insane of medieval times had expressed such beliefs by raving about witches, but by the nineteenth century well-read paranoiacs were adopting a more scientific jargon, insisting that some sort of electrical fluid or Mesmeric current or (later) X-ray was assaulting them. "Influencing machines" operated by their enemies, sending forth electricity or rays, were injecting uncanny feelings and thoughts. They insisted that rays carrying sexual power or willpower, emanating for example from a "black sun" or a "radium tube," were ruining the world or healing it.[24]

Ideas about rays that came up spontaneously in minds operating outside normal limits, ideas that could be traced far back in history, ideas that were common folklore among peoples around the world, whether primitive or civilized—such ideas must not be taken lightly. Some individuals would be impressed by rays that represented illumination or blazing thunderbolts from God, others by rays that projected health or illness, others by rays that altered the mind or peered into secrets, still others by rays carrying sexual symbolism. At least one nineteenth-century fantasy novel, Edward Bulwer-Lytton's *The Coming Race*, featured all these attributes of rays together. However, by that time ray imagery was generally scattered and inconspicuous, part of the anonymous background of culture. It was the discovery of atomic energy that wove the diverse images firmly together again.[25]

Another striking idea about radiation that became linked with atomic energy had nothing do with biology: the idea that radium rays could be used as a physical weapon. It was again Gustave Le Bon who, in 1903, introduced the most imaginative speculations, announcing that radium might emit a sort of ray that could fire gunpowder at a distance. From New York to St. Petersburg the press took up his fantasy of exploding the ammunition in an enemy nation by wielding a small tube. Soon there was newspaper talk of radium rays smashing battleships directly. Magazine fiction followed, such as a story by Jack London about a genius who used ray powers from his island hideout to kill people around the world. Unlike talk of atomic bombs or talk of medical applications, these death-ray stories had not a shred of scientific fact behind them, no relation to any of radium's actual properties. The stories were descended entirely from older myths.[26]

After all, this specific idea of a physical ray weapon was nothing new. The Romans had the legend of Archimedes blasting an invading fleet with giant mirrors that concentrated the rays of the sun, and one ancient author had even associated such devices with a mention of the whole world consumed in flames. Talk of a philosopher's giant burning-glass destroying armies or cities had continued down the centuries. In the early nineteenth century Washington Irving penned a sketch about Moon-men conquering the earth using concentrated sunbeams, and at the end of the century H. G. Wells wrote of Martians blasting soldiers with beams of heat. As soon as Roentgen announced his discovery of X-rays, some people wrote him to express fear of his "death's rays." While most scientists scoffed, the press kept the pot boiling for decades by quoting a few respectable scientists and inventors who thought that ray weapons really could be made, using heat or electricity or perhaps the energy within atoms.[27]

The pulp science fiction magazines that became popular among boys and young men in the United States from the 1920s on, along with radio shows and comic books like the famous Buck Rogers stories, could hardly have made it through an issue without rays. Atomic or "radium rays" would incinerate enemies, transform living creatures, and enslave minds. The ancient heritage of such images was transparent in a 1940 Captain Marvel comic where the villain used rays not only to attack people but to make gold; as he rightly said, these were "powers such as men have dreamed of since the beginning of time." More than half a dozen movies about ray weapons appeared before 1930, for example a Russian science fiction film straightforwardly titled *The Death Ray*, featuring the usual inventor blasting things from an island hideout. At least two dozen more death-ray movies came between 1930 and 1940.[28]

Roughly half of these movies used the rays as a defense against airplanes, the most dreaded menace of the 1930s. Typical was the 1940 American film *Murder in the Air*, featuring Ronald Reagan as an agent who guarded the secret of a projector that could bring down enemy bombers—"the greatest force for world peace ever discovered." The idea seemed so plausible that some citizens called on their governments to build real ray weapons to protect them. In the mid-1930s Japanese authorities set up a death ray research program that continued until 1945; meanwhile the British assembled a secret panel of top scientists to study the air defense problem with ray weapons as the first item on their agenda. The British committee found that they could not build a death ray, but they went on to develop radar, a true farseeing defense ray.[29]

RADIUM AS MEDICINE AND POISON

The public's image of atomic radiation was not built wholly on old myths, for by the 1930s many solid facts were known. Educated people could shrug aside the tales of rays that struck people like thunderbolts or brought them back to life. But as the verified medical and biological facts came into view, they showed a surprising ability to remind people of primitive life-force imagery.

At the turn of the twentieth century use of the new X-rays had brought a new malady, radiation injury. The victims felt weak, suffered vomiting and diarrhea, perhaps temporarily lost patches of hair. Anxiety about this increased when cancers showed up months or years after the victims had been exposed to X-rays; by 1911 doctors had reported more than fifty cases, some of them gruesomely fatal. Equally unsettling to the public was the news that men whose genitals were heavily X-rayed could be left sterile for weeks or longer. By 1912, when Rudyard Kipling wrote a story about a future world government of virtuous Airmen who quelled an unruly crowd by beaming down hypnotic rays, the author could take it for granted that the crowd would fear sterilization.[30]

But public concern over X-rays was muted, for almost all the injured people were physicians or technicians who had exposed themselves to too much radiation. It was an honorable tradition that when a new therapy was plied against deadly disease one would have to accept a certain number of accidental deaths, especially among the physicians themselves.

When radium came along a few years after X-rays, people benefited from the earlier experience. Although some doctors and technicians who used radium came down with radiation injury and cancer, these tragedies were minor compared with the prospective benefits. Physicians devised ways to use radioactive substances to destroy various types of tumors without risky surgery, and radiation became familiar as a part of medicine. During the 1930s more than 100,000 people around the world were treated with radium every year, mostly to their advantage. And over 100,000 were X-rayed every *day*, usually to diagnose their ills but sometimes to cure them. Experts and publicists expected still greater use of radiation in the future; for example, J. B. S. Haldane predicted that someday every neighborhood would have a radiologist's shop, treating clients with prescribed combinations of rays much as a pharmacy dispensed drugs. Meanwhile doctors experimented by poking X-ray projectors and radioactive substances into every part of their patients' bodies. Curiously, these men took a special interest in irradiating the vagina and uterus: gynecological experiments led the way in both X-ray and radium therapy.[31]

An industry arose to meet the demand for radium. The element is rare, a product of the slow transmutation of uranium. So from the jungles of the Congo to the tundra of northern Canada, discoveries of uranium ore raised as much excitement as the discovery of a lode of gold. Indeed, journalists emphasized that radium, gram for gram, cost thousands of times more than other precious metals. Articles about radium often focused on the element's commercial value, strengthening the belief that radioactive matter was valuable beyond compare. How foolish were the alchemists, some writers said, who had sought mere gold!

The greatest transmutational wonder of all was announced in 1927 by Hermann Muller. An independent and combative biologist, Muller had been convinced from his boyhood that humanity must seize control over its own evolution—advancing toward ever-better social organization and even improving our species' biological form. When he turned his mind to radiation in the mid-1920s, it was because people had long been fascinated with its potential to reshape living things. Scientists had subjected the eggs of simple creatures to X-rays and radium, finding that if the damaged embryo did not die it might grow up a monstrosity. People who worked with radiation had naturally begun to worry about their own offspring. What about hidden changes in the genes, they wondered, the kind that might be inherited and show up in a later generation?[32]

Muller bombarded flies with heavy doses of X-rays and found that the more X-rays, the more mutations showed up among later generations of flies. Mutations, such as oddly colored eyes or shrunken wings, were nothing new to fly breeders. They suspected (correctly, as it turned out) that mutant creatures could be manufactured on demand by exposing the grandparents to certain common chemicals, or even by slightly raising their temperature. But Muller was the first to alter genes deliberately, and it was particularly exciting that he did this with radiation.

One magazine now reported that rays could be used to alter creatures "exactly as physicists bombard the atom in order to change its composition." Muller himself, consciously intending to make a connection with nuclear physics, called his discovery the "Artificial Transmutation of the Gene." Although he warned that indiscriminate use of radiation might create unwanted mutations, Muller attracted much more attention when he suggested that his discovery could help scientists guide the evolution of improved plants and animals, not to mention humans.[33]

Making it all seem more plausible, scientists pointed out a genuine connection between radiation and evolution. The world has always been bathed in natural radiation, coming down in cosmic rays, coming up from the traces of uranium and other radioactive elements that are present in ordinary rocks, even coming from the radioactive minerals that all creatures carry within their bones. After Muller's announcement scientists realized that this radiation, along with natural chemicals and heat and other sources of mutation, had provided evolution with its raw materials by altering genes at random. Some science writers, ignoring the slow workings of natural selection, leaped to the conclusion that Muller's rays could speed up evolution. Writers for pulp science fiction magazines imagined that a scientist who could concentrate cosmic rays might compress millions of years of evolution into hours, crafting a monster or a superman. This was one of various ways to introduce the weird being who was born under the scientist's rays in the nuclear energy legend.[34]

Scientists offered still more provocative ideas. Not long after Muller's discovery a professor in Moscow announced "mitogenetic rays," and soon numerous European and American laboratories confirmed and extended the discovery. Life-forces were the essence of these rays, which were emitted directly by living organisms; indeed, a fertilized egg emitted more rays than a sterile one. The rays conveyed life-force itself, for they could make nearby plants grow faster. Many scientists were skeptical of the news but others embraced it, noting that the new rays were no more amazing than any other sort of "radioactivity." The discovery fascinated science writers: weren't the "M-rays" another wonderful tool to help humanity? Alas no, it was the story of N-rays all over again.

Such incidents were not entirely aberrations, for it is the job of science to produce bold hypotheses and test them rigorously, following dozens of false leads for each new truth. What is interesting is the specific ideas projected onto M-rays. The case showed that there remained a minority of scientists who were only too eager to pull archaic radiation myths from the surrounding culture, and then hand the myths back to the public as fact.[35]

In every case, real or imaginary, the hope for benefits far outran any worries. The most famous of radioactivity workers, Madame Curie herself, refused to take precautions that might slow down her work, and regularly got radioactive matter into her mouth and lungs. Workers at her Radium Institute followed her example and fell sick; one of them died, and she herself weakened year by year. Yet she refused to believe that her wonderful radium, which was saving so many lives, was killing her by inches.

The public was little wiser. As late as 1929 one magazine, noting the elaborate safety rules that doctors were imposing on the use of radium, remarked, "Many will be surprised to know the dangers of this healing substance." The doctors themselves were unsure, for a number of experiments on animals and plants gave evidence that very small amounts of radiation sometimes really did stimulate growth and health. In fact, the influence of a tiny dose of radiation was so slight compared with other influences on health that the effects for better or worse were scarcely detectable. The problem came when some people said that radioactivity was so extraordinary that large doses of it were bound to work wonders.[36]

A 1929 European pharmacopoeia listed eighty patent medicines whose active ingredients were radioactive. You could take radioactivity in a tablet, bath salts, liniment, inhalation, injection, or suppository. You could eat mildly radioactive chocolate candies and then brush your teeth with radioactive toothpaste. Manufacturers promised that their nostrums would give relief from any number of ailments, notably rheumatism, baldness, and "symptoms of old age," as if radioactivity were a genuine elixir of youth.[37]

A notable example of the people who sold such medicines was the despicable William J. A. Bailey. He had first come to the public eye back in 1915 when he was arrested in New York City for selling automobiles which, it turned out, he never owned. Subsequently Bailey took up a safer line of work, selling radioactive patent medicines. In his extensive advertising Bailey offered to cure everything from influenza to flatulence, but he particularly implied that radioactivity could bring sexual rejuvenation. His most memorable nostrum was "Radithor," a liquid in little bottles which he sold by tens of thousands. Regrettably, unlike many who peddled supposedly radioactive medicines, Bailey really did put some radium in his liquid.

Radithor's most loyal customer was Eben Byers, well known in Pittsburgh society as a wealthy manufacturer, sportsman, and playboy, approaching fifty years of age. For five years Byers drank down his bottle of Radithor almost every day, as instructed. He believed it was a sort of aphrodisiac and urged bottles on his paramours. At first Byers and his women friends felt better for the tonic, possibly because they were imagining things, or possibly because the radium they drank provoked physical changes that gave them a temporary feeling of well-being. If so, they were following in the footsteps of medieval Chinese alchemists, whose elixirs of arsenic and mercury could stimulate appetite and sexual potency for a little time before the harm set in. Byers continued to take Radithor more desperately each year as his health failed, until in 1931 he entered a hospital, feeble and emaciated, his very breath radioactive. He did not have time to develop cancer but died of direct radiation injury within a few months.[38]

This was the first proven case of death from a patent radioactive medicine, and newspaper headlines spread a ripple of fear across the country. Thousands of people had swallowed their daily dose of Radithor, while hundreds of thousands around the world patronized spas that proudly advertised the radioactivity of their waters. The Byers tragedy brought the usual results of a technological failure: demands that authorities take action, committee investigation, revision of official guidelines. Bailey closed down his factory, protesting that he had done no harm.[39]

The public did not seem to become fearful of radioactivity itself. Radium was only one among many kinds of patent medicines, some of them far more deadly. When journalists warned against such nostrums they took note that, in the hands of competent physicians, radiation remained a great force for health.

Radiation injury was more worrisome when it happened outside a medical setting. Radium had another large-scale use, for it could make certain ordinary chemicals fluoresce. The public was delighted with luminous paint, and everything from children's toys to doorknobs began to glow in the dark. At the center of the industry, in New Jersey, hundreds of young women applied delicate lines to watch dials, tipping the points of their brushes on their lips, unaware that they were ingesting radium from the paint. Radium that is eaten behaves much like a chemically similar mineral, calcium, tending to settle in the bones. Local dentists began to notice an epidemic of problems that they called "jaw rot." In 1925 radium was suggested as the culprit. When the victims sued their employer, newspapers around the world exposed the young workers' travails in mawkish stories about the "Case of the Five Women Doomed to Die."[40]

To the ordinary reader this was just another industrial tragedy, and not the most striking one. For example, the world had heard a few years earlier about the horrid death throes of petroleum industry workers poisoned by the gasoline additive tetraethyl lead. Many tons of this "insanity gas," as one magazine called it, were being spewed into the air in automobile exhaust, but the public took scant interest. Meanwhile, studies in the late 1920s of a small group of uranium miners in Bohemia showed that more than half of them were dying of lung cancer, and radioactivity was suspected as the cause. The miners got even less attention than the watch dial painters. Reformers had more pressing concerns, such as the several thousand coal miners who died every year from accidents alone. By comparison, radium poisoning was not so much a problem as a curiosity.[41]

The public was not easily convinced that radioactivity could be dangerous at all. The enterprising Bailey bounced back to sell an "Adrenoray," a belt allegedly holding radioactive substances and advertised to cure "lowered sex function, general debility, and neurasthenia," much like the electric belts that were still widely used. Doctors of sound reputation continued to use heavy doses of radiation to treat not only serious ailments but also cosmetic problems like warts or excess facial hair. Some even offered men temporary birth control through X-ray sterilization. As late as 1940 many hospital and laboratory workers were casually exposing themselves to radiation at levels far above the official guidelines. If a scientist could not use one hand for a while because it was inflamed by X-rays, or if a doctor lost his appetite for a few months because he had gotten some radium into himself, they were willing to run such risks in the service of humanity.[42]

During the 1930s nuclear physics research become more closely connected than ever with medical marvels. The medium was that universal connector of ideas, transmutation. Devices like Ernest Lawrence's cyclotrons could transmute many kinds of atoms by bombarding them with particles, and the changed atoms would often be artificially radioactive. By the late 1930s physicists could serve up radioactive forms of sodium, iodine, and numerous other elements, which promised to surpass radium in treating cancer.

The development was timely for supporters of nuclear research, for their vision of discovering a source of limitless industrial power had slipped away. As the years kept passing without any breakthrough, Rutherford's and Millikan's skepticism spread. By the late 1930s few physicists or well-informed science journalists were holding out much hope for an atomic industrial revolution. When physicists wanted to justify their work to patrons and to a practical-minded public, they would talk instead about benefits for biology and health. From Paris to Tokyo, physicists seeking to raise funds teamed up with biologists and medical researchers. To impress an audience, some scientists would make a "cocktail" of radioactive sodium which a volunteer or the lecturer himself would drink. A few minutes after drinking the elixir, the subject would wave his hand in front of a Geiger counter. The counter would chatter and the audience would applaud.[43]

Just how far did approval of radioactivity go? One tool to probe opinions of the literate and attentive part of the public is the *Readers' Guide to Periodical Literature*. Although these volumes list American magazine articles only by title, the writers of titles knew what would interest the public, and anyway most readers often got no further than the title. A search in the *Readers' Guide* to see what appeared under such rubrics as "Radioactivity" and "Atomic Energy" found many titles that would awaken little emotion, and a few that balanced positive and negative feelings. These neutral or balanced titles made up three-quarters of the total published between 1900 and 1940, but I set them aside to study the remaining, emotive titles. Some of these titles, such as "Curing Cancer with Radium" or "Harnessing the Atom," would evoke mainly positive, optimistic feelings; others were likely to evoke negative or even fearful feelings: "Handling Invisible but Deadly Rays," "Radium Poisoning," and so forth.

From 1900 until the middle 1920s I found very little negative language; whatever anxieties were mentioned came under neutral or hopeful headings. In the late 1920s the news of radium poisoning brought overt fears, yet positive titles still outnumbered the negative titles nearly two to one. By the mid-1930s radium hazards were no longer news, and the proportion of optimism rose to three to one; by the end of the thirties there were again almost no anxious titles among the many hopeful ones. On the whole the public was exercising its common sense, giving more credence to the successful experience of hundreds of thousands of patients than to isolated misfortunes.

Later in the century, when radioactivity frightened many citizens, experts suggested that it was feared particularly because people could not see it or feel it, or because it brought up instinctive concerns about sexual reproduction and babies, or because it could cause the especially dreadful disease of cancer. However, the public in the 1930s already knew these features of radioactivity yet showed no special anxiety about them. After all, there were other agents such as X-rays, common chemicals, and viruses that also could not be seen or felt and were known to cause sterility, birth defects, and cancer. In the 1930s concern about radioactivity mishaps existed on the same moderate level as fear of these other things.

Yet in fact there was already a special anxiety connected with radioactive life-forces. It was separate from ordinary thoughts of health problems, and scarcely appeared in nonfiction magazine titles. One could detect it only by looking closely into imagery, as a psychologist might notice a minute twitching in a seemingly confident patient with a hidden neurosis.

4

The Secret, the Master, and the Monster

Now I come to the core of the nuclear energy tales. Almost every story had a tremendous forbidden secret; a powerful authority who mastered the secret; and a device, often personified in a robot or monster, whereby the master caused good or harm. These themes seized the public's imagination with a strength that historical traditions and social pressures cannot entirely explain. For a full understanding it is important to see how the influences of history and society were reinforced by tendencies ingrained within each individual mind, what I will call psychological forces.

The theme of the cosmic secret is a good starting place, for here the historical, social, and psychological forces were all plain. The tradition behind the theme was familiar; tales about dangerous secrets were common among native peoples around the world. Western culture had Adam and Eve, Prometheus, Eurydice, Lot's wife, Bluebeard's wife, the sorcerer's apprentice, and a thousand more who came to grief by grasping after forbidden knowledge. The Christian apocalypse itself would begin when the book of seven seals, which no man is worthy to open, is unsealed (Revelation 5:1–4). In this tradition the medieval Catholic church, thinking it sinful to pry too far into God's mysteries, distrusted alchemy, while the alchemists themselves wrote endlessly about their perilous secret.

Alongside this tradition worked social forces that kept promoting talk of secrets, forces that became clear when the talk revived in connection with atoms. People who were nervous about the growing power of science, or upset by discoveries that overturned established beliefs, hastened to invoke the old idea that some things were best left hidden. The supporters of science found even more reason to talk of mysteries. Typical were radio talks given on separate occasions by Rutherford and by a noted German professor, in which each called himself an heir to the quest for the philosophers' stone. They both denied, like the medieval adepts, that they were seeking mere mundane gold; their real interest was the "search into one of the deepest secrets of Nature," the "unveiling of secrets surrounding the inner structure of matter." Many science journalists repeated the theme, as in a Kaempffert article headlined "Ultimate Truths Sought in the Atom." By saying that scientists were questing after cosmic mysteries, these people frankly hoped to increase public support for science. After all, humans are instinctively curious and tend to honor those who can answer their questions.[1]

But raw curiosity was only one of the psychological forces that could be awakened by talk of secrets; critics of science called up a different emotional pattern. Millikan was on target when he said his opponents saw the scientist who meddled in forbidden things as a "bad small boy." For it was common to warn children away from certain adult matters. If talk of harm from probing into secrets was both an ancient tradition and an effective weapon for critics of science, that was partly because everyone knew from an early age that a secret was precisely something forbidden—which meant dangerous. The person who forbade the secret presumably did that because it was dangerous to know, and anyway the person might punish a child who inquired too far.

Of course anxiety about secrets did not come only from childish associations with adult threats. Fear of the unknown was an instinct in all creatures, a valuable instinct necessary for survival. But individual caution toward a specific thing could be taught, becoming a cultural trait; the process was observed even in apes. An important example in Western culture was that children who probed into the mysteries of sex and birth were often forcefully turned away, encountering a boundary to curiosity like a jolting electric fence. Psychologists accordingly noticed that the secret of the womb or of coitus was for many patients the most fearful secret of all. Clinical observations of modern children found that they, like medieval alchemists, associated the secret transmutational life-force not only with hopes but with dire perils.[2]

In sum, instinctive caution, reinforced by widespread suppression of the natural childish curiosity about matters such as sex and birth, combined with traditional myths, which were evoked in modern debates over the social role of science, led many people to view the forces of nature in general and life-forces in particular as forbidden secrets. This view would become central to the public image of nuclear science.

SMASHING ATOMS

In nuclear energy tales, and many other tales too, the theme of the cosmic secret was inseparable from another theme: an attack upon the secret things in search of mastery. Twentieth-century physicists inadvertently fitted themselves into this pattern with their talk of atomic mysteries, above all when they insisted that the way to unriddle these would be to destroy, disintegrate, or "split" atoms. That sounded much like the central work of alchemical transmutation, which also involved breaking down substances before reconstructing them. This idea opened a pathway for strong psychological forces.

The alchemists had specifically symbolized matter as female (the word "matter" itself stems from the same archaic root as *mater*, mother); thus the basic alchemical operation of breaking down matter could sound like a symbolic assault on something female. A few ancient writings bear a resemblence to the modern clinical material collected by the child psychologist Melanie Klein. She observed that children may have fantasies of attacking their parents, striking into their bodies in belligerent attempts to discover the secrets of the womb.[3]

Ever since the invention of the plow, people had spoken of how "man" used technology to exploit Mother Earth and all Nature. Around the time of the scientific revolution, both the actual exploitation of natural resources and the language used to describe it grew bolder. The alchemists themselves had meant to help matter rise back from destruction, and the best scientists continued to respect the mysteries they studied; they hoped to find in themselves the creative principle associated with female Nature, or to join with it in a marriage of equals. Other writers were less discreet. Enthusiasts for science like Francis Bacon described scientists as men who would "master," "disrobe," and "penetrate" a feminine Nature. Such innuendo became habitual. By the nineteenth century it was commonplace to say that a scientist "interrogated Nature" in his laboratory, "unveiled" her, and "forced her to respond to him."[4]

Old metaphors sometimes came into modern times in a single step.

For example, Goethe learned alchemical symbolism firsthand through study of antique writings and his own alchemical experiments, and a proto-scientist's destructive quest for the eternal feminine principle became the central theme of his *Faust*. In the twentieth century Goethe's verses were learned by every German-speaking schoolboy, including some who would become the leaders of world science. If one wished to visit the time and place where modern nuclear physics took shape, a good spot would be the Niels Bohr Institute in Copenhagen in the spring of 1932, and there one would find the physicists amusing themselves by performing a satirical *Faust*. A leading character in this skit was a puzzling atomic particle (the newly discovered neutron, which would turn out to be the key to nuclear energy), representing the mysterious feminine principle which Faust avidly pursued.[5]

Twentieth-century scientists and journalists who wanted to stimulate public interest in physics found their most striking phrases in this old metaphor of aggressive pursuit. Atomic scientists investigated "the most intimate properties of matter," indeed "penetrated" hidden mysteries, "tore away the veils" to reveal inner secrets, and "laid bare" the structure of atoms. Language about breaking apart the indivisible atom could be openly belligerent. Already in 1905 a friend told Rutherford that another physicist was "so anxious to bust atoms artificially that . . . he would have tried it with a cold-chisel before long." A quarter-century later Millikan wrote of the "satisfaction in smashing a resistant atom."[6]

In the 1930s such talk came to a focus on "atom smashers," apparatus built to study nuclear structures. Physicists worked to publicize the devices, if only because they needed considerable money to run them. The press and public were fascinated by the shiny metal towers, massive and phallic, shooting forth powerful rays in what Kaempffert called "violent assaults" on the nucleus. Magazine illustrations showed atom smashers crackling with artificial thunderbolts as visible signs of the energies they invoked. The sparks and silver globes were also put to work to signify weird forces in science fiction illustrations and adventure movies, sealing the transfer of wizardry from electricity to atomic energy.

Of course there were also more lofty associations. The press rarely failed to mention that atom smashers were designed not just to split atoms but to transmute matter, passing through destruction to the miraculous. Once the atom was smashed, wrote Kaempffert, once its "holy-of-holies, its secret shrine" was laid bare, gold would be transmuted on an industrial scale and society would begin anew. Moreover, he said, the nucleus was the epitome of all processes, even of life and thought, a repository of the "secrets of the universe"; he who penetrated the shrine would understand "the whole plan and method of creation."[7]

With so much at stake, some writers found the idea of smashing atoms disturbing. To split an atom, an entity that (as Isaac Newton said) had been made a unity by God at the beginning of time—wasn't that like annihilating matter itself, or even the individual soul, impiously breaking down the basis of all things? The more enthusiastic atomic journalism became, the more it might remind people of the "bad small boy," the person who poked aggressively into forbidden matters, whether scientific, sexual, or spiritual.[8]

THE FEARFUL MASTER

The most obvious problem raised by nuclear energy tales was not that the legendary scientist poked into forbidden secrets, but that he might then use the powers he discovered to victimize others. Desiring to put the world in order, he would build uncanny devices, plotting to master and alter people. This typical domineering streak corresponded to another widespread human problem. From childhood conflicts with parents into adulthood, everyone struggled through relationships with authorities. Often the authority figure seemed dangerously knowing, powerful, and heedless of the wishes of others; these were threats that the scientist was especially apt to symbolize.

The arrogant villain who sought to rule the world, an old stereotype, was increasingly associated with science. Especially memorable were the American movie serials that captivated young people of many nations during the latter 1930s. Gene Autry descended into an underground atom-powered city to battle a prince armed with radium ray devices; Crash Corrigan overcame an undersea ruler who planned to enslave the world using "the atom—the most destructive force known to science"; and the incomparable Flash Gordon sabotaged an "atom furnace" to foil his archenemy Ming, who had boasted that "Radioactivity will make me Emperor of the Universe!" To be sure, Ming and his ilk might not look much like scientists themselves. But the use of advanced science was becoming a key attribute of the totalitarian nightmare.[9]

Historians who studied the genealogy of tales about technological dictatorships traced the ideas back to H. G. Wells, beginning in 1896 with his story *The Island of Dr. Moreau.* The brilliant doctor had strapped down and vivisected animals to turn them into half-humans, then controlled them with hypnosis and crude propaganda; his island held fiction's first society of scientifically dominated under-people. At the conclusion Wells warned, "The degradation of the Islanders will be played over again on a larger scale." He and his imitators fulfilled the promise in many later books and movies about technological slave states.[10]

Political oppression was only part of the point. Dr. Moreau wielding a bloody knife upon a screaming animal-woman: that called up more primeval thoughts. The best way to analyze these thoughts is to adopt the viewpoint that the authors themselves usually chose, that of the scientist's victim. A close look will show that feelings about victimization were basic to nuclear fear, for they are basic to all fear.

The most biologically primitive fear is a shrinking from pain and mutilation, but this is only an entry to the anxieties that psychologists have analyzed. Within mutilation they uncovered thoughts of castration, a fear more common among men than many supposed, while women had a similar and not unfounded fear of attack on their genitals. Both men and women particularly dreaded an attack that could rob them of the ability to procreate. A more general fear was that of losing all pleasure and power; psychologists found that the worst thing some people could imagine was to be strapped down, frozen and helpless. Many imagined that death was just such a paralysis, while others believed in a more explicit Hell, but in either case the fear of being a victim was hardly distinguishable from the fear of death itself. Indeed, children often saw death simply as a punishment. In short, not unlike the primitive peoples who believed that harm was usually due to attack by witches, nearly all people associated their basic human fears with the thought of being victimized by someone powerful.

From these interlinked fears the evil scientist drew his ability to come alive in countless tales. His rays, whether they maimed, sterilized, paralyzed, hypnotized, or simply annihilated people, stood for the power to deprive a victim of life and will.

There was another and perhaps even more important fear, linked to the evil scientist in a very different way. Anxiety about what others might do was not limited to their attacks. According to evidence gathered by a number of psychologists, at the root of many people's anxieties lay something almost opposite: the fear of being abandoned. Many people dreaded death itself mainly because it seemed like a total separation from other people. Such separation can be truly devastating, as in certain orphaned children who never laugh or cry, but sit by themselves uncomplaining and hopeless. In a lesser form everyone has felt if only for an hour the icy touch of that loneliness, the feeling of being cut off from humanity and from our own inner springs of emotion. And if adults have reason to fear abandonment by loved ones, children dread it still more, knowing by instinct how utterly their well-being depends on others. To the nursling, the mother is the world, and separation from such a caretaker can be as devastating as the loss of "Mother Earth" itself. Being abandoned is like being the last person alive in a ruined world.[11]

Such associations may have had special meaning for Frederick Soddy. His mother had abandoned him by dying when he was near eighteen months old, which clinical evidence has shown is a sensitive age for such a loss. As a substitute mother he had only a much older half-sister whom he came to hate. It was this background, Soddy believed, that made him solitary and indifferent to the people around him; modern studies of motherless children suggest that he may have been correct. In both his lonely childhood and his bitter maturity, Soddy had special reason to imagine how the world might be desolated.[12]

But why should thoughts of separation be connected with thoughts of dangerous science in particular? For many people the association probably came in a roundabout way, through the theme of the forbidden secret. A child thinking about punishment for poking into forbidden things might well fear separation, if only because isolation, rejection, or even outright threats of abandonment did serve as punishments in Western families. And this association between forbidden secrets and separation was reinforced by tradition. Long before the overweening curiosity of the protagonist of *The Purple Cloud* plunged him into loneliness as a Last Man (a punishment that took up most of the 1901 novel), Adam and Eve had been expelled from Eden, Eurydice had been torn from her Orpheus, Faust from his Gretchen, and various fictional scientists from their loved ones, all for unwisely seizing secrets.[13]

Separation suggests victimization of a different sort from that inflicted by a simple wielder of rays. The pattern here is too intricate to unravel all at once. For the moment I will only show what sort of associations were widely known; later I will show the surprising way they all fitted together.

The symbolic associations between secret forces and victimization were described most fully by people who specialized in fear, that is, paranoiacs. The classic case was Daniel Paul Schreber, a German jurist who entered an asylum in his middle years, saying that strange powers controlled and tortured him beyond endurance. Freud and other psychologists who studied Schreber's memoirs found that his problem was related to his childhood experience under his father, a doctor. The distinguished Doctor Schreber had apparently regulated every thought and action of his son from infancy on under a psychotically strict discipline, even building contraptions of iron bars and straps to enforce "correct posture." He was a genuine mad doctor.

A curious symbolism was displayed in the son's schizophrenia, as a hidden pattern may be revealed when a metal is stressed to the breaking point. Schreber thought that a world catastrophe had struck, leaving him the sole survivor. Divine "rays," he said, were radiating down upon him, clamping and tormenting his body while at the same time transforming and encouraging him (much as his domineering father had done). He looked forward to rebirth. Imagining himself "as man and woman in one person having intercourse with myself," the jurist insisted that through his suffering and survival he would renew and redeem the human race. Schreber lived within precisely the pattern of symbols—descent into chaos, all-powerful rays, lonely survival of world catastrophe, magical marriage, rebirth and redemption—long associated with transmutation. Was there some sort of connection between transmutation and fears of victimization?[14]

It was in fact typical of schizophrenics to think about the end of the world, to feel a paralyzing loneliness as if they were lost in a frozen waste, to envision cosmic battles and revelations, and at the center of it all to see themselves, chosen for survival and rebirth to restore the life of the universe. The pattern was not restricted to schizophrenia; epileptic fits and hallucinogenic drugs tended to bring forth similar overwhelming images, while much the same visions came to apocalyptic mystics from medieval times onward, and sometimes to more ordinary poets and writers. In all these people, under extreme stress everyday consciousness peeled away to reveal a bedrock charged with mythic imagery.[15]

In their hopes for transmutation, victims and visionaries were attempting to invert the most common human fears. In place of mutilation they looked for health; in place of sterility, fecundity; in place of helplessness, magic power; and in place of loneliness, a new community of universal love. Perhaps with some people the anxiety came first, and hopes for transcendence were a desperate response to the dread of victimization. In healthier people the hopes might come first, with anxiety following from the reasonable fear of losing the good of life. In either case the dread of victimization was as close to hopes of rebirth as a photographic negative is close to the positive picture.

The common psychological tendency to connect all these things together was reinforced by tradition. The most important example, the prototype of all mad scientists, was Victor Frankenstein in Mary Shelley's immensely popular story. Scholars analyzing her book have found that its psychological center was a transmutational urge mingled with anxieties about rejection and victimization. Frankenstein, a mixture of alchemist and scientist, was determined to "penetrate into the recesses of nature" in order to master her secret forces of birth. He meant to create life not in the ordinary way through marriage but all by himself, usurping divine powers. After he made his creature he immediately abandoned it in disgust; then the child-monster, in unbearable loneliness, sought murderous revenge. Scholars have suggested that the novel's theme of vengeance for paternal rejection had something to do with Shelley's own childhood problems with a cold, demanding father. Whatever the origin of her feelings, many people took the images to heart. The most famous of all horror movies, the 1931 *Frankenstein* by Universal Studios, diverged far from the original novel in plot, but this and many other versions of the tale followed Shelley closely in the underlying combination of uncanny birth, rejection, and victimization.[16]

Most nuclear energy tales got their emotional impact from the same set of associations. A telling example was a 1940 movie, *Dr. Cyclops*. At its center was Dr. Thorkel, an archetypical evil scientist with a bald skull and thick round glasses, whose mild voice rarely showed any emotion except contempt for whoever did show emotion. Thorkel's one passion was to work himself to exhaustion at a secret radium mine where he was grasping "the cosmic force of creation." Aiming, as he said, for total control over life, he used rays to tear living matter to bits and put it back together again—transmutation of flesh. When visitors intruded he shrank them to the size of mice, after which the film became a thriller of chases and attacks between the scientist and the shrunken visitors.

As in fairy tales about giants, the movie could easily have reminded people of children's victimization by a parent. Such struggles might have a sexual aspect. The conflict between Thorkel and his visitors began when one of them spied at night on his secret activities, discovering the scientist with a cylindrical, bulbous-ended radioactivity concentrator, straining rhythmically as he raised and lowered it within the vertical shaft of his radium mine. It looked suspiciously like the child's forbidden discovery, the penis in the vagina. Whether or not the scriptwriters meant to hint slyly about raping the earth, they gave a more general and explicit warning: an atomic scientist might seize "powers reserved to God" and use them to persecute the innocent.[17]

The tradition of mad scientist tales, and the universal childhood experiences it drew upon, were far from simple. If the scientist came to stand for a dangerously powerful and uncaring authority, he also stood for something else. In the end Dr. Moreau, Dr. Frankenstein, Dr. Thorkel, and their like died horrible, lonely deaths. That seemed appropriate punishment, for they had been prying into forbidden secrets, which was less the role of the parent than of the bad small boy. Who was really the authority, who the victim? This odd ambiguity points to the central knot in the puzzle of the nuclear energy legend, the least understood and most disquieting feature of all. To explain it I will leave the mad scientist for a moment and focus on his creations, which turn out to share the same ambiguity.

MONSTERS AND VICTIMS

In most tales about a dangerous scientist, the audience's horror focused less on the scientist than on the weird being or device he created. In nuclear energy legends this creation took various forms, from mutant child to atomic robot; each was evidently a complex symbol that could represent a number of interconnected ideas. At the center could be found, yet again, the theme of transmutation. To understand this unexpected association, and so eventually to get a better grasp on public attitudes toward science in general and nuclear energy in particular, I will take one last look into early traditions.

People of many cultures imagined that the life-force could create living beings out of dead matter—another form of transmutation. Early Greek and Arab philosophers spoke of statues brought to life, for example by drawing down influences from the stars, and certain followers of alchemy claimed that in the wombs of their vessels they could create a living homunculus. Loosely associated with such beliefs were legends of the ecstatic rites of certain medieval Jews who reached not only for spiritual perfection, but also for mastery of the secret by which clay could be raised into a living creature: the golem.

Originally the only danger was to the golem's makers, with their blasphemous ambition of displacing God as a creator, but gradually the danger became connected with the creature itself. Ideas of spiritual excess mingled with older, worldwide tales of witches who sent forth uncanny beasts. Not until the early nineteenth century was the story perfected. A rabbi of Prague, said the new legend, brought a clay servant to life, but it broke from obedience and went on a murderous rampage. The tale struck a nerve; particularly in the early twentieth century, stories and movies about the golem flooded Europe.[18]

Such a monster was easily associated with worries over science in general and atomic energy in particular. Reporters liked to say that radioactivity would be a "docile servant of mankind," but people began to wonder whether the servant would stay docile. One citizen wrote about this to the *New York Times* in 1931 after a prominent physicist had announced that he might be on the way to releasing atomic energy. Might not the atom get out of hand, asked the citizen, and "turn into a 'Golem' which could destroy man?" Raymond Fosdick put the new cliché more generally: would all of technological civilization become "a Frankenstein monster that will slay its own maker?"[19]

Even better at carrying such warnings was the golem's cousin, the mechanical robot. After all, it was machinery that actually took away people's jobs, mangled them in wars, and insulted the countryside. Going beyond that, writers from the mid-nineteenth century on used "the Machine" as a compact symbol for threatening new social arrangements, such as political centralization, factory discipline, and bureaucracy. A more specific symbol was the first significant robot in film, a mechanical giant that went about in 1918 killing people with electrical rays. It turned out to be a fake with its inventor inside; the villainous creature was the evil scientist himself.

Scholars who studied the many versions of *Frankenstein* similarly noticed that the writers conjoined scientist and monster. Mary Shelley's scientist himself cried that his creature was "my own spirit let loose from the grave." Not without reason did millions of people from the early nineteenth century on get confused over whether "Frankenstein" was the name of the scientist or the monster. They were two halves of one problem.[20]

The division of the dangerous character into two parts reflected a Western cultural tradition: a distinction between logical reason and emotional feelings. In the nineteenth and twentieth centuries particularly, many people spoke as if humans were split into halves, with a strict and rational side keeping precarious control over murky urges. During the same period this was also a popular image of all of Western society; upright elites supposedly held in check "dangerous classes" wont to break loose in crime and bestial rioting.[21]

This symbolism was evident when fictional scientists tried to suppress their emotions and concentrate on research, only to find their evil urges set loose in their creatures. For example, Dr. Moreau said he found emotion meaningless, yet his lust for pain and blood could be guessed, and such lust was acted out in full when his band of creatures broke loose. Audiences brought up on this tradition, watching a Dr. Thorkel scoff at human feelings, would not be surprised when he himself became tyrannical and murderous. That was the mad scientist's madness, and the result was the same whether it was his monster or his own monstrous urges that went out of control.

Yet there remained a curious reversal, in which the creature or robot could represent not a threat but a victim. From the nineteenth century on workers cried that factory life was turning them into slaves of production, "automata." By the 1930s social critics were warning that citizens would even lose control over their own thoughts and emotions; molded by corporations and bureaucracies, people were becoming frozen, rootless, isolated, and controlled creatures, that is, robots.[22]

The word "robot" itself bore a heritage of more primitive ideas about victimization. Derived from a Czech word meaning compulsory labor, it entered the world's languages through one of the century's most widely performed plays, Karel Čapek's *R.U.R.* The play featured an obsessed scientist who grasped the secret of life and created impassive, sterile, quasi-living servants. Apocalypse followed: the destruction of all humanity when the creatures rebelled like suppressed proletarians, and then, when two of the creatures became human, a marriage and the birth of a new race. This story held personal meaning for Čapek, a shy man who was preoccupied with thoughts of illness, infertility, and death. According to the leading student of Čapek's writings, the author partly modeled his fictional scientist on his father, and the robots' physical and emotional sterility may also have reflected Čapek's intimate concerns. The blank-faced robot—like the clay golem—made an apt symbol for the frozen state of a victim, whether oppressed by society or by more personal problems.[23]

In all such tales a number of associated beliefs and symbols crowded together into a bipolar structure. At one pole were the evil scientist, the corporate authority, the dangers of science and technology, and the coldhearted parent; at the other pole were the creature, the assembly-line worker, the threatened citizen, and the rejected child.[24]

Yet the golem and robot usually seemed to stand not with the victim but with the attacker, hardly separable from the scientist who created them. If the creature embodied the mad scientist's murderous impulses, why did it also sometimes represent a victim? This confusion between the two poles is an important clue, for the central meaning of a bipolar structure is often found right within its apparent contradictions.

The solution was perhaps most clearly revealed in a 1936 movie, *The Invisible Ray* by Universal Studios, leading vendors of horror to the world. The star was Boris Karloff in a switch from his famous role as Frankenstein's monster; now he was the scientific genius, the proprietor of a gloomy laboratory lit by bursts of lightning from a passing storm. Karloff's scientist built a cylindrical radium ray projector with the standard abilities to smash cities and work miraculous cures. But he caught a sort of contagious radioactivity, so that he glowed in the dark and could kill with a touch of his hand. Meanwhile the scientist was so dedicated to his work that he ignored his young wife, and she left him, whereupon he set forth to kill. At the climax the glowing scientist's mother smashed the vial of a chemical he required to stay alive, and he burst into flames, consumed by radioactivity.

The studio claimed that all this showed scientific theory that might soon come true, but in fact there was not a scrap of reality in any of the images. The images came, yet again, from the quest for "secrets we are not meant to probe" (as the mother had warned). Only now the punishment for probing was that the scientist himself became an uncanny creature, rejected by everyone he loved, dying in hell-fire. The basic idea was summarized by the homicidal Professor Radium in a 1941 comic that stole the movie's ideas: "I have made myself a monster!"[25]

Thus the threatening adult merged with the punished child in a comprehensive image of one who grasped forbidden forces. The final goal of mastering secret powers, of transmutation, was to transform oneself. But the mad scientist's attempt stopped halfway, leaving him a monster, his own victim.

It was indeed true that whoever set out to create something new in the world ran grave risks. Such an individual, whether trying to shape a new personal self or attempting a creative act in science, art, or politics, had to abandon patterns of belief that were familiar since childhood and were supported by society. Loneliness, disintegration, and chaos, as the alchemists and mystics said, had to precede rebirth. There was no guarantee that the pilgrim would not become mired in those depths, rejected by society, shorn of personality itself.[26]

On a more childish level, similar perils awaited the fairy-tale hero. Those who studied young children found them associating magic powers with fantasies of aggression. The way to grasp forbidden secrets in the first place might be through an attack, perhaps against a parent. And children knew that aggression or even simple peeping might lead to punishment; the questing hero always risked imprisonment, torture, and death.

There was a still worse fear. Suppose one had come to the secret, and suppose along the way one had been punished and abandoned. The victim, although crushed into a frozen state, would still contain a fiery point of rage. From their first tantrums children knew what overwhelming emotions could erupt, right up to the ultimate desire to kill their parents. If such an outburst came after one had seized ultimate forces, what then? Might unbearable loss come as punishment not from an outside authority, but directly through one's own dire acts?

To the extent that an audience could identify themselves with the mad scientist, the worst peril he evoked was not merely that someone else would act out dreadful urges upon us. The worst was that we might abandon our own self-control, and thereby lose everything we loved. This was not only a problem of children; many unfortunate people, taught to distrust their emotions, feared that a loss of self-control would be disastrous. Therefore they became their own rigid oppressors and their own victims.[27]

For many people this chain of associations would end with a premonition of overwhelming guilt. Therefore the fantasy of power and aggression would be thrust away and almost forgotten, a secret of its own, entangled with the secret life and death forces and dreams of cosmic destruction, all so hard for an adult to take seriously.

Sometimes a careful analyst could excavate the pattern whole. An Italian psychiatrist reported a patient's nightmare in which the sleeper, a chemist, hurled miniature atomic bombs at a world globe until the globe exploded. The terror he felt from this dream reminded the chemist of the terror he had felt when, as a child, he imagined night attacks from a monster. Analysis identified this monster as the child himself, or rather as a ferociously angry part of the child, a part that wished to destroy everything. The cause of the childhood rage also came to light. The chemist's mother had died when he was a year old, and he had grown up feeling abandoned, furious at her and all the world, and terrified of his own fury. This was perhaps the way in which Soddy, that other orphaned chemist, had come to his own vision of atomic doom.[28]

Popular culture showed such things plainly. For example, advertisements for *The Invisible Ray* explained that Boris Karloff "sought to destroy the world . . . because his world of love crumbled." A more elaborate picture came at the climax of the Gene Autry movie serial about an underground empire. The lovely queen rejected the villain's advances, so he prepared to destroy her with a radium ray projector. When the hero rescued her the villain ran amok, working his machine until it too ran amok and melted down the underground city, queen, himself, and all. From attack on a rejecting love object it seemed a short step to triggering the annihilation of one's entire world.[29]

A more thoughtful analysis appeared in a 1924 novel by Čapek. Here again a scientist was tempted to use atomic energy to make himself "uncontrolled master of the world," but he only succeded in accidentally blowing up a town. Most of the book was less about atomic energy than about the scientist's aggressive pursuit of women (at the time he wrote this, Čapek himself was frustrated in his longings for women). Atomic destruction, the novelist explained, was a force that came from within the scientist, expressing his willful passion. "If it had not been in you," a wise man told the scientist at the end, "it would not have been in your invention."[30]

REAL SCIENTISTS

Of course most actual research was not driven by belligerent fantasies, but by other equally fundamental drives—a desire for praise and status, or an urge to help people, or most often of all, sheer curiosity, the hunger for wisdom. Scientists might ponder the questions of life and death, but few if any followed Victor Frankenstein in attempting solitary mastery of the life-force. Creative people in all fields of work, faced with their inevitable mortality, usually sought an alliance with something greater than themselves. Scientific research was a way to identify with the eternal order of things, while putting something of one's own into an imperishable structure of permanent benefit to humanity. Scientists were driven less by aggression and the fear of death than by their opposites, love and the hope for life.[31]

Hope, like fear, has its diseases. The problems of hope are not often seen in the psychologist's clinic, but world literature is filled with figures brought to grief by unattainable dreams. Science's difficulties have often been the diseases of hope. Some scientists became too infatuated with their own discoveries, such as Blondlot with his N-rays and Soddy (at first) with the redeeming potential of his atomic transmutation. Even the most level-headed could slip into exaggerated praise for their particular field of discovery.

Among the public, as among scientists, the image of science up to 1939 was still warped less in the direction of fear than of unreasonable aspirations. Although there were scientist villains in science fiction, they were outnumbered by scientist heroes by two or three to one; meanwhile the ordinary mass fiction of magazines and books, when it showed scientists at all, portrayed them as clean-cut and useful characters. Nonfiction magazine and newspaper articles likewise had little but praise. In truth science was no career for somebody who wanted to have an easy life or make a lot of money, and scientists had the right to say that they were sacrificing their personal comfort in a worthy undertaking. They showed no desire to control society by dictatorial commands, but built up their own diverse, self-regulating communities. It did not seem incongruous when Millikan, for example, told the public that scientists offered an example of right conduct that could improve religion and public life. In his own life he devoted himself visibly and indefatigably to the search for truth and the advancement of civilization.[32]

A popular movie, *Madame Curie,* captured the majority image of scientists as neatly as a butterfly on a pin. In a central scene the young Pierre and Marie Curie bent over a dish containing their first radium, faces bathed in its soft glow, enraptured by the powers for good that they had discovered. Along with their noble vision the movie showed their sacrifices, as Marie drove herself to exhaustion and risked cancer in hopes of conquering death. As an old professor told Marie at the opening and again near the close, the scientist's sole reward was "to catch a star on your fingertips."

Yet the film also suggested there was something odd about scientists. Neither Marie nor Pierre Curie could easily show human emotion. In the impressive closing scenes, when Marie learned that Pierre had died, for days she remained speechless and staring at nothing, frozen, unable to mourn. In historical fact, when the real Marie's friends took her to her husband's corpse she kissed his cold face passionately, then clung to his body, had to be dragged from the room, and wept like any suffering human. However, that was not for the movie-going public. Why not?[33]

By the 1930s the image of the real scientist had much in common with the stereotype of the evil one. The remote scientist in a laboratory, threading intellectual mazes where few could follow, made the sort of half-seen picture that could easily sustain projected images, and some images fitted especially well. To become a real scientist meant years of arduous study and long nights in the laboratory, requiring unusual self-abnegation. Some scientists went beyond that to claim that science could progress only if researchers saw the world in impersonal, emotionless terms. Few groups were as apt as scientists, then, to stand for suppression of human feelings. In fact the best scientists were often warm and gregarious types, much appreciated as colleagues and teachers, but this reality could not overcome the stereotype. To newspaper and magazine writers, the scientist was an odd character who ignored mundane concerns, risking his health and scorning riches (as scientists themselves claimed), an unworldly "wizard" who isolated himself in the pursuit of tremendous secrets. In the movies, both Dr. Thorkel and Dr. Curie worked on radium day and night with superhuman endurance and rigidly suppressed feelings. In short, every scientist was supposed to resemble a robot, if not the atom itself: prodigious energy locked up like a compressed spring.[34]

Nuclear scientists were especially liable to carry this stereotype. For example, in 1936 when the Curies' daughter, Irène Joliot-Curie, entered French politics, lending her prestige to a party that promised to increase funding for research, a satirical cartoon showed her striding coldly among her party's enemies, blasting them with rays from her fingertips. The bad scientist was one who let desire for power get out of hand; the good scientist was one who maintained self-control and renounced social ambition. Scientists tended to agree with that. Joliot-Curie herself soon fled politics, telling reporters that she preferred the calm of her laboratory.[35]

The actual life of the laboratory was scarcely visible to the public. The press talked more about the remarkable personalities of scientists than about their discoveries, and talked about many other things more than it talked about science at all. The public understood little of what nuclear physicists were learning. After all, there were only a few hundred such scientists in the world, and all of a nation's physics laboratories together cost less than one more battleship. Newspapers and radio networks of the 1930s poured forth millions of words about economic predictions and peace conferences, while few people paid attention to what was happening in the laboratories, which as much as anything was the true history of the twentieth century.

In the four decades after nuclear energy was discovered, scientists learned the main facts about the construction of atoms. After Rutherford found the nucleus buried within the atom, other physicists discovered that each nucleus is a clump made up of still smaller particles, the protons and neutrons—simple facts, yet it took two decades to learn them. With help from atom smashers, scientists began to understand the forces that held protons and neutrons together in the nucleus. The hope of discovering something really fundamental, some master key to the universe, began to fade. By the late 1930s some physicists were beginning to suspect that the nuclear forces behind radioactivity held no more cosmic secrets than the everyday electrical forces in a burning match. To be sure, every nucleus remained a tremendous mystery, but only in the same way as every pebble.

The real work of nuclear scientists became less a quest for overwhelming truths than a working out of puzzles about specific types of nuclei. For example, the nuclei of certain metals, beryllium and protactinium, did not seem to be radioactive in the way theorists predicted; during the 1930s much effort went into straightening out just what was happening. Another uncommon metal was uranium. The things that were known by 1938 about the behavior of the uranium nucleus did not quite fit with nuclear theory, so a handful of scientists began to take a closer look. This was routine science, which no newspaper would have dreamed of reporting.[36]

THE SITUATION BEFORE FISSION

Although ordinary people knew little about nuclear physics, by the 1930s they did know the most essential facts. Anyone who could read a comic book or see a Saturday movie serial knew that vast amounts of energy were locked within atoms—energy somehow connected with the transmutation of elements, energy somehow connected with radiation, energy that might possibly be released someday for great benefit or destruction. These were scientific facts, and imagery could never be wholly independent of such facts. Historical events too had made an impression. Newspaper readers knew that radioactivity had killed scores of people and could bring sterility and mutations, but that in the hands of doctors it had helped many thousands. The public, with sound practical wisdom, was skeptical about anything beyond these facts; more fantastic images were something to shrug at. Yet the fantasies would not go away. Around the few simple facts an ancient and mighty array of themes was coalescing.

By the end of the 1930s any mention of atomic energy could remind most people of fears about loss of self-control and dreadful victimization, along with hopes for mastery and rebirth. Talk of radioactive rays recalled old ideas of a life-force, sexual or otherwise, leading to monstrosity or immortality; talk of the mysterious core of atoms recalled the problem of forbidden secrets; talk of atomic energy recalled fantasies of magic power, apocalypse, and a Golden Age. In short, atomic energy seemed to offer what was the supreme promise and at the same time the supreme threat: mastery of the secrets of life and death.

This set of archaic themes had been scattered at the end of the nineteenth century, exiled to various fantastic novels and fairy tales, the original interconnections within the image of transmutation almost forgotten. But radioactivity had characteristics that could be associated with all the themes, and it drew them back together. It was not entirely coincidence when new facts resembled old symbols. People had long been searching for mighty, life-transforming rays; now at last they had found something resembling what they sought.

The connection between transmutation symbols and radioactivity was not only a matter of ancient traditions revived in modern laboratories; it also came through the deliberate use of imagery. Some groups used atomic images to further what has been called the "propaganda of integration"—the work done by everyone from schoolteachers to advertisers to encourage people to play appropriate roles as producers and consumers in modern society. Others invoked imagery to do just the reverse, that is, to promote distrust of the new scientific and technological authorities.[37]

The first to know about atomic energy, scientists were nearly always the first to connect it with ancient myths. Some were proud to evoke the language of transmutation as they publicized their work, even boasting that they would create a utopian future. Science journalists, devoted followers of the scientists' visions, joined in their enthusiasm. As part of their craft the writers tended to emphasize, among all the things that scientists said, the ideas most likely to interest the public; these were precisely whatever ideas recalled traditional fantasics and aroused emotion.

Once the idea of mighty forces was abroad, thoughts of catastrophe naturally followed. These thoughts too were exploited for sensational purposes, for example in movies, and also found expression among people who were already worried by the advance of technology. Yet nearly every warning concerning atoms, like every promise, was first voiced by a scientist. Indeed, the warnings almost sounded like further boasts about the power of science.

Although all these ideas and myths were generally known, this does not mean that they seemed important to the majority of people in the 1930s. The books on atomic themes that I have mentioned were not widely sold, and most of the scientists' pronouncements were not widely heard except among educated and attentive people, a small minority of the public. The most impressive of the archaic images appeared chiefly in pulp magazine stories and horror movies, genres that, despite their impact on youths, seemed of little concern to sensible adults. Nevertheless, even the most bizarre tales had something to say about reality. For images do not always stay quietly within the mind.

The movie *Madame Curie* was released in 1944. The image of the fictional Curies striving to catch a star on their fingertips must have stirred mixed feelings when the movie played in the hastily built halls of America's new secret cities—Oak Ridge, Hanford, Los Alamos. Legendary tales about nuclear energy were turning into realities. Before the process ended it would change our image of science and of modern civilization as a whole.

part two

CONFRONTING REALITY

1939–1952

5

Where Earth and Heaven Meet

At Christmastime of 1938 the fantastic images of nuclear energy began to enter the world of physical reality. European nuclear scientists found that when they bombarded uranium with neutrons, sometimes a nucleus would not do the expected thing and transmute into a similar nucleus, but would split apart into two altogether different atoms. One of the scientists, Otto Robert Frisch, accordingly sought a new name for what was happening. The uranium nucleus, quivering and elongating until it broke into two pieces, reminded him of the mysterious central transformation of birth, the division of a living cell. A biologist friend told him what that was called: fission.

Frisch explained uranium fission to the director of his institute, Niels Bohr, who was just then departing for the United States. In January 1939 Bohr told a meeting of American physicists about the discovery, and one of the science reporters who tended to hover around such meetings immediately put fission into the newspapers. Journalists and scientists everywhere were caught up in the excitement, and through 1939 they wrote hundreds of articles to inform people around the world.[1]

The public quickly grasped the chief meaning of fission. Any physicist could calculate that when a uranium atom split in two it would release a large burst of energy, exactly the energy that had held the atom together until then. Scientists scarcely needed to explain this to science reporters, who for decades had been primed for such a discovery. Now all the hoary clichés about atomic energy, driven back by Rutherford's and Millikan's public skepticism, could march forth once more. As Kaempffert wrote in the *New York Times*, "Romancers have a legitimate excuse for returning to Wellsian utopias where whole cities are illuminated by the energy in a little matter." Similarly, *Newsweek* said that steamships driven across an ocean by a spoonful of matter and other "pseudo-scientific fantasies painted in the Sunday supplements" could be coming true, while *Scientific American* warned, "The tabloids love to write of blowing up the world, and it's not such a sensational idea as one might think." Not only in the United States but from Britain to the Soviet Union, the press carried awestruck stories.[2]

As usual, reporters trumpeted spectacular ideas while scarcely ever remarking that most scientists were more cautious. There was no proof yet that fission would quickly lead to any practical device. Scientists were so doubtful that in all the world only two small groups set out to seek a way to put fission to work, racing neck and neck through the spring and summer of 1939. One group was at Columbia University in New York City, the other at the Collège de France in Paris. Both groups kept journalists well informed about their progress. The stories reflected a subtle shift in the public's view of nuclear energy, as new fears began to outweigh new hopes.

IMAGINARY BOMB-REACTORS

Through 1939 and 1940 fission could promise only vague possibilities of releasing energy. Neither scientists nor the public understood whether the energy might come through controlled burning of fuel, or in a grand explosion, or in some tricky combination of both. Cautionary science fiction tales remained as good a guide as any to what it would all mean.

There was one scientist who had been thinking for five years about nuclear energy as a realistic prospect. This was Leo Szilard, a bumptious physicist whose round face had a baby's look of mischievous innocence and secret wisdom. Since his childhood in Budapest Szilard had worried about the uncertainty of human life; he had been particularly struck by a classic Hungarian poem about civilization degenerating into savagery beneath a dying sun. Perhaps his sensitivity to tragedy was strengthened by his situation as member of a doomed empire and a doomed culture, the Jews of Hungary. Whether because of this or from more personal reasons, he grew up with an ambition, as he put it, for "saving the world." H. G. Wells's ideas about a reforming brotherhood of technocrats especially appealed to Szilard, a moralizing and intelligent loner much like Wells's virtuous Airmen. When nuclear physics caught his attention, it was because the thought of apocalyptic energy fitted his Wellsian visions.[3]

Szilard had figured out in 1934 how one might actually release nuclear energy. Was there some kind of atom, he asked, that when struck by a neutron could transmute in such a way that it emitted two neutrons? If so, these might each strike another atom nearby and cause them to transmute in turn, emitting two neutrons apiece, or four altogether—and so on, doubling the number of neutrons at each step. If you had a block of such stuff, a chain reaction would run like wildfire through the whole mass and energy would rush forth.

Szilard had warned Rutherford and other leading physicists that terrible atomic bombs might be in the offing, and sought funds to try experiments, but his seniors paid little heed. He was forced to turn to other research until the news of fission showed a possible way to nuclear energy: when you hit a uranium atom with a neutron, the process might release enough additional neutrons to make a chain reaction.

Szilard was then in New York, and he set to work alongside a team headed by Enrico Fermi. Within a few months they found that uranium fission did emit some neutrons. Meanwhile a team in Paris led by Frederic Joliot-Curie determined that there would be more than enough neutrons to get a chain reaction going. Both Joliot-Curie and Szilard determined to try it out.

There was a difficulty, however. Fission was easily provoked by slow-moving neutrons; but when a uranium nucleus split in two it spat out fast neutrons, which would usually whip through the next uranium nucleus too swiftly to provoke a new fission. The answer was to slow down the neutrons, using what Fermi and Szilard called a "moderator." Water or graphite might possibly do the trick. The uranium should be mixed in with the moderator, for example by embedding blocks of uranium in slabs of graphite, or simply by wetting down uranium compounds. The fast neutrons that came from a splitting nucleus would bounce around in the moderator, slowing down at each bounce; sooner or later the slowed-down neutrons would straggle back into some uranium and cause the next fission in the chain.

Moderators seemed highly important not only to make a source of industrial power, a nuclear reactor, but also to make a bomb. All through 1939, the world's physicists saw no other way to make an explosion. But suppose they piled up uranium and moderator and started the chain reaction going, and suppose they made the reaction go as fast as possible? Because the neutrons would have to slow down for each step, the chain reaction could not spread very quickly. The pile would heat up, and after some seconds it would melt or blow itself into a gas. That would separate the uranium atoms from one another, ending the chain reaction before more than a tiny fraction of atoms could get involved. As a bomb, a reactor would not make much of a bang.

A runaway chain reaction could still destroy a laboratory, so the French researchers were nervous as they piled up sacks of uranium compounds and drenched them with water. The team member most worried about this, and at the same time the one most convinced that nuclear energy would be a revolutionary development, was the German-born Hans von Halban, Jr. He took exaggerated precautions and did not hesitate to tell journalists of his fears that uranium experiments would explode.[4]

Some physicists suspected there was a chance of getting a real bomb. They began with the fact that nearly all uranium atoms are one or the other of two isotopes, uranium-235 or the slightly heavier uranium-238. Only uranium-235 was easy to split, but this fissionable isotope makes up less than 1 percent of ordinary uranium as it comes from the mine. Suppose you enriched the uranium, increasing the concentration of the uranium-235? The task would be formidable, but if it could be done, and if you built an extremely compact reactor with a concentration of uranium-235, might you not somehow be able to make it explode with real force?

When physicists themselves could scarcely tell the difference between a compact power reactor and a bomb, it was small wonder that journalists did not make the distinction. A French science writer compared the likely result of uranium reactor experiments to a great volcanic explosion, and in 1941 a *Reader's Digest* article asked whether experiments on uranium might start a chain reaction that would unravel the whole planet. Other reporters and even a few physicists warned against playing the "Sorcerer's Apprentice." Only Kaempffert of the *New York Times* was accurate, citing a scientist who said a reactor might "blow up like an overheated steam boiler." Kaempffert went on to point out that such an accident could be worse than any ordinary boiler explosion, for it might spew out hundreds of pounds of isotopes all as intensely radioactive as pure radium. No thank you, wrote Kaempffert; he would stick with a conventional coal-fired power plant, "and view with satisfaction what little smoke floats up from the stacks."[5]

The teams working on reactors in Paris and New York in 1939 were more sanguine about using uranium for industrial power. This did not mean ignoring military uses, for they would be proud if uranium could be used, for example, to propel submarines. Once nuclear reactors were well understood, many years down the road, someone might possibly find a way to build great bombs as well. If so, the scientists hoped the bombs would make war unthinkable.

When the Second World War got under way the teams in New York and Paris began to consider weapons seriously. Fascist anti-Semitism had already forced Szilard, Fermi, Halban, and other physicists from their homelands; they knew a Nazi victory would mean unspeakable things, including their own deaths. Having been brought up in the German scientific tradition, these physicists also felt that if anyone could figure out how to build atomic bombs, it would be the Germans. Their worries were soon repeated in the newspapers as a few anxious reporters warned that nations were in a race to develop atomic energy, a race that might decide who won the war.[6]

If the things said about atomic energy in 1939 and 1940 had the flavor of science fiction, that was no accident. An earlier generation of scientists had inspired science fiction stories, and the stories had inspired scientists in turn. One member of the Paris team said the work reminded him of his beloved Jules Verne novels, while another said it recalled H. G. Wells's *The World Set Free*. That book had been a main inspiration for Szilard on his path to the chain reaction, and he cited it as the initial reference in his first scientific report about building a reactor. More fundamentally, the hopes raised by such stories had already had much to do with the public funding and the recruitment of young scientists that had sustained nuclear science for decades.

Nobody among the public was as intrigued by the possibilities of uranium fission as the readers of *Astounding Science-Fiction*. The magazine's editor, John W. Campbell, Jr., was building up a stable of writers who were lifting the genre above the level of boyish pulp adventure. A male college student majoring in science or engineering, eager for new ideas, was the typical *Astounding* reader in 1939; the magazine was as close as anyone could find to a probe sunk into the back brain of American technology. Like most science fiction writers, Campbell had thought about atomic energy during the 1930s. In one of his stories the hero used it to save the universe, while in another, an atomic experiment destroyed a vast area. In June 1938, six months before the discovery of fission, Campbell had recklessly declared in one of his editorials, "*The discoverer of the secret of atomic power is alive on Earth today.* His papers and research are appearing regularly; his name is known." Soon he could boast that his prediction had been precisely true.[7]

Astounding's stories did more than any factual article to tell the meaning of fission. Most striking was a story by a new writer, Robert Heinlein. Heinlein had been trained to see ideas from the engineer's practical viewpoint, but in 1939 that training seemed useless, for he was ill with tuberculosis and had little visible chance for a career or even for survival. His early stories were often about loneliness and oppression by sinister conspiracies. Usually his heroes overcame their problems and opened a new world of hope, but some of his stories ended with a frankly paranoid vision as the world literally disintegrated into chaos. His first atomic energy story echoed that personal theme.[8]

"Blowups Happen" appeared in September 1940, heralded by a Campbell editorial announcing that the story was based on the latest discoveries. Heinlein had indeed rewritten the story several times as new information came from the laboratories. The tale revolved around a future uranium power plant which the scientists called, with simple candor, "the bomb." Could any human being be trusted to operate such a plant, Heinlein asked, when one mistake might cause a catastrophe that would devastate thousands of square miles? Might not the pressure of the job drive an operator mad? In the story, tension rose when an error was discovered in the calculations, proving that if the plant did blow up it could leave the whole earth as barren as the moon. The power company's board of directors, convinced that the plant was necessary for the nation's economy, refused to admit that any risk existed; they agreed to make changes only when worried scientists threatened to mount a damaging public relations campaign.[9]

If the story seemed to anticipate newspaper editorials of later decades, that was not because Heinlein had some uncanny gift of prophecy. Exploding atomic plants, world devastation, and arrogant power companies too, had all been banal ideas since the 1920s. Heinlein's story had nothing to do with the facts of nuclear engineering, a subject entirely unknown when he wrote his tale, but dealt only with the fact that great energy might be released by human hands. The rest followed directly for a man like Heinlein who could get to the core of common ideas.

His next atomic energy tale, again accompanied by a note from

Campbell announcing that it was scientifically plausible, also applied plain thinking. Heinlein suggested how radioactive substances could be used as a devastating weapon. With that discovery, one of the characters exclaimed, "The whole world will be comparable to a room full of men, each armed with a loaded .45. They can't get out of the room and each is dependent on the good will of every other one to stay alive." In Heinlein's story the good will failed, and war left cities as uninhabitable as any in *The World Set Free.* At length order was restored, in H. G. Wells style, by an airborne international Peace Patrol. But the future remained uncertain, for Heinlein and Campbell lacked Wells's trust that such guardians would stay wise and virtuous. Their opinion of what could be done about atomic weapons was summed up in the story's title: "Solution Unsatisfactory."[10]

Through the war years, fiction writers and journalists continued to speculate now and then about atomic weapons. For example, in late 1944 *Time* magazine reported that London was buzzing with speculation that Hitler's next deadly surprise, after his V-2 missile attacks, would be an atomic bomb. But Heinlein's "Solution Unsatisfactory" remained unique in the way it stripped down the consequences of such weapons to a basic logic. When the story appeared in May 1941 only a handful of people knew that, in our closed room, the guns were already near to hand.[11]

At this point the story divides into two streams. One stream was the public history. The few facts that scientists had explained about fission in 1939 had simply revived the talk common for decades, spurring a few people to think seriously about nuclear energy but otherwise only reminding everyone of familiar fantasies. Through the war years this public imagery remained stagnant. The other stream was the secret history known to a few hundred American and British leaders. Here ideas ran ahead swiftly, going through all the main arguments that would later appear over everything from reactor safety to missile warfare. Yet at the end of the war when the two streams rejoined, the fantastic and the scientific, they would merge with unreasonable ease.

REAL REACTORS AND SAFETY QUESTIONS

In a few years of frantic secret effort, scientists figured out all the major facts about nuclear technology: how to build atomic bombs, how a reactor accident could happen, how radioactive wastes could spread into the environment. With the help of government and industry officials they devised strategic plans and safety practices that would guide all future nuclear endeavor. Much of this followed from the technical facts, which now brought a new element into nuclear energy thinking; the laws of physics would have as much to say about events as all the forces of history, society, and psychology. But the thinking of scientists and officials was not based entirely on the new facts, for like everyone else they had older images in the back of their minds.

The history of real nuclear weapons began with a secret calculation made in England in the spring of 1940. It was two refugee physicists, Otto Robert Frisch and his colleague Rudolph Peierls, who figured it out. Exactly what would happen, they asked themselves, if you could cull from natural uranium a mass composed purely of the rare uranium-235? Bohr and others had told the public that there could be enough energy there to blow up a city, but nobody had worked it out as a serious technical possibility. Now Frisch and Peierls realized that with fissionable uranium-235 atoms all crammed together, there would be no need for a moderator to slow the neutrons down, since even the fast neutrons emitted in each fission would have a good chance to provoke another fission. The whole chain reaction would go so swiftly that, before the mass had a chance to blow itself apart as a runaway reactor would do, many of the uranium-235 atoms could split and release energy.[12]

Unlike many scientists, Frisch could envision an industrial effort great enough to separate a few kilograms of pure uranium-235, so he and Peierls calculated what would happen if you put together such a mass. The appalling answer started a chain reaction of its own. First British scientists were convinced, and then they convinced their government. Yes, atomic bombs could be built, and in time to affect the outcome of the war. There was important physics involved here, of course, but that is not the subject of this book; I am concerned with some of the human forces that brought the physics into history.[13]

Frisch and Peierls, who started it all as much as anyone did, were unusually shy, sweet people, with many warm friends and no enemies. They were only afraid of what Germany could do with such a weapon. Both men had been driven into exile by the Nazis, and Frisch had recently been through the grim process of extricating his father from a concentration camp. It was with good reason, in sorrow and fear, that these gentle men turned their minds to weapons.

In a democracy the final decision on such matters rests with elected officials, and in due course the problem worked its way up through committees to the desk of Winston Churchill. Churchill held his position as wartime Prime Minister not least because he was one of the best-qualified people in the world to contemplate strategic bombardment. His taste for new weapons was well known. Back in 1914 he had ordered the first pre-emptive strike in the history of air warfare, a bomber raid on German zeppelins that were preparing to bomb London. The public could easily have predicted how he would view nuclear weapons.

Churchill had indeed been involved with those weapons years before—in fiction. A 1932 novel had pictured a pacifist British cabinet faced with the invention of atomic bombs in 1939. "It would be impious," cried a trembling liberal, "for us to dabble in the Satanic potentialities." Nothing made the fictional politicians so nervous as the thought that the opposition leader, Winston Churchill, might learn of the weapons, for nobody could doubt he would insist that they be built and used.[14]

In the real year 1941, after German bombers had tried their hardest to smash his cities flat, the real Churchill addressed the question of atomic bombs with his usual display of good cheer. "Although personally I am quite content with the existing explosives," he told his chiefs of staff, "I feel we must not stand in the path of improvement." The British accordingly set up a program to build factories for separating uranium-235. Their resolve spread across the Atlantic, helping to push the Americans to set up their own nuclear energy project. Another push came from a discovery, made separately in Britain and America, that reactors could have a close connection with bombs after all.[15]

The uranium-238 that would make up most of the uranium in a reactor seemed only a nuisance, soaking up neutrons that otherwise would reach the fissionable uranium-235. But when a uranium-238 atom absorbed a neutron, the atom transmuted. By mid-1941 some scientists understood that the result would be a metallic element never before seen: plutonium. The name was suggested independently by several people who remarked that the new element stood in the same relation to uranium (two spaces outward in the table of elements) as the planet Pluto stood to the planet Uranus. They also understood the crueler implications of naming an element after the Lord of the Underworld, for it was clear from the start that the metal would be fissionable.

A reactor would transmute ordinary uranium into plutonium, and once you had a few kilograms of plutonium you had the makings of a bomb. That might or might not be easier than separating out pure uranium-235 from natural uranium, so the American authorities decided to try both routes, isotope separation factories and reactors to make plutonium. The nation had a scientific community strong enough to sustain such an effort, for Americans in the 1930s had enthusiastically supported atom smashers and the like, paying for more nuclear physics than the rest of the world put together, precisely because people had hoped for amazing things from such work.

The burden of building American reactors fell on the square shoulders of Arthur Holly Compton. His upbringing had been deeply religious, and in his youth he had planned to become a missionary until his father advised him that as a scientist he might perform "a more valuable Christian service." Only by trying to improve the lives of others, Compton felt, could his own life have meaning. When American reactor work was gathered together at the University of Chicago in 1942 under his direction, he took the responsibility most seriously.[16]

Compton hesitated at the prospect of building the world's first nuclear reactor, a pile of uranium and graphite, right on the university campus. To be sure, by this time physicists understood that such a pile could never explode with any force greater than that of an ordinary TNT bomb. But Compton knew that whenever a uranium atom split, each fragment would be a new, intensely radioactive isotope. Considering the care that had to be taken over just one gram of radium, it was a serious matter to create kilograms of radioactive matter in the middle of Chicago. After difficult meditation, he decided the need was so urgent that the risk must be accepted "as a hazard of war."[17]

All the project workers were impressed by the thought that they were on the brink of unleashing cosmic forces. General Leslie Groves, the career Army officer who had taken control of the entire Manhattan Project, was shaken when Compton told him the pile would be started up where it stood. Physicists were equally concerned; some had recently read "Blowups Happen" and the rest could recall earlier warnings.

Fermi, chief builder of the pile, set up strict precautions for its first run. To control the chain reaction was simplicity itself: into the pile Fermi sank a rod containing cadmium, a metal that greedily absorbs neutrons. As the rod was gradually pulled out, the number of neutrons bouncing around within the pile would increase, and if something began to go wrong the man holding the rod only had to shove it back in. But scientists volunteered additional safeguards. One rigged a rod that would go in at the touch of a button. A second added a rod to drop in automatically if there was trouble. Yet a third hung a rod overhead by a rope and stationed a reliable colleague alongside with an ax. Not satisfied, still another physicist organized a "suicide squad" with buckets of cadmium solution, ready to soak the pile at any sign of danger. Since an accident with the buckets would set back the work for months, some complained that this last precaution went too far, but Fermi let it stand. In short, the team members were as nervous as if they were attending the birth of a Frankenstein's monster.[18]

The watchers who crowded into the room saw only a tedious inching out of the control rod while Fermi made calculations, until he announced that the chain reaction was self-sustaining. It should have been about as exciting as watching someone tune a radio, yet for those present it was unforgettable. Some, like Szilard, left the room troubled by thoughts of weapons, while others thought of the potential of reactors for bringing a shining industrial future. Compton believed the event was a "miracle," the birth of a new era.[19]

After all, a vocabulary of birth was embedded even in the new theoretical descriptions of the chain reaction. American physicists spoke of neutrons "reproducing" in successive "generations." Such terms were made explicit by other physicists, trying to build a reactor in Germany, who said that neutron reproduction was an exact analogy to people in a population, each fission being like a "marriage" with neutrons as the children. Although nobody in Chicago said it, in a symbolic way their pile of uranium had come to miraculous life.

Now that they had a living reactor, or at any rate a working one, the Chicago scientists had to face those kilograms of radioactive isotopes. Compton personally knew scientists who had been injured by X-rays or radium, and everyone had heard of the doomed radium dial painters. "Because of the urgency of our task," Compton later reported, "these unusual hazards have been taken as military risks." The Manhattan Project workers could hardly complain when they were spared the dangers facing soldiers at the front and even the high death rates among workers in some other war industries. It turned out to be familiar problems such as construction work rather than radioactivity that caused nearly all the Manhattan Project's casualties. Nevertheless, radioactivity remained, in the Project's code phrase, the "special hazard."[20]

Even before building the first reactor, Compton had set up a team to study radiation hazards, led by a feisty, no-nonsense physician, Robert Stone. The team's first problem was to define how much radiation the Project's workers should be allowed to receive. They worked out a unit of measurement, the "rem," which roughly speaking was one hundredth of the energy that would be needed, delivered by an X-ray machine in one concentrated dose, to irritate a person's skin. Prewar experience in hospitals and the radium industry suggested that a person could regularly absorb a rem or so spread across a few days without suffering observable damage. Building on guidelines that had been set up by committees of experts in the 1930s, Stone's team decided to allow Manhattan Project workers no more than one tenth of a rem per day.[21]

Nobody could be sure that even such low levels of radiation would not cause subtle damage, perhaps showing up only years later. The way prewar experts had approached the problem was explained, with telling exaggeration, by Robert Heinlein in his 1941 *Astounding Science-Fiction* story. "Not all the heroes are in the headlines," he wrote. "These radiation experts not only ran the chance of cancer and nasty radiation burns, but the men stood a chance of damaging their germ plasm and then having their wives present them with something horrid in the way of offspring—no chin, for example, and long hairy arms. Nevertheless, they went right ahead." Stone was not satisfied with such heroics, and he insisted that the one-tenth rem limit serve not as an everyday "safe" dose but as a ceiling. Actual radiation doses should be kept as low as could reasonably be achieved without halting the work.[22]

Stone's team went on to study the hundreds of different radioactive isotopes produced in a reactor, searching out the particular dangers of each. Plutonium was the biggest worry. Gram for gram it would be roughly as dangerous as radium, and workers would be machining kilograms of the metal into bomb components. The problem turned out to be manageable, because plutonium metal taken in through the skin or mouth is quickly excreted out of the body. A worker who picked up a piece of the metal bare-handed or even swallowed a speck of it would be foolish, but with ordinary luck he should suffer no harm. (Decades later, studies of the health records of Project workers confirmed this.) There was one glaring danger signal, however. The deaths from lung cancer in the Bohemian uranium mines warned against inhaling even a tiny bit of radioactive dust. When Stone had to decide what amount of plutonium it would be permissible for a worker to inhale, he gave a characteristically firm judgment: "None at all."

To enforce their rules the physicians devised elaborate procedures, which would set the tone for the future nuclear industry. Each work site had a health team, which might, for example, take urine samples to test for plutonium. They set up special laundries for contaminated clothing and assigned special garbage cans for radioactive waste, which they buried in temporary disposal areas, leaving final disposal for the future. These measures went far beyond anything that was done for toxic products in the chemical industry, if only because the project leaders wanted to reassure workers who felt that nuclear energy was somehow uniquely dangerous. However, not everyone felt that way. Health teams found some scientists skipping the urine tests, scorning precautions, and getting far more radiation than the rules would allow. Still, in the end the teams could boast that not one Manhattan Project worker had received as much radiation as careless doctors in hospitals routinely inflicted upon themselves. Even those who had inhaled some plutonium (a closely monitored group over the following decades) never suffered any ill effects.[23]

The worst radiation worry was not overconfident American workers, but Germans. The Chicago scientists suspected that Germany was ahead of them in the race, which raised a problem explained in Heinlein's 1941 story. Even if you could not make a bomb, once you got a big reactor going you could make tons of radioactive fission products. Heinlein had described airplanes dropping radioactive dust on cities until citizens perished or fled in panic. American and British scientists toyed with the idea of such a weapon but never laid serious plans, for it seemed like a needlessly difficult and odious variation on ordinary poison gas. Yet the Germans might try it, so Stone's team had to figure out the dangers of fission products spread across a wide territory.

They calculated that one of the worst hazards would be a little-known element. When a uranium atom split, one of the halves was likely to be a radioactive isotope of strontium, a metal in the same family as radium and calcium, easily taken up by living bones. After feeding tiny amounts of radioactive strontium to mice, Stone's team found bone tumors. A still graver danger was radioactive iodine, often the other half of the split uranium atom, an element that concentrated in thyroid glands.

If nobody ever used fission products as a weapon these substances would still be a problem, for every nuclear reactor would be stuffed with them. In February 1943, soon after Fermi finished his first pile, he sat down with Compton and Stone to decide what sort of hazard they had built. The answer was reassuring. If the pile started to heat up, then the neutrons in it would bounce around faster; since fast neutrons are not as effective as slow ones in causing fission, the chain reaction would falter. Fermi's pile had what all engineers love: automatic self-regulation. About the time the pile got too hot to touch, the chain reaction would halt of itself.

They were less confident about larger reactors, the kind that could produce significant amounts of plutonium. Everyone agreed that such devices should be kept away from populated areas, so the Chicago team's second pile was built at Argonne, out in the Illinois countryside, while the next and more substantial reactor went among the isolated hills of Oak Ridge, Tennessee. Although these precautions were expensive, Groves welcomed them, for the general was more worried than any scientist about atomic experiments going catastrophically wrong. He set the project's final and largest reactors at Hanford in the remote wastelands of Washington state.[24]

But could the Hanford reactors endanger the nuclear workers who would have to live nearby? Compton made some rough calculations. The conditions were set by an elementary and crucial fact: you get one gram of plutonium when you operate a reactor that produces one megawatt of heat (a million watts) for one day. To get the kilograms per week wanted for bombs, the Manhattan Project accordingly planned three 250-megawatt reactors. These would be blocks of graphite moderator piled three stories high and laced with rods containing uranium metal. To carry away the hundreds of megawatts of heat, the engineers planned to pump 30,000 gallons of water a minute through channels in the graphite. What if an accident or sabotage cut off the flow of coolant? The chain reaction would stop once the temperature rose, but that would still leave a problem. The big reactor would be packed with fission products so intensely radioactive that they would give out megawatts of heat. With no coolant, the isotopes would heat up the reactor until the uranium melted or perhaps caught fire.

To see what might happen next, there were two possible approaches. First, Compton looked into "the extreme case in which all the fission products are scattered and deposited uniformly over an area," as in the schemes for poisoning a city with radioactive dust. A simple calculation showed that an area 60 miles square might be made temporarily uninhabitable. The calculation seemed unreal, however; a reactor that got overheated was not the same thing as a reactor that exploded. Even in an explosion the heavier dust would fall back down on top of the reactor, while most of the lighter material would rise and drift far away, diluted into insignificance by the atmosphere. The uniform spreading of isotopes over an area was conceivable in principle, in the sense that any event is conceivable that is not excluded by the laws of nature, but it was the sort of event that any knowledgeable person would find flatly incredible.

Compton, like all subsequent nuclear reactor builders, decided on a second approach: assuming the worst credible result. After an accident some fraction of the fission products, particularly those that were already gases, might drift downwind in a slowly spreading cloud. Rough calculations showed Compton that seven miles from the reactor was the farthest that the cloud would kill people through direct radiation injury, even if things were ten times worse than he assumed. As a result of these and later, more detailed calculations, Groves decided to set the Hanford reactors six miles apart and to keep the workers' housing several times farther away.[25]

These decisions disturbed Leo Szilard. It was all a "question-answer game," he complained in an angry memorandum. The scientists were asked to set a completely safe distance, so they made "the most pessimistic conceivable assumptions." Then an official took that answer and acted upon it. If the scientists had been asked specifically to advise how far apart the reactors should be, taking into account not only barely credible accidents but also the cost and delays of building miles of roads and railroads, Szilard thought they would have decided to build the plants closer together. In a number of other areas too, scientists began to criticize their superiors for taking precautions to an extreme, needlessly delaying the Project.[26]

Unprecedented caution nevertheless continued to mark the operation of the reactors and of the chemical plants that extracted plutonium. Project leaders feared that with so much radioactive material around, some might leak out. Health teams measured wind patterns near the reactors, monitored radioactivity for miles around, exposed rabbits to gases that came through a reactor, and studied the effects of radioactive waters on fish. No toxic industrial product had ever been watched so closely.

Less care was taken by those in Germany who were meanwhile working on nuclear energy. Foremost among these was Werner Heisenberg, a master of theory but one who lacked Fermi's taste for experiment and solid engineering. Heisenberg expected to build the world's first reactor, but the only landmark he achieved was the world's first reactor accident. It was actually just a small sphere with some powdered uranium metal, far from anything that could sustain a nuclear chain reaction. One day in June 1942 the uranium accidentally got wet. As any chemist knew, wet powdered uranium metal would generate a bubble of hydrogen gas, which could be explosive when mixed with air. The hydrogen did explode, scattering burning uranium around the room and nearly killing Heisenberg. In Berlin, as in Hanford, it was less the fascinating "special hazards" of nuclear energy that threatened lives so much as dangers like those that engineers had faced for generations.[27]

PLANNED MASSACRES

Unaware of how disorganized and hapless the German atomic project was, scientists in Britain and America pressed their work forward with the haste of fear. Compton had put the worry succinctly when he drafted an official report to President Roosevelt in October 1941. The first nation to get fission bombs, said Compton, would have its enemies at its mercy and could conquer the world within months. As late as the summer of 1944 Compton still believed that he and the Germans were racing neck and neck, and that only an American bomb might prevent a future of Nazi horrors.[28]

These fears reflected a growing understanding of a new fact: in modern war nobody would hesitate to massacre people by the million. This ghastly reality pressed on the Manhattan Project scientists, slowly reshaping their thinking. The same reality would also have to be taken into account by everyone else, sooner or later, in their attitudes not only toward nuclear energy but toward all modern life.

The issues can be seen most clearly by considering the advance of technologies other than nuclear energy. Some scientists had refused to join Manhattan Project laboratories, but not out of moral qualms. They disliked the science-fiction atmosphere, the chaotic living conditions, and especially the extreme security measures, the spying and censorship that General Groves had imposed. It seemed unlikely that atomic weapons could be built under such circumstances, and these scientists preferred to work on other projects—something more certain to destroy the enemy. Like most of their fellow citizens, the scientists were committed to total war.[29]

Germany and Japan had begun with war aims that, however grandiose, were specific and limited, but the United States, Britain, and the Soviet Union had adopted the unlimited aim of annihilating their enemies. The only thing they were not willing to use was poison gas, because that had acquired a taint of moral horror and because nations hesitated to set a precedent that might someday be turned back upon themselves. Besides, they had means of slaughter more efficient than gas, perhaps even more efficient than atomic bombs.

Blockade was one proven method. In the First World War, Britain and

Germany had tried to strangle each other by cutting the sea lanes. In Germany beginning in 1916, the result was one or two hundred thousand civilian deaths each year above the number that would have died in peacetime, and starvation was a main reason for the uprisings that ended the war. In the Second World War Britain and Germany again tried to starve each other, but it was Japan that was most vulnerable to blockade, and the United States pulled the noose tight. By 1945 submarines and mines were so effective that not even fishing boats could go to sea. Supplies of food began to run out and tuberculosis struck with deathly force, so that the Japanese could have been crushed by blockade alone. It would not necessarily have been the kindest way to end the war; a blockaded nation feeds its soldiers while its old men, women, and children die lingering deaths.[30]

The quiet deaths of children through economic warfare were largely out of sight for the public, who took more interest in the spectacle of bombardment. In the 1930s many had thought that no democracy would bomb civilians, and when the Second World War began all sides meant to use threats of retaliation to deter the enemy, restricting bombardment to military targets. The result was a textbook case of what would later be called "failure of deterrence through escalation." German bombs meant for the London docks went astray and fell on houses; Churchill decided to "raise the stakes" and bombed Berlin. In modern war, everyone becomes a military target.[31]

There was nothing new about destroying cities. Possibly the worst single massacre in the Far East, not excepting the bombing of Hiroshima, had already set the precedent back in 1937 after the Japanese Army captured the city of Nanking. Japanese soldiers had gone on an atrocious rampage, literally putting to the sword Chinese men, women, and children. The death toll from these old-fashioned methods was never established, but newspapers reported that some quarter of a million perished. A still worse slaughter, also done with traditional methods, befell Leningrad in the winter of 1941 under German blockade. The besieged citizens, prostrated by starvation, lacked strength even to clear their streets of the countless frozen corpses, some of which were eaten; children were driven mad with privation and fear. At least a million Russians died, the worst military massacre at one location in history.[32]

Spurred by Russian demands for help, by 1943 waves of British bombers were unloading at random over German cities. Their first big success was a fire storm that engulfed Hamburg, turning the city's streets into a hell where shrieking mobs were reduced to charred heaps of corpses. British and American military men called it a matter of destroying the enemy's "industry" and "housing," of smashing their "will to resist" and "morale"—their bodies were seldom mentioned. The Americans in particular, through close relations with journalists and filmmakers, worked to portray an image of bombers dropping their loads with precision on strictly military and industrial targets. In reality most of the commanders on all sides, together with their nations' citizens, had crossed a great moral divide with little reflection or debate.[33]

By 1944 citizens of the Allied nations approved any kind of bombing. As news of enemy atrocities arrived, many decided that the Germans were naturally cruel and the Japanese even worse, cunning and apelike barbarians. Roughly one American in ten thought the nation's war aim in the Pacific should be simply to exterminate all Japanese. The military men and civilians who did think in moral terms mostly expected that merciless bombing would shorten the war; they were groping for Douhet's vision, the white flags springing up over bombed cities. What was achieved, however, was something else Douhet had predicted: the ancient boundary between soldiers and civilians was erased.[34]

The war in the air, according to one of its historians, "assumed a significance in the popular mind out of all proportion to the contribution of air forces to victory." Airmen became the war's most mythicized heroes, or in the case of enemy bomber crews, villains. The supreme power of bombers was often described in the press and even in a Walt Disney movie, *Victory through Air Power*. People exclaimed that another war after this one would surely end civilization, mainly through bombardment. Vice President Henry Wallace offered the H. G. Wells solution of an international bomber force that would smash aggressors. In secret the United States Joint Chiefs of Staff agreed, studying the idea of an "International Police Force." The Joint Chiefs trusted that they would be top policemen, and sketched out a plan, which Roosevelt approved, for the United States to scatter bomber bases around the world after the war ended.[35]

Meanwhile, in March 1945 General Curtis LeMay sent some 300 bombers over Tokyo in a low-level incendiary attack. Fifteen square miles of the city went up in a fire storm, a far larger area than would burn at Hiroshima and Nagasaki together. The heat in Tokyo was so intense that small waterways boiled, bad luck for the people who dove in to escape the flames. The night's work caused fewer deaths than would come at Hiroshima but many more burns and other injuries—a million casualties. LeMay repeated his tactics week after week, methodically burning down cities. By August 1945 the Japanese people, between fire raids and blockade, were unable to go on much longer.

Such results could not be achieved without research. Groups at the Standard Oil Company and Harvard University had worked on incendiaries, devising napalm in early 1942. This and other substances were studied at the Massachusetts Institute of Technology to see how well each would ignite various types of wood. Elsewhere, researchers built copies of typical German and Japanese houses and test-bombed them. One result was the M-69 bomb, designed to go through the roof and spray its burning chemicals into the attic or living quarters. The M-69, like the German V-2 rocket deployed in the meantime, came close to being a strictly genocidal weapon.[36]

In warfare, as in other areas of modern life, research and industry had combined to do with stunning efficacy whatever people wanted to do. By mid-1945 the U.S. Army Air Force was destroying Japanese urban areas at an incremental cost of about three million dollars per square mile—substantially cheaper than the first atomic bombs.[37]

The nuclear scientists, however, were convinced that they were the ones who would revolutionize international relations. Since 1941 they had been insisting that atomic bombs were an ultimate weapon that could end the war in a few weeks. Political leaders up to Churchill and Roosevelt, impressed more by the familiar old atomic imagery than by any new facts, began to agree. In the war and in the maneuvering against the Soviet Union that they expected would follow, these leaders came to see atomic bombs as the solution to all their problems. When, they kept wondering, would the bombs be ready?[38]

Nobody felt the pressure more intensely than Robert Oppenheimer, who stood at the junction between science and mass death. Since childhood Oppenheimer had been troubled, and as a young man he had dismayed his friends and parents with bizarre actions, feeling so inadequate that he thought of suicide and murder. What helped him to overcome his problems was the fact that he could hope for a future, accepted among the best, by becoming a physicist. In later years Oppenheimer would describe with matchless eloquence the way science could give value to mortal life. There was the straightforward craftsmanship of the daily work, and the "clear and well-defined community whose canons of taste and order simplify the life of the practitioner," and beyond all that, physics itself, the "brilliant and ever-changing flower of discovery." Such beauty and order, said Oppenheimer, in the end sustained the life of scientists like himself.[39]

Reconciling this order with violence became Oppenheimer's career when General Groves chose him to head the new laboratory of Los Alamos, New Mexico, where atomic bombs would be designed and assembled. Oppenheimer traveled up and down the United States to recruit physicists, warning them that Hitler must not be allowed to get atomic weapons first. Opposition to the chaos and death that the Nazis represented was the main reason many scientists followed him to Los Alamos.

But there were other reasons, as Oppenheimer reflected later, that brought people to that secret city on top of a mountain. "Almost everyone realized that this was a great undertaking," a summons to a historic task, what another scientist called "the culmination of three centuries of physics." The fantasy of setting the world free with atomic energy seemed about to come true. Once atomic bombs were made they would quickly end the war, and presumably would spell the end of all great wars. Then the other long-promised benefits of atomic energy would be just around the corner. Alice Kimball Smith, the historian of these physicists' consciences, said that such were the motives that attracted her to Los Alamos with her scientist husband. Years earlier they had seen *Wings over Europe* and were so impressed that they had bought a copy of the script, but this play was only one of a hundred dramatic images that had prepared such people to go to Los Alamos.[40]

Once there the scientists had little time to think about future history, whether of utopia or massacre, for they were finding it extremely hard to make a bomb. They faced the problem that made it impossible for a reactor to undergo a mammoth explosion: once a chain reaction got well underway the mass would blow itself apart and halt the reaction. The best solution they found was to make a fist-sized lump of pure fissionable metal such as plutonium, not quite the critical size, and pack ordinary chemical explosives around it. If the chemical explosives were molded with exquisite and unprecedented care, then touched off in precisely the right way, the shock wave could squeeze the central lump almost instantly down to the size of a walnut. Inside that knot of molten metal, neutrons would multiply in a frenzied chain reaction during the split second before it all blew apart.

The whole process was so precarious that nobody expected it could work very efficiently. The best guess at Los Alamos was that one atomic bomb might bring destruction equivalent to an ordinary air raid of perhaps a hundred bombers. In the eyes of leading American scientists and statesmen, the first atomic bomb would raise no different moral questions than would a normal air raid. Yet nobody imagined that something so ingenious, so costly, indeed so fantastic, could not be used to great effect. In physical terms an atomic bombing might equal a night's work with incendiary bombs, but the psychological impact would be another matter.

Some leaders thought of the impact on the stubborn Soviet government; others considered the impact on the United States Congress, which would want to know where two billion dollars had gone; but the historical record taken as a whole indicates that most leaders thought mainly of using atomic bombs to send a message to Japan. In view of the literally suicidal defense of Saipan, where more than half the *civilians* (women and children included) killed themselves rather than surrender, in view of the waves of kamikaze airplanes that were hurling themselves with deadly effect upon Allied ships off Okinawa, the Japanese looked like a race of fanatics almost incapable of submitting. That was why Oppenheimer and other project leaders, brought together to advise on the use of the first bomb, suggested dropping it on "a vital war plant . . . surrounded by workers' houses." They hoped the tremendous spectacle of flash and blast would bring the Japanese to their senses. The first atomic bombings would be an act of rhetoric, a science fiction image aimed less at the enemy's cities than at his mind.[41]

A bomb test in July would teach the scientists what their achievement could really do, but by then it would be too late to reconsider. Two bombs would be ready to drop in August, and from early September the Manhattan Project was expected to deliver roughly one bomb a week. The Los Alamos scientists labored to fulfill these plans.

A different mood spread in Chicago. Compton's laboratory had no urgent tasks once the Hanford reactors were finished, and the scientists had time to think, perhaps even to read *The World Set Free* (someone had put a copy in the laboratory's library). They could also put together the realities of fission with the realities of slaughter that newspapers were reporting. By the summer of 1945, a fearful vision came to some as they walked the streets of Chicago. They imagined the sky suddenly lit by a fireball, the steel skeletons of buildings bent into grotesque shapes, masonry raining into the streets. A number of Chicago scientists, with the irrepressible Szilard in the lead, petitioned the United States government to set off a bomb on a deserted island as a demonstration. They thought the apparition of nuclear energy all by itself might make the Japanese break out white flags. But their main concern was for the postwar world.[42]

The Chicago scientists saw that there would be no safety from atomic bombs without revolutionary changes in domestic and international arrangements. The bomber would always get through, or if not, the V-2 showed that soon any city on earth could be bombarded by missiles. Yet the American public, ignorant of atomic bombs, still believed itself secure. In a democracy there was only one way to proceed, as Compton had written on a memo in 1944: "public education." The Chicago scientists made ready to explain the new facts of life once wartime secrecy was lifted. Some wanted to convince people to welcome an international peace force, a United Nations backed by virtuous Airmen with atomic bombs; others wanted to warn their countrymen to disperse industry so as to be less vulnerable in war. Compton felt the public should learn about nuclear energy especially so that people would willingly pay for physics research, and not only on weapons.[43]

Already in April 1944, some rough calculations had suggested that uranium reactors might produce power for industry as cheaply as coal or oil plants could do. The Chicago scientists got excited about the prospects for a new field of work, dubbing it "nucleonics" to show that it would be as important as electronics. Besides helping humanity toward a prosperous future, the rise of nucleonics would conveniently solve their problem of where to find jobs when the war ended. Provided, of course, the public could be taught the supreme importance of nuclear energy.[44]

The wisest comment on these plans was offered by Karl Darrow, a senior physicist respected for his caustic insights. "I take it that there are two main objects," he wrote to one of the Chicago scientists, "apart from inspiring interest in science for its own sake. One is, to please the public with the prospect of beneficial uses of atomic power, and the other is, to scare it out of its boots by threatening it with new weapons." He doubted that both goals could be reached at the same time.[45]

"THE SECOND COMING"

While the Chicago scientists thought over what to tell the public when secrecy was lifted, General Groves laid his own plans. He gave one journalist, and one only, entry to the Manhattan Project. The man he chose was ideally qualified to exalt the work that Groves had accomplished, becoming the world's foremost prophet of atomic miracles.

William L. Laurence had been born in the fifteenth century. At any rate the Jewish hamlet in Lithuania where he had grown up, with its cramped wooden huts, muddy streets, and plodding horses, had nothing to do with the twentieth century. There was the same timeless simplicity in one of Laurence's earliest memories, a story that offers a key to his personality. His white kitten grew weak, mewed, and died. Devastated, the little boy decided to seek out God, who seemed close by, over at the horizon where heaven came down and met the earth. So he set out to walk to the horizon and ask, "God, why did you kill my little kitten?" A long walk later, heartbroken, he gave it up. Yet when Laurence in his old age remembered that day, it seemed to him that he had never really stopped trying to reach the place where heaven met the earth.[46]

A more material heaven came into view when Laurence came upon some modern books. Reading about Mars, he swallowed whole the theory of an elder civilization. If only we could communicate with those wise creatures in the sky, humanity could skip ahead over millennia of agony to reach its Golden Age! His reading made him an atheist, and he disrupted his Hebrew school until his father sent him to study elsewhere. In the Russian Revolution of 1905 he stampeded through the streets with other young students, singing revolutionary songs. The apocalyptic message stirred him for the rest of his life: "After the bloody battle," he would sing, "there will rise the sun of truth and eternal love."

When the revolution collapsed Laurence decided to go to America, the land of the newly invented airplane and radio. He had not forgotten the dream of communicating with Mars, and he meant to find the technology to do it. He wanted, as he later recalled dryly, to learn "the secret of life—how to produce life synthetically. That was one of the little questions I was going to ask the Martians to tell me." It was not so different from the question he had once meant to ask God.

Laurence's mother helped smuggle him out of Russia to Germany in an empty barrel, and he took a passage to America. When he saw New York harbor at night he felt he was already entering the white city of his dreams. Abandoning his unpronounceable Lithuanian name for the patrician "William L. Laurence," he worked his way from penniless immigrant to a scholarship at Harvard. His body filled out, energetic and squat, topped by a flat, homely face with a thoughtful look around the eyes. In 1930 he joined the *New York Times* as a science reporter.

Laurence believed that science was the religion of the future. Like any religion, it needed missionaries, and that was what a journalist like himself should be. "True descendents of Prometheus," he declared, "the science writers take the fire from the scientific Olympus, the laboratories and universities, and bring it down to the people." Along with Compton (who compared his own group with Prometheus), Oppenheimer, and many others, Laurence took part in science as a way to counter human weakness and rise above mortality.[47]

Atomic energy had always fascinated Laurence as much as any science reporter, and the news of fission stirred him to write articles which the editors of the *Times* and the *Saturday Evening Post* almost rejected for sounding like science fiction. In 1945, when General Groves asked Laurence to come on a secret mission, the reporter felt more than ready.

But he was not prepared for the Manhattan Project. At Oak Ridge they took Laurence to a mountaintop and showed him the valley below covered with a staggering secret city. One Oak Ridge plant for separating uranium-235 was the largest factory ever built, four stories high and stretching half a mile; as a senior administrator told Laurence, the Project had outdone Jules Verne. Laurence warned his editor that he was onto a story bigger than anyone could have imagined, "a sort of Second Coming of Christ yarn."[48]

At Hanford Laurence was shown a nuclear reactor. All he saw was a big wall with thousands of holes for fuel rods, operating soundlessly, but he already had an image of atomic energy, and to stand at a real reactor seemed overwhelming. Jotting down his impressions, he wrote that "mighty cosmic forces" were at work "in the presence of the very act of elemental creation of matter," an event comparable only with the biblical Genesis. Laurence took special note of the elaborate controls "devised to keep this man-made Titan from breaking his bonds. Left without control for even a split second, the giant would run wild." Nuclear energy was a sort of golem, then, although one securely chained.[49]

Finally Laurence reached the most secret of cities, Los Alamos. He called it "Atomland-on-Mars." If the Los Alamos Martians could not reveal the secrets of life and death, they held secrets that seemed almost as important.

The scientists tended to agree that almost nothing lay beyond their cosmic powers. Back in 1942, when a team of theorists led by Oppenheimer had first begun to work out the details of atomic bombs, a thought had struck them. Could such a bomb set off a chain reaction in the oceans or the earth's atmosphere? Rutherford had already explained that a world so unstable would have vanished long ago, but such logic did not satisfy the Manhattan Project scientists. Over the next years one or another of them would come back to the question, nervously calculating proofs that they did not have the power to destroy everything. Fears persisted up to the eve of the first test, when Fermi joked that he would hold bets on whether or not the bomb would ignite the atmosphere.[50]

Before dawn on July 16, 1945, the leaders of the Manhattan Project assembled in a valley deep in the desert. Stretching from the barren earth toward the heavens was a tower carrying their first device. When the fireball illuminated the valley, some scientists thought for a moment that the world was starting to blow up, while an Army engineer, General Thomas Farrell, cried, "Jesus Christ, the long-hairs have let it get away from them!"

Surely the sight was awesome, but Farrell was bringing older images to the scene. General Groves said that Farrell caught the feelings of almost everyone when he wrote that the sound of the bomb, a long hard thunder echoing around the distant mountains, "warned of doomsday and made us feel that we puny things were blasphemous to dare tamper with the forces heretofore reserved to The Almighty." At another site an observer found witnesses reacting with a "hushed murmuring bordering on reverence." It was as if they had seen something beyond the mortal realm.[51]

When Laurence asked Oppenheimer what he had felt at the moment of the explosion, the answer stunned the reporter. As Oppenheimer watched the cloud boil miles up over his head, multicolored and radiant, he had recalled a passage from an ancient Hindu scripture that described a vision of the divine Krishna suddenly growing to reach from earth to heaven, taking on a dazzling multicolored form with numberless arms and eyes, numberless flaming jaws that swallowed whole armies: "I am become Death, the destroyer of worlds."[52]

As always, death was matched with hopes of rebirth. Farrell reported that the witnesses felt they were viewing something not only threatening but tremendously promising. To Laurence, the rumbling echoes that reminded Farrell of doomsday seemed "the first cry of a newborn world." He reported that some scientists broke into a gleeful jig like "primitive man dancing at one of his fire festivals at the coming of spring."

Laurence believed that he and Oppenheimer and probably many others there had shared a profound religious experience. He said later that witnessing the explosion was "like being present at the moment of creation when God said, 'Let there be light,'" or indeed like seeing the Second Coming of Christ. It was exactly what he had always been looking for.[53]

The language of birth and apocalypse flowed outward from the Trinity test site. Coded messages to the American Secretary of War described the success as a miraculous birth, with "Doctor [Groves] . . . most enthusiastic and confident that the 'Little Boy' is . . . husky." Groves sent Farrell's histrionic description as an official report to the new President, Harry Truman, who promptly remarked that the scientists might have the most useful thing ever discovered and certainly the most terrible, perhaps the "fire destruction" prophesied in the Bible. Meanwhile Churchill exclaimed, "This atomic bomb is the Second Coming in wrath." All this before they had received any solid facts about what such a bomb could do. For those who witnessed the Trinity test and for those who only heard about it, the explosion served less to introduce new ideas than to bring ancient thoughts of apocalypse to new and vivid life.[54]

The scientists returned to Los Alamos to assemble bombs, and Laurence flew to Tinian Island in the Pacific. All that week fleets of ordinary bombers went back and forth from Tinian and other bases, smashing and burning Japan. Laurence saw the bomb destined for Hiroshima, code-named "Little Boy." It was loaded aboard the bomber *Enola Gay*, named for the pilot's mother (her first name was taken from "alone" spelled backwards). The bomb was scribbled over with people's signatures and, following an Air Force custom, with vengeful and ribald messages to the Japanese.[55]

6

The News from Hiroshima

"It is an atomic bomb. It is a harnessing of the basic power of the universe. The force from which the sun draws its power has been loosed against those who brought war to the Far East." President Truman's announcement astonished the world, for few had expected that atomic bombs would come in their own lifetime. The statement itself sounded like something William Laurence might have written for a 1930s Sunday supplement. In fact the President had received the statement from a committee, who had adapted a draft forwarded by General Groves, who had gotten it from none other than Laurence. The reporter had drawn up the draft for Groves two months before the Trinity test.[1]

The public, like Laurence, could only understand the news in terms of what they already had in their heads. Some intellectuals promptly revived the complaint of the 1930s that science was advancing too fast, that the humanities and human thinking in general could not keep up. But in fact there had been solemn warnings and imaginative treatments about atomic energy for decades; further delay would not have led to a deeper understanding. Thought waited upon actual experience. However, there was not much solid information about nuclear war, and even that had to be fitted in with people's hard-won stock of existing ideas. Most people began with the myths that had grown up long before the first bomb burst.

CLICHÉ EXPERTS

The War Department gave the press a number of reports written weeks or months earlier by Laurence, all full of millennial awe, and the reporter's language appeared in newspapers around the world. At first this was almost the only material available. When other writers began to describe the emergence of nuclear energy, they either followed Laurence's lead or reached into the same stock of familiar images that he had used.[2]

A few bare facts were available. Planes flying over Hiroshima and then Nagasaki told of square miles aflame, while Japanese radio confirmed that the cities had become graveyards. Much could be made of that, given that atomic energy had done it. A typical American newsreel exclaimed that Hiroshima was annihilated by a "cosmic power . . . hell-fire . . . described by eyewitnesses as Doomsday itself!" This idea of apocalyptic power cropped up everywhere at once, like dormant seeds sprouting under a sudden rain.[3]

The idea was immediately connected with the question of control over dread secrets. The bombs were a "revelation of the secrets of nature, long mercifully withheld from man," according to Churchill's official statement. Statements by Truman and by Clement Attlee, the new British Prime Minister, likewise spoke of incalculable forces, control, and secrets. The leaders hoped that the new power could be used to "serve" humanity, made into a helpful rather than a wrathful creation.[4]

Was the bomb a golem then, a mad scientist's monster? Certainly the War Department's report on the Trinity test suggested a scene right out of the 1930s Universal Studios films: "darkening heavens, pouring forth rain and lightning immediately up to the zero hour" had set the stage for the perilous secret experiment. In reality the storm had passed hours before the test. But newspapers, drawing on the thunderbolt imagery and on General Farrell's histrionic report, did not fail to exclaim that scientists had tampered with an uncanny creation.[5]

From the outset nuclear energy was personalized. There were not many "atomic bombs" in all the talk during the next few years; everyone spoke of *the* atomic bomb, or just The Bomb, capitalized like a creature from myth. Cartoonists drew muscular genies escaping from bottles or brutal giants labeled "atomic power," looming over hapless scientists or politicians. Particularly explicit was H. V. Kaltenborn, a famous NBC radio pundit, in his broadcast on the day of Hiroshima. "For all we know," he concluded in hushed tones, "we have created a

Frankenstein." That term was soon heard everywhere from street corners to the United States Senate.[6]

Usually a large fraction of the public paid scant attention to world news, but a survey found that 98 percent of all Americans had heard about atomic bombs. Most had talked about them with family and friends, and surveys found that ordinary people were speaking much like the journalists. Western Europe was swept by the same excited talk, for not only did the Europeans share the full burden of prewar atomic imagery, but they now got almost all their new atomic information from the United States.[7]

Instead of parading more examples of such talk, I will defer to an authority. Mr. Arbuthnot, the fictional creation of humorist Frank Sullivan, admitted in 1946 that he was the leading "cliché expert" on the atom. No question could stump him. For example, he was asked,

> Q. Where do we stand, Mr. Arbuthnot?
> A. At the threshold of a new era . . . Will civilization survive? Harness.
> Q. Harness, Mr. Arbuthnot?
> A. Harness and unleash. You had better learn to use those two words, my boy, if you expect to talk about the atom.

The cliché expert knew what had been harnessed or unleashed, namely, "the hidden forces of the universe." He easily identified whose stone atomic energy was (the philosophers' stone); and whose dream (the alchemists' dream); and of course whose monster. He had an easy time summarizing talk about the first bombs, for he could have given exactly the same answers to questions about atomic energy a decade earlier.[8]

Mr. Arbuthnot, and the real-life cliché experts who staffed newspapers and radio networks around the world, insisted above all that atomic energy might open the way to a utopian civilization. Laurence told a huge radio audience that with the new energy, "We can air condition the jungles and make the arctic wastes livable . . . and we can lick disease." And of course ships would cross the ocean with only a little fuel—a lump the size of a pea, said Mr. Arbuthnot; "the pea is the accepted vegetable in these explanations." Nuclear energy could even solve the problem of war. In mid-1945 most people had expected many more bloody months of battle, and now two bombs seemed to have magically brought surrender and peace.[9]

But the main words to keep at hand, according to Mr. Arbuthnot, were "vast" and "doom." A poll found that a quarter of all Americans thought it likely that atom-smashing experiments would someday destroy the planet. In early 1946 Paris radio broadcast a tale in which an announcer, imitating news flashes, cried that "atoms from radium used for research in America have broken loose" and a doomsday atomic storm was sweeping around the world. Parisians went into a panic, jamming switchboards with terrified calls; the scare cost the director of the French national radio his job.[10]

Some fundamentalist Christians began to speak of true apocalypse. Atomic bombs, they announced, proved that the day foretold in biblical revelation was at hand. There had always been people at the margins of society preparing for Armageddon, but from 1945 on the most sober leaders, from presidents to popes, spoke in language that could evoke such thoughts. As sociologist Edward Shils remarked, atomic bombs made a bridge across which apocalyptic fantasies, marching from their refuge among fringe groups, invaded all of society.[11]

During the initial shock, that is, for the first few years, few people in any nation had specific fears. Attack by the United States or the Soviet Union or some other particular enemy was no immediate prospect. The public simply felt that the ground had fallen away under them. One element in this was the realization, which struck many people right from the first news, that at some point in the foreseeable future no city on earth would be safe. A related element, harder to pin down in factual concerns, was best expressed in a famous editorial by Norman Cousins: "The fear of irrational death . . . has burst out of the subconscious and into the conscious, filling the mind with primordial apprehensions." The old sense of security was lost; something unimaginable had come into the everyday world to stay.[12]

Mr. Arbuthnot and his colleagues had done something important. As late as the 1930s the "public" who were regularly exposed to the various images of nuclear energy—counting, say, everyone with a high school education—had been a minority of the population in every nation, and even they had given the subject little attention. From August 1945 on, the whole tangle of fantastic nuclear imagery, mingled with the very few facts then available, was a significant part of the mental equipment of everyone within reach of a radio.

HIROSHIMA ITSELF

There was one nation that had something original to say about atomic bombs: Japan. Imagine that you close your eyes and count to five, and when you open your eyes everything around you to the horizon is a smashed wasteland inhabited by wounded crowds. In other bombed cities like Hamburg or Tokyo, some thousands of people had experienced the supreme terror of a near-miss, the tremendous crash of light and sound, the view into a hell. But only at Hiroshima and Nagasaki did that happen to hundreds of thousands all at once. It mattered how the people of Hiroshima and Nagasaki felt, for some were determined to bear witness.[13]

When the psychologist Robert Lifton interviewed Hiroshima survivors he found they tended to merge the bombing experience with childhood images of victimization and the end of the world—images of separation, helplessness, and annihilation. To have been in Hiroshima or Nagasaki was to know a terrifying "coming together of inner and outer experience," of death fantasies and reality. The bombings seemed less like a military action than a rupture of the very order of nature, an act, as the wartime Japanese government had declared, of sacrilege.[14]

People in many nations questioned the morality of the deed. In the United States the great majority felt that the bombs had been justly used (and a substantial minority were sorry that more had not been dropped before the Japanese had a chance to surrender), but there too questions of guilt arose instantly. A few critics, notably some thoughtful ministers and priests, had already condemned the fire-bombing of helpless cities. However, Hiroshima inspired more debate than the rest of the war's destruction put together. It was as if all the other recent massacres could be set aside and the entire moral problem of modern war could be concentrated in this one question.

The feeling that the bombs went beyond some human boundary strengthened as solid news arrived from Hiroshima. The public was stunned by photographs and newsreels that swept across square miles of burnt-out wasteland. To be sure, sections of Tokyo and dozens of other Japanese cities looked little better, but it was atomic force that caught the imagination.

In August 1946, one year after the bombings, the *New Yorker* devoted an entire issue to "Hiroshima," by the young journalist John Hersey. For days it seemed that everybody was talking about Hersey's report, and soon in the United States and elsewhere newspapers were running the entire text while radio networks read it aloud. It became a best-selling book, with pirate editions turning up in France, China, Bolivia. Within a few years many high schools were using it in reading programs, and for decades it would remain a favorite with teachers.[15]

Hersey's approach was shaped by his tour as a war correspondent in the Pacific, where he had faced death time and again. Once he was ambushed with a jungle patrol; another time his airplane went down in the sea and he had to fight his way out of the sinking wreckage. Obsessed (as he said) with the human will to survive, Hersey was unusually qualified to probe Hiroshima's encounter with death. When he heard Truman's radio announcement of the atomic bomb, like many others he felt dread for the world's future. What he saw in Hiroshima redoubled the dread. But rather than express such feelings outright, Hersey faded into the background like a good reporter, building his book on the stories of a few sympathetic individuals, giving the bomb survivors a voice.[16]

A typical reader who finished "Hiroshima" at midnight said that the rest of his night was fitful, heavy with dreams. Hersey's work, along with other similar reports, gave people a taste of the Japanese experience. The atomic bomb had not, after all, been used against an "enemy" or a "city" but against people like ourselves; from now on the threat would be personalized. Moreover, like the famous shadows of humans scorched into concrete, a set of horrifying images was burned permanently into the public mind. These images became new emblems of nuclear energy, and at the same time new emblems of modern war—and of death itself.[17]

When it came to linking nuclear energy with death, the most disturbing news concerned radiation injury. At first Tokyo radio hinted that lingering radioactivity was killing relief workers who entered Hiroshima. Meanwhile, an American chemist and science writer announced that for seventy years radioactivity would leave the city as desolate as the moon. The report made a sensation, and the American War Department was deluged with anxious telephone calls. Japanese began to tell one another that radiation would poison Hiroshima and Nagasaki for decades, that not even plants would grow there.[18]

Radiation held as much mythical meaning in Japan as anywhere. An ancient legend told how the Japanese had lost a battle when they fought with the sun in their eyes, but would be victorious so long as they had the rays of the Rising Sun striking from behind them, as on the Japanese battle flag. Medieval Japanese Buddhist icons depicted halos shooting forth rays, and even rays projected from a divinity's eyes. A leading Japanese scientist reportedly had a wartime dream in which he saw a deep cavern containing a great atom smasher from which emerged luminous rays, striking around the earth to destroy Washington. The cavern might sound like the underground laboratories of Western science fiction, but it also sounded like a cave in Japanese mythology that had hidden the radiant Sun Goddess. In short, to the Japanese as to everyone else, rays had long stood for a mysterious, divine force of life and death.[19]

The feeling of uncanny attack that perturbed people at Hiroshima and Nagasaki was redoubled by real radiation injury. The Japanese came to know the symptoms well, the vomiting and diarrhea in the first days, and in later weeks the loss of hair and the bleeding gums, sometimes ending in recovery but sometimes in death. Such injury was featured in the first news story out of Hiroshima, by a British correspondent who had wandered through the rubble stupefied by vistas of death. Determined to issue a warning, he wrote that "people are still dying, mysteriously and horribly . . . from an unknown something which I can only describe as atomic plague."[20]

All this General Groves dismissed as Japanese propaganda, and American officials severely harassed scientists and reporters who exclaimed about radiation from the Hiroshima bombing. Most of the casualties, the authorities pointed out, had resulted from ordinary blast and fire. Yet it was also true that many thousands were injured or killed by radiation, and that news could not stay hidden. Indeed, reporters followed the Japanese in emphasizing radiation casualties above more familiar sorts of wounds and burns. Hersey's calmly gruesome description of radiation injury, coming as the climax of "Hiroshima," showed everyone how strange and horrible it seemed to the victims.

Radiation anxieties were reinforced by the world's only other experience with atomic bombs during their first years: tests held in the Pacific at Bikini atoll in 1946. The United States meant to see what bombs would do to surplus battleships, and when the test plans were first announced, few thought of radiation. People expected an image out of Jules Verne, a fleet instantaneously pulverized near a secret island, if not Crookes's picture of battleships hoisted thousands of feet into the air. Around the world there were some who worried about tidal waves, clouds that would bring a year of rain, or even a chain reaction that would consume the oceans. More than a hundred journalists came to Bikini, and to many of the public (according to a French poll) the test was the most important news event of the year.

Bikini's lesson was one of anticlimax. The first test bomb, exploding far off target, sank only a few of the many ships. Soviet newspapers minimized the results; an Argentine radio announcer said he would broadcast the sound of the explosion, then gave a ludicrous peep. A second bomb test was more impressive, but the spell of magical dread had been broken. In the end the Crossroads exercise reassured the public, dispelling fears of vast explosions. The initial shock of living in a world with atomic energy now began to die away, and people began to hope they could deal with the new bombs as weapons like any other.

The Bikini experience did promote one type of nuclear fear, however. The Navy took strict precautions against radioactivity, and the sailors and reporters who entered the lagoon after the explosions were not reassured when technicians kept testing their bodies with Geiger counters, until the formerly unfamiliar crackle of the instruments became a sound of dread. Nervous jokes and rumors about sterility began to circulate. Newsreels reported the official caution and unofficial anxiety, for example with film of a sign reading "Danger Radioactivity—Keep Out." Tales of Bikini reached the public at almost the same time as Hersey's "Hiroshima," redoubling the impression of bomb radioactivity as a special horror.[21]

Hiroshima and Bikini came together, for example, in a special Paramount newsreel released to movie theaters in August 1946. Shortly after the Hiroshima bombing, courageous Japanese cameramen had shot film in the ruins. The American War Department had confiscated their footage but released snippets for the newsreel along with spectacular shots of the Bikini explosions. The *New York Times* reported accurately that the most stunning image was not the blast but the Japanese victims, who looked "as though they had been seared by an acetylene torch." Although it was not nuclear rays but heat radiation from the bomb's fireball that caused such wounds, the *Times*'s reviewer thought he was seeing "the deadly radioactive effects of the bomb."[22]

The facts from Hiroshima and Bikini were becoming tangled up with older images of magical attacks. There were stories that turned radioactivity into a horrid contagion, such as rumors that the bodies from Hiroshima were so radioactive that they had to be buried using long poles as if they were plague victims. David Bradley's *No Place to Hide*, a 1948 book on the Bikini tests that was read very widely, told how bombs had contaminated the idyllic atoll until it became a land of ghostly peril, focusing on radioactivity as if that were the bombs' chief danger.[23]

On the other hand, many people accepted a remark in Hersey's book that when vegetation sprang up lushly in Hiroshima, that was not because of the rich ashes and suddenly open sunlight but because the bomb had somehow "stimulated" the plants' "underground organs." Of all the mysteries of radiation, the most fascinating had always involved life and procreation. In the first significant novel to invoke atomic bombs, Aldous Huxley's widely read *Ape and Essence* of 1948, the radiation left from a future nuclear war not only generated grotesque babies but also made women uncontrollably lascivious for one month of each year.[24]

Medical experts and others began to worry about a new phenomenon, what William Laurence and others called an "unreasoning fear" of radiation. The experts' worries were premature, for most people were still not fearful of all radiation; they continued to get X-rayed with few qualms, and in 1946 when four physicians died as a result of too much exposure to X-rays, the press took scant notice. Yet reports that one man from the Oak Ridge uranium works had died of radiation exposure stirred a minor newspaper sensation. (The press promptly forgot about the worker when it turned out he had died of hepatitis.) It was not radiation alone that was dreaded—not yet—but the combination of radiation with nuclear weapons. Fears of explosions had subsided after the Bikini test. But the feeling of sacrilegious violation that the Hiroshima victims experienced, and to some degree all the abhorrent results of modern warfare, had been neatly symbolized by one particular physical effect.[25]

SECURITY THROUGH CONTROL BY SCIENTISTS?

Some groups were not satisfied to let feelings of awe and horror develop by themselves, but worked to shape the world's image of nuclear energy for their own purposes. Most of these groups used scare tactics. However, laboratory studies have shown that if fear sometimes aids in persuasion, at other times the results may be the exact opposite of what is planned.[26]

First to know of atomic bombs, the scientists of the Manhattan Project had also been the first to work out a public relations plan. As soon as the war ended the Chicago group launched their campaign, and scientific colleagues in many countries joined them. The Chicago scientists had some specific political steps to recommend, but first they meant to instruct the world in the dangers and opportunities created by their discoveries. Like most scientists, they believed that in any impasse the first step should be a full exchange of information.

Leo Szilard saw the dreams of his youth coming true. Always convinced that scientists should band together to help reform society, the bustling Hungarian took the lead in politicking. His first concern was that the United States government was planning to leave postwar atomic energy work in the hands of the Army. Many who had labored under General Groves's rigid secrecy rules would fight that plan tooth and nail, and when Szilard sounded an alarm scores of scientists hopped on trains for Washington, formed a Federation of Atomic Scientists, and plunged into political battle.

The "atomic scientists" caught everyone's attention. Physicists were asked to attend countless private dinners, club gatherings, and government meetings, to write articles for all sorts of magazines, to speak everywhere from the Quiz Kids radio show to the White House. A sociologist who studied congressional hearings on nuclear energy concluded that even senators looked upon atomic scientists in much the way that primitive groups looked upon their shamans, as beings "in touch with a supernatural world of mysterious and awesome forces whose terrible power they alone could control."[27]

The association between awesome nuclear force and awesome men was embodied in one of the most common nuclear pictures: a face with wise, sad eyes and a chaotic shock of white hair. Often the talismanic sign $E = mc^2$ was added, as in a 1946 *Time* magazine cover that superimposed the face and equation on an atomic bomb cloud. This was the man whom the press had already in the 1920s made into a symbol of esoteric thought, an eccentric sage who grasped secrets beyond ordinary mortal reach. It was not the Albert Einstein who had actually discovered the equation in 1905, a cocky young man with a trim dark moustache. The real Einstein had made no discoveries in the physics of the nucleus; he had never had anything to do with fission, aside from an incident in 1939 when he had served as a figurehead to help draw the government's attention to uranium. To be sure, his famous equation expressed the conversion of nuclear mass into energy when a nucleus split, but only in the same way as it expressed the conversion of molecular mass into energy when a match burned. If Einstein had never lived, the date of the bombing of Hiroshima might not have been altered by a week. When scientists and the press collaborated in making the old man a prime symbol of nuclear energy, they were using the symbolism to insist that scientists, however wise and well-meaning, had released forces far beyond the ordinary ken.[28]

Some physicists were taking a different view of nuclear energy. By the late 1940s theorists recognized that this energy was no supreme mystery, no key to the universe, but something on a level with any other property of matter. They began to suspect (although it would not be proved until around 1980) that the force revealed in radioactivity is simply one aspect of the same force otherwise revealed in ordinary electrical and chemical processes. However, nobody explained to the public that making a nuclear device work was neither more nor less magical than wiring a house for electricity or building a bonfire. Atomic scientists were willing to be seen as almost like Einstein himself, masters of ineffable knowledge.

Combining their towering new prestige with earnest effort, scientists became the ones who gave citizens most of their basic information about nuclear energy. Information rarely comes all by itself, however; it usually comes as one ingredient in a mixture of ideas, feelings, and images. What emotional messages were scientists passing along, blended into their lectures on neutrons?

Time magazine suggested that the atomic scientists flung themselves into politics because they were "guilty men." Einstein agreed, saying that those who had worked on the bomb were driven to work for peace as an atonement. Oppenheimer put it best, writing that atomic bombs "touched very deeply man's sentiments about the evil of having too much power." In 1947 he captured the idea in an instant cliché: the atomic scientists had "known sin." It was as if they had eaten the forbidden fruit of the Tree of Knowledge—a comparison that some physicists made explicitly. These thoughts came not just from anxiety about cosmic secrets but even more in response to the news from Hiroshima. Oppenheimer himself, when he met Truman, blurted out that he had blood on his hands.[29]

Inquiries found that a majority of Manhattan Project scientists, like most other Americans, said they did not feel guilty, for they believed the bombs had shortened the war and saved many lives. Yet guilt could be subtle. Psychologists have observed that people might feel secretly guilty when someone near to them died; one study of terminal cancer patients found that relatives and victim alike might think they were being punished for an unknown sin. On seeing a photograph of Hiroshima, even people whom nobody could blame for the destruction might feel a gnawing anxiety not unlike guilt.[30]

After all, everyone recognized that when the United States made atomic bombs it had opened up the possibility of being bombed that way in turn. Scientists thought more than anyone about being paid back in their own coin. For example, in the summer of 1946 two brilliant young physicists, Freeman Dyson and Richard Feynman, were driving across the United States. Dyson had not been in the Manhattan Project, but Feynman had served as a Project theorist, meanwhile letting off steam as the wild jester of Los Alamos, playing ridiculous pranks on the Army guards. Now Dyson was struck by the bleak way in which his companion saw the future. "As we drove through Cleveland and St. Louis," Dyson recalled, "he was measuring in his mind's eye distances from ground zero, ranges of lethal radiation and blast and fire damage."[31]

This was the emotional message the world swallowed along with the atomic scientists' information: mortals had loosed forces too great for them. Whenever Oppenheimer tried to give a cool and balanced speech, he noticed that his listeners went away anxious. Other scientists communicated their anxieties deliberately. "I write this to frighten you," the chemist Harold Urey told millions of readers of *Collier's* magazine in an article ghostwritten for him by the Federation of Atomic Scientists. "I'm a frightened man, myself," he continued. "All the scientists I know are frightened." Other atomic scientists conveyed their feelings especially by describing what bombs could do, explaining the range of blast damage and so forth. "Just a Wreck! That's Chicago Under A-Bombs," said the *Chicago Tribune* after a typical scientific speech. Especially ingenious was the scientists' lobbying group at Los Alamos, who sent lumps of fused sand from the Trinity test site to forty-two mayors as a hint of what could happen to their cities.[32]

Scientists spread fear not only because that represented their own feelings but also because fear would move the listener. Technical lectures drew yawns, but warnings of doom brought audiences to the edges of their seats. The principle was explained by a committee of eminent psychologists who, answering a request from the Federation of Atomic Scientists, made a study of atomic bomb anxiety. The psychologists warned that fear could be unhealthy; it could lead to passive despair, to panic, or to aggression. But they said fear could also be helpful, a way to get people moving. "Our first objective," the committee reported, "must be to mobilize a healthy, action-goading fear for effective measures against the real danger—war."[33]

Both the psychologists and the atomic scientists expected that cries of atomic bomb dangers would move society toward certain specific political goals. Already before Hiroshima was bombed the Chicago scientists had worked out their program, as clear and sharp as a diamond. Now they drummed their reasoning into the public. Never again—and this was the hardest to explain, for it went against all habits of thought—never again could a nation win security through a commanding military lead. For once each side had enough atomic bombs to wipe out the other's cities, more bombs would make little difference. And there could never be a defense, no matter how far technology advanced, when only a few small bombs would suffice to wreak unthinkable damage. The scientists warned that "security against aggression will rest on the fear of retaliation." But that would mean "a world of fear, suspicion and almost inevitable catastrophe." The only way to avoid doom would be for nations to abandon some of their sovereign rights; to be specific, they must put all nuclear energy projects under international control.[34]

It was plain logic, based squarely on the facts. Nobody since has offered a better permanent solution. But how could international control be set up in our suspicious world?

Many hoped that the very existence of atomic bombs would automatically impose the solution. To a humanist like Raymond Fosdick, the earnest League of Nations supporter who had repeated Soddy's warnings back in the 1920s, atomic bombs simply turned forebodings into a demand for action. Sooner or later, wrote Fosdick in a phrase already becoming famous, there would be "One World or None." On BBC radio, Bertrand Russell explained that "no middle course is any longer possible" between annihilation and a utopian age of peace. The fictional Mr. Arbuthnot, with his unfailing knowledge of current clichés, said humanity was at a "crossroads": a metaphor that many editorial cartoons made visible. Once everyone was shown the two paths leading from the crossroads, "World Destruction" or "World Control," surely they would choose rightly.[35]

It was as if nuclear energy were such a cosmic force that it would sweep away history, instantly replacing the web of international tensions with a millennial age of peace! The idea was so powerful that it became the official policy of the United States and many other governments. Truman announced that atomic energy was "too revolutionary to consider in the framework of old ideas." On behalf of the United States, Bernard Baruch enunciated a precise and ingenious plan for international control of atomic energy, a plan accepted in principle by a large majority of the United Nations. The chief developer of the scheme had been Oppenheimer.[36]

"Oppie's plan," as a sympathetic official put it, "would set scientists up as social policemen." The keystone was free exchange of information among international teams that would inspect every nuclear facility. To avoid technical surprises the international agency must also, as scientists particularly insisted, employ scientists in research; these workers could meanwhile turn the new force to peaceful uses. The scheme was modeled on the open, self-regulating, peaceful community that scientists already enjoyed, and at the same time resembled H. G. Wells's vision of virtuous technocrats enforcing peace and bringing a Golden Age.[37]

Scientists were joined by many other groups that tried to promote the plan by spreading fear of the alternative. Foremost in the United States was the National Committee on Atomic Information, which brought together leaders of important American organizations such as the National Farmers' Union, the American Federation of Labor, and the League of Women Voters. Through materials like the short film "One World—or None," which began and ended with a picture of the globe consumed in flames, the Committee taught that only international controls offered safety. The Federation of Atomic Scientists provided much of the material that the Committee distributed in tens of millions of copies.[38]

Other groups such as world federalists worked still more directly to implant fear. One broadcast over ABC radio began with a frightened little boy wailing, "What's happening to my arm?" His mother shrieked in horror, "Look! . . . his left arm is gone! And his right arm is slowly—*Eeeee!*" The terrified scream faded into a program featuring celebrities calling for world federation. Such shows could have a permanent impact. For example, a little girl who listened to an atomic bomb radio special called *The Fifth Horseman* recalled, decades later, that "it frightened me more than anything I can remember." The impression that lingered was one of apocalyptic danger. "Because of that radio show," Jane Fonda continued, "I've always thought that we were tampering with things . . . We were playing God by creating such weapons, such material."[39]

Leading newspaper columnists and radio commentators likewise emphasized atomic horrors in an attempt to spur the public into action. For example, Joseph and Stewart Alsop wrote an article for the *Saturday Evening Post* with the title, "Your Flesh *Should* Creep." Master cartoonist Herblock worked up a scary caricature that newspapers and magazines and the National Committee on Atomic Information put in almost every mailbox in America. His famous symbol, rapidly imitated by other cartoonists, personified the atomic bomb as a brutal and uncontrolled giant, a true golem. "I felt a good deal of satisfaction," Herblock recalled, "when people used to tell me . . . that the Atom character gave them the shivers."[40]

These expert communicators could make people afraid but they could not create a new world order. Indeed, the more powerful the bombs seemed, the more reluctant most Americans were to turn them over to an international agency. The negotiations broke down, partly because of this reluctance and still more because Soviet officials frankly scorned the Baruch plan, which would leave the Americans armed for an indefinite period while denying everyone else the chance to build bombs of their own.

None of the processes I have described so far did much to influence opinion in the Soviet Union. The Russians had no orgy of atomic bomb fear and felt no drive to establish practical international controls, for official Soviet policy worked actively to suppress any such feelings. The day after the Hiroshima bombing, Radio Moscow had broadcast the full text of Truman's announcement and everyone in Moscow had talked about it anxiously, while the Soviet leadership became more determined than ever to advance with desperate haste their own work on fission, begun in 1943. But outwardly the leaders showed only confidence. Soon Stalin was declaring that the American atomic bombs were "intended for intimidating the weak-nerved, but they cannot determine the outcome of war."

While a torrent of facts and speculation washed through the democratic nations, the Soviet public received only a trickle of generalities, mostly about the beneficial uses of atomic energy. What information they did get could be misleading. By combing the Soviet press a reader might discover that the dead at Hiroshima and Nagasaki had numbered in the "thousands," a trivial number to a nation that had survived the siege of Leningrad. There was little chance for Russians to get better information. For example, most of the "radio sets" in the Soviet Union were wired like a telephone to central stations, so the only choice that typical citizens had was between listening to Radio Moscow or listening to nothing. Sometimes they did not even have that choice, for tendentious "news" was announced over loudspeakers in factories, lunchrooms, and public squares.[41]

Despite the vigilance of the leaders, nuclear fear did seep into the Soviet countries, thanks to the Communists' own propaganda apparatus. During the years when most of the world's nuclear weapons were held by their enemies, Communist organizations began working to set such weapons apart from other devices. From 1948 on the pivot of international Communist propaganda was a "peace movement" which cried that using atomic bombs would be a crime against humanity and that they must be outlawed. Stalin had good reason to push forward the idea. If by chance the United States agreed to a paper ban without inspectors, Russia might feel a little more secure; meanwhile, provoking disgust toward atomic bombs would teach the world to despise the Americans who owned them. Thus revulsion against nuclear weapons went on the agenda of the Communist propaganda network—the newspapers and magazines in many languages, the millions of Party activists around the world who in those years followed Moscow's lead.

The agitation reached a peak in 1950 with the Stockholm Appeal, a petition demanding that atomic bombs be outlawed as "weapons of terror and the mass destruction of whole populations." Although it proposed a moral ban rather than practical controls, many people found the Appeal attractive. First to sign the petition was Frederic Joliot-Curie, followed by other Communist and leftist scientists and notables. Communist parties around the world ordered their cadres into the streets to gather signatures, getting tens of millions from ordinary citizens in Western Europe and still more elsewhere.[42]

Averell Harriman, who knew Russia well, believed that the Stockholm Appeal was directed as much inward to Soviet citizens as outward to the world. "No Russian government," he explained, "could keep itself in power if the people did not trust it as the champion of peace." A campaign to gather signatures for the Appeal inside the Soviet Union rose to a frenzy in July 1950. Hundreds of news items flooded the press, while a large proportion of citizens attended mass meetings and signed.[43]

As often happens with a propaganda campaign, few people were as strongly persuaded as those who carried it out. The conviction that nuclear weapons were uniquely horrible became seated in leftist thought. In the heat of the campaign, *Pravda* let slip remarks by Soviet scientists about "total destruction" and "mass extermination." Most outspoken was T. D. Lysenko, Stalin's favorite biologist, who described how enemy warmongers were laying plans to "wipe from the face of the earth . . . all that has been created through the centuries by the genius of mankind." Although the rush of talk vanished from the Communist press almost as swiftly as it had come, it left behind a residue of frightening thoughts. Anti-American propaganda had joined the other political forces that promoted nuclear fear.[44]

In every language people were called upon to choose between the road to doom and the road to a festival of peace. Few noticed that there was another path leading from the symbolic crossroads. Between the two highways shown in cartoons lay an imperceptible trail, striking out across the countryside toward a future where nuclear energy would neither incinerate humanity nor redeem us.

SECURITY THROUGH CONTROL OVER SCIENTISTS?

While scientists, world federalists, and Communists thought to use nuclear fear in international politics, others wielded the same weapon in the domestic politics of individual nations. The power of the atom was so staggering, they said, that radically new institutions must be built to handle it. At home as abroad, the only route to security would lie through the strictest controls. This attempt to solve the problem of bombs, even more than the international propaganda, would only lead to increased fears.

The atomic scientists and a number of other commentators warned that without international control, or perhaps even with it, the social changes to be expected would involve greater government power, especially in the military and executive branches, with a corresponding reduction in individual freedom. Perhaps most insightful were the Alsop brothers. They wrote that henceforth the United States must be on a permanent war footing, ready for battle at an instant's notice. Civil liberties must weaken when saboteurs could smuggle in atomic bombs, while "security police" with Geiger counters must have the right to come snooping into any home. Worse, the constitutional rule that war must be declared by Congress would disappear, with authority over the very life of the nation placed in the hands of the President alone. Worse still, since Washington might be destroyed instantly, this ultimate authority would have to be at hand to unelected military officers. All this seemed unbearable; the Alsops and others felt that a nation with such practices would be no true democracy. Nevertheless, the predictions came true.[45]

The United States Congress created a Joint Committee on Atomic Energy with unprecedented privileges, sole overseer of a government monopoly that likewise held sweeping powers, the Atomic Energy Commission. The AEC legislation threw out traditions as old as the nation. For example, basic scientific information, which traditionally had been freely available to all, became secret government property.[46]

Control of nuclear energy was turning out in practice to mean control over secrets. Nuclear scientists themselves, beginning in 1940, had violated their own tradition of open publication and clamped down private censorship on fission research. General Groves pushed this secrecy much farther, teaching the scientists a new meaning for the word "security." To Groves, security meant hundreds of miles of fences with armed guards and special passes, censorship of private letters, and Army counterspies complete with hidden microphones—all far beyond what was found in other wartime civilian work. Of course the word "security" also kept its old meaning of being secure, safe from danger. When the war ended, Groves sent telegrams to warn Manhattan Project workers that "loose talk" still "jeopardizes the security of the nation and must be controlled." He had the ideas of control, secrecy, and safety all twisted up together, all knotted around and concealing the facts of atomic energy.

It was psychologically fitting that the idea of "security" came to include the idea of controlling secrets. The General felt personally responsible for the entire Manhattan Project, and he thought that such responsibility meant keeping everything, even his own feelings, under tight discipline. This was the approach he had learned from his father, an Army chaplain and strict taskmaster. Like his father, Groves neither drank nor swore, and he revealed no self-doubts even to his wife. Only once, in his exhilaration when the Trinity test succeeded, did he confess that for three years beneath his mask of confidence he had been as afraid of failure as if he were crossing Niagara Falls on a tightrope. Groves' personality, in short, was built for keeping secrets. That was one reason he had been picked to run the Manhattan Project, for everyone from President Roosevelt down had demanded extreme secrecy.[47]

When the project was revealed its secrecy impressed the public. They were astonished to learn of immense factories where tens of thousands of workers had had no idea of what they were making; the Manhattan Project became known as "the best-kept secret of the war." When Groves told the public that the Soviet Union could not come up with atomic bombs for many years, most people thought that was because only Americans knew the secret of building such bombs. Actually, the only important secret was the fact that atomic bombs were feasible, a secret lost forever at Hiroshima, but Groves never admitted that. His intense personal desire to control what he called "vital secrets" found echoes around the world.[48]

I was struck by the way secrecy and control were emphasized in the *New York Times'* index of the articles it published in the autumn of 1945. Of all the articles about atomic energy, roughly two-thirds were mainly about international or other "control," and of these, nearly half were largely about "secrecy." When I came across a collection of mid-1947 digests of American network radio comments on atomic energy, these turned out to have the same preoccupations in roughly the same proportions. Even President Truman thought that his government owned an exclusive treasure, and one of his first postwar cabinet meetings degenerated into a quarrel over whether the "secret" of atomic bombs should be given away to the world. Countless newspaper columns, radio shows, magazine stories, and entire novels were written about "the secret," as if it were a formula on a piece of paper in a safe somewhere.[49]

Secrecy became a mania. A government document stamped "Top Secret" turned out to be a copy of an article that had been published in a London newspaper. In 1949 when a press exposé announced that a bottle containing uranium was missing and perhaps stolen from an Atomic Energy Commission laboratory, the Joint Committee on Atomic Energy summoned commissioners and top scientists to explain the loss. In fact the bottle had not held enough uranium to drive a steamship across the Atlantic, nor scarcely a toy boat across a bathtub, but a desperate search overturned the laboratory until a bottle with uranium was produced. (Decades later, scientists smiled when asked if it was *the* missing bottle, or *a* bottle.)[50]

In other ways secrecy was not funny. At the war's end congressmen began writing bills to impose the death penalty for revealing atomic secrets; if this became law then a scientist who told a friend even the results of pure research, done privately at home, might be legally executed. In early 1946 a Canadian spy ring was revealed that had indeed passed a few atomic secrets to the Soviet Union. Groves and others used the revelation to convince Congress that spies imperiled America. The legislation went through, death penalty and all. New spy stories, such as the treason of Klaus Fuchs and the trial of Julius and Ethel Rosenberg, kept the pot boiling for years.[51]

The concern about loss of secrets harmed scientists more than anyone. Although not one American atomic scientist was ever shown to have turned traitor, no group was more closely inspected or forced so often to prove their loyalty. Because of this special attention, physicists and mathematicians made up more than half of the people who were identified as Communists in congressional hearings. Hundreds of scientists were mercilessly pursued, often losing their jobs, some of them ending in exile or suicide.

Americans came to accept something so undemocratic that they would have found it unthinkable a decade earlier: an elaborate peacetime system of guards and fences, locked safes, visitors making detailed inquiries about the personal lives of friends, and plain spying. This system was strongest in the Atomic Energy Commission; by the end of the 1950s the government had investigated in detail some 150,000 people in connection with AEC employment. The system spread into many sections of government, industry, and even the universities. It remains fully in force today.[52]

The same passion for controlling secrets beset every other nation that planned a nuclear energy program. France too set up an Atomic Energy Commission with unprecedented privileges, and leftist scientists were hounded from their jobs regardless of their patriotism. The British government set nuclear energy even farther apart from democratic traditions, keeping the work so secret that Parliament had no idea how much money was being spent on it, and even the scientists in charge did not know when the decision to build bombs was made.[53]

Atomic bomb fears aggravated "security" demands most severely of all in the Soviet Union. The Soviets had customarily imprisoned scientists and other intellectuals, but in the late 1940s these attacks grew worse. Meanwhile fission research pushed ahead at a feverish pace under the command of none other than the secret police chief, Lavrenty Beria. Beria was a sadist, with a passion for seeking his personal security through control over others. Cruelty was only one of his means; even more important was secrecy, for Beria sought to control people and to protect himself by concealing everything, including his hostile feelings. Beria's combined mania for domination and personal inscrutability was so extreme it exceeded even that of his master, Stalin. Both men smiled at their comrades while planning sentences of death, and their underlings survived only by concealing their own anger and fear.[54]

The fact that such men held power said much about the Soviet system. Some Western observers believed that the Russian character itself was fixed in a vicious circle of insecurity and secrecy because of a long history of invasions. However, Germany had been invaded more often, and now an arbitrary line divided two Germanies, one increasingly alive with open discussion, the other as cramped as the Soviet Union itself. It was the domestic political institutions that determined how personal urges for control and security would be expressed in practice.

Soviet nuclear laboratories were fenced off as a separate part of Beria's empire of prisons, which in turn was fenced off from the rest of the Soviet Union. Prisoner scientists worked alongside free ones, although the latter had scarcely more choice. At these hidden installations, as at Los Alamos, fences and guards restricted the activities of everyone including the project leaders, but in Siberia it was plain that the fences were less to keep spies out than to keep scientists in. Everywhere, control over atomic energy was in practice coming to mean control over scientists.[55]

The anguish over spies—what Americans called McCarthyism—obviously involved more than nuclear fear. Everything that happened in these years was conditioned by battles over the rise of Soviet-style regimes in Eastern Europe and Asia, the set of historical events that opened the Cold War. Not only physical but mental territory was in dispute, for discourse everywhere was saturated with warnings that the Communist powers, or, as the case might be, the capitalist ones, were out to enslave the world. Aside from the manifold political and social causes, already analyzed in many books, what role did nuclear energy play in this tide of fear and hatred between rival blocs? The question cannot be answered in full. The American and British archives show that leaders there had nuclear weapons much in mind, but the Soviet archives remain closed, and in any case such enormous events cannot be reduced to a simple formula. Yet historians of the period agree that, at the least, the enormous threat of the new bombs intensified the feelings of suspicion and aggression on both sides.

One feature of the political trends was in turn particularly important for the history of nuclear imagery. Secrecy was central to the Cold War, if only because of the Iron Curtain—a new creation which contemporary thinkers found deeply disturbing. For the first time in history a large part of the civilized world was not allowed to know what people elsewhere were saying; nothing could have done more to promote confusion, anxiety, and universal concern over secrets.

This international concern was strengthened by domestic forces within each nation. In the United States, accusations of treachery were useful, for example, to certain Republicans who meant to discredit New Dealers, and to primitive conservatives who meant to discredit any upstart group that threatened prevailing social patterns, not only the newly powerful scientists but also Hollywood writers, State Department homosexuals, and so forth. In the Soviet Union, similar functions were served by the similar if more deadly wave of accusations.

Moreover, in complex societies possession of information was a touchstone of political power. For example, only a few top leaders in any nation were allowed to know how many nuclear weapons their nation owned. Others, left in the dark, could not offer rival policies with the same confidence. Even Nikita Khrushchev, when he was Soviet Premier, complained that his military fended off criticism by keeping things secret. In short, secrecy served as both an instrument and a talisman of political authority. Meanwhile, what politics drew from public anxiety about bombs it returned with interest, pouring new fuel on the smoldering fires of nuclear anxiety.[56]

So much for the pressures of historical events and social structures in this period, matters that many have analyzed. The forces that I want to look at here, though less often studied, were actually more obvious, standing at the forefront of public thought: things that not only officials but newspaper reporters, movie directors, and common citizens were saying. Wasn't there something familiar about those scientists and spies with their incalculable secrets?

Fiction brought the theme of atomic "security" into the realm of the personal. A number of movies made a profit from the spy craze by showing traitors at work, while children's television shows such as "The Atom Squad" brought similar stories into the living room. Even Little Orphan Annie was kidnapped by villains seeking to extort atomic information from her seniors. A new popular stereotype arose, the insidious Communist who endangered people through stealing secrets. This meant atomic secrets more often than every other kind put together.[57]

The stories became one more force promoting mythical imagery. For example, in the detective film *Kiss Me Deadly* a woman who lifted the lid of a stolen "Pandora's box" was consumed by hellish rays; the secret was literally radioactive, not to say demonic. In each of three other films where spies tried to kidnap a man for his atomic secrets, the man in question had been in an accident that made him weirdly radioactive, like Boris Karloff when he played a glowing mad scientist. In yet another spy film the camera panned along a Los Alamos fence from a sign reading "Contaminated Area" to another reading "Restricted Area"—the peril from radioactivity merging with the peril from those who sought to know forbidden things.[58]

Scientists had always been seen by many as queer, single-minded, powerful beings working outside normal society, and the stereotype was repeated in journalists' stories of atomic scientists. Whether dedicated and brilliant Manhattan Project workers or dedicated and brilliant traitors or saboteurs, they seemed only too inclined to secretly inflict violent change upon society. People put their finger on the emerging pattern when they likened McCarthyism to a "witch hunt." Sometimes talk descended to a level even more primitive than the stereotype of an antisocial sorcerer. Everyone from congressmen to commissars declared that scientists were "babes in the woods" when it came to politics, and that was not the only way physicists reminded people of babies. A woman wrote a United States senator begging him to "try and stop those crazy scientists . . . from playing with those atomic bombs before we are all blown to bits. They act like children with a new toy." Many leading citizens agreed. Mr. Arbuthnot, the cliché expert, may have the last word: atomic scientists were like "little boys playing with matches."[59]

A sophisticated British film, *Seven Days to Noon*, featured just such a scientist. Dr. Willingdon, a childishly naive physicist who helped make atomic bombs, began brooding over them until his mind cracked. Stealing away a bomb in a satchel, Willingdon announced that he would blow himself up along with all London unless Britain promised to renounce nuclear weapons. Despite the crazed-genius stereotype, in its overall atmosphere the film resembled nothing so much as an ordinary police thriller; the mad scientist had stepped into the world of realism.[60]

A few observers remarked that mythical thinking was overpowering rational discussion, but even they barely glimpsed all the symbolism at work. Official secrecy was obviously connected with the desire to control others, the wish to be an insider superior to ignorant outsiders; but why then did so many citizens on the outside insist that officials should keep atomic secrets? There was, Stewart Alsop remarked, "a sort of Victorian reaction to the whole subject," as if atomic energy were best not discussed. Some British members of Parliament were more specific. One complained that when he asked about bombs the government reacted as if he had "asked about something indecent"; another said that the press treated atomic energy in the cramped way they had once talked about sex.[61]

The old innuendos had indeed broken out of the enclaves of science fiction and science journalism. For example, a United States Department of Defense film about the Bikini tests exclaimed, "Man has torn from nature one of her innermost secrets!" President Truman himself, in his final address to Congress, spoke of the decision to "probe . . . innermost secrets." More pointedly, everyone from Soviet writers to Mr. Arbuthnot spoke of nuclear energy as an advance in "man's mastery over nature."[62]

One poet recognized a meaning in such language. Immediately upon hearing the news of Hiroshima over the radio, Dame Edith Sitwell wrote a poem in a more disturbed tone than she had ever used before. The bombing made her think of Man as a tyrant "that conceived the death / Of his mother Earth, and tore / Her womb, to know the place where he was conceived." She had recognized the psychological pattern that I described in Chapter 4, a pattern symbolized by the mad scientist seizing a cosmic secret.[63]

Worries about "security" were commonly associated with that theme. People as different as Groves and Beria believed that security could be maintained only through control of treacherous feelings, which must be suppressed not only in others but also in themselves. For many people the prototype of such dangerous desires was the bad small boy's aggressive urge to probe into forbidden things, and it seemed vital to keep such urges locked away, making them a secret of their own. Of course security also seemed to require controlling and keeping hidden the catastrophic forces that the urges aimed to uncover—forces that extended from real technical facts to the magical powers that children often associated with the mysteries of sex and dreaded punishments.[64]

Atomic energy, a truly great and hidden power, neatly fitted this complex psychological pattern and its long tradition of imagery. Atomic bomb anxiety became a condensed way of thinking about more than the forces of science and technology in general, as had happened occasionally since the turn of the century. It meant more than an epitome of all the horrors of modern war, which had been a focus of concern since 1945. It also stood for the cruelest secrets of the heart: forbidden aggressive prying; treachery; the drive to master others; and the urge to destroy, like Dr. Willingdon, even one's own city.

These associations did not come of themselves, but were forged by individual people. The ordinary citizen who repeated mythical themes was using familiar ideas in an attempt to comprehend the news from Hiroshima. Others used imagery more deliberately to impress others, although the results were not always as intended. Talk about sorcerous atomic powers and scientist-magicians could encourage people to follow the lead of liberal scientists—or to fear them and bring them to heel.

In the end the second tendency won. By 1949 the atomic scientists' movement, its goal for an international scientific police having come to nothing, was reduced to a tiny remnant. Meanwhile citizens in the United States, Europe, and the Soviet Union pushed aside their initial dismay and confusion to focus apprehensions on a specific enemy, be it the domestic traitor or the foreigner, Russian or American. After all, it was reasonable to take the fearful hostility provoked by nuclear weapons and displace it onto those who might someday use the bombs, or who aided those who would use them, or who had already used them. Some observers of the time felt that the warning of the psychologists consulted by the Federation of Atomic Scientists was confirmed: propaganda that attempts to frighten people toward peace may instead rouse aggression. The historian Paul Boyer, studying the United States in this period, concluded that when the atomic scientists worked to intensify emotions, they "created fertile psychological soil for the ideology of American nuclear superiority and an all-out crusade against Communism."[65]

There is no final way to prove that nuclear fear, or any other combination of emotions and imagery, played a specific role in history. What can reliably be said, based on a large sample of writings, films, polls, and so forth, is that this closely knit structure of associated emotions and images did pervade Cold War and McCarthyist discourse.

The groups of the left and right who spread nuclear imagery understood that they were provoking emotions, but they did not recognize how powerful and ambiguous a force they were using. All the talk about secrets, control, and security was less likely to reassure citizens than to remind them of their most intimate problems. Nuclear fear was like a Chinese finger-trap: the harder people tried to pull out of it, the tighter it gripped them.

7

National Defenses

After atomic bombs were revealed, much of the talk about security was specifically about "national security," that is, the search for safety through military means. Surely the bombs, as physical weapons, could be countered with physical defenses? Generals, admirals, and other authorities up to President Truman, along with a large part of the world public, insisted that every weapon must eventually meet its match; the scientific geniuses who invented the bombs would doubtless make a counter-weapon. In trashy movie serials, atomic bombs were neutralized with Cyclotrode X or Meteorium–245, and Little Orphan Annie heard about Eonite, "the defense against atomic energy." The scientists' movement devoted much of its effort to informing the public that no such discovery could possibly turn up to save them. Yet most people believed that the road to security lay through technology: if not Cyclotrode X then bomb shelters, or perhaps one's own nuclear devices. These forms of defense, ranging from shelters through novel weapons to entire branches of the military, mobilized millions of citizens and billions of dollars, with correspondingly powerful effects on nuclear imagery. The results would leave people more insecure than ever.[1]

In the United States and Western Europe, thoughts of defense became urgent in late 1949 when Truman announced shocking news: a Soviet atomic bomb test. Evidently international control, and even internal control over secrets, had failed. The search for some other solution became desperate. During the next decade the United States, and still more the Soviet Union, would spend huge sums of money on interceptor airplanes and far-flung radar systems. These were well publicized, and most Americans and Russians trusted their armed forces to preserve their nation from extinction and even to win a victory. Yet polls in the United States found a majority admitting that if war came, their own city at least might be smashed, and Russians seem to have felt the same way. The bomber would always get through.[2]

Some scientists and their followers said the solution was to disperse industry and population. Concentrated skyscraper cities were obsolete, mere sitting ducks, so the new cities must stretch out along narrow strips; besides, that would give people more access to parkland. Evidently nuclear energy would revolutionize even town planning. Meanwhile huge sums of money must be spent on shelters. Mr. Arbuthnot and his fellow cliché experts explained that atomic bombs might "drive cities underground," a fantasy of hiding from atomic bombs in the most literal way. Others began to discuss a more plausible defense: civil defense.[3]

CIVIL DEFENSES

The American press began talking about bomb shelters after the Soviet test, and the occasional talk rose to a clamor around mid-1950 when the Korean War broke out. What if the Soviet Union decided to drop atomic bombs on America? The government came under pressure to do something, if only to calm the public's fears. Truman created a Federal Civil Defense Administration which soon had a staff of more than a thousand people, while local agencies added thousands more, coordinated by the FCDA. This effort was still far too small to do much against a real atomic attack, but the government was unwilling to pay the tremendous cost of a full-scale program. Civil defense officials decided the only way to do their job was to find and train millions of volunteers through a massive public relations campaign. They knew they would have to adjust their campaign carefully if it was to ease fears rather than stimulate them.

Experts felt that the handful of bombs the Soviet Union owned in the early 1950s might cause more damage by inducing panic than by direct destruction. Most people assumed that a bombing would mean shrieking mobs and cities disintegrating into chaos; by 1953 the American press was using the word "panic" fourteen times more often than in 1948. In fact, wild mobs were a rarity in disasters. At Hiroshima, for example, most people had reacted with stunned inactivity, meek random actions, or attempts to help one another. But people had always expected that atomic bombs would signal the apocalypse, and that traditionally included collapse of the social order.

The first civil defense instruction was accordingly aimed at reassuring citizens. The aim was to forestall panic if war came, not to mention soothing public qualms in the meantime; the basic civil defense message was "Keep Calm!" The FCDA's first booklet, *Survival under Atomic Attack*, insisted that atomic bombs were "only" another sort of explosive. It said most citizens could survive if they simply learned a few facts, which the booklet conveniently provided.[4]

Reassurance had limits. To prepare citizens for war a program would have to instruct them in what atomic bombs really could do. The civil defense agencies accordingly taught their millions of volunteers some raw facts about blast, fire, and radiation injury. Beyond simple training, officials hoped that exposing citizens to selected images of atomic war would "inoculate" them, getting them accustomed to horrid sights so they would not run around screaming when the bombs came. Accordingly, civil defense materials showed carefully adjusted scenes of destruction. No actual corpses or gore appeared, but the images of wreckage and of first aid practice on mock wounded told a story that was, deliberately, about as frightening as the public could bear.

Besides training citizens and inoculating them against panic, civil defense agencies had a third reason to spread images of atomic war. While congressmen often said civil defense was important, they never voted sums of money to match the enormous task; through the 1950s the FCDA's budget appropriation was typically about one-fifth of the amount that presidents requested for it. Worse, not enough volunteers stepped forward. Civil defense leaders constantly worried about what they called a "criminally stupid" lack of support. If only the public could be made to understand the terrible dangers, wouldn't they be more supportive?[5]

For all these reasons, civil defense agencies spread images of nuclear disaster more efficiently than even the atomic scientists had done. In 1951 the FCDA and local agencies handed out twenty million copies of *Survival under Atomic Attack*, while a film of the same name sold more prints than any film had ever sold before. Traveling "Alert America" exhibits reached over a million people with pictures of atomic bomb destruction, while slide shows and lectures came to every community in the United States.

Other agencies joined in. For example, the U.S. Air Force helped radio networks to produce dramatizations of Russian attacks on American cities—frightening demonstrations that the bombers would always get through, at least unless the Air Force received far more support. Within the White House itself, beginning in 1953 James M. Lambie, Jr., an aide to President Eisenhower, got free advertisements for civil defense placed everywhere from newspapers to bus placards. Like many of Eisenhower's supporters, Lambie wanted to stir Americans to do their utmost in the Cold War. He worried that unless he made citizens face their peril, they would demand "dangerous reductions in our expenditures for armaments."[6]

Government propagandists got whole-hearted cooperation from the American press and other institutions. Newspapers printed tens of thousands of civil defense items in the early 1950s, while all major magazines published articles. Every radio listener from 1953 on was periodically jolted when Conelrad, the emergency warning network, interrupted broadcasts. Towns set up air raid sirens and tested them at intervals. Spurred by Lambie's advertisements, hundreds of thousands volunteered for the Ground Observer Corps to watch the skies for enemy bombers sneaking in. Most impressive of all were a series of "Operation Alert" exercises held from 1954 on. As Russian bombers supposedly approached, citizens in scores of cities obeyed the howl of sirens and sought shelter, leaving the streets deserted. Afterward, photographs of the empty streets offered an eerie vision of a world without people. The press reported with ghoulish precision how many millions of Americans "died" in each mock attack.

Eisenhower's cabinet spent many hours discussing the details of the exercises. To the former general it was simple prudence for a nation, like an army, to hold what he called "war games." On only one occasion did the cabinet discuss how the games might affect public attitudes. In 1956 Secretary of Defense Charles Wilson suggested that the upcoming Operation Alert would "scare a lot of people without purpose." A fiscal conservative, Wilson warned that such drills might ultimately harm the economy by spurring citizens to inflate the military budget beyond all reason. For the drills would "strengthen and confirm the views of what might be called the fear lobby here in America—the people who might be said to have a vested interest in massive preparations for war." Wilson was overruled and the exercise went forward.[7]

Most successful of all the civil defense advocates who sought to increase public fear was a famous novelist, Philip Wylie. He was glad to serve as a consultant to the FCDA, for he feared that Communist attack might come at any moment. Since his childhood Wylie had been outraged when he met smug self-confidence, and he particularly hated to see anyone looking confident about nuclear war; he wanted to shove people's noses into the facts. When FCDA officials turned down his memos, he set off on his own and wrote a stunning civil defense novel.

Tomorrow! was a story of Russian bombs falling on America in the near future, with unforgettable scenes teaching civil defense lessons. Descriptions of the look and smell of burnt flesh, or a woman flayed by a burst of shattered glass from a window, no doubt instructed some readers. But overall Wylie's novel, along with realistic radio shows, magazine stories, and so forth, written by other authors in the same cause, did less to instruct the public than to frighten them. As Wylie later admitted to a group of fellow writers, "We have taught the people to be afraid—because most of *us* are afraid."[8]

A still more impressive exercise in civil defense instruction took place in the Nevada desert. In 1953 the government allowed hundreds of reporters to witness an atomic explosion. Crouching in a trench was a network pool reporter, young Chet Huntley, who described the blast as "the most tremendous thing I have ever experienced." According to a survey, three-quarters of the nation heard about the test or saw it on television, coming away with a new feeling for the bomb's destructive power. Of special interest were houses constructed nearby with mannequins in the living rooms and kitchens. Among those who had promoted this idea was Wylie, who held that a grisly demonstration would help awaken the public. After the test, television and magazines showed the mannequins lying twisted amid broken glass and collapsed lumber, as macabre a scene as Wylie could have wished. Most Americans would have agreed with the name reporters gave the houses: Doom Town.[9]

Year after year the Nevada tests continued, until millions of Americans had witnessed them directly. More than two hundred miles from the test people would be awakened and jump from their beds, aghast, as their bedrooms were lit by an uncanny false dawn. The light was sometimes seen as far away as Los Angeles. Over the years a quarter of a million men experienced the tests much closer in. The Army hoped to learn whether troops would panic in face of atomic bombs, and claimed to be heartened by the results. But many young soldiers who felt the blast shake their bones knew terror and nightmare images of apocalypse.[10]

Still more susceptible to civil defense messages during the 1950s was an even larger audience: children in nearly every school in the

United States. Thirty years later I asked groups of American adults if they remembered what to do if a teacher sounded the alarm, and about a third would put their hands up. The procedure recommended by the FCDA for a sudden attack was "Duck and Cover," and in tens of thousands of schools the children (I among them) practiced ducking under their desks and covering their heads. To practice for attacks with advance warning, students filed into a basement or evacuated to their homes.

A typical program was in Detroit, where schools held surprise drills and used the *Survival under Atomic Attack* booklet as a fourth-grade text. Since the children might be separated from their homes, Detroit parents were asked to put names on clothing with indelible ink, and about half complied. But experts frowned on identification by marking clothes, since "clothing can be destroyed by blast and fire." Some cities therefore handed out metal identification tags to hundreds of thousands of school children. Occasionally people kept the tags to adulthood, tucked away in a drawer, almost forgotten.[11]

The training materials for children, and for the several million adults who took civil defense courses, deployed strong images. The tone was set by confident, calm voices uttering reassurances, but in an undertone the materials whispered a different message. For example, *Survival under Atomic Attack* noted cheerfully that "lingering radioactivity . . . is no more to be feared than typhoid fever." The film of the same name showed a window bursting and plaster raining down on a table set for a family dinner. Another short film, aimed at schoolchildren, pictured a surprise attack where a boy was lit by a sudden flash, hurled himself to the ground until the blast wave roared past, then ran frantically for shelter in a neighborhood suddenly converted to rubble. Fictional television shows and movies of the 1950s showed no scenes of nuclear attack as shockingly realistic as that. Some children began to understand that no matter how well they obeyed instructions, their chances of surviving an attack might not be high.[12]

The children seemed to take it all in stride, and few adults asked what was really going on in those young heads. Not until long after, in the late 1970s, did some people begin painfully to describe their childhood thoughts. A woman remembered how as a fourth grader she was sometimes frightened at night if she heard an airplane drone overhead, and would sit up begging, "Please don't let them drop the bomb on me!" A man recalled a night in his childhood when he heard a siren and believed that the attack had come, "the most terrifying thing I can remember." A college student said that as a little girl she had had nightmares of running all alone for shelter from bombs; this nightmare of separation, of desperately seeking home or a hiding place, had been particularly common. Other adults would still occasionally shrink at the sound of a passing airplane or a siren, wondering for an instant, as in their childhood, if the moment had finally come.[13]

Michael Carey and other psychologists, interviewing young American adults in the 1970s, found that claims of confidence about nuclear war soon gave way to recollections of terror. As children these Americans had shifted back and forth between fear of atomic bombs and more general anxieties, dreams and fantasies of dying or being separated from their families. Nuclear weapons had become almost inseparable in the minds of many young people from overwhelming death itself.[14]

In the 1950s some adults had similar feelings. One official wrote Eisenhower to complain that civil defense exercises would "cause more heart trouble than this country has ever had," since "older folks now become frightened every time they hear the siren on an ambulance . . . They fear it is a raid."[15]

How widespread was such anxiety? The age of public opinion polls had begun, and pollsters tried to answer such questions. Within weeks after the bombing of Hiroshima they found severe concerns. A majority of Americans foresaw a real danger of their own families dying in atomic attacks, along with most of the people in the world's cities. Evidence from polls is scanty outside the United States, but it seems that a majority of citizens in every industrial nation believed what their newspapers said: a dreadful atomic war was all too likely.[16]

Yet once the initial shock was over, that is, by the end of 1946, most people had decided the problem was not immediate but in a nebulous future, and had stopped talking about it. After the first great burst of attention the media gave less and less space to nuclear weapons, which was natural enough, for once a magazine has carried several stories about a particular topic it must turn to something newer. More surprisingly, most people around the world said they trusted in their scientists to devise a defense, or trusted their government to keep the peace, or simply trusted God. Americans in particular felt that at present their nation's military defenses could stave off total devastation, and through the early 1950s that was true. Of course you might die under an atomic bomb on some distant day, but you were bound to die anyway some day. And your nation as a whole would survive. Typical was the Texas rancher who admitted that nothing could protect him against atomic death in some future war, but who added, "Most of my friends are more interested in this year's calf crop."

The unconcern struck contemporary observers as peculiar. People who confessed that they were in deadly peril declared with the next breath that they weren't worried. Asked questions about foreign affairs, most people did not even bring up atomic bombs at all unless asked. Everywhere, many people said that they just didn't know. The whole question was set aside as incomprehensible, something only scientists could deal with.[17]

Close observers suspected that the seeming confidence hid severe anxieties. Like death itself, atomic energy seemed so overwhelming that most people shied away from the subject. Such was the conclusion of various sociologists pondering survey results, and likewise of a committee of newspaper editors who puzzled over the lack of public demand for atomic information, and also of a psychologist who speculated that everyone was so afraid of apparently "supernatural" atomic forces that "we hide our heads in the sand."[18]

In 1950 a survey team tested the idea. The people they interviewed gave the usual answers: they were not worried about atomic bombs right now, for God or the government would take care of things. However, the team noticed subtle signs of insecurity. It took only a few probing remarks to overturn the nonchalant answers. For example, an aircraft worker insisted that the Russians could never bomb the United States, but when his confidence was gently questioned he abruptly reversed himself and confessed his fear. "Their planes could get through," he admitted tensely. "I hate to think of what would happen."[19]

Other interviewers and individual observers found that most people likewise hated to think about the danger. American surveys in 1954 found nearly all the respondents saying that civil defense was a sensible precaution, yet very few of these individuals had taken the least concrete action to safeguard their own families. At a dinner party in Baltimore, Philip Wylie heard citizens joke about a news report of the black and yellow signs that New York City was posting to identify bomb shelters. Everyone was taken aback when Wylie said that Baltimore too had such signs. He found that only one of a hundred people he questioned could recall perceiving the shelter signs that were all around them. Their fear, Wylie concluded, was so great that their minds refused to recognize the signs. This may have been caused not only by awareness of facts about bombs but also by the images, even more disturbing than facts, that bombarded everyone. Laboratory studies of propaganda have shown that when people believe there is no effective action they can take, statements designed to inspire fear may instead lead them to ignore or minimize the threat.[20]

In 1956, soon after the cabinet argument over what Operation Alert might do to public thinking, Eisenhower asked for a "thoroughgoing study of the effect on human attitudes of nuclear weapons." A blue-ribbon panel of social science experts was called together, and after some months they delivered a top secret report. The distinguished panelists confessed themselves baffled. They suspected that most people were so afraid that they wanted to avoid nuclear war at almost any cost. Beyond that, the experts just could not say what the bombs would ultimately mean for the citizenry.[21]

Outside the United States civil defense efforts usually produced the same disarray. The British government vigorously promoted civil defense, but pamphlets describing the power of bombs seemed only to heighten the public's doubt that an individual could survive. Elsewhere in Western Europe and in Japan, among people still digging out from the rubble of the last war, thoughts of another war were entirely unbearable and civil defense was scarcely attempted. Only Sweden and Switzerland, prosperous and long untouched by war, began leisurely programs to build shelters. According to a report assembled by the *New York Times*, most Western Europeans "turned to other matters not so terrifying to contemplate."[22]

Communist countries also ignored atomic civil defense during the early postwar years, but for a different reason. With Stalin insisting there was nothing to fear from American atomic bombs, defense authorities disdained to mention them, and most Soviet citizens had no clear idea what such a bomb could do. Only after Stalin's death did reality begin to seep in.

In 1955 the Soviet government launched a civil defense program, organizing mandatory groups in factories, offices, and apartment houses. Through lectures, articles in newspapers and magazines, radio programs, posters, and films, civil defense officials pounded in a few basic facts. By the end of the 1950s most Soviet adults had sat though ten hours or more of teaching, and in the early 1960s instruction was extended into the schools. This practice never ceased; in questioning Soviet émigrés around 1980 I found none who had not received civil defense training.[23]

The civil defense program saturated the Soviet populace with thoughts of nuclear attack. The overall message was designed to reassure citizens that a nuclear war could be survived, just like the Second World War. But for Russians the Second World War had been so terrible that this was scant comfort. And the facts they learned about nuclear blast, heat, and radiation, although scanty and presented in a bland way, were enough to show that the next war would be still worse.

Around the world people were coming to realize that civil defense could never guarantee their safety, no more than international controls could, nor spy hunts, nor scientists devising some Cyclotrode X, nor even plain soldiers with antiaircraft cannon. But such insecurity was intolerable. A solution had to be found.

BOMBS AS A PSYCHOLOGICAL WEAPON

By the early 1950s many people in the United States, the Soviet Union, and elsewhere had concluded that the only defense against the threat of nuclear weapons lay in nuclear weapons themselves. The result was a race to command the most apocalyptic devices. But this began to look like something apart from previous arms races, where common sense had called for tanks and battleships to bar the approach of enemy tanks and battleships. Atomic bombs could not stave off the enemy's bombs in that way. How, then, were they to be used?

This was the beginning of a great debate, fought with increasing intensity over the next decade, a debate that would eventually catch the attention of nearly every intellectual everywhere in the world. I will chiefly follow the debate in the United States, for that was its center. This was only partly because at that time the United States government made many of the key decisions in nuclear as in other world affairs; another reason was that Americans were in almost every case the first to bring basic nuclear facts into public view, and the first to raise basic questions.

The starting point of the debate came when farsighted people suggested that the bombs might be less important as physical weapons than as mental influences. The concept was gradually developed into an elaborate theory of deterrence. Foreshadowed by some thinkers starting in the 1920s and publicized by atomic scientists in 1946, the idea was precisely stated by an American academic strategist, Bernard Brodie. "Thus far the chief purpose of our military establishment has been to win wars," he said. "From now on its chief purpose must be to prevent them." Many American and European leaders were convinced that atomic bombs had already deterred the Red Army from marching into Paris. During the 1950s the United States government adopted the policy of threatening "massive retaliation" for any incursion: without ever being used, the bombs would quench the enemy's aggressive urges.[24]

Rather more common during the early years was a more traditional idea—that nuclear bombs were plain military tools, things you could use directly to defeat the enemy in war. A particularly famous description of what that meant was the October 27, 1951, issue of *Collier's*, devoted entirely to "The War We Do Not Want." Articles datelined a few years in the future told how, when Soviet troops went one step too far in central Europe, the United States launched a nuclear attack. American troops invaded Russia while exchanges of bombs wrecked a few cities on each side. Almost at once the Soviet order fell apart, and Russians with tears of joy blessed their liberation from godless Communist tyranny.

Many people in the United States and abroad were dismayed by the story, for they believed that this and similar writings reflected government policy. They were right. The commander of the American bomber forces, General Curtis LeMay, had no qualms about striking first, and he believed that bombing a few Russian cities would set off a popular revolt that would end Communism. The highest American policy committee, the National Security Council, secretly decided that if the United States went to war it should feel free to use nuclear weapons "as other munitions," and set forth grandiose war aims: international Communism would be dismantled and the Soviet Union would be divided into powerless segments. The Air Force expected nuclear weapons to achieve this more or less unassisted, by destroying the Russians' "will to fight." Once a few bombs went off, white flags would flutter over the enemy's remaining cities. Thus the theory that atomic weapons were plain munitions, almost as much as deterrence theory, placed faith not only in bombs themselves but in the idea of bombs.[25]

Deterrence, which worked by pointing to a hell of uttermost destruction, might seem a world away from calculations of how to encourage surrender with the measured application of military tools, but in fact strategic plans bore a resemblance to visions of boundless punishment. Where the two pictures overlapped an ugly image appeared: an enemy neither deterred nor defeated, but obliterated. Thus in Philip Wylie's novel *Tomorrow!* the United States answered a Russian attack with a grossly oversized bomb. A radioactive flame scorched the Soviet Union from end to end, leaving not an insect alive.

The combination of fearful and hostile urges underlying such a picture—in fact underlying nuclear bombs and all weapons—is surely obvious. I will go on to a more subtle matter. Wylie thought all would be well if only the evils of Communism could be erased, and his novel predicted that once Russia was gone universal peace would prevail, a "brand-new" and "infinitely better" future. Like those who waged earlier wars to end war, he wanted to attack evil itself, passing through Armageddon to world rebirth.[26]

American war plans came as close to such a vision as the weapons at hand allowed. Within a month after the bombing of Hiroshima, an Air Force study proposed that on the first day of the next war every available bomb should be dropped on the Soviet Union, aiming for "immediate destruction of the enemy centers of industry, transportation, and population." This all-out spasm became official strategy. That was barely enough for General LeMay, who reportedly told weapons designers that he wished they would build him a bomb (like Wylie's) that could destroy all Russia in one blast.[27]

American strategy, this mixture of deterrence, atomic munitions, and fantasies of obliteration, was questioned by only a few penetrating thinkers. The clearest-headed was a British physicist and expert military analyst, P. M. S. Blackett, who calculated in a 1948 book that the Red Army, sprawling across millions of square miles, could not be destroyed with fewer than several thousand atomic bombs. The United States would not have so many bombs until the mid-1950s, and by then both sides would hold many weapons. At no point, then, could anyone safely launch either a military invasion or an annihilating spasm. In short, according to Blackett, no sustained military advantage and indeed no sane human purpose at all could be achieved if generals actually took to exploding atomic bombs.

President Truman said privately that Blackett was making a serious mistake in analyzing atomic bombs as instruments of war. After all, the President exclaimed in a wild but significant exaggeration, the United States was building enough bombs to "blow a hole clean through the earth." Atomic bombs were "not a military weapon" at all, according to Truman and many others; their real use was as a threat during disputes, that is, as a diplomatic and political weapon.[28]

Yet Truman found that talk of bombs did nothing to turn the Soviets from their insulting rejection of American plans for the postwar world. Veiled threats only made the enemy more intransigent than ever. By 1950 a small number of leaders were coming to a surprising conclusion: nuclear weapons might be useless in every way.

The Korean War brought the first clear test of whether atomic energy could be an irresistible force in world politics. In late 1950, as Chinese troops routed Americans, talk about the use of atomic bombs grew loud, and Americans from factory workers to senators spoke of launching a preventive attack to destroy Red China and Russia together. The Communist press in Europe and the Soviet Union seized on the scattered statements and used them to stir up fear of war-crazed Americans. Then, at the end of November 1950, Truman casually remarked in a press conference that he was, of course, always thinking about the possibility of using atomic bombs in Korea.

A journalist in New York City reported an ominous mutter rising from crowds that pressed about newstands to read the tall black headlines. Most Americans were appalled by the thought of a nuclear war, and outside the United States the war scare was even stronger. Indian newspapers, for example, shocked by Truman's remark, warned that Asia would erupt in fury if the bombs were used again on a nonwhite people. The shock was strongest in Western Europe. Since the Berlin blockade of 1948 many Europeans had expected that their countries might soon be bombed, and Truman's remark brought scary headlines, street demonstrations, and frantic governmental debates. The British Prime Minister, backed up by messages from the French Premier, flew to Washington to make Truman see reason.

Truman did his best to soothe people, promising he was far from ready to use atomic bombs. In fact not even the Joint Chiefs of Staff wanted an exchange of bomber raids; such an exchange might have done less damage to the huge Asian armies than to the fragile American expeditionary force in Korea, dependent on a few ports. Once the United States' few hundred bombs were expended, the nation would be like a hunter with an empty rifle facing a wounded and enraged bear. Commentator Drew Pearson further warned his national radio audience that using atomic bombs would leave the United States to "fight a world war without allies."[29]

If the bombs could not be used as weapons, they were no better as threats. Their existence did not prevent the Korean War from starting, nor keep it from dragging on for years, until Stalin's death allowed Communist policy to change. In other cases too, such as the fighting in French Indochina, the Joint Chiefs felt the bombs would make poor military tools, while veiled diplomatic threats only brought cries of defiance from the enemy, threw sand into the gears of relations with allies, and frightened Americans themselves.[30]

Yet leaders and the public alike could scarcely believe that nuclear weapons were useless. Surely they could at least deter nuclear war itself? Even if true, that idea turned in a circle, leaving the bombs without meaning except in their own hell. Perhaps there was some way, people thought, that this great energy could at least prevent or win small wars, and otherwise help nations exert their will in ordinary affairs. Experts decided that since hundreds of atomic bombs were of little help, a nation must possess thousands, or tens of thousands, or something mightier still.

Nuclear imagery and actual weapons had begun to feed on one another, each helping the other to grow. To explain this crucial process, it is not enough to talk abstractly about public attitudes; the history was made by specific individuals who accepted nuclear imagery and then gave it back to the public, redoubled, in the form of action.

Nobody showed the historical effect of nuclear images so plainly as Brien McMahon, the "Atomic Senator." McMahon was the most elegantly dressed man in the Senate, although as his solid jaw and heavy eyebrows hinted, he was also a tough political brawler. All that was in service of visionary aims. He had entered the Senate in 1945 bearing an idealistic commitment to peace and freedom, learned from his liberal mother and his professors at Yale. When he heard the news from Hiroshima over the radio his first thought was that something unimaginably monstrous had gotten loose. Cutting short a vacation, he rushed back to Washington to begin work on the problem, getting a head start on more senior men. The senator was particularly afraid that technocratic "experts" would now come to power, shouldering aside elected representatives. He decided his only defense was to become an atomic expert himself. Beginning with William Laurence's articles, he went on to long hours of lessons from the atomic scientists who came lobbying in Washington. What he learned, McMahon told a reporter, "frightened him to an exquisite degree."[31]

The law that eventually set up the Atomic Energy Commission and the Joint Committee on Atomic Energy was his bill, the McMahon Bill, so when the Democrats won control of the Senate in 1949 they made him chairman of the Joint Committee. The senator crammed more working hours into his day than seemed possible and devoted about half of those hours to the single issue of atomic energy. According to a close observer, McMahon felt that "the whole world revolves around the exploding atom."[32]

McMahon's restless energy was partly plain political ambition. Only the magic of the atom could have brought an obscure freshman senator to a position of such prominence that some expected him to run for President before long. But such mundane goals intersected with nuclear terrors and visions: McMahon's root ambition was to save democracy and indeed all the world. In 1948 he declared that the United States would shortly have the means to annihilate human life over the entire planet, which at that moment in time was a fantastic exaggeration. He also hoped to see atomic energy bring about a fabulous new age when "every man can be in private possession of such energies as hardly any men had dreamed of."[33]

McMahon's work exemplified the search to control cosmic force. He took charge of domestic legislation and personally supervised the Atomic Energy Commission. He also spoke out from the start for international controls, but he distrusted Communists and soon recognized that negotiations were failing. Urged on by his friend Philip Wylie, he took an interest in civil defense, but found no solution there either. The Senator's fears became acute after the Soviet Union tested its first atomic bomb. Toward the end of 1949 he announced his solution.

Half of the solution was an idealistic plan to cut the military budget drastically and use the saving to help people around the world. The other half of his plan was more likely to be put into effect: McMahon thought the military budget could be cut because "obsolete" ordinary weapons could be replaced with cheap atomic munitions. He wanted to put atomic bombs on every warship, in every airplane, almost in the hands of every soldier. "No matter how many we might come to possess," he believed, "we would need and could profitably use far more—in the event we were attacked." Only atomic power, he explained, could meet what he saw as the central issue, "survival."[34]

Close to the Senator's way of thinking was Edward Teller. At first sight the physicist was only a stubby man with big brooding eyebrows, yet his intelligence and diffident charm made him the most persuasive of all the scientists McMahon knew. Teller's background had encouraged an interest in cosmic, personal, and social catastrophe. There was his enthusiastic reading of science fiction. There was his mother, who was abnormally fearful for her children's lives (she had even reportedly held a string attached to young Edward when he went swimming). Above all, there was his experience as a Jew of Budapest in a community threatened time after time with extinction at the hands of Russians, Communists, and Fascists. If Teller was concerned with insecurity it was because he felt, more personally than most, how horribly precarious life can be.[35]

Faced with the forces of death, many people yearn to join with something beyond the mere individual self. A politician like McMahon might strive to represent liberal ideals, a scientist like Teller might commit himself to scientific research, and those paths could run together. At the beginning Teller believed even more strongly than McMahon in the need for a new international political order, while the politician believed even more strongly than the scientist in the power of science to reform civilization. The atomic scientist and the Atomic Senator went still farther. They sought to defend their nation through mastery of the greatest power within human ken: hydrogen fusion.

There is no stranger story of how pure thought can invade the world of action. In the 1930s, of all the problems of science the most truly unworldly was the question, What makes the stars shine? Scientists heightened the celestial glamor of nuclear energy when they suggested that it was the source of starlight. Around 1939 they figured out the mechanisms. Deep within a star where the heat and pressure are high enough to smash the nuclei of hydrogen atoms together, the nuclei can fuse, and in that fusion release energy. But the details remained obscure, making a fine scientific puzzle.

At Columbia University and later at the University of Chicago during the war, the puzzle came to the attention of Enrico Fermi. When astronomers happened to join Fermi at lunch they could not talk about his own work, the top-secret uranium fission project. So the talk might turn to something infinitely remote from military matters, the topic that was most on astronomers' minds at that time—the conditions for fusion in stars. Fermi loved to help people solve puzzles, and he gave the problem some thought. One day it came to him that a fission bomb might for an instant create heat and pressure as intense as any at the center of a star and set off a fusion reaction.[36]

After lunch Fermi mentioned the problem to Teller. Already before the war Teller had given some thought to fusion in stars, simply as an intellectual game. Fermi's question struck a resonance, and soon Teller was working single-mindedly to invent something a thousand times more powerful than fission bombs. At Los Alamos he kept worrying at the problem whenever he could get time off from ordinary fission studies, and finally he abandoned work on fission altogether in order to pursue vaster explosions. There was only one other problem that really excited Teller, he recalled later: the question of whether the first atomic bomb test could trigger a chain reaction that would wrap the earth in fire. That was the kind of brain teaser he found "really delightful."[37]

After the war Teller remained preoccupied with fusion. In later years when people called him the father of the hydrogen bomb, he demurred with his unfailing politeness, but in truth he kept after the problem for years like someone fiercely intent on conceiving a child. Later, when his ideas were proved with a great explosion, he sent a prearranged message to tell Los Alamos of his success: "It's a boy."[38]

In his personal struggle for the means to create a fusion weapon,

Teller's break came with the first Soviet bomb test. Senator McMahon, urged on by Teller and by more prominent scientists, wrote a long, fervent letter to Truman. The United States must leap ahead, the Senator insisted, for if the Soviets got a hydrogen bomb first it would mean certain catastrophe.

In secret, Oppenheimer and others disagreed. They argued that ordinary fission bombs could be made big enough to destroy any military target; fusion weapons were not needed to attack anything short of an entire metropolitan region. Some physicists consulted by the government added that such a massacre could never be justified "on any ethical ground which gives a human being a certain individuality and dignity even if he happens to be a resident of an enemy country." A hydrogen bomb, they said, was "necessarily an evil thing considered in any light."[39]

These top-secret debates involved not only ethics but technicalities, particularly the question of how to use the small amount of uranium the nation then possessed. Such technicalities were not revealed for decades, and many insiders doubted that the public, ignorant of the facts, could play any useful part in the discussions. But anyone who carefully followed the press could learn many of the essential facts. For example, the ultimate secret was the limited number of bombs the United States held—a secret so tremendous that the President was never given the number in writing but only verbally—yet in 1950 *Life* magazine published a guess that the number was then about 300, and it was an accurate guess. Besides, within a few years the nation got enough uranium to use in every possible way; over the long run the secret technical facts meant nothing compared with political and ethical standards.[40]

These the public considered from the start. Everyone who paid attention to public affairs understood that weapons far more powerful than fission bombs might be developed. In late 1949 the news slipped out that the government was just then considering whether to make the attempt. Immediately the news media filled with vigorous debate over whether, as McMahon and others declared, such weapons were urgently needed, or whether, as others said, to develop them would be unwise and immoral.

The most thoughtful debaters, both in public and in secret, agreed that a hydrogen bomb would not be so much a thing built of metal as an uncanny force, a "Frankenstein" as one scientist called it. The Joint Chiefs of Staff themselves said they wanted hydrogen bombs not for actual warfare but as a "psychological" weapon to keep the Soviets mentally on the defensive. Truman too protested that he only wanted to have the weapon in his pocket when he bargained with Stalin. A more immediate psychological use was a domestic one, for the administration was under acute pressure to do something, anything, to answer the menace of the new Soviet fission bombs. Urged on not only by McMahon but by more senior government leaders, and with the backing of a majority of the public, the President decided to press at full speed toward hydrogen bombs.[41]

McMahon and others pushed Congress to heap money into the lap of the Atomic Energy Commission, and immense construction works got under way on a scale beyond even the Manhattan Project. Reactors were built to make plutonium by the ton, while new factories covering dozens of acres began to separate more uranium-235, which is required for efficient hydrogen bombs. Even before the new bombs were deployed, the relevant laws of physics, as much as any other historical force, were having an effect on the nation. Within a few years the American nuclear program was using about one-tenth of the electricity produced in the United States, more electricity than was used in all of Britain.[42]

Other nations made similar decisions. The Russians too believed that they had to possess thousands of nuclear weapons of every variety, and zealously pushed ahead immense construction projects. During the next decade they were joined by Britain, then China, and finally France, all building their own reactors and isotope separation factories. Other major nations stood aside, most notably those defeated in the Second World War. Yet by the 1960s Germany and Japan, despite their lack of bombs, would seem no weaker in the international arena than Britain, France, or China. It would not be in world politics, but in the development of civilian nuclear energy, that the gargantuan factories would one day exert a pressure.[43]

THE AIRMEN

As the number of bombs in the world climbed into thousands, the fantasy of control over apocalypse became a matter of daily practice within institutions. The most famous of these was the Strategic Air Command of the United States Air Force.

Air Force public relations officers deliberately promoted an image of SAC with open houses at airfields, kits for journalists, aid to filmmakers, and mock bomber raids staged for radio and television. On the surface the image this promoted was flattering; the Air Force wanted the public to feel proud and confident with SAC. Besides, they wanted citizens and Congress to agree that the heavy cost was worth paying in return for security. However, reassurances were mixed with warnings, resulting in contradictory images. For the logic of deterrence meant convincing the enemy that war would mean utter destruction; making SAC look deadly was Air Force policy of the highest order. Furthermore, there were domestic reasons for warning that the enemy bombers were likewise formidable.

An example will show how the process worked. SAC commander Curtis LeMay invited Arthur Godfrey, perhaps America's most popular and trusted radio personality, to visit Air Force bases, and took care to befriend him. Godfrey was soon praising SAC's prowess on his radio show. In 1955 he published an article in the *Saturday Evening Post* that called for more bombers, warning of the "desperate need for air power." Godfrey especially endeared himself to his Air Force friends in 1957 when, at LeMay's instigation, he asked his listeners to write Washington to demand higher military pay.[44]

Soviet officers and political chiefs were similarly concerned to inspire their people with respect for fleets of bombers and at the same time to daunt foreigners. Everyone particularly watched the flyovers during military parades in Moscow. Most memorable was a 1955 show that stunned Western intelligence experts with the number of advanced bombers; only years later did anyone realize that the Soviets had put every modern bomber they possessed into the air and perhaps had even circled them around for a second pass, like a stage army.[45]

American generals and senators cried "bomber gap" while demanding funds for the Air Force—deliberately making the public feel insecure so they would approve buying more weapons. Of course nobody could ever expect conscientious military officers and their supporters to say that they had enough arms to guarantee perfect security. Officials wanted the public to feel safe behind their shield, but they were like many of the actors in this story: because of their position in society they felt duty-bound to remind people of mortal dangers.

No one took such duties more seriously than SAC's creator and chief public symbol, General LeMay. He was the picture of toughness with his bulldog face and scowling eyes, a cigar clamped between compressed lips, and anybody who doubted that the general was ready to destroy cities had only to recall that he was the man who had designed the Tokyo fire raids. When the Air Force wanted to show the world that it could drop bombs anywhere, it sent bombers on intercontinental flights with LeMay as pilot. It was hard to say whether that was comforting or frightening.

Popular magazines described LeMay, and by implication SAC, as "more machine than man," a creature of "irrefutable logic." It was true that the general set discipline above all else, concealing whatever human feelings lay behind the compressed lips. Yet the feelings were there. To his biographer LeMay revealed his childhood fascination with flying, which had seemed like a mysterious and almost godlike power, a fascination that persisted in his feelings toward war planes and their gadgetry. Somehow that reminded him of the trick camera he had pointed at little girls in his boyhood, with a toy snake that popped out so the girl would scream and run away ("usually she liked to be scared," he added). In the same way, LeMay said, electronic snakes were packed into every inch of a bomber's body, monstrous treasures "throughout the stiff flesh." Most impressive of all was the nuclear bomb, that "baby . . . clinging as a fierce child against its mother's belly."[46]

It would take a psychologist like Melanie Klein to appreciate what all that might have to do with Air Force imagery. Klein had noticed how young children might imagine that powerful phallic material, somehow identified with babies, was hidden within a mother's body. After all, that was not far from the truth about birth, and such fantasies might be only a symbolized first stage of the universal search for the secrets of life. Like one of Klein's small patients, LeMay used a language of babies and phallic symbols within a mother to express desires that many humans shared.[47]

I am speaking of the desire to master the secret forces of life and death. Of course that drive, and the more explicit sexual imagery connected with it, had always intermingled with ideas about armed force and weapons. When the Hiroshima bomb was called the "Little Boy," cradled within an airplane named after the pilot's mother, this followed a military symbolism that had been rife during the Second World War. Ordinary bombs were called "eggs," while bombers, called not "it" but "she," were often painted with pictures of women and given names like "Laden Maiden." Such symbolism continued in postwar stories and movies. Once military men became custodians of nuclear energy, which had itself long since become enmeshed in the imagery of sex and life-force, the whole tangle became more interwoven than ever.[48]

Not only airplanes carrying bombs but nuclear bombs themselves could be seen as something female impregnated with aggressive force. The first bomb tested in 1946 was reportedly named "Gilda" and the second, "Bikini Helen." In a film instructing soldiers about atomic attack, an officer remarked that the bomb was "like a woman . . . never underestimate its power." Of course, calling a dangerously seductive woman a "bombshell" had been common since the 1930s; behind that lay a still older tradition that saw both strong mothers and sexy women as having appropriated dangerous male force and hidden it within themselves.[49]

Nuclear energy itself was more commonly shown as directly masculine. When William Laurence described the Trinity test as a birth, when Groves's coded message used the same metaphor, and when Teller announced his successful fusion test with "It's a boy," they were each hinting at pride in male generative forces. More explicit was the first published description of atomic bombs, in H. G. Wells's 1913 novel, where a pilot preparing to drop the weapons took pride in "the thought of great destruction slumbering in the black spheres between his legs." From 1945 on when editorial cartoonists personified The Bomb as a mythical giant, it was invariably male. The cartoonists usually showed cigar-shaped devices; so did the first movie about the Manhattan Project, *The Beginning or the End* in 1947, although the real Trinity device and Hiroshima bomb had been more nearly spherical. Dr. Thorkel's cylindrical ray generator rising and falling in the radium mine shaft in the 1940 horror movie had something in common with 1960s missiles in silos within the earth. The phallic significance of such devices was never a secret, least of all to military officers, who joked about "emasculating" the enemy by destroying his missiles.

In short, although nuclear mythology had always hinted at life-force, by the 1950s there was a specific, widely felt association between nuclear weaponry and aggressive sexuality. I will say nothing about war psychology in general, that great, knotty problem. For the history of nuclear imagery, my point is that henceforth all thinking would be warped by this poorly understood association with sexual aggression. It would make any manifestation of nuclear energy unreasonably attractive to some, and unreasonably repellent to others.[50]

Military officers themselves began to shy away from provocative imagery. SAC felt uncomfortable about the girls painted on its bombers, and during the 1950s the paintings gradually disappeared, while war planes became not "she" but "it." Cute names for nuclear weapons also faded away as officers began calling bombs by technical designations, Mark 17 or whatever. A new, dry word for a weapon spread through SAC: "nuke."

LeMay took care to select only officers like himself, men who kept their feelings under strict control. This was in agreement with established Air Force thinking, for even before the Second World War the officers (according to historian Ronald Schaffer) "had an extraordinary closeness with death," as their friends died in air accidents and they had narrow escapes themselves. So they steeled themselves against death, a reaction easily extended to indifference about the deaths of enemies. Such an attitude was, moreover, a strategic necessity. Officers would deter the enemy only if they were visibly ready to bomb without hesitation; they must forget even their own families, who were quartered near airfields and would be first to die when the enemy struck back. Besides, as a writer explained in the *Saturday Evening Post*, bombing had to be done "mechanically, with swift, sure precision, undisturbed by emotion, either of fear . . . or pity."[51]

The public was frequently reminded that such professionalism had developed during the Second World War, when a cool head was needed to pilot a bomber through the terrifying antiaircraft fire. For example, around 1950 millions of people read or saw the Second World War bomber epic *Twelve O'Clock High*, about a heroic general who stifled his personal feelings and demanded that his men too be "harder than the metal of their B-17's." Magazines served up LeMay as a paragon of such thinking, and they were right. When he had flown bombers over Germany, LeMay recalled, his imagination had caught a picture of a little girl down below, horribly burned and crying for her mother. "You have to turn away from the picture," he said, "if you intend to keep on doing the work your Nation expects of you."[52]

LeMay reinforced such attitudes in his men with remorseless training, and he let the public know it. Less well known was the effect on the officers. Alcoholism was a widespread problem in SAC for decades. The officers' relations with their wives tended to be unusually formal and unromantic, while extramarital sex relations became popular. As the rate of divorce and other disturbances climbed, as officers resigned in large numbers to save their marriages or their sanity, LeMay became worried. One of his responses reached the public.[53]

A Hollywood screenwriter close to LeMay proposed to film the story of the first atomic bombing, and the General, hoping that a movie would improve SAC's internal morale as well as its public image, gave Air Force help. Film critics could not understand why *Above and Beyond* took time away from its semidocumentary theme for a syrupy love story, but the real key to the film was the strained relationship between the pilot who bombed Hiroshima, Colonel Paul Tibbets, and his sentimental wife. In the film the steely officer lost his self-control only once, when his wife remarked that children just like their own might be killed by bombs; Tibbets furiously commanded her never to say such things. He clamped down not only on feelings but on facts, ordering his wife to stick to the family laundry and keep her nose out of his secrets (which was the role LeMay assigned to his own wife). At the end of the film Tibbets' rigidity proved correct, for his work ended the war and his wife gave him her love. (In real life the couple got a divorce.)[54]

If the public did not catch the undertones in *Above and Beyond*, Hollywood and SAC cooperated to give them other chances. Tibbets had been torn between his wife and the beautiful new B-29 warplane with its atomic bomb. The widely seen 1955 movie *Strategic Air Command* was different: an officer was torn between his wife and the beautiful new B-47. For a change, in a 1957 movie a crew chief was torn between his daughter and the beautiful new *Bombers B-52* of the title. Since the great flashing airplanes stood for patriotism and resolution, the men always chose them, and eventually their women all agreed they were right. Most insightful was the 1963 film *A Gathering of Eagles*, written by a former SAC colonel in close contact with LeMay. Here a commander was torn between his wife and the device that could summon him at any moment, what movie advertisments called "The Red Telephone . . . His Mistress . . . Her Rival." In the concluding reconciliation he took wife and telephone into the bedroom together.[55]

The theme was familiar. At Fort Apache or Guadalcanal, according to John Wayne movies, men did have to leave weak sentiment back with the womenfolk. As another cowboy actor, Ronald Reagan, put it in a television Western, the officer was a "divided man," torn between love and official duty, who had to "deny himself the feelings that people have." Such stories drew upon an old symbolism, for the contrast between tough-minded logic and tender emotion had been represented for centuries as a dichotomy between masculine and feminine, the two searching for harmony.[56]

The public of the 1950s yearned to hear that sweet domestic peace could be maintained despite the presence of shocking weapons, that dutiful officers could keep aggression under control. The SAC movies, with their dedicated pilots, obedient wives, and soaring airplanes, were designed to reassure. According to a press release for *Strategic Air Command*, the movie would "build up our confidence in our protective devices and allay some of our uneasy feelings." SAC was to be, at last, a defense against nuclear fear.[57]

Countless publicists and politicians said that the bombers gave their nation "security." From 1957 on a motto went up on signs at the entrance to every SAC base and even on the SAC stationery: "Peace Is Our Profession." A publicist wrote that SAC "may be the forerunner of the world force for peace-in-being often visualized." Some SAC officers felt they almost had more in common with their Russian counterparts than with all the squabbling politicans, and indeed the Soviet armed forces served up identical promises of "security" and international guardianship. The virtuous Airmen had arrived.[58]

Perhaps the truest emblem of SAC was its official coat of arms, selected by LeMay and painted on the nose of every bomber where the pinup girls used to be: an armored fist in the clouds grasping one olive branch and three thunderbolts. The physicist and military analyst Blackett was the sharpest observer of these officers. "Spiritually intoxicated by flight at 50,000 feet in a jet bomber . . . ," he wrote, they "sang of the ease with which they could keep erring mankind in order by threatening them (as if they were Jove himself) with atomic thunderbolts."[59]

How reassuring was this, really? Talk about steely logic, denial of feelings, and authorities controlling superhuman force could cast a chill shadow. Watching the SAC movies, a film buff with a good memory might have recalled scenes in *Frankenstein* or *The Invisible Ray* where the scientist ignored his woman as he relentlessly pursued secret powers. Russians wrote frankly of the American "robot-soldier" who could "drop atom bombs on civilian towns without shuddering."[60]

Although nobody noticed at the time, nuclear defense efforts were slowly moving society toward a bitter confrontation between logic and feelings. Over the next decades people would increasingly find less safety than horror in the image, promoted by Air Force publicists, of unsentimental males commanding nuclear energy. This failure of psychological security became all the greater because the search for military security was itself releasing new forces far greater than fission.

part three

NEW HOPES AND HORRORS

1953–1963

8

Atoms for Peace

The island that held the device was well over the horizon from the ships where observers waited, yet the explosion turned their tropical night to dazzling noon. A fireball heaved itself up from the sea, growing and growing, much larger than anyone had expected; sailors thirty miles away felt the heat sear their skin as if a furnace door had been opened; some scientists thought that this time they had finally gone too far, that they had set off the last experiment. Then the familiar cloud began to mushroom upward, but enormously larger than any atomic cloud ever seen. November 1, 1952: the first fusion device.

Within weeks the public knew that something prodigious had happened, for when sailors described the apparition in letters home, American newspapers printed melodramatic stories. Although there was little information in the stories, there was enough to remind people of talk that had already been circulating for several years. Back in 1950, for example, Drew Pearson had told his radio audience that a scientist said 2,000 hydrogen bombs could "blow up the world." A few scientists had voiced well-publicized speculations that even a single such "Hell Bomb" might set off the fabled chain reaction that would consume the earth; although most experts denied that possibility, others had secretly made doomsday calculations just to be sure, and built a special atom-smasher to measure data for the calculations. In short, from the outset everyone saw fusion as something that went farther even than fission bombs into the realm of apocalypse.[1]

Most leaders met the appalling news of the fusion test with intensified effort along the lines established for fission—more negotiations for international controls, fences around scientific research, civil defense exercises, and fleets of warplanes. Some began to explore different responses: perhaps only a tremendous change of mind could meet the new reality. Two such movements arose. One was a campaign against radioactive dust, which I will come to in later chapters; the other, which started first, was a crusade called Atoms for Peace.

This crusade began in the White House. The ground was prepared by the Eisenhower administration's conviction that in holding off Communism, public relations and "psychological warfare" were as important as bombers. To wage this propaganda war Eisenhower had brought in C. D. Jackson, a mature, balding man who looked like an energetic vice-president of a major corporation; and so he was, for he had come to the White House on leave from Time Incorporated. Jackson believed his job was to think of a way to "go on the moral and ideological offensive against the Communists . . . give it a bite and punch which would really register on both sides of the Iron Curtain." But no such idea came to mind until atoms were offered.[2]

The initiative came from a special panel appointed in 1952 to take a new look at disarmament now that hydrogen bombs were on the way. Oppenheimer dominated the panel, as he dominated most groups he sat with. The group's recommendations came to Eisenhower's desk as a secret report in January 1953, and Oppenheimer explained the essentials in public. The physicist began with the premise that only open debate could lead to wise decisions and therefore to peace. The first step, then, must be "candor," telling people just how much destruction was now possible.[3]

Demands for nuclear information were meanwhile raised by the press and others, so the Eisenhower administration grudgingly took up Oppenheimer's idea. The President put Jackson in charge of drafting a speech and a publicity campaign under the title "Operation Candor." But in nuclear matters nothing ever stayed simple, for people sought to use the powerful images for their own purposes; Jackson's drafts reflected his zeal for scoring points in his worldwide propaganda contest with the Communists. He was backed by James M. Lambie, Jr., the White House aide in charge of civil defense publicity, who laid plans to turn Operation Candor into a blitz of domestic advertising that would use the threat of Russian hydrogen bombs to instill discipline and moral fiber within the United States.[4]

After reading Jackson's drafts for a speech that told of the awful damage Soviet bombs could inflict, Eisenhower reportedly complained, "We don't want to scare the country to death." Jackson went back and redrafted the speech to emphasize America's enormous power to retaliate. That sounded frightening too. As Jackson admitted, the proposed speech boiled down to "bang-bang, no hope, no way out at the end." [5]

Eisenhower himself had been disturbed about hydrogen weapons ever since he learned the results of the first test. Most impressive was a top secret film, called *Operation IVY* after the test's code name. It was screened in the White House on June 1, 1953, for the Cabinet, the Joint Chiefs of Staff, and others like Jackson; the few dozen people on that level were the entire audience for whom the Air Force and Atomic Energy Commission had made the film. Nobody who saw it was likely to forget its picture of an entire atoll vanished into a crater, or the fireball with a dwarfed New York City skyline printed across it in black silhouette.[6]

Worse news arrived in August: the enemy too had tested a fusion device. It was in fact only a crude experiment, and the Soviets were a year or more behind the United States, but the evidence available to Eisenhower in 1953 left him unsure whether the Soviets might not be ahead. Americans would have to be told something about all this, yet little good would come of any speech that offered only "bang-bang." [7]

A way out was offered by an idea that Jackson's drafts had mentioned only in passing, the "peaceful uses" of the atom. The idea came to Eisenhower's attention more pointedly in a long letter that Sterling Cole, the new chairman of the Joint Committee on Atomic Energy, wrote after hearing of the Soviet test. Cole wanted to speed up the building of bombs, but at the same time he wanted to "assure the world we stand ready to share the benefits of peacetime atomic energy." He said he was about to propose legislation on civilian nuclear industry that would "pool our resources and talents" with allies.[8] Eisenhower wrote back that his thinking was almost identical with Cole's, and the President asked an assistant to draft a memorandum. The key proposal, in the second paragraph, was later published in the memoirs of Eisenhower's supporters. "Suppose," the President suggested, "the United States and the Soviets were each to turn over to the United Nations, for peaceful use, X kilograms of fissionable material . . . "

On the surface it looked like a halfway version of the international atomic energy organization that scientists had advocated since 1945. But sharing some materials seemed such a modest step that it might actually be taken, setting a precedent for cooperation and meanwhile promoting civilian uses of nuclear energy. More subtly, at a time when fissionable materials were more rare and magical than gold, the donation of "X kilograms" should kindle sympathy for the entire American nuclear program.

Psychological warfare with the Kremlin was paramount. This became clear in the third paragraph of the memorandum, which Eisenhower's men neglected to publish: "The amount X could be fixed at a figure which we could handle from our stockpile, but which it would be difficult for the Soviets to match." The offer to give the United Nations a substantial amount of uranium-235 or plutonium would embarrass the Soviets if they held back, and would seriously deplete their small stockpile of bombs if they joined.[9]

On December 8, 1953, Eisenhower delivered the much-revised speech on hydrogen bombs before the General Assembly of the United Nations. His opening was a relic of the original Operation Candor, describing with awe but with little new information the power of modern bombs. The American nuclear armory, he said, far exceeded the total force of explosives spent by all sides in the Second World War—the first time such a comparison was made officially. Then came the main point. The President, his eyes shining with emotion, offered to support an International Atomic Energy Agency to develop the new power for life rather than death.

The speech had an unexpectedly great impact on the world public. The final result would be something the President scarcely intended: to promote a nuclear power industry far more rapidly than ordinary business practices would have advised. This happened partly because of the universal demand for an answer to hydrogen bombs, and still more because the hopes for civilian benefits had been prepared well in advance by certain elite groups.

A POSITIVE ALTERNATIVE

Since 1945 cliché experts had never ceased to tell the world it stood at a crossroads, and many journalists had said less about the road to doomsday than about the one to an atomic White City. The best known of these was William Laurence ("Atomic Bill" as some called him), now more of a missionary than ever. He explained that atomic energy could turn deserts and jungles into "new lands flowing with milk and honey," and, in sum, could "make the dream of the earth as a Promised Land come true in time for many of us already born to see and enjoy it." He admitted that fission reminded him of his adolescent daydreams about Martian civilization, and even of a new Tree of Knowledge that would return us to Eden. Not everyone went so far, but many journalists, and scientists as well, spoke as if ancient millennial dreams were almost established fact.[10]

The titles of articles in the *Readers' Guide* suggest what Americans were reading. In the 1930s about one article in 3000 had been mainly about atomic energy, and for the years 1945–1953 that proportion rose to about one in 300. Many of the articles after 1945 were about military uses, and many of these naturally had titles that might provoke anxiety. But writings on civilian uses were at least as numerous as those on weapons. When neutral titles are left aside, the emotion-laden titles on civilian atomic energy are seen to be almost entirely positive. Similarly, for the first postwar decade wondrous promises about nuclear energy may be seen as frequently as stories about weapons in the index of the *New York Times*, in the titles of articles from various nations listed in the *Internationale Bibliographie der Zeitschriftenliteratur*, and in a study of the typical Paris newspaper *France-Soir*. The only medium I surveyed that gave most of its attention to bombs was the newsreels, a major source of simplistic half-truths until television arrived. Newsreels loved to show the spectacular bomb tests, yet they too included many items explaining that the atom would bring marvelous benefits.[11]

The scientists and science journalists who originated most of this publicity offered only vague explanations for their visions. At most, some noted physicists would remark that uranium was such a compact fuel that power plants could be set down even in the middle of a jungle, with industry decentralized at last. That was why Kaempffert told *New York Times* readers that Africa "could be transformed into another Europe." Other scientists said that nuclear power could pump water to irrigate deserts, rivaling the engineers of Mars. After all, projects already under way, such as the monumental dams of the Tennessee Valley Authority, were scarcely less astonishing. Some Americans looked forward to a government-operated civilian atomic energy program, an "atomic TVA." Not to be outdone, in 1949 a Soviet spokesman boasted that his government "right now" was using atomic energy to irrigate deserts and so forth. The forces of traditional imagery were encouraging scientists, journalists, and officials to make gorgeous promises that went far beyond fact.[12]

Nobody was as well qualified to understand those forces as David Lilienthal. His enemies sometimes made the mistake of seeing Lilienthal as a simple man, for his face, whose drooping features might have seemed melancholy, carried a quirky grin that gave him a gentle and elfin look. Yet he had risen like a rocket through Harvard Law School and government service, and by the age of forty he was running the TVA. Here he had faced all the contradictions of his progressive beliefs, the desire for both compassionate government plans and efficient private competition, for both community spirit and individual initiative. In the Tennessee Valley he succeeded, as well as anyone ever did, in combining the White City of technology with the green hills of Arcadia. If anyone deserved to be called Chief of the Martian Engineers, it was Lilienthal. President Truman was making a natural choice when he picked him in 1946 to become the first chairman of the Atomic Energy Commission.[13]

Lilienthal was as susceptible as anyone to nuclear imagery. Briefings from Oppenheimer and other scientists were a "soul-stirring experience," he told his private diary. He felt as if he had been allowed "behind the scenes in the most awful and inspiring drama since some primitive man looked for the very first time upon fire." Lilienthal was an introspective man, and he would spend much of his life reflecting upon such imagery.[14]

He noticed that experts were looking for a single miraculous answer to the apocalyptic problem of nuclear bombs. From his experience in law and government he doubted that such a single answer would come; people would have to fumble with all sorts of partial measures, gradually working through the complexities. Nevertheless, even Lilienthal temporarily fell for the illusory promise of an international inspectorate. He soon became disillusioned with that idea, and he did not believe in SAC and hydrogen bombs either, but he kept looking for a way out. He later confessed that he and his colleagues "became emotionally committed to the search for an Answer."[15] The only real solution, he finally told Truman, would be to work toward civilian benefits—atomic TVAs all over the world. Lilienthal often spoke in public about that. When he left the Commission in 1950 he insisted that "my theme of Atoms for Peace is just what the country needs."[16]

Most other nuclear authorities agreed. For example, Lilienthal's replacement on the Commission, Thomas Murray, was a very different man—a hardheaded engineer, millionaire industrialist, and devout Catholic—but he was equally fascinated with civilian uses. Murray had asked Truman for a post on the AEC because of what he later called a "secret, unconscious fascination" with atomic energy; it became almost a personal religion of Murray's that this energy was a gift from God which the nation was morally bound to develop. In later years Lilienthal wryly recalled such missionary fervor. He and the rest had been convinced, he admitted, that anything that could yield such a dreadful weapon simply had to have important peacetime uses. At the atomic crossroads, all that was necessary was to set people on the correct road.[17]

When men like Lilienthal and Murray as well as scientists and journalists all talked about the coming economic revolution, industrialists wondered what it would mean for themselves. Business magazines regularly printed information and speculation on atomic energy, optimistic or pessimistic by turns. Most businessmen were cautious, with no inclination to invest much money yet. A few plunged in for reasons that had little to do with immediate financial returns.

Some feared that the United States government would block the development of a private nuclear industry. A few enterprising men, chiefly from the electrical industry, resolved not to be left behind when the atomic revolution arrived. Demand for electricity was doubling every decade; sooner or later the most economical fuel might be uranium, and when that day came they were determined that the uranium would be in the hands of forward-looking businessmen like themselves, rather than controlled by a socialist atomic TVA. In the early 1950s they began to urge Congress toward new legislation that would lower secrecy and other barriers.[18]

In some other nations, leading groups were still more strongly impelled toward nuclear power. In Europe and Japan the most easily mined coal was gone, the best hydroelectric sites were already in use, and almost every barrel of oil was imported. The lesson of fuel vulnerability had been driven home by the Second World War when lack of gasoline forced most citizens to get about on bicycle or foot and when the German and Japanese armed forces, despite a military grand strategy largely aimed at seizing oilfields, had been virtually immobilized by late 1944 for lack of fuel. The war was followed by painful coal shortages. During the winter of 1946 the citizens of Paris often found themselves with neither fuel to heat their rooms nor electricity to light them, and into the 1950s Japanese cities suffered brownouts. Atomic scientists of every nation promised to solve the problem with nuclear power.

The United States government began to worry that other nations would seize the lead in this grand new industry. The British were visibly leaping ahead, while the Soviets voiced total enthusiasm and seemed bound to join the race. In March 1953 the National Security

Council endorsed a policy submitted by the AEC: "economically competitive nuclear power" must become "a goal of national importance." A healthy American industry could capture the world market, and meanwhile the threat of Communism in impoverished countries could be met with the prestige and economic blessings of atomic energy in jungle and desert. And of course a civilian industry would make a solid foundation for America's military programs. The Joint Committee on Atomic Energy swayed Congress in the same direction through extensive hearings in which industrialists and others agreed that civilian development was urgent. It was in view of these pressures, as much as in search of a response to hydrogen bombs, that Chairman Cole had written Eisenhower and helped to launch the Atoms for Peace speech.[19]

All these discussions were held among elites in the absence of any well-formed public opinion. Most people, when asked, would agree with such statements as atomic energy was "wonderful—think of all the new things that will come of it." Yet in the early postwar years the idea was not uppermost in their minds. A 1946 survey concluded that few citizens thought of any constructive uses when atomic energy was discussed; "to the general public atomic energy means the atomic bomb." A 1950 survey noted that, despite the flood of extravagant promises in the media, "involvement with the atomic energy process is restricted to the upper socio-economic and relatively well-educated groups in the population." Even among opinion leaders such as newspaper editors and businessmen, one-third could name only military uses for atomic energy. As of 1953, the drive toward a civilian nuclear industry was confined to some nuclear scientists, their followers in journalism and government, and a small minority of industrialists. But those groups could be persuasive, fired as they were not by dry facts but by a vision of saving the world and leading it to atomic utopia.[20]

ATOMIC PROPAGANDA ABROAD

The day after Eisenhower's "Atoms for Peace" speech to the United Nations, Jackson reflected that it was "a direct challenge to the Soviets' near monopoly of 'peace' propaganda." He exploited the speech to the hilt with a blizzard of press releases ("Era of Atomic Power is on the Way," and the like) that got into newspapers around the world, along with countless magazine articles, pamphlets, radio broadcasts, and films prepared by the United States government. Particularly effective was a set of exhibits made by the AEC, which over the next few years displayed beneficial uses of atomic energy to huge crowds in Karachi, Tokyo, Cairo, São Paolo, Teheran, and many other capitals. It was a psychological warfare triumph beyond Jackson's dreams. A secret report to his office in November 1955 boasted that the campaign "detracted popular attention away from the image of a United States bent on nuclear holocaust," diverting the public eye to "technological progress and international cooperation."[21]

The campaign did more than Jackson intended. It gave people everywhere the idea, which until then had seemed convincing only to a small elite, that atomic utopia could become a reality within their lifetimes. Eisenhower had opened a floodgate, and most other governments rushed to show that they too could use atomic energy for something besides destruction. Indexes such as the *Readers' Guide* and the *Internationale Bibliographie* show that the volume of publication on civilian uses of atomic energy doubled or tripled between the five years before Eisenhower's speech and the five years after. The card catalogs of major American and European libraries show a parallel wave of books classified under "Atomic Energy," more than at any time before or since. Even the newsreels found ways to dramatize civilian uses.[22]

The United States and to some extent Britain continued to dominate world opinion. A majority of the European books were translations from English, and of the more than 100 short films on peaceful uses offered on loan worldwide within the decade after Eisenhower's speech, nearly half were American and nearly half the rest British (France came in a distant third, followed by the Soviet Union). When it came to information and images, most nations looked to the United States.[23]

The Atoms for Peace crusade broke through to a new level of credibility at an international conference proposed by the United States and convened in Geneva in 1955. Three thousand scientists and their followers gathered to hear representatives from various countries declare that nuclear industry was close to being commercially profitable. Until then, most nuclear scientists and those who paid close attention to them had warned that as an economic venture, nuclear power was still decades away. At Geneva those cautions were cast aside.

More than money was at stake, as might be seen in the working nuclear reactor that was the centerpiece of the American exhibit at Geneva. Magazines around the world ran photographs of this deep cylinder of crystal-clear water with its rods of uranium, where visitors could see the water glow with a ravishing blue light, as if vitalized by sorcery. Guides at the exhibit and other scientists at Geneva waxed enthusiastic. An onlooker told how one young scientist spoke with such intensity "that he seemed the priest of a mystical religion, to which his listeners could not but become converts."[24]

Many came home eager to buy or build reactors. Characteristic was a West German representative, Karl Winnacker, a sober and respected leader of the chemical industry. He had talked with the euphoric scientists and gazed down into the American reactor, whose glow seemed to him (he later said) like "the light of Aladdin's magic lamp." Inspired, Winnacker fought successfully to bring atomic industry to Germany. In many other nations too, leaders abandoned their caution and began to reach for nuclear energy as a panacea.[25]

These plans took on unexpected urgency in 1956 when war over the Suez Canal plunged Europe into its second postwar fuel crisis. As cars lined up for rationed gasoline and economic ruin threatened, everyone saw the weakness of an energy system built on the shifting sands of foreign oil. Officials in Europe and elsewhere drafted plans to build dozens of nuclear electric plants with the greatest urgency.

To spread the enthusiasm as widely as possible, the United States offered to give reactors to almost anyone, along with the necessary uranium-235 (the "amount X" to be distributed abroad was set after long discussion at 20 metric tonnes). The Americans hoped to make future exports easier by embedding their technology in foreign nations, and meanwhile they expected to make the world safer by driving back poverty and Communism. There was also a more specific way in which the program could bring national security: the United States would help a nation build reactors only if these were subject to inspection, for example by an International Atomic Energy Agency, to make sure that no plutonium was diverted to make bombs. This was the same idea proposed in Eisenhower's 1953 speech and before that in Oppenheimer's 1946 plan for international inspectors, an idea that would remain at the core of all future international nuclear energy negotiations. Atoms for Peace was officially promoted above all as a step to controlling weapons.[26]

Leaders in many poor nations grabbed at the chance, as a Pakistani representative put it, to "condense decades of progress into a few years." Besides, even the smallest reactor could serve as a badge, certifying that the owners were up-to-date. By the end of 1957 the United States had signed bilateral agreements with forty-nine countries from Cuba to Thailand, and American firms had sold foreigners twenty-three small research reactors more or less like the Aladdin's lamp of Geneva.

The results of this campaign were peculiar. It set down reactors in nations that lacked the economic structure and skills to exploit them, nations that had far more need for fertilizer or high school teachers. From his vantage point as a private citizen, Lilienthal called the program an "absurdity" driven only by "the desire to prove somehow that atoms were for peace."[27]

That desire was even stronger among Soviet leaders. Presumably they were as concerned as anyone with psychological warfare, international commercial rivalry, and the advent of hydrogen bombs, but what was most obvious was their straightforward faith in atomic paradise. *Izvestia* explained in 1950 that anyone who denied the economic promise of atoms was a dupe of "American monopolists," coal and oil trusts that would do their utmost to block nuclear power. At Geneva in 1955 Soviet authorities claimed that in some regions of their country nuclear power was already economically sound. After all, the idea that improved technology drove history toward a perfected future remained a central Marxist tenet, as well as one of the sources of legitimacy for the Soviet government, justifying its demands for sacrifices. Along with Marxism, a fierce national pride urged Russians to stand second to none in modern technological projects; huge reactors would join huge dams, rockets, and steel mills as proofs of pre-eminence.[28]

Not only in reply to American propaganda, then, but also from native drives, Soviet officials showered the world with their own optimistic messages. They too sent exhibits on international tours and offered research reactors to friendly nations. Still more did the Soviets teach nuclear optimism to their domestic audience. To counterbalance the dread of weapons and to instill confidence in government plans for the future, officials released a swarm of magazine articles and radio speeches and put up a pavilion in Moscow where millions of visitors thrilled at the underwater glow of a reactor. As psychological warfare, atomic imagery could have its greatest impact at home.[29]

ATOMIC PROPAGANDA AT HOME

Eisenhower, noting the success of Atoms for Peace exhibits overseas, wrote the AEC that he "would favor additional exhibits being prepared and displayed to a large number of our own people." The AEC scarcely needed encouragement; the Commission and its laboratories had been deep into public relations from the start. Back in 1948, a poll at one laboratory's exhibit in New York City had found that visitors came out feeling less fear about atomic energy and more hope—precisely what was intended. After 1953 the AEC stepped up its efforts in every area, for example arranging many interviews for journalists and distributing copies of hundreds of speeches, nine-tenths of them dealing with peaceful uses. It offered film and television producers quantities of footage and made free loans of its own movies, such as "Atoms for Peace," which was originally assembled by the United States Information Agency for distribution abroad. Nuclear officials in other countries made similar efforts. In each nation the government efforts were seconded by numerous groups, until almost everyone was half convinced that civilian nuclear energy could soon be as real and important as the bombs.[30]

A main force behind the public relations work was Lewis Strauss, a Wall Street financier who was now chairman of the Atomic Energy Commission. With his easy Virginia accent and a grin somewhat like that of his good friend Eisenhower, he seemed master of any situation. But beneath his courtly manners Strauss easily became indignant, for he could never admit that he might be in the wrong. Devoutly religious, he centered his life on a constant battle against what he called "the powers of evil and of godless atheism." In his childhood Strauss had avidly read Jules Verne and dreamed of learning the secrets of nature, and in high school he had been excited by Millikan's textbook with its question about whether man could "gain control of this tremendous store of sub-atomic energy." In later years Strauss thought mainly of using that energy against atheistic Communism, doing much to speed the development of hydrogen bombs. He was not at first enthusiastic about Atoms for Peace, although he told Eisenhower that at least it "might have value for propaganda." But gradually Strauss became wholly converted to civilian nuclear energy. "Beneficent use of power which the Almighty has placed within the invisible nucleus," he declared, "will prevail over the forces of destruction and evil."[31]

The press was his most important ally in this struggle, as Strauss explained to a meeting of the National Association of Science Writers in 1954. He was concerned, he said, that the AEC's funds might be cut back unless the public approved of its programs. Therefore he asked the science writers to tell about the bright part of the AEC's work. Our children, he said, will enjoy "electrical energy too cheap to meter," an "age of peace," and so forth. Laurence and others relayed such claims to the public verbatim.[32]

Journalists had their own reasons to laud Atoms for Peace. Besides honestly reporting what experts said, science writers liked these stories, and not only because tales of an astonishing future were their stock in trade. After the bombing of Hiroshima the number of science writers mounted rapidly, and the new generation, like their elders, were either trained in science or otherwise inclined to view it with respect. As one of them explained, the first atomic bomb explosions drew him into science writing because "I realized then that science was going to rule us." Such journalists were the last to doubt that atomic energy was going to do wondrous things.[33]

Next to nuclear authorities and science writers, it was certain businessmen who were finding the biggest personal stake in Atoms for Peace. They were impressed when Strauss's claims were repeated by experts like the authors of a 1956 book, *Atomic Energy for Your Business: Today's Key to Tomorrow's Profits*. A vice-president of General Electric, noting the ambitious British and Soviet plans to export reactors, warned that "already the contest is on." He said courageous entrepreneurs must march forth on behalf of the nation and democracy, not to mention their companies.[34]

Few were as enterprising as the president of the Detroit Edison Company, Walker Cisler. An engineer by training, Cisler had risen to the top of this respected electric utility because of his ability to make things work. With a strong jaw and silver hair framing his impassive face, he looked a model of industrial leadership, and *Fortune* held him up as an example of the engineer-administrators who were replacing the crude old tycoons. Cisler had a crusading spirit, throwing himself energetically not only into industrial management but also into volunteer civic work and government service.[35]

Nuclear energy caught Cisler's attention in 1947 when he joined a committee that the AEC set up to make connections with private industry. He soon began persuading Detroit Edison and other companies to promote an atomic age under free enterprise. In 1952 it was Cisler's lobbying as much as anyone's that brought business concerns about nuclear energy to the attention of Congress and the President. Meanwhile an acquaintance suggested that in view of the upcoming legislation, business needed a new institution to coordinate its efforts. Cisler took up the plan with his unfailing energy, inviting dozens of industrialists to meet for a lunch shortly after Eisenhower's Atoms for Peace speech; soon he was president of an Atomic Industrial Forum. By 1956 the Forum had some 400 companies as members.[36]

The Forum was a main agent for spreading nuclear visions within the business community. Its specialty was meetings that brought representatives of corporations together in a hotel to sit through lectures or join discussion groups. Since most lecturers were AEC commissioners, employees, or contractors, men whose careers were devoted to atomic energy, it was not surprising that their message was enthusiastic. Typical was a 1955 meeting with the topic, "Atomic Energy—A Realistic Appraisal." The chairman opened the proceedings by joking, "Perhaps it is a little early in the morning to work up a keynote mood of wondering and of passionate awe. Nevertheless . . ." Nevertheless, such was the mood that speaker after speaker evoked. Even Cisler rose above his normal pedestrian prose to compare nuclear energy with the greatest religious developments of history.[37]

In addition to such meetings, the fledgling nuclear industry developed trade journals, newsletters, and other mechanisms to share its information and hopes. An international network was growing, a community of people who drew their knowledge and ideas largely from one another, people who had decided to stake their companies' money and their own careers on nuclear energy.

These people worked to tell not only their fellow businessmen but the whole world about Atoms for Peace. The Atomic Industrial Forum did so much public relations work that in the White House Lambie and Jackson praised it as a valuable "cold-war weapon," although the Forum chiefly addressed the domestic rather than the international public. Individual companies did still more. Foremost was General Electric, which ran the AEC's Hanford reactors and by 1957 was employing more than 14,000 people in its nuclear divisions. GE had been acutely concerned with public acceptance of its products since the turn of the century, when the company had faced widespread fears about electrocution. The public had come to accept electricity, and GE meant to help them accept nuclear power too. Its efforts began in the late 1940s and continued in high gear for over a decade. GE public relations experts praised nuclear energy through their "opinion leader advertising series" in American magazines and promoted it to the tens of millions of viewers of the GE Television Theater. Similar efforts by other American and foreign firms helped spread similar messages around the world.[38]

The optimistic pronouncements of national nuclear agencies, scientists, science writers, and industrial companies confirmed other leadership groups in attitudes that already attracted them. For example, in 1955 the World Council of Churches called for energetic development of atomic energy for peace. Meanwhile the American Federation of Labor and individual unions formed study committees that endorsed the swift development of nuclear power as a source of jobs and prosperity.[39]

Large majorities of ordinary citizens around the world, and especially the young, caught the optimism. A 1957 world survey of college students from Poland to Brazil found them aware of radioactivity hazards and worried about bombs, but on balance pleased that fission had been discovered. Most enthusiastic of all were American students. The AEC received an increasing stream of letters from children asking for information about civilian applications.[40]

Since 1945 teachers had insisted, in accordance with their professional mission, that the primary response to bombs must be "education." Some American teachers also said they hoped to counter the anxiety that civil defense drills aroused in their pupils. From state education organizations to local advisers of high school Atomic Energy Clubs, everyone wanted students to face the future not only with knowledge but with good cheer. The teachers were supported by the AEC, which sent traveling exhibits to high schools and developed teaching materials. General Electric helped too, distributing millions of copies of its comic book "Inside the Atom," and reaching about two million students a year with its animated color film *A is for Atom*.[41]

Perhaps most effective of all was Walt Disney's *Our Friend the Atom*, shown on television and in schools beginning in 1957. The great storyteller naturally introduced the subject as something "like a fairy tale," indeed the tale of a genie released from a bottle. The cartoon genie began as a menacing giant much like Herblock's bomb-monster. But scientists turned the golem into an obedient servant who wielded the "magic power" of radioactivity, symbolized as glittering pixie-dust.[42]

An example of the results of all this work was an essay written by a second-grader:

> *Good Atoms.* Everything is made of atoms. When we learn more about how valuable these atoms are, people will be very happy . . . the business man will have machines and better things to sell. Every body will be happier.[43]

9

Good and Bad Atoms

Scholars studying fairy tales find that the stories can be classified, hundreds at a time, into one or another traditional pattern. In the same way, there was much in common among the various Atoms for Peace productions: *Our Friend the Atom*, the American newsreel series *The Atom and You*, the corresponding newsreel by Actualités Français, the U.S. Information Agency's *Atoms for Peace* film, the long Russian documentary by the same name, the scores of books in various languages, the exhibits, and so on. As a scholar might explicate the structure of a set of folk tales, so I wish to unfold the pattern of common themes within Atoms for Peace productions. For these themes became basic to public attitudes toward the nuclear industry, toward science and technology in general, and eventually toward our entire civilization.[1]

At the center of the structure was the polarity of weapons versus peaceful uses, the atomic genie who could be either menace or servant. By the mid-1950s everyone was familiar with both extremes, although different people would view the set of images differently; some would focus on wonderful benefits while others would see death in the foreground. Atoms for Peace productions were built on the principle of setting one extreme against the other. The producers typically gave only a glimpse of an atomic explosion near the beginning or a brief, banal warning against catastrophe near the end, but that mention of bombs was scarcely ever omitted. Even the Soviet works, which avoided talk of weapons at all costs, by repeatedly emphasizing "strictly peaceful purposes" showed that the other pole was never forgotten. Since it was precisely because of the bombs that so many spotlights were focused on peaceful uses, in the end the Atoms for Peace publicity only gave greater prominence than ever to the whole ancient tangle of transmutation imagery, the bright and the somber together.

MAGICAL ATOMS

Public anxiety about nuclear weapons was so strong that Atoms for Peace productions could not hope to balance it except with the most wondrous visions. The core vision, offered by Soddy decades earlier, was rephrased by Laurence in 1946 for the new generation: atomic energy was "a philosopher's stone that not only could transmute the elements and create wealth" but could also provide an "elixir of life" and "mastery over time and death."[2]

Nuclear healing was the most prominent theme of all. No production on Atoms for Peace seemed complete without a patient lying calmly on a table and gazing up at a gleaming white radiation mechanism, as in the Russian documentary film, which reached its dramatic climax when radiation therapy erased a disfiguring tumor from a child's face. Medical radiation was so popular that United States officials had to issue a warning against "atomic" potions peddled by quacks, while respected physicians dosed thousands of people with radioactive substances for diseases like tuberculosis even though less risky therapies were now at hand. During the 1950s X-rays were often used to kill unwanted body hair; thousands of fluoroscopes in shoe stores across the United States and Europe showed people the bones in their children's feet; some hospitals routinely X-rayed infants simply to please parents with an inside view of their offspring. The accumulated radiation sometimes reached severely hazardous levels, yet the public continued to trust in rays.[3]

Medical hopes were not all unfounded, for even small reactors like the Geneva model could make artificially radioactive isotopes that could reduce tumors and aid diagnosis. Soon the isotopes were successfully treating hundreds of thousands of patients every year. Already by the late 1950s the number of lives taken by radioactivity, including those in Hiroshima and Nagasaki, was surpassed by the number of lives saved by radioactivity.

Still more important in the long run was the fact that the most minuscule amounts of isotopes could be detected with Geiger counters, and thus could be used as "tracers" to follow biological processes. Isotopes became an invaluable tool for studying everything from physiology to the way heredity works. Much of the tremendous progress in biology and medicine since the 1950s would have been impossible without radioactivity.

Popularizers exaggerated even that, promising everything up to immortality. Tracer isotopes would unveil the secrets of life itself! And, as a narrator explained in a CBS radio program, "when you get deeper and deeper into the secrets of life, you find them so fascinating you sometimes forget that the atom can kill." (This 1947 program was meant to bring just such a change of mind, and testing showed that it did leave listeners less fearful.)[4] The most common specific boast was that, as a Soviet Academician put it, tracers would lead to the "discovery of the mystery of photosynthesis," with immense benefits for agriculture. Atomic rays could also be applied more directly to promote life through increasing the supply of food. Fantastic growth was the property most often stressed, as in a film by the United States Chamber of Commerce showing a little boy gaping at oversized peanuts "created by radioactivity." The idea of the life-force within rays, embodied in the elixir of isotopes, was so appealing that biology, medicine, and agriculture together took up between a quarter and a half of typical Atoms for Peace productions.[5]

In terms of hard cash, however, the life sciences took up only a tiny fraction of the budgets of the various national nuclear establishments. The U.S. Atomic Energy Commission, for example, spent less than a tenth of its budget on any civilian uses in the mid-1950s, and the life sciences had a small share even of that fraction. The nuclear industry dealt with uranium-235 and plutonium by the ton, while Atoms for Peace imagery relied upon a stock of isotopes that could have been stored in a closet.[6]

Atoms for Peace publicists also spoke of industry, but still with particular attention to tracers: "Experiments with Radioactive Piston Rings Hold Promise of Benefits to Motorists," and the like. Substantial money was going into reactors, but early Atoms for Peace productions often gave nuclear electric plants less than half of the total space devoted to industrial applications. An example of this pattern was a high school film series entitled *The Magic of the Atom*, with segments on *The Atomic Alchemist*, *Radiation—Silent Servant of Mankind*, and so forth. Only one of the film segments was on electrical energy (*Power Unlimited*), while there were three apiece on isotopes in industry, in medicine, in biological research, and in agriculture. The emphasis would change rapidly, for by 1960 electrical power was recognized as the center of gravity of civilian atomic energy. But on its first look the public beheld not prosaic electric generating stations so much as wonders akin to magical elixirs and talismans.[7]

Popular thought was already in the realm of fantasy. After 1945 even more than before, storytellers made use of atomic rockets, atomic rays, and the "atom-powered two-way wrist radio" that Dick Tracy wore from 1946 on. Children everywhere from America to Russia played with "atom-powered" toy cars and airplanes. The most prolific fictional inventor was Tom Swift Jr., who powered aircars and the like not with electricity as his father had done, but with atoms. Real authorities strove to outdo him. In 1951 Benson Ford, an officer in his grandfather's company, announced that atomic cars were on the way; four years later David Sarnoff, chairman of RCA, wrote in *Fortune* that by 1980 every home would have its own atomic power plant. Scientists began speaking out to refute such fantasies, and journalists who followed the scientists' lead began to repeat their cautions, until gradually most people understood that suitcase-sized reactors were not on the horizon.[8]

Large reactors remained plausible, yet using uranium simply as a replacement for coal-burners to generate electricity did not seem exciting enough. Scientists, journalists, and officials claimed that big reactors were also ideal for propelling ships or rockets. In the 1950s, engineering teams in the United States and the Soviet Union even studied nuclear-powered railroad locomotives. From 1945 until the late 1950s propulsion got almost as large a fraction of all atomic energy publicity as did electricity generation.

The height of organized fantasy was a plan for an atomic aircraft to go farther and faster than mortal man had ever flown. Here the duality underlying Atoms for Peace was unmistakable, since civilian transport would surely come only after the military version—a bomber that could stay aloft for months on end. Magazines printed diagrams of the warplane complete with snack bar and lifeboat, diagrams with a striking resemblance to the ones drawn by Tom Swift Jr. and his adolescent readers. Spurred by warnings from experts and senators that the Russians had already seized the lead in building this "ultimate weapon," the United States threw more than a billion dollars into the struggle until President Kennedy put the program out of its misery in 1961.[9]

The only nuclear propulsion schemes that actually worked involved ships, for in truth a kilogram of uranium-235 could get the proverbial steamship across the Atlantic. In 1959 the Soviet Union launched a nuclear-powered icebreaker with great fanfare, and later came an equally well publicized American cargo vessel and more modest German and Japanese ships. Here, as so often with new technologies, a good idea proved to be not quite good enough for commercial success. It was in another area, where cost meant little, that atomic ships would become important.

REAL REACTORS

Since 1939 the scientists working on fission had thought of using it to propel submarines. Captain Hyman Rickover made it work. Contemptuous of his Navy superiors and cruel to his subordinates, punishing with stinging scorn anyone who fell short of his standards for hard work and competence, Rickover came to be despised as much as respected. Even his formidable bureaucratic skills would never have lifted him to high command, but for atomic energy—the one field everyone from contractors to congressmen held in awe. Rickover's *Nautilus* (named after Jules Verne's submarine), the world's first nuclear-propelled vehicle, hit the water in 1955, perhaps five years before a normal technological program would have produced such a thing. In retrospect it made little military or political difference that the United States had such a submarine so early, but at the time it impressed everyone. Rickover's achievement exerted a curious influence on the growing nuclear industry.[10]

There are many ways to build a reactor, and during the 1950s engineers experimented with dozens of designs, each as different from the next as a steam-powered locomotive differs from a diesel truck. For example, a group sparked by Teller took safety as their prime criterion and designed reactors for which a major accident would be physically impossible. The prime criterion for Rickover's reactor, on the other hand, was that it must fit inside a submarine. His engineers solved that problem by using highly concentrated fuel, the proportion of uranium-235 raised from the natural 0.7 percent to 90 percent or more. They had no problem getting this exotic fuel, thanks to the enormous plants the AEC was building for its hydrogen bomb program. A number of related choices culminated in the elegant and compact "light water reactor," named for the ordinary water it used as both coolant and moderator.[11]

The light water type was not optimized either to turn a profit or to make accidents impossible, but it was the only developed reactor at hand, and that attracted the AEC when the agency decided to build a civilian power reactor. In the summer of 1953 Rickover took on the project. Meanwhile Commissioner Thomas Murray gave an enthusiastic speech to a group of utility executives, inviting them to join the work, and the speech sparked the interest of Philip Fleger, chairman of a Pittsburgh utility. Fleger later said he was attracted by the prospects for cheap nuclear energy in the hands of private industry rather than government, by Murray's call for a Cold War triumph, and by a shrewd understanding that the first company to master nuclear power would gain matchless publicity. Besides, Pittsburgh was suffocating in smog, and citizens who had united against coal burning would welcome reactors as a clean and modern substitute. In short, when Fleger teamed up with the AEC in building a reactor, nuclear imagery meant more to him than direct profits.[12]

This is not to say that economics played no role in nuclear energy development; I am leaving discussion of economics aside, only pointing out that it was not the sole factor, nor at the start was it the main one. Not even the engineering details were driven strictly by commercial needs. The reactor's design by the Westinghouse company was simply a scaled-up version of the submarine engines they were building for Rickover. Constructed at Shippingport, Pennsylvania, the plant came on-line in 1957 to universal acclaim. As the first large reactor whose technology was free of secrecy, the light water reactor made a strong impression on engineers everywhere. The design would eventually dominate the nuclear industry in most nations. Yet the reactor's technical design, like its government-subsidized funding and the abundant supply of uranium-235, had originated less in civilian needs than in military programs. The same was true for British and Soviet prototype power reactors, built and widely praised during the same years.[13]

Businessmen and nuclear officials knew that commercial nuclear power still rested largely on noneconomic factors, yet they agreed that uranium might well be the fuel of the future, if only because other fuels were running out. To be sure, centuries' worth of coal lay in the ground, but it was getting ever more expensive to dig it out, and now the cheap fuel was oil. In the 1920s three-quarters of all American power had come from coal, but by the end of the 1950s three-quarters came from oil. Estimates of oil resources were far from certain. For the United States they ranged from the official speculations of the Geological Survey, which by the end of the 1950s was saying there was enough oil for a long time, to the calculations of the iconoclast geologist M.

King Hubbert, who warned that crude oil production in the United States would reach its peak in the 1980s or even earlier. Even that deadline seemed a long way off. Some experts took comfort in the huge resources of the Middle East and elsewhere, while others said that solar energy or newer inventions would doubtless be developed in time.[14]

Other people insisted on planning many decades ahead, relying only on whatever technology was already visible, and not trusting Middle East politics. Those people included electric utility executives like Walker Cisler, government officials like Strauss, and military officers like Rickover. They all insisted that to play it safe, a large-scale nuclear power industry must be in place by the 1980s. Their views prevailed. By 1960 many of the glittering trappings of Atoms for Peace had been set in the background while nuclear reactors, generating civilian electricity or driving submarines, stood more and more in the fore. These reactors were prosaic devices, simply sources of hot steam. Yet from their birth they had been seen as part of a grand mythical structure—and that was one reason, as strong as economics, why they were built. In that mythical structure they were fated to remain.

THE CORE OF MISTRUST

An ethnologist studying Atoms for Peace publicity might be reminded of the many folk tales in which there is a virtuous brother and an evil one. The anthropologist Claude Lévi-Strauss pointed out that in such tales the main point may be the bipolar structure itself—the fact that the world is divided, rather than precisely who is virtuous or evil. Sometimes, in related stories gathered from different tribes, the sides changed places while the structure persisted. Often myths and symbols made an impact precisely because they contained opposite meanings simultaneously, gathering contradictions into a package that made some sort of sense, meeting the human need to reduce life's paradoxes to meaning. In just this fashion the Atoms for Peace tales built their message on the division between what *Time* called "Good & Bad Atoms." It was the structure already seen in the editorial cartoons that showed one bright and one dark path leading from a crossroads: the point was that extremes of power now lay within human choice.[15]

As in sets of folk tales, the sides in nuclear tales could switch. For example, while some saw nuclear bombs as wholly evil, the Strategic Air Command saw them as the mainstay of peace. Similarly, the more some people spoke of the wonders of peaceful reactors, the more others thought of danger. If civilian nuclear energy could turn deserts into gardens, how easily could it turn gardens into deserts!

When polls showed that a majority felt optimistic about civilian nuclear energy, the same polls showed that a minority did not. The rising nuclear industry realized that, as a pollster told an Atomic Industrial Forum meeting, a "hard core of about one-fourth" were "dominated in their thinking by fear." The expert knew what lay behind such fear. He had asked Americans what came to their minds when they heard the word "atom," and two-thirds immediately spoke of bombs and destruction. European polls gave similar results. The Atoms for Peace vision had entered almost every mind, but it seldom managed to screen out more fearsome images.[16]

The nuclear industry urgently wanted to dissociate civilian products from bombs. Publicists encouraged everyone to speak not of "atomic" but of "nuclear" power; aside from being scientifically more accurate, they hoped that would disentangle reactors from atomic bombs. Usage began to shift, with the London *Times,* for example, adopting "nuclear" in 1957. However, the replacement word was immediately applied to bombs as well as reactors, and the word "nuclear" in turn soon had a frightening tinge.[17]

Atoms for Peace publicity itself cast inadvertent shadows. For example, countless exhibits, films, and photographs showed concrete shields with thick glass windows, workers hidden in white protective suits, and mechanical "slave" hands for manipulating radioactive substances from a safe distance. Such pictures were used if only because few other images of civilian work could grab an audience's attention. The images were supposed to show how carefully experts protected everyone, but they carried a more archaic message. A French author put it precisely in his caption to the typical picture of a face peering intently through glass at robot hands opening a bottle of isotopes: didn't that symbolize "an ever more mechanized Humanity, avid to tear from Nature her eternal secrets, and at the same time fearful of the unknown that can escape from a simple bottle, a modern representation of Pandora's box?"[18]

The publicity itself hinted at the mad scientist and his golem. In addition to all those robot hands, everyone from advertising men to presidents talked about making atomic energy an "obedient and tireless servant"—a formidable creature indeed, perhaps barely under control. With civilian energy as with weapons, the word "control" was on every tongue. This reached an extreme whenever reactors were mentioned. Some of the most widely seen Atoms for Peace productions, seeking to dramatize the immense power of civilian atoms, abandoned the facts and spoke as if a reactor were ready to explode exactly like a Hiroshima bomb, held back only by a few control rods.[19]

Talk of precariously controlled powers was joined by images of victimization. When Atoms for Peace productions featured a patient prone on a hospital table, surrounded by gleaming machinery and waiting to be penetrated by rays, the picture looked strikingly like the mad scientist's victim in comic books. As if to drive home such ideas, the AEC's stock of free film was especially generous with footage of helpless white rats that scientists inspected and injected. Yet at first hardly anyone made public complaints against the new industry; the nuclear community only knew that overt public confidence was somehow accompanied by obscure fears.

The way the lines of future conflict were sketched out may be seen in the case of the Brookhaven National Laboratory. This was the flagship of the AEC's fleet of laboratories, a set of buildings sprawling across newly planted lawns behind miles of chain-link fence, out in the middle of rural Long Island. Brookhaven was not a weapons laboratory; rather, its hundreds of scientists and engineers were dedicated to driving away war with nuclear benefits. A poll of staff members found them more concerned than other citizens that atomic bombs would be used against the United States and at the same time more optimistic that industrial use would soon begin.[20]

Brookhaven's neighbors had mixed feelings. The laboratory's public relations head warned, "We know there is a lot of thought—conscious and subconscious—and we know there is a serious potential public relations problem." Although the local press had nothing but praise and the laboratory received few open complaints, the staff felt surrounded by anxiety. Airline pilots whose routes passed nearby inquired if they might be in danger of sterility; a Long Island woman wrote to ask if the radiation could make her pregnant; a man accused the scientists of impiously tampering with God's elements; neighbors insisted that weird gases were making them ill. And all this was in early 1947, before the newborn laboratory had on hand any radioactive materials at all.[21]

Since it seemed that worries about civilian atoms came only from ignorance, some laboratory experts began to take a condescending attitude. Brookhaven mounted a publicity campaign "to assure," as its director said, "that the Laboratory would not be hampered in its operations by the uninformed fears and suspicions of its neighbors." The staff gave tours of the laboratory to high school teachers, went on local radio to explain that "installations dealing with nuclear energy are actually safer than almost any industry," and spoke at hundreds of Parent-Teacher Association or Rotary Club meetings. A scientist would describe the five-foot-thick concrete walls and elaborate safety precautions around the reactor that Brookhaven was building, and would then wearily face the inevitable question from the audience: Couldn't the reactor blow up like an atomic bomb?[22]

The Brookhaven staff's obvious competence and the fact that they had their own homes nearby were powerful arguments. Here and around every other nuclear installation, public speeches and personal contacts with scientists encouraged trust. Whether for that reason, or simply because of the normal desire of people to insist that their homes and jobs are safe, polls found that those who lived near nuclear facilities tended to express less anxiety about them than did the public elsewhere. But such personal contact could never touch the great majority of citizens.

A 1950s poll found that a significant fraction of Americans associated civilian nuclear energy with things "dangerous to touch or be near." The category of perilous things might include human beings. When two workers in Texas were accidentally contaminated with a little radioactive material in 1957, neighbors shunned both men's homes and even shied away from their children, as if fearing a weird plague. It became a widespread joke that anyone who took in radioactivity would glow in the dark, like Boris Karloff playing the mad scientist who killed with a touch. Despite the Atoms for Peace campaign (or even because of it), people felt increasingly nervous about anyone who dealt with atomic energy.[23]

TAINTED AUTHORITIES

When Atoms for Peace publicity turned a spotlight on the Atomic Energy Commission and similar organizations in other countries, people did not like everything they saw. Often enough they found arrogance and foolishness. In fact most atomic officials behaved neither more nor less arrogantly and foolishly than officials in any other large organization, but they were subjected to a more intense and agitated scrutiny. Atomic authorities seemed like no ordinary mortals: they were men controlling unthinkable power, and whatever mistakes they made could not be forgiven.

The first public complaints were about secrecy. At first the AEC and its foreign counterparts had been trusted and indeed required to act on their own behind impenetrable walls. Officials often came to feel that only they knew enough to make decisions about the cosmic secrets for which they were responsible, and adopted what critics called a "father knows best" attitude. It made no difference when scientists like Oppenheimer and Teller objected on principle to the restrictions on information, nor when journalists began to cry out against censorship.

Not only scientists and journalists, those natural enemies of secrecy, but many others yearned to peek at the inner mysteries of the atom. Even the Joint Committee on Atomic Energy complained repeatedly and bitterly that the AEC was barring them from important facts. Industrialists brought similar complaints, demanding more technical information as their involvement in nuclear energy mounted. The AEC, constrained by strict laws, could never satisfy these critics, and at any rate it was inevitable that people would suspect whoever was a keeper of secrets. But some of the AEC's political decisions reinforced the suspicion. For example, in 1950, when *Scientific American* began printing an article about hydrogen bombs, the AEC forced the magazine to halt and to burn 3000 copies that had already come off the press, although in fact the article contained nothing secret. The publisher, Gerald Piel, promptly denounced the AEC in a speech to the American Society of Newspaper Editors, warning that the agency was bent on intimidating anyone who opposed building hydrogen bombs.[24]

The AEC's greatest political mistake was the highly publicized removal of Oppenheimer's security clearance in 1954. The decision only meant that the physicist was expelled from the sanctum of military secrets, but the impact on public opinion was as great as if he had been condemned for treason. When it became clear that Oppenheimer had fallen partly because he had opposed Air Force plans for large bombs, liberals began to call him a martyr in the cause of peace. Further, since Teller and Strauss had been trying for years to undercut Oppenheimer, they were branded as villains. Scientists who had once been colleagues refused to shake Teller's hand, plunging him into a bleak depression, while columnists like Joseph Alsop (who had cultivated a friendship with Oppenheimer) savaged Strauss in print. The AEC had begun as the great hope of atomic scientists and their liberal allies. But its security policies had increasingly alienated them, and the Oppenheimer affair was the last straw. The AEC had forever lost the trust of many in those groups.[25]

The Oppenheimer stories sounded like tales of a hero in cosmic battle against a villain, and such tales could be read either way; many saw Oppenheimer as the traitor while Teller and Strauss played the heroes. Subsequently the division between good and evil was projected into Oppenheimer's own soul. For example, a 1964 play that reached large audiences in Europe and the United States portrayed the physicist as a flawed martyr who confessed that in building bombs he had been "doing the work of the Devil," and a 1981 film saw him as a "Faust" who had sold his soul. Whether it was Teller or Oppenheimer or both who were called morally sick, the message was the same: the world's foremost nuclear authorities had been penetrated by evil.[26]

Other mistakes followed, above all in the battle that had raged since the 1920s between government-owned and private electric utilities. The fate of the nation's economy was still thought to be at stake in this battle, fought without quarter until leaders on each side customarily spoke of their opponents as "those bastards." Lilienthal had avoided the minefield, but Strauss marched right in bearing the flag of free enterprise. He was backed up by Eisenhower, who was determined to reverse what he called the "creeping socialism" of enterprises like the Tennessee Valley Authority. Strauss found an opportunity in the fact that the AEC's isotope separation factories made heavy use of TVA electricity. To block the TVA's growth he set up a complicated deal involving the AEC with two private utility men, Edgar Dixon and Eugene Yates. This was in 1954, an election year, and Democrats made the "Dixon-Yates affair" into a stick for belaboring the Republican administration. The issue quickly became personal as congressional hearings brought out the fact that Strauss had acted without the full knowledge and consent of his fellow commissioners. Among these Thomas Murray in particular, despising Strauss's high-handed ways, roundly criticized him.

Since Congress was just then writing legislation to make Atoms for Peace an industrial reality, a passionate battle broke out over public versus private control of nuclear power. It climaxed in a record-breaking thirteen days of Senate filibuster, until the Republicans forced through an Atomic Energy Act that encouraged privately owned reactors. Meanwhile the Dixon-Yates affair grew ever more complex and acrimonious as secret deals were uncovered. The AEC had originally seemed a haven of idealism above party squabbles, but now nuclear energy had lost its political virginity.[27]

The self-righteous Strauss had permanently alienated from the AEC many physicists, their followers in liberal circles and the press, and Democrats. He could have avoided this. In other countries such as Britain and France, nothing happened to provoke such intense suspicion of nuclear agencies. But the AEC's trustworthiness was important in every nation, for most of the world's information on nuclear matters, ranging from reactor design to Soviet bomb tests, still came from the United States. Now the AEC chairman had provoked widespread suspicion even within the nuclear community itself. To the public his activities brought constant reminders of the disquieting structure that underlay every Atoms for Peace production—the paradox of an all-powerful force for good that in arrogant hands could work great harm. It was with this burden of ambiguous imagery that the AEC would face its most severe test.

10

The New Blasphemy

The fireball heaved itself up from the sea, growing larger and larger; then it froze still. Across the hemisphere of flame appeared tiny skyscrapers in silhouette. "The fireball alone," said the narrator, "would engulf about one-quarter of the Island of Manhattan." On April 2, 1954, the public was watching the *Operation IVY* film. American television stations played it repeatedly all day, and it was soon riveting audiences around the world. I remember staring in dread fascination at the television that afternoon; I was twelve years old. Just at the time that Oppenheimer was falling, his hope for candor triumphed.[1]

Since 1945 most people had shrugged aside warnings of Armageddon as fantasies of some fairly remote future. The future had leaped closer in late 1952 with rumors about the first fusion test, and everyone had begun to hound Eisenhower for information about the new weapons. It was the *Operation IVY* film that forced his hand. For one thing, civil defense officials wanted to use it in their campaign to spur citizens to action. For another, its information was leaking into the press, as when a magazine reported that the IVY test had dug a "submarine crater several hundred feet deep and wide enough to swallow a dozen Pentagons," which was almost a direct quote from the film. Eisenhower reluctantly agreed to release a censored version.[2]

Everyone grasped what the news meant. The Hiroshima fission bomb had expended energy roughly comparable to the conventional explosives dropped on Japan in a typical week during the summer of 1945; it had been more in method than results that the explosion differed from earlier massacres. One hydrogen bomb was a thousand times mightier. Churchill told Parliament in 1955 that fission weapons "did not carry us outside the scope of human control," but with the coming of hydrogen bombs, "the entire foundation of human affairs was revolutionized."[3]

Many people only now gave serious attention to nuclear energy, and in the mid-1950s the number of publications on the subject increased sharply around the world. Part of the increase resulted from the Atoms for Peace literature that I have already mentioned. In the *Readers' Guide* and other indexes, articles dealing with civilian uses of the atom increased, as a proportion of articles on any subject, by more than half between 1948 and 1956. But still more striking was the rise in articles on nuclear weapons—a fourfold jump. A panel that surveyed 1957 press clippings on nuclear energy from around the world found five times more about bombs than about civilian uses.[4]

Bomb fiction also proliferated. In the eight years from 1946 through 1953 some fifteen significant novels had been published whose central element was nuclear energy, nearly always as a weapon; in the next eight years twice as many of the same sort were published. For the same periods the number of atom-related feature films rose from approximately twenty to eighty, and again nearly all focused on weapons.[5]

Nevertheless, the public would not give its main attention over the next decade to the fact that a hydrogen bomb could reduce several square miles to incandescent gas. What came to attract the most concern was a fact barely hinted at in the *Operation IVY* film. It turned out that a hydrogen bomb could kill people not only nearby, but hundreds of miles distant. That news opened the way for a dismaying idea, one that had long been buried within nuclear imagery but that only now emerged as a major influence on public attitudes: the idea that releasing nuclear energy was a blasphemous and guilty horror, a violation of the entire planet.

BOMBS AS A VIOLATION OF NATURE

Already in planning the first bomb test scientists had worried about the dust that the explosion would hurl into the air. Passing through the swarm of neutrons and fission fragments in the fireball, the dust would become dangerously radioactive, then drift downwind to eventually "fall out" on the desert. The problem was publicized in the 1946 Bikini tests and in such places as a 1947 *Reader's Digest* article which told how an atomic bomb could shower a city with a "mist of death." Civil defense experts warned children that after an atomic attack they should not eat a sandwich or drink a glass of milk until authorities had checked for radioactive dust.[6]

When the AEC started bomb tests in Nevada, a few citizens began to wonder about risks. In 1952 some NBC radio reporters made quiet inquiries. "AEC sources uniformly deny there is danger from these tests," an NBC man wrote privately, "but other sources, equally reputable, say that while the danger is remote, it does exist." Tests in 1953 produced so much fallout that citizens of nearby towns were ordered to stay indoors while the clouds passed, and thousands of sheep, weakened by bad weather and then exposed to fallout, died. Songwriter Tom Lehrer suggested that visitors to the Wild West should bring along lead underwear.[7]

The AEC was acutely concerned about the latent public anxiety, for the agency was determined to let nothing impede its tests. Despite the doubts of some of its own scientists, the AEC committed itself to a policy of reassurance, and mounted a public relations campaign insisting that there was no chance whatsoever of harm. To keep local citizens from worrying, some overzealous officials failed to warn them of precautions they should have taken. The press repeated the reassurances. That was not only because most reporters believed the experts who said there was little danger, but also (as the NBC man wrote) because the press too did not want to interfere with tests that were important for the national security. Pointing to the cloud of radioactive dust that drifted off from the televised 1953 Nevada explosion, network anchorman Walter Cronkite assured the audience, "It's not dangerous." Most people trusted that they would be safe so long as they were not within a few miles of a bomb explosion.[8]

Attitudes began changing after an American fusion test at Bikini on March 1, 1954. The BRAVO explosion was more than twice as powerful as planned, and the fallout went far beyond what had been predicted. At Rongelap Atoll, 115 miles downwind, gritty dust drifted from the sky. The natives were hastily evacuated but nevertheless fell sick, and for decades afterward some would have medical problems. Somewhat closer to the blast was a Japanese fishing vessel, the *Lucky Dragon*. Gray ash coated the crew, and when they got back to Japan two weeks later they were showing the classic signs of radiation injury.

The incident profoundly shocked Japan. As the *Asahi Shimbun* reported, the Japanese were "made to feel acutely once again the horrors of an atomic bomb." Adding to their dismay was Strauss's refusal to admit that the AEC had made any mistake. The fishermen's skin sores, he announced, were "due to the chemical activity of the converted material in the coral rather than to radioactivity." When the crew member who was most weakened by radiation died, the AEC insisted he would have lived if he had received perfect medical care, which was beside the point.[9]

Respected journalists like the Alsop brothers accused the AEC of feeding the public pap. These were the same months when the AEC was losing the trust of many Americans through Strauss's handling of the Oppenheimer and Dixon-Yates affairs, but for the rest of the world it was chiefly evasions about fallout that poisoned the Commission's reputation.

Meanwhile fine radioactive dust, catapulted into the stratosphere by the explosion, drifted on the high winds. Independent scientists who had already been worrying about the effects of radioactivity in the near vicinity of a bomb explosion now warned that a test might cause birth defects among people hundreds or even thousands of miles away. But month after month the American government issued no more than scraps of ambiguous information mixed with reassurances. The fact that hydrogen bomb fallout could devastate entire territories was treated as a military secret, for it undercut prevailing ideas on nuclear war and civil defense; the government scarcely knew what to say to itself, let alone to the world public. In the absence of reliable facts, indeed with the world's main source of atomic information largely discredited, anxieties could only grow.[10]

Open fear erupted among the Japanese when they learned that a few tuna fish caught in the Pacific had excess radioactivity. Millions stopped buying fish; in terms of diet and emotions it was as if Americans did not dare to eat beef. Japanese government agents with Geiger counters selected fish for destruction by the ton, and fishermen came near bankruptcy. At the peak of the scare some parents refused to let their children swim in the ocean.

The Japanese had reason to feel endangered. Thanks to the exquisite sensitivity of Geiger counters, their scientists could track faint radioactivity from the BRAVO explosion for thousands of miles through the wandering currents of the Pacific. In some cases the isotopes became concentrated as they worked their way up through food chains, so that a fish could be much more radioactive than the surrounding water. A person who ate many such fish might have an increased risk of contracting cancer. Scientists began to warn the world of a new reality: a human act at one locality could physically affect the environment across vast distances.[11]

Uneasiness took some curious turns. For example, a month after the BRAVO test hundreds of citizens in Seattle telephoned the police to complain that a mysterious agent, probably from the bomb, was making little pits in the windshields of their cars. Many covered their windshields with newspaper for protection, and the mayor made an emergency appeal to President Eisenhower. Few believed the real explanation: windshields ordinarily collect tiny pits over time but usually nobody notices. Anxiously scrutinizing their surroundings, the citizens were projecting into their perceptions a novel fear.[12]

The new fear was this: bizarre nuclear damage could strike anywhere in the world. In 1956 presidential candidate Estes Kefauver declared that hydrogen bombs could "right now blow the earth off its axis by 16 degrees"; in 1957 Soviet Premier Nikita Khrushchev reportedly boasted of owning a bomb that could "melt the Arctic icecap and send oceans spilling all over the world"; a British screenwriter planned a movie about nuclear tests that set the planet adrift toward the sun. Many people imagined that a bomb test could somehow cause earthquakes thousands of miles distant, and even the AEC took that seriously enough to seek reassurance from professional geologists.[13]

The most persistent rumors were about weather. In 1951 the AEC began receiving letters from all over the world saying that bomb tests were upsetting weather patterns, and beginning in 1953 the world press often discussed such complaints. Farmers blamed the AEC for heat waves or cold spells, cloudbursts or drought. People in Moscow said American bombs caused the late arrival of spring; a poll in France found a third of the public suspicious that recent bad weather was due to the tests, and another third flatly convinced of it. In April 1955 the Joint Committee on Atomic Energy, in its first public hearings devoted to global fallout, gave as much attention to weather as to health effects.

The Joint Committee joined the AEC in denying there were problems with either weather or health, and the press repeated the reassurances. Well-informed citizens tended to set aside problems of earthquakes, weather, birth defects and all the rest together as fantasy. Yet many remained uncertain. The irrepressible rumors pointed to thoughts stirring below the level of informed discourse.[14]

Whether true or false, the stories taken as a group made symbolic sense. To say that nuclear bombs were polluting fish or causing birth defects, to say that they could disrupt the weather or set the planet's axis askew, in each case was to say that nuclear energy violated the order of nature.

This idea was bound up with one of the strongest of primitive themes: contamination. In most human cultures the violation of nature, and forbidden acts or things in general, have been directly identified with contamination. According to the anthropology theorist Mary Douglas, whatever is "out of place," whatever goes against the supposed natural order, is called polluting; dirt seems right in a vegetable garden but disgusting in vegetable soup. People explained their taboos against such things by saying they would cause grotesque harm. For example, a woman who carried out an action inappropriate to her sex would bring down sickness on herself or even on her whole clan. This fitted a familiar psychological pattern: as everyone knows, the bad small boy who defies the proper order of things is especially liable to make a filthy, damaging mess, defiling himself and perhaps others too.

The concept of contamination had an important social dimension. Some anthropologists noted that the disgust and fear associated with taboos would fasten especially on a person who was "out of place" in the accepted social structure, such as a person who broke caste rules or made unreasonable demands. It was common to call that sort of person a witch. The most straightforward belief was that witches attacked the inner purity of others by projecting a deadly substance right into their enemies' bodies. More generally a witch was someone who could prevent the conception of babies or bring an unseasonable storm, someone who above all violated the proper scheme of things, perverting the community and nature itself.

Much the same was said of nuclear bombs and the men who made them. Scientists and nuclear officials made particularly apt targets for suspicions about the disruption of childbirth, the weather, and so on, for at least some of that sounded plausible. Equally important, there was a long tradition of accusing science and technology of violating the order of things—which came close to saying that they brought pollution.

A striking literary expression of these feelings, largely independent of scientific fact, had been published back in 1908 by H. G. Wells in one of his most widely read novels, *Tono-Bungay.* The protagonist, a technology-minded young man corrupted by greed, manufactured fraudulent patent medicines for the new mass market. Toward the end of the story he set out to steal a stock of radioactive ore. The ore was surrounded by a lifeless zone extending for miles; worse, it exerted a baleful influence on all who approached it, afflicting them not only with illness but also with feelings of hatred and futility. Wells used the "incalculably maleficent" ore as a metaphor for what was the main subject of his book: the disregard for time-honored values and customs that was undermining England's social structure. Radioactivity itself he called a "disease" of "debased and crumbling atoms . . . It is in matter exactly what the decay of our old culture is in society, a loss of traditions." Thus the corruption of the protagonist was reflected in the minerals he used.[15]

In sum, the powerful theme of contamination was historically, psychologically, and socially in a good position to link up with nuclear energy. Of course the linkage does not explain why people were horrified by nuclear bombs; it only indicates why the emotion was expressed in particular ways and linked with particular images. If none of these connections had existed the public would still have been horrifed because of the plain facts of what the bombs could do. But it happened that some of these facts strengthened the association between nuclear energy and uncanny pollution.

Most important was the fact that radiation could cause genetic defects. This fact resonated with certain old and widespread ideas about contamination. Traditionally, defective babies were a punishment for pollution in the broadest sense, violations such as eating forbidden food, looking at something that should not be seen, or breaking a sexual taboo. Thus deformed children, according to a sixteenth-century book, "may come by reason of inordinate or unkindly copulations, when the seed is not conveyed into the due and right places." Incest in particular, according to many groups both primitive and civilized, would almost inevitably bring unnatural progeny. Sometimes it was not human acts but the radiated power associated with sexuality itself that created prodigies; the Maya thought that eclipses sent down demonic forces to deform babies in the womb. This primitive belief that contamination, perhaps involving forbidden secrets and rays, could cause strange births seemed to be confirmed in the facts of radioactivity.[16]

The association between nuclear energy and pollution was strengthened by another fact, that radiation could cause cancer. By the 1950s the word "cancer" had come to stand for any kind of insidious and dreadful corruption. Demagogues labeled Communists, prostitutes, bureaucrats, or any other despised group as a social "cancer," while people afflicted with real tumors came to feel shamefully invaded and defiled. Thus the symbolism of cancer coincidentally reinforced the idea that nuclear weapons meant odious violation of the proper order of things.[17]

It was such coincidental facts, reinforced by social and psychological mechanisms which also happened to fit the case, that forged the association between nuclear weapons and the complex theme of contamination. Nobody set out deliberately to make these associations; they came to many people at once, following readily from all that had been said earlier about atomic energy. Later, of course, many groups would exploit the associations to give emotional force to political propaganda, but at first the context was only generalized disquiet about the way the world was going.

On occasion the ideas were openly invoked. As early as 1950, liberal newspaper and radio commentators had exclaimed that hydrogen bombs, wrongfully exploiting the "inner secrets" of creation, would be "a menace to the order of nature." On receiving news of the BRAVO test, the conservative publisher William Randolph Hearst told millions of readers that such explosions "could cause dangerous changes in the orderly processes of natural law." Even Pope Pius XII, in Easter Sunday messages heard over the radio by hundreds of millions on every continent, warned that bomb tests brought "pollution" of the mysterious processes of nature.[18]

All the themes could easily mingle, as shown in a letter from twenty-one pastors of Boise, Idaho, petitioning President Eisenhower to stop testing bombs on moral grounds. They spoke of the Seattle windshield pits (seen also in Boise), the risk of setting off a chain reaction that would engulf the world, the apocalypse of the Book of Revelations, and the forbidden fruit of knowledge. The mixture might seem confused, but in fact it summarized an ancient and tightly structured cluster of ideas.[19]

Officials dismissed all such talk as superstition, and at first most educated people followed their lead. When columnists, statesmen, or clergy called bombs unnatural, they offered no more than a scattering of vague phrases which seemed to have little connection with scientific reality. Besides, during the early 1950s the specific idea of contamination was scarcely visible amid all the talk about bombs triggering storms or earthquakes. It was not in newspapers or speeches or sermons that attitudes about contamination were most plainly emerging, but in another form of discourse, a more popular form where myth became ferociously visible.

RADIOACTIVE MONSTERS

What was it that bomb tests brought from the deeps of the Pacific in 1954, radioactive and threatening, to terrorize millions in Japan? That is, aside from tuna fish? The answer was Godzilla, the 400-foot prehistoric reptile that stomped Tokyo flat in the first Japanese movie to achieve international financial success. The world had seen dozens of earlier films about cities wiped out in a natural disaster or a war, or about prehistoric beasts, or about monstrous results of scientific experiments, but *Godzilla* combined all these themes. The most disturbing elements of modern imagery were coalescing into solid form and connecting up with bomb tests. Besides reflecting anxiety about hydrogen bombs, the imagery would load new meanings upon nuclear energy.[20]

The connection with uncanny nuclear energy became entirely clear in the 1954 movie *Them!* Killer ants the size of buses crawled out of the desert near the Trinity test site, "a fantastic mutation," as the film's scientist explained, "probably caused by lingering radiation from the first atomic bomb." Moviegoers found that plausible enough to put them into a cold sweat. After the Army exterminated the creatures, an official in the film worried that if such horrors followed the first test, what would come from all the bombs exploded since? The answer soon appeared in theaters.[21]

The financial success of *Them!* inspired a crowd of movies about oversized creatures engendered by radiation, including a giant radioactive squid, giant leeches, giant scorpions, a tarantula the size of a house, and a pair of 25-foot crabs. Never was so much faith shown in the idea that radiation could promote growth. As if to verify the Atoms for Peace promises, one low budget thriller described a radioactive substance invented to enlarge vegetables; insects ate the produce and an army of gigantic grasshoppers advanced on Chicago.[22]

These were no detours of the popular imagination but well-traveled highways. Millions of people would pay to see the cheapest production if it had a radioactive creature in it. Meanwhile comic books also impressed millions of youngsters, and adults too, with tales of monsters as well as of contamination, world destruction, and perilous rays, all usually associated with atomic energy. For example, after 1946 the immensely popular Superman found himself susceptible to weird "Kryptonite" rays, and his authors revealed that he and Kryptonite had both originated on a planet blown up in an atomic apocalypse. Similar themes proliferated in science fiction stories, now breaking out of their earlier limited market. A sociologist estimated the American science fiction readership in 1953 at six million people, mostly well-educated, and the editor of *Astounding* calculated that his magazine was read by a third of all young American technical personnel. Although the fountainhead for science fiction was the United States, Europeans and Japanese translated great quantities and began to write their own; only in the Soviet Union were scientific monsters strictly excluded.[23]

For the specific new genre of atomic monsters, the central symbolism appeared right in the first minutes of the first film, a 1953 American production where scenes of an atomic explosion faded into scenes of *The Beast from 20,000 Fathoms* as it began to emerge from arctic ice. Critics quickly recognized that such creatures were personifications of nuclear weapons—or rather, in the popular imagination the creatures filled a vacant space where the public declined to see real weapons.[24]

The films scarcely referred to nuclear war itself, except indirectly. Not one of the countless futuristic movies of the 1950s dared to show an audience the realistic results of atomic bombing. At most a film might mention nuclear wars on some faraway planet, or tucked out of sight as a past event. Scriptwriters preferred symbolic war. For example, an American film treated *The Deadly Mantis* exactly as if it were an approaching Russian bomber, tracked by radar and Air Force interceptors; similarly the Japanese film *Rodan*, about flying monsters hatched as usual by bomb tests, showed citizens anxiously scanning the skies for bomber-like contrails, then running in panic as air raid sirens howled. Almost every monster film showed mass evacuation and collapsing cities, as well as troops mobilizing, officers issuing commands, and scientist advisers hovering about, just as in the real Cold War.[25]

In a famous essay Susan Sontag suggested that audiences, not daring to think about nuclear war, projected their fear onto a monster in order to be able to defeat it vicariously.[26] However, the movies did more than simply deal with people's fear of being bombed. The creatures introduced a new and peculiarly horrible imagery of their own. The most disturbing new thing about the radioactive monsters was their remoteness from humanity. Unlike the Wolfmen and Draculas of an earlier generation, most 1950s creatures were literally inhuman, devoid of intelligence and feelings. Now the world was menaced by things as impersonal as *X The Unknown*, no living animal at all but a sort of glittering sludge that oozed about in blind search of isotopes.[27]

Curiously, a reciprocal gap appeared in the delusions of insane people, for they too failed to connect nuclear energy with humanity. Psychologists noted with surprise that postwar paranoids, unlike earlier ones with their X-rays and radium rays, rarely used atomic rays in their fantasies about "influencing machines." It might now be television that sucked thoughts from their heads and injected strange ideas (which seemed almost plausible), or perhaps radar, but rarely atomic devices. Nor were nuclear bombs themselves, despite their notoriety, a common fantasy object among psychiatric patients.[28]

Did filmmakers and insane people alike see nuclear weapons as altogether remote from human emotion—or did the weapons have a human meaning too overpowering to face? The answer can be suggested by putting together a number of stories in order to discover a common structure in the places where they overlapped. Consider the sludge in *X the Unknown:* it was inspired by a disgusting mass in an immediately preceding film from the same studio, and there the blob had indeed once been a man, an astronaut who degenerated after probing into forbidden realms of radiation. Many other films began with people scientifically victimized, turned into inhuman creatures rejected by all the world; even beasts like Godzilla could seem like pitiful victims at the end. The tradition behind this theme could be traced in a pioneering 1951 movie, *The Thing,* where a radioactive vegetable creature lurched about looking remarkably like Boris Karloff playing Frankenstein's monster. In short, the new radioactive horrors were descendants of the clay golem and the alchemist's chaotic mass. Therefore, they may well have held the same terrifying meaning: violation of forbidden secrets punished by loss of true life and feeling, a takeover by the most inhuman and bestial parts of oneself, a wrong turning in the attempt at rebirth.[29]

The likelihood of a connection with these old themes is strengthened by the fact that most of the films combined victimization with references to abortive rebirth and forbidden secrets. Godzilla, said the script, was temporarily "resurrected" by hydrogen bombs; more graphically, the Thing, the Beast from 20,000 Fathoms, and the Deadly Mantis came to doomed life when they were melted out of a literally frozen state. Meanwhile giant ants or blobs emerged from dark, secret holes in the earth. Womb symbolism was most explicit in *Rodan,* where scientists went down a tunnel to an underground room and saw a monster emerge from its egg, what the actors called a forbidden sight. If moviegoers did not already associate radioactivity with such primal problems when they went into the theater, they would do so by the time they came out.

A related meaning could be connected directly with radioactive birth: in an old sense of the word a "monster" was a malformed offspring. If bomb tests violated the natural order, then perhaps it was only to be expected that monstrosities would emerge from the womb of the contaminated earth. Movies did not dare touch the question of defective babies directly, but since the 1930s science fiction magazines, less prudish, had run stories about children deformed by radiation. By 1952 the idea was so hackneyed that the editor of a major pulp magazine begged authors to stop, for his desk was buried under tales of atomic bomb children with too many fingers or heads. Nobody suggested that the film monsters were likewise a symbolic progeny of the breaking of taboos, but the idea was clear on a more general level. As the film critic John Brosnan put it, the 1950s monsters symbolized a retribution for "man's technological tampering with nature . . . the same message propounded by the 'mad scientist' films of the 1920s, 1930s and 1940s."[30]

The echoes of mad scientist themes in postwar movies added yet another idea onto the imagery associated with nuclear weapons, an idea that at this time was rarely expressed directly. With Senator McCarthy's admirers scrutinizing every production for hints of disloyalty and with Cold War heroes triumphing daily on television, science fiction allegories were one of the few ways to express doubts about the technological defenders of democracy. The science fiction hero would defeat his monster in the end, but the end often carried a question mark, such as the suggestion in *Them* that more bomb tests might well create further monsters.[31]

Some prewar mad scientist stories could be interpreted as showing victimization not just by science but by cold-hearted authority in general, and that theme too became common in the 1950s movies. Aimed chiefly at teenage audiences, they might show adolescents struggling less against the monsters they discovered than against unmindful parents. "Dad, you've got to *listen* to me or everybody will die!"—a plea that was also a threat. Critics noticed that the monsters themselves, raging and lonely, seemed like rejected children. An especially direct pair of movies featured an officer irradiated at a bomb test who grew into precisely a gigantic, outcast, violent baby.[32]

The violent victim did not have to be connected with nuclear energy at all. Indeed, some oversized spiders, caterpillars, and so forth were produced not by radiation but by biological or chemical tampering. However, atomic rays were by far the most common plot device, and even when other pseudo-scientific explanations were invoked, the film might suggest, or the audience might spontaneously assume, that radioactivity was somehow involved. As a critic remarked, one could no longer believe in werewolves and vampires, but "I am still prone to terror at authentic-looking young atomic scientists talking what may well be scientific nonsense." In consequence, when the Geiger counter began to clatter and the Professor whispered, "It's radioactive!" audiences learned to expect something slimy and vicious to crawl out. These stories not only exploited, but thoroughly reinforced, the association between nuclear radiation and monsters, dragging in all the old implications of obscene contamination and victimization.[33]

On the surface most people continued to respect atomic scientists. The postwar movies scarcely ever featured a mad scientist in person but almost always had a role for a clean-cut (if rather odd and emotionless) scientist hero. If the Beast from 20,000 Fathoms represented nuclear weapons, then it was easy to see what was represented by the white-suited scientist who destroyed the monster with a new isotope—good atoms mastering bad ones. But the neat plot endings were less impressive than the central image. Overshadowing the clever scientists loomed their bastard offspring, unkillable despite all, 400 feet high and growing.[34]

BLAMING AUTHORITIES

If the natural order was violated somebody must be to blame, and authorities could not escape that blame forever. Censure was all the more certain to settle on them when bomb tests and fantasy movies brought pangs of anxiety. For anxiety is normally accompanied by guilt and shame, sometimes redirected against oneself but sometimes directed angrily against others. Such denunciations could distort political discussions, but the same emotions could also connect with a moral impulse akin to hope, a feeling that things should have been otherwise in the past and must be changed in the future. All these feelings became part of nuclear thinking after the advent of hydrogen bombs.

The pioneers in converting blame into action were certain survivors of Hiroshima and Nagasaki. Like people anywhere who helplessly watched family and neighbors perish, they had tended to feel in the wrong, unworthy of surviving when others did not. Moreover, from the outset the Japanese survivors had felt themselves to be participants in a uniquely horrible violation; simply by being present when a city changed instantly to a charnel house, they felt shamefully contaminated. Those who tried to forget the experience were forcibly reminded in the early 1950s when dozens of survivors came down with symptoms reminiscent of radiation injury: bleeding under the skin, weakness, death. This was leukemia, which arises after a few years in a small fraction of irradiated individuals. Many Japanese confused leukemia with miscellaneous other pathologies, including normal feelings of weakness. By the mid-1950s, when one or another survivor of the atomic bombings died, perhaps of some ordinary illness, newspapers in Japan and elsewhere might announce the cause of death as a new and uncanny affliction, "A-bomb disease." Survivors feared that they were too contaminated to have healthy children, and hesitated to admit they had been exposed to radiation. Robert Lifton reported that every one he interviewed showed a "sense of impaired body substance."[35]

So many thousands in Hiroshima and Nagasaki felt the same way that they were able to group together, finding hope in mutual emotional support and practical action. Some groups began to demand special medical aid or financial compensation from the American or Japanese governments; others campaigned for international peace. Both the people seeking a stipend and those seeking universal amity had reason to emphasize the special horror of their experience. Besides explaining their sense of shame, that could give them the right—more than the other millions of war wounded—to make demands on the world.

When atomic bomb survivors directed their feeling of shameful violation outward, one of their first targets was American scientists. Some resented the American and other doctors who came around to study their illnesses dispassionately. A few suspected that such people were not only unfeeling but utterly inhuman: had the bombs been dropped not for military reasons but as a scientific experiment to see what they would do to people? Most Japanese denied that, seeing the bombings only as an inevitable tragedy of war; such was the official view of the Japanese establishment, which had immense atrocities of its own to bury in the past and an alliance with the United States to maintain. The Japanese Communist Party and the Left in general had more reason to launch a campaign of moral censure. American guilt for the atomic bombings became a theme of their numerous speeches, movies, and books, and it was soon common for anti-American writers around the world to call the bombings "a horrible experiment." In most nations this polemic of blame could be dismissed as a propaganda tool, but in Japan there remained the bomb survivors themselves, speaking with a conviction that plainly went beyond politics.[36]

The *Lucky Dragon* incident and the tuna contamination scare unlocked anti-American feelings that since the war had lain hidden within most Japanese. Now not only the survivors of Hiroshima and Nagasaki but all Japanese could see themselves as victimized by a nuclear experiment. When American scientists came to study the *Lucky Dragon* crew, Japanese newspapers cried, "We are not guinea pigs!" To underline the nation's vulnerability, fallout from Soviet bomb tests too came drifting across. "I hear Japan is a sort of valley into which all radioactivity flows," complained a character in a 1955 film. Meanwhile descriptions of the 1945 bombings, which had been suppressed during the American military occupation, came into the open, while the *Operation IVY* film and other hydrogen bomb news convinced many Japanese that they would perish if war came. In an initiative that was only loosely connected with the bomb survivors, Japanese housewives concerned about fallout began to circulate petitions against further hydrogen bomb tests and soon had tens of millions of signatures opposing all nuclear weapons. A mass political movement was taking shape. A large rally in Hiroshima on the 1955 anniversary of the bombing drew dozens of representatives from abroad and caught the eye of the world press, marking a new departure in world politics.[37]

Meanwhile talk of hydrogen bomb tests revived American and European interest in the Hiroshima and Nagasaki victims and the moral questions they stood for. At the time of the rally in Hiroshima, groups in New York City held a sympathy rally. Around the same time the veteran world federalist Norman Cousins cooperated with Japanese survivor organizations to bring twenty-five "Hiroshima maidens" on a well-publicized visit to the United States where they got plastic surgery for their disfiguring scars—a token reparation. Although the majority of Americans continued to insist that they felt no guilt over the atomic bombings, an important minority were making it plain that *somebody* should feel guilty.[38]

The press seized on stories about one or another remorseful atomic scientist, even when the scientists themselves insisted they felt little guilt. Widespread rumors held that the pilot of the plane that bombed Hiroshima was frantic with remorse, that he had committed suicide, that his entire crew had gone mad. Paul Tibbets, the pilot, remarked that on each anniversary of the bombing since the mid-1950s he would get telephone calls from reporters checking up on his sanity. The reporters seemed disappointed when they learned that Tibbets and his crew were all leading ordinary lives with no regrets.[39]

From 1957 on, respected newspapers and magazines around the world told of Claude Eatherly, the "Hiroshima pilot" who fell into insanity and crime because his conscience tortured him for helping to destroy the city. Intellectuals in many nations wrote articles, poems, and plays about Eatherly, and reviewers endorsed his moving book, *Burning Conscience*. In fact, Eatherly had piloted only a weather plane and never witnessed the bombing. At first his only regret was that he had not won fame by dropping the bomb himself; only in 1956, when his burglaries got him into trouble and he sought special consideration as a disabled veteran, did he begin to claim that guilt had driven him mad. Eatherly's case showed less about the unstable airman than about a curious tendency of people to believe what they heard about the bombing—provided it showed guilt.[40]

The same tendency turned up in arguments about the first decision to use atomic bombs. Because of endlessly repeated allegations, millions around the world came to believe it was proven that Truman had slaughtered defeated Japanese for no reason but to cow the Soviets. Debate over this issue continued for decades, backed up by scholarly books and documentary films. The arguments were remarkable for their passionate tone and limited scope. All but a few debaters ignored the decisions behind the earlier massacres which had set the key precedents; scarcely anyone asked about the morality of the deadly American blockade. The debate had relatively little to do with the real history of 1945—the decision to destroy a few cities by atomic bombs rather than by fire bombs.[41]

Few Americans and Japanese who had personally experienced the war saw the bombing of Hiroshima as an exceptionally guilty act; the arguments of blame were most attractive to people who showed little interest in the war as a whole. The real debate was over whether or not authorities, or at least American authorities, were moral monsters, and whether or not nuclear weapons were immeasurably more loathsome than other ways of attacking people. The question on every mind was not, "Should atomic bombs have been exploded in 1945?" It was the much more interesting, "May hydrogen bombs be exploded now?" Fear of hydrogen bombs was at the root of all the anxious blame, and of the rising tendency to see authorities as people who would not scruple to contaminate the world.

11

Death Dust

If it was wrong to set off bombs, then why not just stop doing it? This idea, like almost every idea involving nuclear energy, was raised first in the scientific community. The issue was brought to the highest official level in 1954 by AEC Commissioner Murray, who privately asked Eisenhower to call a moratorium on tests of large fusion devices. For one thing, Murray wrote, an international moratorium might forestall the development of usable hydrogen bombs. For another, halting tests could be a small first step toward mutual trust and disarmament. Soon physicists and others made the same proposal in public, for the same reasons: a test ban would serve as both a practical and a symbolic obstacle against the spread of weapons.[1]

There was another argument for halting bomb tests, one that struck at a different level of consciousness. A few experts warned that fallout from the tests might cause leukemia or other proven forms of radiation harm, not only near the explosions but around the earth. Some people began to suspect that the tests might cause "A-bomb disease" everywhere. On hearing about the *Lucky Dragon*, India's Prime Minister Jawaharlal Nehru exclaimed, "How can we be sure that our children may not gradually go blind or contract some internal disease?" Joined by other leaders of emerging nations, he called for an immediate halt to the tests. Communists endorsed the idea as an adjunct to their long-standing campaign for a ban on nuclear weapons.[2]

Citizens who followed the news with care began to understand that a bomb brought three sorts of radiation hazards. First were the direct rays from the fireball; these were what had injured thousands at Hiroshima and Nagasaki. Second was the fallout of heavy radioactive ash that an explosion could scatter hundreds of miles downwind; so far such ash had visibly injured only the Rongelap Atoll natives and the *Lucky Dragon* crew, but in a hydrogen bomb war it would be a major killer. Last was the tenuous dust that a hydrogen bomb lofted into the stratosphere to drift about the globe for years, settling almost indetectably into people's air and food. It was this last type of radiation—the faintest, most widespread, and least understood—that became the focus of world debate.

At first most Americans, Europeans, and Russians were reluctant to believe that the bomb tests endangered world health, but when scientists began to speak, attitudes began to change. A crusade gathered against the tests and the authorities who sponsored them. As a response to hydrogen bombs this movement would become more powerful even than Atoms for Peace, if not in immediate practical terms then in its permanent effect on public imagery.

CRUSADERS AGAINST CONTAMINATION

The first risk of global fallout that scientists consistently held up for debate involved the most primal of threats, genetic damage. Scattered early warnings converged at the 1955 Atoms for Peace conference when biologists attracted attention by insisting that even the tiniest amounts of radiation could injure human genes. The most famous paper prepared for the conference, however, was never delivered there, for the AEC had refused to give Herman Muller a forum. He therefore made one for himself in the world press.

Muller, a Nobel Prize winner for his 1920s experiments with X-rayed flies, looked like a respectable family doctor with wire-rimmed glasses, thinning white hair, and a sensible voice that seldom revealed the passion of his commitment to furthering the human race. As early as 1933 he had reminded X-ray specialists that the future of the race was embodied in minuscule quantities of germ plasm within each person, scarcely totaling one spoonful for the entire world; the people of each generation were "custodians of this all-important material," he said, and should not "contaminate" their germ plasm with indiscriminate use of radiation. Fallout from nuclear weapons brought a new source of contamination, and Muller used his fame to drive the point home. He warned that each bit of radioactivity would add to the "erosion" of our precious genetic material.[3]

The message was reinforced by a study that the National Academy of Sciences of the United States published in June 1956. The Academy's group of experts denied that the weak radioactivity from bomb tests could cause cancer, and pointed out that in any case the fallout was slight compared with all the other radiation that bathed the public, such as the widely misused X-rays. However, the group's Genetics Committee, with Muller among them, announced that any amount of radiation, no matter how small, would cause some genetic damage. Newspapers displayed that conclusion on the front page and in many parts of the world the warning seemed more convincing than all the reassurances in the rest of the Academy's report.[4]

Many other scientists, especially those working for the AEC, disagreed with Muller, finding no evidence that the extremely dilute global fallout could cause harm. The Eisenhower administration in general and Strauss in particular, convinced that national security depended on bomb tests, suspected that complaints about radiation perils reflected nothing but Communist propaganda. AEC officials therefore issued a barrage of reassurances, and during the mid-1950s the American press tended to go along. The only opponents were a handful of scientists and their followers, writing mostly in small intellectual journals like *The Nation* and Norman Cousins's *Saturday Review*. Mass-circulation magazines showed their loyalty with an occasional soothing article, as when *U.S. News & World Report* devoted six of its slick pages to insisting that there was "not a word of truth in scare stories" about deformed babies. More commonly during the mid-1950s, the press scarcely mentioned fallout at all amid the flurry of Atoms for Peace articles.[5]

Under the calm surface, however, strong forces were aligning with each side of the bomb test issue—political forces. When Strauss clashed with Democrats over private ownership of nuclear power and other matters, it became clear that the AEC chairman scarcely distinguished between factual statements and special pleading for Republican policies, until even magazines that supported bomb tests sometimes questioned whether Strauss was telling the truth about hazards. One reporter remarked in 1955 that the AEC's critics might be right about fallout—"Death Dust I like to call it, because I think people understand it a little better." Although the controversy had scarcely begun, in his phrasing the reporter was already choosing sides.[6]

The sides were sorting themselves out along traditional political lines. As early as 1952 a sophisticated survey had found Americans diverging on whether it was "fair" to use atomic bombs, with conservative Republicans mildly in favor, liberal Democrats mildly opposed. It was no coincidence that the leftist *Nation* worried about bombs and fallout while the rightist *U.S. News* did not. Somehow the issue touched upon the same fundamental beliefs that determined political choices.[7]

Within the AEC itself Murray, the commissioner most sympathetic to the Democrats and most outraged at Strauss's high-handed ways, openly declared that hydrogen bomb fallout could eventually endanger the human race. In April 1956 he made public his call for a moratorium on tests of large fusion devices. Nine days later the idea was endorsed by Adlai Stevenson.

Stevenson's presidential campaign was then in full swing, so his endorsement brought the bomb test issue into the limelight. In harsh language he warned the nation against radioactive isotopes—"the most dreadful poison in the world." However, polls showed that Americans supported their government and its bomb tests by a heavy majority, and it did not help Stevenson when the Soviet government openly took his side. Nevertheless he kept returning to the issue, impressed by the outcries of Hiroshima victims and convinced that hydrogen bombs were of central importance for the election and beyond. When Muller and other prominent scientists, a dozen here and a dozen there, backed his warnings, some other citizens became convinced. Eisenhower was showered with mail opposing bomb tests.[8]

The hundreds of citizens who wrote the President typically said that the scientists' warnings about harm to children raised a moral problem. The letter writers were not only concerned for their own children, but condemned the immorality of AEC "Frankensteins" (as some put it) who endangered innocents around the world. The writers often mentioned their Christian convictions, and indeed many were ministers; the rest were largely from the professional and intellectual classes. Two-thirds of the writers were women. These people did not have enough votes to elect Stevenson, but they would not forget what he had told them.[9]

The next call rang out from a wholly unexpected quarter. It was the first political act of Albert Schweitzer, the saintly physician-philosopher who had set aside his worldly career to minister to the sick in Africa. Norman Cousins, on his way back from Japan and the Hiroshima Maidens, had made a pilgrimage to Schweitzer's jungle hospital and persuaded the sage to speak out. In April 1957 Schweitzer's attack on bomb tests, highlighting the threats of cancer and genetic damage, was read over the radio in many nations.[10]

A number of other humane and strong-willed men began raising the banners of protest. Most effective of all was Linus Pauling, a biochemist at the California Institute of Technology. Pauling's power came partly from his stature as an outstanding scientist, but still more from his personality. With his finely turned sentences and his rubbery features that stretched into an infectious smile, he was an irresistible persuader. Constitutionally disinclined to trust officials and at the same time deeply concerned about nuclear war, Pauling began to think about fallout. He concluded that it was definitely causing many deaths already.

His colleagues often looked askance at Pauling's self-confidence, for while other scientists found the perils of fallout uncertain, he cast intellectual caution aside and presented his conclusions as obvious truths. The AEC's insistence that global fallout was perfectly safe seemed no better, however, and some of Pauling's colleagues were becoming as disgusted as he was with official reassurances. They encouraged him to speak more and more boldly. In May 1957 he drew up a petition against bomb tests and within two weeks got the signatures of 2,000 American scientists.[11]

Pauling's skill as a publicist became apparent in his calculations, which boasted a precision that conventional scientists would never dare to claim. In early 1958 he announced through newspapers, magazines, and television that the tests conducted so far would probably cause a million cases of leukemia, and also perhaps five million genetically defective children. In a scholarly article that immediately became famous he predicted that each year of continued testing could mean another 55,000 defective births and 100,000 stillbirths. Others took up this imagery of numbers, speaking with all the authority of statistics about thus and so many thousands or millions of deformed babies. The numbers suggested a dreadful picture, already popularized by numerous science fiction stories: a world infected with grotesque mutants.[12]

Although the consensus of scientific opinion did not go nearly so far, it was turning against fallout. Influential journals close to the scientific community, such as *Science* and *Scientific American*, gradually became more willing to call bomb tests a risk to health. The shift among experts was displayed to a wider audience by well-publicized hearings that the Joint Committee on Atomic Energy held in 1957.

The hearings deflated AEC optimism, for even the Commission's own scientists admitted that there was at least a chance of cancer from fallout, and that in theory some level of genetic harm was certain.[13]

Nevertheless, many experts insisted that any harm from fallout was too slight to worry about. The most widely heard statements came from Edward Teller. He explained over and over that global fallout gave very low levels of radioactivity, and that harmful medical or genetic effects of such faint radioactivity had never been detected. For all anyone knew, he said, a little extra radiation might even be good for people. Either way the effects would be insignificant. Teller noted that brick houses were more radioactive than wooden houses, emitting ten times more radiation than people got from global fallout, but nobody was proposing to tear down brick houses. Radiation was all around us, for example as minute traces of radium in drinking water; some locales had water supplies that gave residents more radiation than they got from bomb tests, yet nobody was demanding a change in water supplies. Teller could not understand how any rational non-Communist would object to bomb tests unless from ignorance of the facts.[14]

The soothing words of Teller and various other scientists failed to smother public worries. Eisenhower got more and more letters opposing bomb tests, not only from pacifist and religious groups but from labor union locals, businessmen, mayors, and informed people in general. Authors joined the campaign, producing dozens of stories and novels specifically opposed to bomb tests. In 1955 polls had found that only one-sixth of Americans could explain what "fallout" meant, but by 1961 the fraction had risen to more than half, indeed to four-fifths of the college-educated, a high level of knowledge for a new technical term. By the late 1950s a majority of informed Americans and still more in other nations believed that fallout held real dangers. In 1958 Lilienthal wrote privately to another former AEC leader, "I don't remember any instance in which a major public body has lost public confidence in its integrity to the extent that has happened to our old friend the AEC." He and his friend agreed that "we are really in for trouble."[15]

Doubts about government bomb tests were stirring even in the Soviet Union. Although Soviet officials discouraged all talk of technological danger, some biologists privately discussed genetic risks. Backed up by the statements of foreigners such as Schweitzer and Pauling, they won quiet support from leading nuclear physicists and especially from Andrei Sakharov, who wrote Khrushchev to warn against the tests. These private concerns became public when the Soviet government backed up its peace propaganda with a 1958 announcement that it would cease all tests. Now the Supreme Soviet openly stated that testing meant "the poisoning of human organisms." At once Sakharov and other scientists began to publish, in the Soviet press, Pauling-style warnings about the thousands or millions of babies who would be born defective if other nations continued to test bombs.[16]

Ordinary Soviet citizens never saw the more detailed and scarifying articles that were published elsewhere, still less the radioactive movie monsters, but in the abstract they grasped the main ideas. In 1961, when the Soviet government abruptly announced it would resume tests, it admitted that this would "instill alarm in people, causing their hearts to ache." To spare people more heartache, Soviet authorities said nothing more about the mammoth explosions they were setting off, but most Russians heard the news through foreign broadcasts or by word of mouth.[17]

By the end of the 1950s anxiety over fallout had became a powerful force around the world, with the full support of many governments. India was frightened by Nehru's outcries against fallout, which he said "might put an end to human life as we see it." Scandinavian officials issued alarming emergency warnings whenever fallout from a test drifted near. Most anxious of all was Japan, where the government responded to Soviet tests with loudspeaker trucks in the streets telling citizens to wash their fruits and vegetables. When an accident threw dye into the air of a Tokyo park in 1962, leaving blue spots on everything, passersby became terrified that they were victims of fallout. Never before in history had there been such unanimous worldwide concern, official and unofficial, about a technical issue.[18]

Even within the United States, the most widely publicized and convincing of all warnings came from the government itself. After the BRAVO test civil defense authorities had realized that they would have to completely reshape their programs. It was hopeless to try to save anyone within a city struck by a hydrogen bomb, but citizens might at least be protected from the heavy fallout that would drift a few hundred miles downwind. Beginning in the mid-1950s Americans were bombarded with officially sponsored pamphlets, magazine articles, and films with instructions on shielding their families from radioactive dust. Such local fallout in a possible future war had little to do with the far weaker global fallout from current bomb tests, but many people failed to notice the distinction and came to fear all sorts of death dust together.

In some regions of the United States, actual local fallout became a source of gnawing anxiety. Beginning in 1954 the explosions in Nevada were increasingly criticized. Soldiers who had been near the tests quietly began to seek financial compensation for all sorts of maladies; over the years some of them began to feel the same worries about lingering "A-bomb disease" known to Hiroshima survivors, and the same resentment. Some Americans who lived within a few hundred miles of the tests also began, like the citizens of Hiroshima, to demand attention. For example, in 1958 Edward Murrow gave a Nevada critic a national television audience for bitter attacks on the AEC. A little boy had played in radioactive dirt, the critic said, and later died of leukemia. Was he not a fallout victim?[19]

Fallout was perfectly suited to induce anxiety, as psychologists define the term: something that rests upon helplessness and uncertainty, on the feeling that a threat cannot be escaped nor perhaps even comprehended before it is too late. Certainly fallout was inescapable. Still more unsettling, it was cryptic. Part of the problem was that authorities disagreed on certain points and that government pronouncements in particular seemed untrustworthy; but that was not the only source of uncertainty. Nor was it just that radiation was invisible, for so were many other hazards from chemical poisons to viruses, and indeed Geiger counters could detect radiation at lower levels of danger than the levels at which almost any other hazardous agent could be detected. The worst uncertainty came at the next stage, when you knew that you had absorbed some radiation but did not know what the effects might be. Even if you fell ill with leukemia, no physician could tell whether your particular case had come from fallout, or from natural radiation such as cosmic rays, or from some other cause altogether. The mother whose boy had played in radioactive dirt and a few years later died could never prove that the AEC had killed him; the AEC could never prove that it had not.[20]

Americans exposed to bomb tests began, like people in Japan, to call themselves "guinea pigs." The words recalled the irradiated animals in Atoms for Peace films, ignorant of what was happening to them and helpless to prevent it—the model of a scientist's victim. It did not help when AEC spokesmen and loyal journalists insisted that bomb tests were merely "scientific experiments." In 1957 a Japanese physicist spoke for many when he declared that with global fallout, "the whole population of the world is being used as guinea pigs." Contamination brought on by overweening scientific authorities was no longer a vague phrase or a horror movie myth; it was real dust drifting into every home.[21]

A FEW FACTS

Was worldwide fallout an undetectably small addition to the natural hazards that everyone normally endured, as Teller and the AEC said? Or was it the possible doom of millions of infants, as Pauling and other critics said? In fact it was both. The scientific controversy was not as complicated as it seemed; the public was confused because they were projecting traditional myths, personal feelings, and social conflicts onto the invisible fallout. Such projections could be distinguished from reality by those who understood a few basic facts.

A radioactive atom is only a particle of matter, neither a magical friend nor a monster of corruption. Radioactivity is not a supreme secret but a simple physical process going on in every lump of dirt, as unromantic as the rusting of iron. There remains the question of how the process affects living creatures.

That question was burdened by facts so unhappy that few cared to discuss them. One fact is that at least one out of every ten children conceived will have a significant congenital defect. Another is that, among healthy people, roughly one out of five will die from cancer. These common tragedies have always been poorly understood, for the statistics vary in unexplained ways. If radiation raised the rate of birth defects or the rate of cancer in some group by a small fraction, say from 10.0 percent to 10.1 percent, the change would be invisible among the normal fluctuations in the rate. Thus Teller and the AEC were correct: harm from global fallout had never been detected, and at worst it was small compared with ordinary ills which the public rarely discussed. Yet if such a slight percentage increase did occur, among a hundred million deaths in the world it could cause a million more. Thus Pauling too was correct: nobody had proved that fractional increases were *not* adding to the total burden of human suffering. Amid all the fighting over whether radioactive atoms should be called friendly or monstrous, the only scientific question was whether such a fractional increase did occur.

There was never any doubt that radiation in bursts of several dozen rems or more causes injury. Most sensitive are fetuses; by the mid-1960s it had been proved that a few dozen fetuses exposed to heavy radiation from the Hiroshima bomb had been born mentally deficient as a result. A more important proven effect of large doses of radiation is cancer. A given individual could never know whether the cancer would have come anyway, but the statistics for entire groups gradually became clear. At Hiroshima and Nagasaki, on top of the tens of thousands of cancer and leukemia deaths that were normally expected, by 1980 there were a few hundred additional ones among people who had been exposed to some dozens of rems. Similar risks threatened those who ingested a large amount of radioactive dust, like the radium dial painters of the 1920s, some of whom died, or the Rongelap atoll natives, some of whom suffered nonfatal thyroid abnormalities.[22]

Genetic defects from heavy bursts of radiation were far harder to detect among all the defects that occurred anyway, and for humans a genetic effect was never detected, not even at Hiroshima. But ever since Muller's work on flies, nobody had doubted that this was another real risk. The result would not be outlandish mutants but simply more of the usual defects such as muscular dystrophy, mental retardation, and occasional deformities like extra fingers. Perhaps most likely were minor changes such as scarcely recognized allergies or lowered resistance to some disease.

What about smaller doses of radiation? An average person is exposed to nearly two-tenths of a rem per year. Roughly half of this comes from cosmic rays, from traces of radioactivity in ordinary rocks, and from other natural sources; the other half of a person's normal irradiation comes chiefly from X-rays and other medical procedures. In the early 1960s, global fallout added about a hundredth of a rem per year to the normal two-tenths. The question was whether that addition made any difference.

According to one hypothesis, there was a threshold. Five hundred rems given in one burst will kill most people, but cancer patients are routinely exposed to several times that amount in total, spread out over a long period; the patient is weakened but the cancer cells may die. Many experts believed that if the dose was spread far thinner, for example to less than a tenth of a rem per year, it would cause no harm at all. According to a different hypothesis, which only became scientifically respectable in the 1950s, there was no such threshold, and all that mattered was the total quantity of radiation no matter how it was divided up. Under this no-threshold hypothesis, whether a hundred people get one rem apiece or a hundred million people get one millionth of a rem apiece, in either case the radiation would produce the same total number of cancers. That was how Pauling calculated that the many rems' worth of fallout from tests, however thinly spread around the world, would still harm a great many people.

Scientific evidence gave no way to choose between these two hypotheses in the 1950s, and it still gives no way in the 1980s. The tremendous numbers of normal cancers and birth defects conceal any added effect of small traces of radiation. Hundreds of studies suggest that radiation will indeed cause some types of damage even at rather low levels. Hundreds of other studies on plants, animals, and humans, accumulating steadily from the turn of the century into the 1980s, have found minute amounts of additional radiation causing no harm—or actually encouraging growth, strengthening immunity to disease, and stimulating mechanisms that heal genetic defects. Thus Teller's remark that a slight addition of radiation might possibly be beneficial to the world's health, a remark that his critics took as an obvious lie, was just as good in terms of scientific knowledge as the claim that fallout might possibly kill millions. The uncertainty was less a debate between opposing schools than a reflection of plain scientific ignorance.[23]

By the late 1950s most experts realized they had no sure knowledge about the effects of low levels of radiation. But just to play it safe, they began to choose the no-threshold hypothesis, and demanded that exposure to radiation be kept as low as could reasonably be achieved. This was explained in a report on fallout prepared for the United Nations by American, Soviet, and other scientists and published in 1958. The tests held so far, said the report, might have doomed as many as 2,000 people a year to die of leukemia (or none if there was a threshold). Fallout might cause up to 100,000 genetic defects around the world (or possibly none). No doubt the problem was insignificant compared with other health problems that killed hundreds of millions every year; it was even small compared with public exposure to natural radioactivity in water and rocks; but it was a possible tragedy nevertheless. Experts today would arrive at much the same conclusions.

By the late 1950s AEC staff members were admitting that their tests might possibly bring occasional death or genetic damage. Yet few of them suggested the tests should cease. Meanwhile knowledgeable critics understood that the risk to an individual from test fallout was less than the risk from living in a brick house, and far less than the damage done by the widespread X-ray hazards. Yet few except Muller spoke out against misused X-rays. Both advocates and critics of tests were in fact much less interested in fallout than in the bombs themselves.[24]

CLEAN OR FILTHY BOMBS?

As time passed the fallout debate grew more explicitly political and more fervent. Seeing that the future of the entire world lay in the balance from the threat of nuclear war, both sides reached for the most emotion-stirring arguments. In a prime example of the power of imagery, the question of what to do about nuclear weapons became almost indistinguishable from the question of whether or not radioactivity was an uncanny horror.

On one side, a formidable public-relations apparatus responded to the critics by proclaiming ever more loudly the trustworthiness of government plans and the safety of radioactivity itself. Within the testing nations (the United States and the Soviet Union, joined in 1952 by Britain and in 1960 by France) the highest levels of government labored to convince people that a little fallout radiation was tolerable. For example, Eisenhower's cabinet got into an extended discussion of how to word official radiation guidelines not only honestly but also "in such a form as not to scare people to death." Hardest working of all were the AEC staff. Arguing their case in the many magazines and television shows that offered them a forum, they persuaded a majority of Americans that bomb tests were essential for national security.[25]

The American government urgently wanted to test new warheads for use against enemy troops, for the ballistic missiles that were under development, and for various other weapons, all presumably to deter the Soviets from starting a war. In a 1957 top secret address to Department of Defense leaders, the Chairman of the Joint Chiefs of Staff explained that "the use of atomic weapons is the very foundation of our present defense program. The Russians . . . are doing their best to take them away from us." In short, halting tests would mean halting a vital project: the creation of good bombs to defend against evil Communism and to deter nuclear war.[26]

Teller more than anyone devoted himself to this project for the survival of freedom and of life itself. As he struggled to brush off the personal attacks of his critics, he rose to a sort of optimism. Surely the AEC weapons laboratories could create truly useful explosives? In June 1957 Teller and other physicists visited Eisenhower to say that tests must continue so they could perfect a device free of all but a minimum of fallout. Eisenhower, impressed, promptly announced that the government was working to devise bombs as "clean" as possible. Such bombs might be morally tolerable weapons, a sort of "death ray" (as one senator put it) that could strike down invading bombers or troops without harming nearby civilians. The *New York Times* editorialized that the invention "restores the scales again in favor of the free world."[27]

A clean device might grant civilian benefits too. Nuclear explosives had occasionally turned up in Atoms for Peace fantasies ever since 1945, but beginning in 1957 Teller and his AEC colleagues made the idea sound realistic. Taking their motto from the Bible's finest passage about a future Golden Age, they proposed to beat nuclear swords into plowshares. Project Plowshare would "make deserts bloom" (yet again) as well as gouge out harbors and carve a sea-level canal across the Isthmus of Panama. "We will change the earth's surface to suit us," Teller exclaimed. He met with the President's cabinet to present colored drawings of the proposed canal, and the government laid serious plans. For a few years nuclear explosives became part of the worldwide Atoms for Peace excitement. Russian scientists were the most optimistic of all about shearing off mountainsides or excavating reservoirs, and the Soviet government claimed that such schemes were a chief aim of its bomb test program.[28]

Typical of the enthusiasts was Theodore Taylor, a bright young bomb designer who around this time gave up on weapons and left Los Alamos. He dreamed of making nuclear explosives that would burn subway tunnels through bedrock, an idea already patented in 1954 by Tom Swift Jr., who had used an Atomic Earth Blaster to drill tunnels and tap the earth's riches. Other scientists proposed to heat underground chambers in order to generate steam for power, or even to create diamonds with a blast deep in the womb of the earth.[29]

Not everyone was attracted by the idea of such "control over nature," as Teller called it. The most striking fictional picture of a Plowshare program was a 1965 movie in which an overweening scientist fired an atomic bomb down a shaft into the earth, hoping to release "limitless, clean" energy. He only broke open the disastrous *Crack in the World* of the film's title.

There must have been some particular meaning in that image of dangerous probing down a deep shaft, for exactly the same image was important in many other tales, from the original *Last Man* book to the popular movie *Beneath the Planet of the Apes*. Indeed, the secret tunnel is one of the most common images I have found connected with nuclear energy in fictional books and movies. There is no way to say whether many people made the further association between this image and infantile fantasies of forbidden excursions toward the mother's womb. Certainly, in more general terms many people feared that scientists pursuing peaceful benefits would end up, as a well-known science fiction story put it, "digging too deep and blowing up the whole shebang."[30]

From the beginning many people rejected "clean" explosions. The

AEC's critics pointed out that digging a canal with nuclear devices, no matter how ingenious they were, could scarcely fail to produce radioactive fallout, and a nuclear war without lingering radioactivity was still less conceivable. Foremost among the critics was Senator Clinton Anderson, a strong-willed member of the Joint Committee on Atomic Energy who had often clashed with Strauss. Anderson decried the argument that tests were needed to develop clean bombs, announcing that in fact the American military was strengthening its stock of bombs in a way that would make them not less but more radioactive, or as he put it, "dirtier." Soon many people were agreeing that a clean nuclear explosion was a contradiction in terms. Teller's attempt to associate bomb tests with the positive feelings of Atoms for Peace was a failure; the tide began to run in the opposite direction.[31]

It became common for critics deliberately to associate ideas about test fallout contamination with images of war. An example was a five-minute film called "Falling Out," a collage of images of rain falling, charred bodies lying in a bombed street, a red telephone, and charts of ever-larger bomb tests, all without commentary. Such indiscriminate association of images dismayed scientists whether or not they favored tests. As one biologist put it, there were many ways in which the United States affected people's lives around the world, for better or for worse, more than with traces of radioactivity. He thought arguments for and against the tests were a diversion from the real issue, which was nuclear war itself.[32]

A similar conclusion was reached two decades later by Robert Divine, a scholar who studied the history of the fallout debate. He concluded that the hope of banning tests became a "magic talisman, a way that the nation could confront a real and present danger without coming to grips with the true reality of the 1950s—the possibility of total destruction."[33]

That is what psychologists would call displacement, a hostility that shrinks away from what is too threatening, directing itself onto some other target instead. A familiar example is the man who suffers silently through a bad day with his boss, then goes home and scolds his children. Even a bird, frustrated by a stronger bird, may turn upon a weaker one or attack a pebble. A combination of fear and hostility can turn up far from the original source, as in children who cannot admit their feelings about angry parents or death itself, and instead become fearful and hostile toward animals, ghosts, or nightmare monsters. Such displacement is one of the very few defense mechanisms whose existence is admitted by even the most rigorous critics of psychological theory.[34]

All the classic elements of displacement were present in the fallout debate. There was a threat which some people (as they admitted) found too great to face directly, namely the threat of war; there was a connected hostility; and there was a convenient nearby target. Displacement en masse can never be proved for certain, but it seems more than likely that many of the feelings expressed toward fallout from bombs were feelings originally meant for the bombs themselves.

That was not necessarily foolish. If a problem seems all but insoluble it may be wise to go after a piece of it, not so much displacing activity as concentrating it at one point. Pauling remarked that although his petition against bomb tests had stressed fallout, it was the weapons themselves that most worried him and his fellow critics. Norman Cousins, Schweitzer, and other leading opponents, even as they pressed their fight against fallout, agreed that their primary goal was to reduce the likelihood of future war. They said they were agitating for a test ban because this drew attention to the entire problem.[35]

The opponents of testing found they could draw attention above all by talking about milk. The U.S. Public Health Service had begun to monitor fallout in foods, and samples of milk happened to be particularly easy to gather over a wide area for laboratory tests. By the spring of 1959 the laboratories noticed a sharp rise in amounts of the isotope strontium-90, already identified during Manhattan Project days as a problem because it tended to settle in bones, and especially the rapidly growing bones of children. Within a few years the concentration of strontium-90 in American children had doubled.

In fact, radioactive milk was rarely the worst hazard from bomb tests. Only a small fraction of the strontium-90 that a cow ate with its grass was passed along into milk; the radioactive content of vegetables or wheat, for example, was often much higher. Nevertheless it was worries about milk that made thousands of citizens write Congress. Senator Anderson and others attacked Strauss for evading the problem, and elite magazines like *Consumer Reports* and *The New Yorker* carried frightening warnings. Even *Playboy* published an editorial attacking "The Contaminators," exclaiming that all children "may die before their time . . . or after having spawned grotesque mutations." Editorial cartoonists and groups favoring disarmament propagated a new image, a milk bottle labeled with a skull and crossbones.[36]

The clamor became so serious that dairy companies and government agencies feared the public would buy less milk. Authorities insisted there was no hazard; in January 1962 President Kennedy publicly drank a glass of milk and announced he was serving the delicious beverage at every White House meal. The majority of citizens accepted such reassurances, but there remained many who felt anxious and even guilty every time they gave their children milk. A few psychologists speculated that the attention given this hazard—by no means the worst problem of fallout, let alone of nuclear weapons generally—reflected infantile reactions. The authors of a newspaper advertisement against bomb tests were more precise when they explained why they chose this issue: "Milk is the most sacred of all foods." Since the time of witchcraft trials, accusations of obscene attacks on milk had been a powerful weapon against enemies.[37]

Contamination, poison, impurity, pollution, obscenity—more and more people were applying such words to fallout. The imagery was epitomized by a 1961 Herblock cartoon about the resumption of Soviet bomb tests, showing Khrushchev flying over the world on a witch's broom trailing noxious black clouds. All the feelings that nuclear bombs were filthy corruptions of nature had spread into feelings about radioactive isotopes as well. The underlying metaphor was best expressed by a five-year-old who, cautioned by his mother, passed the warning along to a little friend: you shouldn't eat snow "because there is a piece of the bomb in it."[38]

By 1962 the AEC had lost its most important battle. Between the scientific evidence and the impassioned imagery directed against bomb tests, most people had come to see "death dust" as a condensed representation for nuclear weapons or even as a displaced substitute for the weapons. By extension, all radioactive isotopes, whatever their origin, were openly and permanently associated with the most distressing contamination. Revulsion against radioactivity, a new attitude resembling a primitive taboo, like fallout itself was settling invisibly into every home.

12

The Imagination of Survival

In 1961 President Kennedy delivered an address to the United Nations with his customary eloquence, declaring, "Every inhabitant of this planet must contemplate the day when this planet may no longer be habitable . . . The weapons of war must be abolished before they abolish us." The slogan came to Kennedy through a series of three speechwriters, each of whom made changes; the first draft had read, "If mankind does not end war, war will end mankind." A still earlier version was, "If we don't end war, war will end us"—a line in H. G. Wells's 1936 movie, *Things to Come.* Was there nothing new to say? In 1955 when two animators produced a cartoon film of animals discussing how humans had exterminated themselves through war, it was only a remake, little changed, of a film made in 1939.[1]

Yet new meanings were creeping into the old idea of extinction. Up to the mid-1950s, despite dire warnings from some scientists and journalists, it had seemed no more than a theoretical possibility for some distant and hazy future. But as the number of hydrogen bombs mounted into the thousands, citizens began to suspect that the end of humanity was something that a leader like Kennedy could actually bring about, and soon.

Probably the first person in high authority to advise the public that hydrogen bombs could "render the earth uninhabitable to man" had been AEC Commissioner Murray in 1955. A few American senators and other presumed experts joined him, and the idea was spread around the world that same year by a Communist appeal asserting that "the use of atomic weapons will result in a war of extermination of the entire human race." Over half a billion signatures were said to be gathered on the related petition to abolish the bombs. Further pressure came from news of events. From time to time an American president hinted that he was thinking of using nuclear weapons, as happened when Eisenhower wanted to defend the islands of Quemoy and Matsu against Communist Chinese invasion in 1955 and 1958 and to defend Berlin in 1959. The American military in fact doubted that nuclear weapons could profitably be used in these cases, and the effect of the veiled threats on diplomacy seemed slight, but the effect on the world public was clear: each event triggered an avalanche of warnings about doom.[2]

There was no single moment when everyone began to feel they were living on the edge of universal death. Even in the 1960s a substantial minority around the world felt that although a war might kill them personally, their nation would survive handily, while an equally substantial minority had feared since the early 1950s that all civilization might perish. For many Europeans the turning point came in the summer of 1955 when the North Atlantic Treaty Organization held realistic war games. In two days 335 nuclear bombs "dropped" on Germany while journalists, flown about in NATO airplanes, were shocked by briefings that tallied millions of mock deaths among civilians. The following year came a Soviet threat to use rocket bombs unless Britain, France, and Israel halted their invasion of Egypt. About half the Germans surveyed afterward said that the threat might have been carried out, and most believed that in such a war "Germany would become a desert."[3]

The possible destruction of Germany did not particularly worry Americans; for many of them the turning point came in October 1957 when the first Soviet Sputnik orbited overhead. Unstoppable missiles had seemed like something for a remote science fiction future; suddenly they seemed like something that could drop on the United States next year. The American press erupted in an almost hysterical alarm.[4]

The immediate issue now was the survival of nations or even the whole human race. That was bound to inspire intense thinking. In these early years of the hydrogen bomb, more than at any time before or since, intellectuals threw themselves into logical analysis of nuclear destruction. However, behind all the analysis lurked more primal images. Every train of thought started off from one basic question: What would the world look like if the bombs fell? Would we all die? Or if not—and here came the most revealing images—what would survival mean?

VISIONS OF THE END

In the imagination, nuclear war often meant no future but an empty one. That was the image carried, for example, by the statement that the world's armament included "enough explosive force to equal ten tons of TNT for every man, woman, and child on the face of this earth," as President Lyndon Johnson said on television in 1964. In fact a typical 10-megaton hydrogen bomb would be far less devastating than a million separate 10-ton explosions would be. But the latter image persisted, assigning each of us a sort of personal bomb. That future was a void.[5]

Still more purely symbolic was the "clock of doom," reportedly suggested by Teller, featured on the cover of every issue of the *Bulletin of the Atomic Scientists.* The clock first appeared in 1947 set at seven minutes to midnight, and the editors put it forward to a hair-raising two minutes in 1953 when hydrogen bombs arrived. The image, ignoring what would happen after the hour struck, had a clear symbolism: nuclear midnight meant the end of time.

The image was developed to its greatest effect by an aeronautical engineer, Nevil S. Norway. With his weary eyes and long nose, he had the look of a bloodhound sniffing after something unpleasant, and indeed he was no optimist. His autobiography, *Slide Rule,* began and ended with talk about death and was centrally concerned with an airship disaster. In 1955, when he started writing a new book, a series of heart attacks had sunk him in thoughts of mortality—it would in fact be his last book. However, the engineer never said much about his preoccupation with disaster and death; he was reticent about personal matters and did not even say why he adopted as a pseudonym his mother's maiden name. For whatever reason, it was as Nevil Shute that he published *On the Beach.*[6]

England's best-selling novelist, Shute had particularly attracted attention with a 1938 book about a coming war where the bombers got through and civilian society dissolved. He took the idea to the limit in *On the Beach.* The novel imagined that in 1963 a Third World War wiped out all life in the Northern Hemisphere, while Australians, untouched by bombs, tried to carry on normal lives while awaiting extermination as fallout crept inexorably south on the winds. The book was serialized in more than forty newspapers in 1957, and by the 1980s the paperback edition had sold over four million copies, the highest sales of any novel centered on nuclear energy.[7]

The veteran film director Stanley Kramer saw a chance not only to market a good story but to caution the world, and he won the bidding for a Hollywood version. Judging from its gross rentals, his movie sold more tickets than any other nuclear movie with the possible exception of *Strategic Air Command,* and its impact was powerful. Audiences from New York to Tokyo left the theater stunned or weeping; I have found that even two decades later people of various nationalities would remember *On the Beach* with strong emotion. (In Moscow, a select audience applauded but felt it was "too stark" to give to the Russian public.) The audiences were drawn into Kramer's busy world of decent, ordinary people, then watched that world gradually die away. At the end there were only miles of empty streets. It was as if the pictures of unpeopled cities in civil defense drills were to become a permanent reality.[8]

Most film critics, supporting Kramer's overtly antiwar message, did not question his images. Questions were raised rather by American government officials, who feared the movie would weaken the nation's will and promote "extreme pacifist and 'Ban the Bomb' propaganda." Eisenhower's cabinet discussed confidential actions they might take to undermine the movie, and the State Department and the AEC distributed comments. According to the government view, voiced in public through various mouths, *On the Beach* was seriously in error.[9]

The first criticism made by government experts was that the story's premise was scientifically ridiculous, for all the weapons in existence could not possibly spread enough radioactivity to sterilize the planet. This was true at the time, but it was not true that the idea was out of the question. Already by 1947 Teller's group at Los Alamos had secretly concluded that a hundred enormous bombs could raise the planet's radioactivity to "a dangerously high level," and around 1954 atomic scientists like Szilard publicized the more specific idea, adopted by Shute, of thousand-megaton infernal machines encased in cobalt metal to generate many tons of fallout. In fact, real weapons designers preferred one-megaton explosions with as little fallout as possible drifting over their own homes. As Teller remarked, "The cobalt bomb is not the invention of an evil warmonger. It is the product of the imagination of high-minded people who want to use this specter to frighten us into a heaven of peace." Yet if doomsday bombs did not exist, what mattered psychologically was that such things *could* exist. For the first time in history it was not humanity's limited abilities that prevented us from committing racial suicide, but only our good sense.[10]

This made a second criticism of *On the Beach* the crucial one. Would people really act as the film showed them, exploding cobalt bombs and then living on almost normally, finally taking suicide pills? Teller, for once agreeing with *Pravda,* cited a Russian comment: "What manner of men will accept the end without resistance?"[11]

The implicit reply was that nearly everyone was already passively accepting a gradual approach of what the film called "the time," the nuclear midnight. In ignoring this peril, was not the world public like the fictional Australians who refused to acknowledge imminent death? If that was Shute's and Kramer's message, it eluded most people. For the book and film gave most of their space to love stories about steadfast military officers who insisted on doing their duty, and blindly sentimental women who eventually agreed that their men were right—the theme of *Strategic Air Command.* Only now the officers' moral duty lay in giving their families poison pills to hasten the inevitable end. The book and film, by showing none of the physical agony and demolition that a real war would bring, made world extinction a romantic condition. All the characters died in ways of their own choosing. For example, the central figure was a lonely Navy captain cut off from his family, who at the end resolutely sailed away from his lover to drown, saying that death was like going home. *On the Beach* said far less about how to prevent nuclear war than about a more personal problem that Shute, a dying man, felt keenly: acceptance of separation and of death.[12]

Numerous other stories of the period also terminated their nuclear wars in the peace of the grave. If there was nothing beyond nuclear midnight except the quiet nullity of death, then there was nothing more to say about it. Audiences came away from *On the Beach* and similar works not with questions about military policy but with a sense of inevitable tragedy.

SURVIVORS AS SAVAGES

Many people looked for more than blankness in the space following a future nuclear war. But that was a space outside authentic experience, a space visible only to the imagination, so people who peered into the fog ended up projecting images already within their heads. The question raised by a future after the bombs was the ultimate question of human survival, so the images people reached for were ones addressing that question.

"Who survives?" asked a newspaper reporter in 1947. At best a few remnants, he answered, "too weary to rebuild, too weak even to dream, dully scrabbling like their Neanderthal foresires in the primordial ooze." This idea of postwar savagery was spread by authors such as Upton Sinclair and Aldous Huxley along with many lesser writers. In the first widely seen movie to show a world after atomic bombs, the 1950 *Rocketship X-M,* an expedition to Mars found "crazed, despairing wretches" shambling through a radioactive desert. A ridiculously cheap production, the movie made enough money to found a genre. In 1952 *Captive Women* brought the story down to earth with barbaric tribes of survivors, more or less warped by radiation, battling one another in the ruins of New York. In the next dozen years *World Without End, Teenage Cave Man, The Time Machine,* and *The Time Travelers,* and beginning in 1968 the six *Planet of the Apes* movies and two derivative television shows, won an enormous audience with variations on the same theme.[13]

These savages had a heritage. People who nodded wisely at Herblock's editorial cartoons of cavemen shuffling through the postwar ruins had forgotten that his first famous cartoon, also about the technological end of civilization, was drawn early in the Second World War before atomic bombs were devised. Earlier still, the nineteenth century had appreciated the image of ruined cities. The most famous example was Thomas Cole's series of huge paintings depicting a white city rising out of wilderness to glory, then collapsing amid the flames of invasion and civil war, until broken marble columns stood covered with vines in a resurgent wilderness. Other romantic artists showed the proud landmarks of London or Berlin in some future age, fallen like the broken temples of Greece.

People found such scenes attractive. Farmers squeezed by urban creditors and clerks ground down by mass society listened to certain politicians and intellectuals who said that the world would be better off without its cities, those garbage-heaps of grasping bankers and criminals. Traditionalists everywhere longed for the demise of cities where "the good habits and simple customs of the rural communities are gradually being destroyed" (as explained in a 1913 report by the military commander of the Hiroshima district). The image of a savage clad in skins gazing at the wreckage of Wall Street also had an Arcadian appeal. For example, a popular novel of 1885, *After London: or Wild*

England, used the metropolis, decayed into a literal poisonous swamp, as a foil for the saga of a hardy barbarian hero roaming an England restored to verdant woodlands.[14]

Americans with their frontiersman legends were particularly drawn to landscapes shorn of cities. The prototype was Jack London's 1912 novel *The Scarlet Plague,* with spear-carrying savages living at ease in the California forest after civilization fell to an epidemic—a metaphor of decadence. Scholars of London's works noted that when he described society collapsing in riot and flames, that was just what his radical political speeches declared ought to happen. After the death of the mechanical mass civilization, London could imagine people like himself as amoral barbarians, proving their worth by the very act of survival. By 1941 the theme was so common that a reader complained to *Astounding Science-Fiction,* "Why must we always have Rousseau's Noble Savage, with his biceps, his stone ax and his mate, crawling around the ruins of mighty Nyawk or Chikgo?" After 1945 authors only needed to add atomic bombs to transform this minor pulp genre into a universally recognized myth.

The myth became so familiar that sometimes it was only necessary to mention it in order to make a point. An example was Ray Bradbury's *Fahrenheit 451,* a perennially popular novel (it went through forty-four printings in the quarter-century after its 1953 publication). The author's main concern was a detailed attack on modern social trends; he brought in nuclear war only in the final pages, to erase the corrupt cities so a better society could rise.[15]

More often, however, the center of attention was the savages themselves. The symbolic meaning of this became clear when the survivors were pared down to a few individuals, as happened for example in two important American films: *Five,* in 1951 the first serious film on the aftermath of nuclear war, and *The World, the Flesh, and the Devil* in 1959. Each put a handful of survivors into postwar cities, which somehow lacked shattered buildings and rotting corpses. By the end the survivors seemed happy enough. "I hated New York," the young hero of *Five* confided to his girl; "I'm glad it's dead." Now they could enjoy one another without inhibitions, as in the private fantasies that some teenagers had about sex play in bomb shelters. And when villains showed up they could be hunted and shot like animals. Movie audiences or science fiction readers, to the extent that they identified themselves with the heroes and heroines, could picture themselves indulging in sexual adventures and gunplay that would have been forbidden in the old society.[16]

Even the thought of being sole survivor, the Last Man, had the broad appeal that A. A. Milne recalled from his childhood when he imagined that everyone was dead and he could freely loot the candy store. Survivors in after-the-bombs movies might scorn civilization, but they all used its conveniences, free to drive any abandoned car and enter any empty house. A more subtle liberty was noted by an adolescent girl who told her therapist that she liked to imagine herself alone on the earth, free from her anxieties about relations with other people. The therapist asked, But who would love this sole survivor? "In that case I should love myself." Such dreams of narcissistic escape haunted many a nuclear war story.[17]

Best of all, after-the-bombs stories usually ended with survivors setting out amid sunlight and twittering birds to begin the world anew. When a Last Man roamed through the hackneyed stories of science fiction magazines and television shows, audiences would hardly be surprised when the lone Adam met an Eve. Narcissism and unfettered barbarism led to renewal, perhaps to another turn of the Wheel of Time as civilization prepared to rise once more. "How many times will it happen again?" asked the narrator at the conclusion of *Teenage Cave Man*—a question that the audience, groaning at the cliché, might have thrown back on the scriptwriters.[18]

Nuclear weapons brought nothing new to the theme. The author-director of *Five* had conceived its plot in 1939. Going farther back, *The World, the Flesh, and the Devil* was inspired by a property that the studio had bought in the 1930s and pulled out of storage immediately after the bombing of Hiroshima: *The Purple Cloud* of 1901. *Five* reached still farther back for its ending, quoting from the Revelation of St. John, "I saw a new heaven and a new earth." The archaic fantasy within such tales was well put by a four-year-old who thought that "the bombs would kill Daddy, Mommy and everybody, and then God would have to make new people."[19]

The whole cycle might be shortened, according to civil defense training films of the early 1950s. Millions of Americans watched actors portray families that took shelter from bombs, then emerged a little later with their clothes only slightly rumpled. The next step, the father might explain, would be to "wait for orders from the authorities and relax." Some fictional stories of the 1950s likewise suggested that the world could be put back together again almost at once.[20]

Only slightly more realistic was Pat Frank's *Alas, Babylon,* next to *On the Beach* probably the best-selling book about a world after the bombs; it was published in 1959 and remained in print for decades.

Frank had the usual pious intent of warning against war, but he also said he disliked the "suicidal" tendencies of other writings. He depicted a community in rural Florida, far from the explosions, suffering little worse than attacks by a few marauders who were hunted down by vigilantes in an exciting adventure. At the conclusion the town had become admirably self-sufficient. This was not a postwar society of squalid camps filled with sick refugees, but of sturdy individualists with fishing poles, determined (as an advertisement on the paperback put it) "to build a new and better world on the ruins of the old."[21]

Survival of the fittest, Frank suggested, would kill off the weak and the evil: a world not just restored but improved. Other American storytellers agreed that nuclear war would at least rid the world of Communism. Marxists, for their part, insisted that nuclear war was bound to destroy capitalism. This view, the official one in the Soviet Union, was carried farthest by the leaders of the Chinese Communist Party in the late 1950s, who suggested that after a nuclear war their system would prevail almost intact amid the ashes of its enemies. Everywhere it was hinted that society might be purified by its passage through flames and chaos, like lead transmuted into gold.

Social transmutation was traditionally associated with spiritual renewal of the individual. In *On the Beach, Alas, Babylon,* and other stories, disaster did indeed leave protagonists more strong and loving. Nuclear war novels and films also showed a remarkable affinity for straight religious feeling, inserting quotes from the Bible with an obsessive frequency not found in most other genres. And the only after-the-bombs novel of the period to win some respect from literary critics was frankly centered on religious salvation.

The work of an ordinary science fiction writer and devout Christian, Walter Miller's *A Canticle for Liebowitz* sold over a million copies in the two decades following its publication in 1959 and was translated into five languages. Its subject was an order of pious monks who preserved scientific knowledge through a new Dark Ages following nuclear war, although they suspected that overweening humanity would misuse the knowledge and rebuild a wicked secular civilization. Their only solution was the traditional one, to accept the crushing burden of original sin and look for redemption by divine grace. At the end of the story an irreligious civilization was indeed rebuilt, only to destroy itself in another nuclear war. Many saw this complex book as pessimistic. But critics who read it closely found that within its doomed cyclical history lay the possibility of time closing not with death but salvation. As the last abbot died in his blasted monastery he witnessed a saving miracle, a new and pure birth, possibly the Second Coming itself. It was as if nuclear war, just like individual martyrdom in pain and guilt, were a gateway to spiritual redemption.[22]

No idea could have been more dangerous. Perhaps it was useful for literature to use social change as a metaphor for personal change; and perhaps an individual personality does need to go through chaos before it can hope for full rebirth. However, to suppose that not just an individual but an entire civilization can be redeemed by passing through catastrophe—that is a dangerous fallacy. Wars leave most people not more trusting and reasonable, but less so.

THE VICTORY OF THE VICTIM

The nuclear war survivor as a wretched victim could merge with the survivor as a pilgrim moving toward renewal. The idea joining them was simple: escape from victimization equaled rebirth. This idea led to a chilling logical conclusion. If a person who already felt victimized believed that catastrophe would end in release, perhaps even in a rebirth of all the world, should he not want to drop bombs? The question was seldom discussed, for it led into murky questions of human motivation—disconcerting questions that few people cared to consider in others, let alone in themselves.

Deliberate destruction had already been central to the old mad scientist stories, and it reappeared in the newer movies about radioactive monsters who in their anguish stamped around smashing everything. The idea could also be uncovered in the clinic, for some angry paranoiacs felt that the way out of their misery was not merely to escape the attacks of their oppressors, but to vengefully abandon self-control and bring down wholesale destruction. To do that yet still survive would mean winning the greatest power of all, the power to live while the enemy died. A victim who became both a destroyer and a survivor was seizing from the oppressor the magical keys of death, life, and rebirth itself.

Did these turbulent fantasies appear in widely known stories of nuclear war? The answer is yes, only with a crucial omission. Before showing the omission I will show what was clearly present.

Every after-the-bombs book and film included obvious victimization. All the usual threats were displayed: mutilation, disability, lack of healthy progeny, separation from loved ones or perhaps from all human society, and plain death. A number of pulp science fiction magazine stories, and two of the best-remembered episodes of the *Twilight Zone* television series, showed a Last Man wandering hopelessly through desolation, acting out the worst nightmares of paranoia with no final rebirth. In the more numerous stories where survivors eventually found happiness, most of the attention still went to their loneliness and other woes.[23]

The tale that deployed the most complete set of victimization themes before proceeding to rebirth was a widely read English science fiction novel of 1955, aptly titled *Re-Birth.* John Wyndham described a primitive postwar village bordered by a forbidden radioactive Badlands infested with degraded mutants. The villagers, to maintain their racial integrity, killed off any "deviation" who had too many toes or whatever. But they failed to notice that a few children could exchange thoughts by telepathy. When the protagonist's father discovered this new mutation in his son, he condemned him and his friends to death. Just in time a group of adult mutants showed up, a new and superior species who slew the adult villagers and carried the mutant children across the Badlands to a new White City where thoughts mingled in communal friendship and wisdom. These themes were not the idiosyncrasies of one writer, for each of them—degenerate creatures, radioactive forbidden zones, murderous fathers, telepathic prodigies, and harmonious new communities—appeared again and again in one or another book or film connected with nuclear war.[24]

Most typical were the horrid creatures. Entire tribes of them kept the action moving in many a science fiction film and paperback novel. The filthy creatures, like the radioactive rubble they inhabited, put the bombs' blasphemous pollution of the earth into visible form. Supposedly descended from irradiated humans, they were in fact descended from mad scientists' monsters, and like those monsters, even as they threatened people they were themselves miserable victims. Of course an increase in random birth defects could not in fact cause the evolution of a grotesque new race within a few centuries; the authors were simply combining old stories of savages amid ruins with still older stories of monsters. The combination succeeded because both types of story spoke of victimization, not just by nuclear energy but by the tyranny of technology and authority in general.[25]

Stories of monsters and victims traditionally required something more: the investigation of weird secrets. Secrets were indeed common in after-the-bombs tales, where survivors ventured into forbidden zones or even crawled down tunnels into the earth in search of the mysteries of the old civilization. The search might be opposed by an oppressive tribal leader, the one who had made the zone forbidden and otherwise played the dangerous parent.

At this point the argument runs into a difficulty. Explorers in after-the-bombs tales usually found not dire punishment but revelation and renewal, while the tribal authorities were left behind, weak and ignorant. Something essential was missing. The real violation, the blasphemous act that had created the monsters and other evils in the first place, was invisible, somewhere back at the time the bombs fell. If after-the-bombs stories resembled mad scientist stories, then where was the mad scientist himself?

A few stories such as *On the Beach* did show scientists, yet none was insanely destructive. These characters acted more like general representatives of human weakness and aspiration. Virtually no author of the 1950s offered a character who was directly responsible for any of the political or technical acts that caused nuclear catastrophe: the Man Who Started the War did not show his face. It became a convention to avoid naming even the nation that triggered the fictional war. When writers did name someone, Americans of the 1950s took it for granted that Communists would be to blame, while Communists emphatically asserted the reverse; if anybody was responsible, it was nobody the writer knew. Even *The World, the Flesh, and the Devil* entirely omitted the theme of personal responsibility for the end of the world that had been the chief concern of its prototype, *The Purple Cloud.* In this central omission, the nuclear war stories seemed very distant from the mad scientist tradition.[26]

Curiously, in leaving out the obsessed and guilty man himself, nuclear war stories were not diverging from the radioactive monster films. Godzilla and his ilk were not created by any identifiable villain. Like the mutants who scuttled through radioactive ruins and like the primitive human survivors themselves, they were victims of distant and nameless acts.

During the 1950s mad scientists did continue to flourish in their own movies and stories, remote from nuclear war. A revealing example was the most highly praised science fiction movie of the decade, *The Forbidden Planet,* in which the traditional island stronghold of the sorcerer-scientist turned into a faraway planet. There Dr. Morbius meddled with an underground nuclear energy device, left behind by an alien race that had mysteriously vanished. The device could make wishes materialize, and Morbius inadvertently created a murderous monster—what he finally had to admit was his own "evil self," a condensation of his "subconscious hate and lust for destruction." Then he realized that the alien race too, long ago, had unknowingly loosed atom-powered "monsters from the Id," and they had exterminated each other in a single night. This might have been the most apt metaphor yet for the true problem of nuclear war—but neither the scriptwriters nor the film critics mentioned that.[27]

It was not just the normal human lust for destruction that people somehow failed to connect with nuclear weapons; there was a related idea even less often brought up. Morbius exposed it in the end when he got rid of his monster by blowing it up along with the entire Forbidden Planet and himself too. He was simply following his prototype, Jules Verne's Captain Nemo of *20,000 Leagues Under the Sea,* who (in the highly successful Walt Disney film of 1954) ended his quest for a better world in an atomic blast. The picture of a Nemo destroying his enemies along with his own island, like the scientific terrorists who threatened their own cities in other stories, and like the mad scientists who tried to blow up the whole earth in still other stories, stood for a most unsettling idea: mass destruction combined with suicide.[28]

Suicide was no fantasy; every year hundreds of thousands did perish that way, far more than were murdered. And suicide-murder combinations were not unknown. Indeed many psychologists were convinced that the combination was a fundamental one—that suicidal depression was an inward turning of destructive desires. The would-be suicide cherished a hatred of inner evil, a self-loathing, and that might easily be projected outward to inspire attacks on others. There were even a few authentic cases of disturbed individuals who worked on scientific plans for the destruction of all humanity.[29]

Psychologists also found that the urge to annihilate evil was connected with hopes of rebirth. Many suicides held fantasies of death as the doorway to an afterlife, or said that at least their death, like the mad scientist's, would make the rest of the world better. Psychopaths who longed to destroy not only the wrongness within themselves, but everything around them, also imagined destruction as a route to control over death, magical survival, and a better world. Even Hitler in his last bunker, exclaiming that he wanted the German nation that had failed him eradicated, said he meant to clear the way for stronger races. In sum, many of the motives underlying suicide—the desire for revenge and the overthrow of evil, the longing for peace and an end to anxieties, the urge to be decisive at last, the hope for clarification and a new beginning—could also be motives for starting a nuclear war.

Nevertheless, the theme of deliberate global suicide scarcely appeared in after-the-bombs stories. *On the Beach* and *A Canticle for*

Leibowitz did show crowds lining up to get poison pills, but audiences and critics did not seem to notice any connection with the forces that led toward war. Some people declared that if the bombs dropped they wanted to be underneath one and die instantly, but that sounded like merely a personal and passive suicide. From the 1950s on the word "suicidal" became a favorite polemical description of nuclear war, but merely as an abstraction whose meaning few explored. Only a very few psychologists suggested that war might in fact start from a sort of "Samson complex," the deliberate desire to destroy one's enemies and oneself, to wipe the slate clean.[30]

Was there really no connection in anyone's mind between thoughts of nuclear war and the tangle of themes, murderous and suicidal, featured in mad scientist tales? Perhaps all of us were like the gentle folk of *On the Beach,* never dreaming of personally starting a nuclear war. Perhaps even the motives that had burned down a thousand real cities since ancient times—the desire to survive one's enemies and build a better world on their ashes—had evaporated with the arrival of hydrogen bombs?

There is a different way to explain the remarkable omission of a major traditional character, the mad scientist, from a collection of tales where he should have been perfectly at home. Perhaps mass destruction had become so plausible and dreadful that people did not dare to connect it with the desire to bring it about. If so, then at least some of us did resemble the chemist I mentioned earlier, who dreamed of vengefully destroying the world with atomic bombs but who hid his anger even from himself.

To study this matter I will take a few authors, and for each I will note some significant stories, including the author's own life story. This method can suggest what web of associations lay within a particular mind. When that mind originated influential tales of world doom and survival, the exercise can also show something about what underlay the images impinging on the rest of us.

I have already mentioned that the author of the original *Last Man* book was a lonely man who killed himself soon after it was completed; the book itself included the idea of a relentlessly probing researcher working in a laboratory deep within the earth. Evidently an embryonic form of the mad scientist theme shared space with world doom and suicidal urges in the writer's thoughts. I have also noted that biographers of the author of the next major *Last Man* book, Mary Wollstonecraft Shelley, found she was troubled by conflicts with her cold-hearted father and by the loss of her mother at birth, problems that had already helped her to create Dr. Frankenstein. The world catastrophe novel and the self-destructive mad scientist seem to have had a common foundation in her anger and loneliness.

For other authors who developed the mad scientist stereotype, from Nathaniel Hawthorne through Jules Verne to Karel Čapek, biographers have in each case suggested a connection with unhappy relationships with the author's father. But to combine the mad scientist with world doom required further personal complexities.[31]

An especially important case was Jack London, whose *Iron Heel* and *Scarlet Plague* brought tales of collapsing civilization into the twentieth century. Born a bastard, London hated the father who had abandoned him and also nursed an angry distrust of his unloving, neglectful mother. He hated himself as much as anyone and died of an overdose of morphine. Most critics agreed that these psychological pressures propelled London toward his stories of catastrophe. The same pressures were even more evident in the mad scientist fantasies he also wrote. Most telling was a story of a young man who (like London himself) ran away to sea, and who later fell in with his father. This father was an evil scientist, who experimentally froze and revived his son in a parody of victimization and resurrection, until the son managed to kill him.[32]

London's most prolific successor was Philip Wylie. In 1932 he coauthored the most widely read science fiction book of the decade, *When Worlds Collide,* a story of survival after a cosmic collision destroyed the earth. In 1954 he offered a new cataclysm in his civil defense novel, *Tomorrow!,* and during 1963 he showed millions of *Saturday Evening Post* readers a nuclear war of the 1970s that left America empty of life. His scenes of mobs trampling one another were so gruesome that the magazine refused to print the details, but Wylie restored them in the published book.[33]

In all these writings Wylie saw himself as a crusader struggling to purge the world of every evil from Communism to hypocrisy, but he was best known for his vituperative attacks on the modern "Mom," a far from maternal creature. Wylie's biographer suggested some motives. Not only was the novelist's father a hypocrite, but his mother had died when he was five years old, leaving him with "an abiding sense of irreparable loss" and perhaps also "a smoldering rage." Such anger may have given Wylie a particular interest in the gory deaths he vividly and repeatedly described. Adding to his problems were anger directed inward and obsessive anxiety about death. When he wrote nuclear war novels he was concerned not only with the world's fate but his own, for he was ill and despondent, despising himself as he killed himself with pills and alcohol. The annihilation of everything he feared and hated, accompanied by survival and a new birth of virtue, was something Wylie often wrote about but never achieved in himself.[34]

A particularly familiar symbolism appeared in Wylie's first nuclear doom story, which was also the first such fantasy to reach a huge general audience when *Collier's* published it in January 1946. The story was about scientists who sank a deep shaft in the earth to conduct atomic experiments, and accidentally set off a chain reaction that burned up the globe.[35]

For these key writers, and others I will mention later, the end of all things on the one hand, and dangerous scientists on the other, while not necessarily linked in any single story, were bound together by a common origin. That origin lay in problems near the core of the writer's personality. All these writers guessed the secret of the mad scientist—his homicidal or even patricidal or matricidal lust for destruction, often mingled with suicidal thoughts, and accompanied by a desperate desire to seize the forces of rebirth. When these writers, and the public who read them, failed to discuss such things in the same breath as nuclear war, the reason was not that the ideas were far from every mind. They were only too close.

In one popular medium where the images were more impersonal and thus easier to face, the Man Who Started the War did become visible. Drawing on an old pictorial cliché, editorial cartoonists drew the globe as a round bomb with a fuse, and sketched in alongside a politician or a general playing with matches. The planet could be recognized as a bomb because generations of cartoonists had pictured such a spherical device in the hands of wild-eyed anarchists out to blow up society and probably themselves too. The threat of nuclear weapons—the threat of global doom—the threat of dangerous authorities—the threat of the small boy meddling with forbidden things—the threat of an explosion of rage against modern society and against oneself: all merged into one compact symbol.

The symbol was no fiction but reflected a genuine problem. According to a later report, at the peak of the 1962 Cuban missile crisis when the world's nuclear forces were poised for instant action, an American spy in Moscow sent a false warning that the Soviets were about to launch their weapons; having learned that he was about to be arrested, the spy "evidently decided to play Samson and bring the temple down with him."[36]

THE GREAT THERMONUCLEAR STRATEGY DEBATE

A few people strove to look behind the mythical imagery. How, they asked, might future wars actually start? How might they actually end? Surely reason could be brought to bear on the matter!

From the mid-1950s on, the best-known center for such work was a flat, nondescript building packed with little rooms like professors' offices, fenced off among the palm trees of Santa Monica, California. This was the RAND Corporation, devoted to rational analysis, such as calculations of how many millions of people might die in this or that eventuality. Sometimes the analysts felt that their worst enemy was not so much the Soviets as their own sponsors, veteran Air Force generals who sneered that logic could never analyze the chaos of war. Such disagreements were only to be expected in an organization trying to weld together two very different groups, scholars and military officers—a new linkage that was altering both sides.

Similar studies also went ahead on blue-ribbon academic panels assembled by government and private institutions; among physicists publishing in the *Bulletin of the Atomic Scientists* and statesmen publishing in *Foreign Affairs;* within the French and German armies, stirred by the intellectual ferment that sometimes follows military defeat; and in secret seminars that from 1958 on brought nuclear war questions before every high-ranking Soviet military officer. The strategy debates began when hydrogen bombs and rockets came over the intellectual horizon and lasted for a decade of clever and vigorous argument.[37]

The keystone concept in the debate was deterrence. The first national leader to explain the idea clearly to the public was Winston Churchill, speaking to the House of Commons in 1955. He said that by the end of the decade both the United States and the Soviet Union would be able to guarantee total destruction of an enemy, and he called that a stable and even a welcome situation. In one of his last unforgettable aphorisms, Churchill predicted that "safety will be the sturdy child of terror, and survival the twin brother of annihilation."[38]

Deterrence rested on the ability to eradicate the enemy within a matter of days, as in the actual war plans drawn up by the Strategic Air Command and no doubt by its Soviet counterpart. The most effective plan, all strategists agreed, would be the most extreme: a First Strike. Many past wars had started with an all-out sneak attack, and the threat was particularly familiar to Americans with their memories of Pearl Harbor and to Russians with their memories of Hitler's 1941 surprise invasion. Now the goal of strategy was to deter such a First Strike. For this a nation must possess, in the analysts' new jargon, "survivable" and "credible" forces that could wreak "assured destruction" in a "second strike."[39]

This doctrine was one of the many abstract intellectual concepts that could be built in steel. An American fleet of submarines with Polaris missiles, a survivable, credible, assured-destruction, second-strike force, went to sea beginning in 1960, later joined by similar Soviet, British, French, and Chinese ships. The missiles were barely accurate enough to find an urban area. They were not aimed at traditional military targets, but at the minds of enemy leaders.

Deterrence was so far from normal military strategy that, according to some, it did not even require a nation to possess a stronger force than the enemy had. By the early 1960s this concept of "sufficiency" was mentioned in official United States policy, while in practice the British, French, and Chinese built only enough weapons, as the French delicately put it, to tear off an enemy's arm. American and particularly Soviet planners were not content to stop there, however. Each wanted to be the nation with the most weapons, if only to keep the enemy from hoping that after a war he could rule over the survivors.

The desire to have a multitude of weapons was reinforced from the late 1950s on by strategists and statesmen who pointed out that it might not be easy to deter a nation from grabbing a little slice of territory. Would Kennedy really risk the destruction of Washington to keep Russians out of Hamburg? The alternative was to blast invading troops directly with "tactical" nuclear weapons, that is, ones not more than a few times larger than the Hiroshima bomb. Amid fierce international debate, thousands of such weapons were emplaced in Europe and the Far East. Many strategists hoped that war with such armament would be a step down the ladder from total destruction.

For the Western Europeans across whose countryside the tactical bombs would explode, it was not far down. Their strategists advised backing down another step, building a defense that could stand without nuclear weapons. Perhaps a mental cordon could be raised against "first use" of the bombs; perhaps the moral horror of nuclear energy had become powerful enough all by itself to keep war inside traditional limits?

Soviet military doctrine discouraged such talk. The Soviets were sure that an army fighting for its life would use its most decisive weapons almost from the start, and they openly planned to do that. NATO itself never felt certain it could stand off the Red Army without "escalation" up the ladder to nuclear weapons. And in any case Europeans, recalling the deserts of rubble they had witnessed in 1945, found a conventional war scarcely more attractive than a nuclear one. NATO therefore stated what was the simple truth as a matter of policy—that any invasion might bring them to use nuclear weapons.

NATO's soldiers might still look like the brave warriors of yore, but they actually served less as traditional defenders than as a trip-wire, rigged to set off tactical bombs upon any serious incursion. And many thinkers, especially Soviet ones, insisted that on the day tactical nuclear weapons were used, all restraints would be dropped; the wise warrior must therefore begin with a First Strike. No matter how far strategists tried to climb down the ladder, the specter of total destruction followed.

The most famous student of this strange ladder was a young RAND physicist, Herman Kahn. He seemed the opposite of a military man with his exceedingly round stomach and thick glasses, but he could pour out sparkling ideas and neatly packaged historical anecdotes in a way that kept senior officers fascinated through hours of lectures. In 1960 he piled together his notes and published them as a volume entitled *On Thermonuclear War.* At once controversy erupted over Kahn's unfamiliar logic, as exemplified above all by his "Doomsday Machine."[40]

This notional device would be set automatically to destroy the world if ever bombs exploded on the owner's territory: the ultimate in deterrence through assured destruction. Kahn was quick to admit that building a Doomsday Machine would be stupid and immoral. He insisted that nevertheless people ought to analyze it—for the United States and the Soviet Union, with their trip-wire troops and missile submarines and all, were building something ever more like a Doomsday Machine. Kahn had put his finger on a central problem. It had already been hinted at by Churchill, whose landmark 1955 speech noted that beyond a certain point deterrence meant "the worse things get the better." Churchill frankly called that a "paradox" he could not resolve.

In 1962 Kahn got to the heart of the paradox by borrowing an analogy from Bertrand Russell. The philosopher had drawn attention to the adolescent game of "chicken," in which two cars drive headlong toward each other until one driver swerves aside, losing the game. Kahn accepted the analogy to nuclear diplomacy. He added that the best way to win such a showdown would be "to get in the car drunk, wear dark glasses, and throw the steering wheel out of the window as soon as the car has gotten up to speed"; the opponent would surely give way. The game of chicken went even further than the Doomsday Machine in demanding that players give up responsible control.[41]

Unable to refute the logic of this simplified system, some strategists decided that the only solution was to make the system complex, to keep as many rungs as possible on the escalation ladder between a border scuffle and total destruction. A nation must have plenty of everything—troops with rifles and tactical warheads and so on up to forces bordering on a Doomsday Machine. As practical advice this policy was publicly adopted by Kennedy and all later American presidents, and more quietly by every other nuclear-armed nation. Eventually certain leaders like Richard Nixon took Kahn's advice to its logical end, working to convince their enemies that they might use bombs irrationally. The whole system was bizarre but it seemed to work, precisely insofar as leaders, aware of the crazy uncertainty of all human affairs, feared to test even the first rung of the rickety ladder. In short, the system's success rested upon its very lack of sense.[42]

This insoluble paradox in deterrence theory, this precarious logic of unreason, was not clearly understood until the 1980s, and even then the majority of military and diplomatic thinkers remained mired in its deliberate confusion. From the 1950s on the sharpest analysts left ambiguities, internal contradictions, and blind leaps of logic in their writings. Most writers changed their position from one year to the next and sometimes, it seemed, from one page to the next.

An example of the muddle was the failure of most writers to define clearly even the key term "deterrence." Sometimes it meant, as the French translated the term, "dissuasion." That meant arranging things so that enemies would deduce, like chess players, that they should not launch an attack because it was clear they would not win the game. Other times deterrence meant what the Russian translation frankly called "terrorization," which did not address the intellect at all. Of course, military logic on the one hand or an appeal to raw fear on the other might well require different strategies and even different hardware. But most thinkers mixed the two approaches, evading refutation in one mode of thought by shifting indiscriminately to the other. An even more surprising gap in the debate was that, caught up in the details of their argument, many strategists forgot to mention that nuclear weapons could never be used on cities in a way that would satisfy simple morality.

The debates over ambiguous and dehumanized concepts bewildered the public. Popular thinking about strategy took on an Alice-in-

Wonderland quality, amounting to a set of inconsistent clichés that reflected the internal contradictions of more sophisticated discussion. At the deeper level of belief and symbolism, the countless clever articles and books published in the late 1950s and early 1960s scarcely influenced how most people imagined nuclear war. Quite the reverse: the already widespread images of nuclear war influenced the strategists.

One example was that central theoretical concept, the all-out First Strike. A surprise attack had been a reasonable idea when Pearl Harbor was bombed, but in the nuclear age it would be less a military tactic than the slaughter of an entire people. Any child could see that only a uniquely powerful, totally evil, and downright insane person could commit such a crime. The First Strike of the theorists resembled nothing so much as the mad scientist's outburst.

The same theme infected discussion of the alternative policy, graduated deterrence. Theorists concentrated on how to keep escalation from climbing to full-scale warfare, or as they put it, from going out of control. To call what would be a deliberate decision of national authority, prepared long in advance and dutifully carried out with the cooperation of millions, "going out of control"—that too was less military logic than mad-scientist thinking. And the keystone of policy, mutual assured destruction, was still closer to old tales, as the Doomsday Machine made plain. Yet these fantastic ideas almost displaced the concept that had dominated traditional strategy: war as a continuation of politics. The genuine, and far from drastic, local conflicts of interest between the United States and the Soviet Union tended to disappear behind a fog of exclamations about an apocalyptic suicide pact.[43]

Still more curious, most analysis halted at the point where deterrence failed. What would the political situation of the world be like a year after that? In other words, what rational human purposes would be served by a nuclear war? Almost every thinker tended to ignore this question. Even in professional strategy analysis the time beyond full-scale use of bombs usually resembled the space beyond midnight on the doomsday clock, that empty zone the mind could never enter. As one analyst later remarked, "My mind just stops there."[44]

Other mythic themes, for example survivors as wretches or heroes, also permanently infected official thinking. One example showed up in an important debate that the United States Senate undertook in 1968. At stake was a costly new weapons program with grave implications. The issue was whether to settle for an armory sufficient to wreak total destruction, whatever that might mean, or to work toward the ability to win a war, whatever that might mean. Two Senators, familiar with the complexities of military thinking, addressed each other:

> *Mr. Clark:* There comes a time when the tens of millions of casualties are so enormous that civilization is destroyed, and if there are a few people living in caves after that, it does not make much difference.
>
> *Mr. Russell:* If we have to start over again with another Adam and Eve, then I want them to be Americans and not Russians.[45]

THE WORLD AS HIROSHIMA

Most adults suspected that real nuclear war would have little to do with the empty streets of *On the Beach,* or the mutant monsters of *Teenage Cave Man,* or the heroic survivors of *Alas, Babylon,* or even the calculations of *On Thermonuclear War.* People who tried to think seriously about war pushed colorful fantasies and ingenious theories into an out-of-focus background, leaving actual experience in center stage. This experience was limited. The public was familiar with some basic bomb effects described by atomic scientists, some tales of Japanese victims, and a few dozen photographs and newsreels of the Hiroshima wreckage. Many people were like a woman who said, "What's imprinted on my brain is those photographs that I have seen, so if I ever did think of it I'd see a city destroyed and blackened and burned." That became the world public's fundamental image of a world after the bombs.[46]

It was a dangerously incomplete image. Although some exceedingly gruesome photographs had been taken of the wounded at Hiroshima, the United States government released very few of these through the 1950s and 1960s; the commonly seen Hiroshima pictures mostly showed vast landscapes of rubble, empty of victims. This impersonal image was reinforced by newsreels of empty houses flattened in bomb tests and newspaper drawings of skyscrapers snapped in two. The destruction was usually viewed from an Olympian distance, as in the frequently published maps that showed with concentric circles how many square miles of a city would be pulverized. The best examples were paintings done by the master illustrator Chesley Bonestell for magazines in the 1950s, views gazing down from a great height upon a city lit by a nuclear fireball. Earlier illustrations of cataclysm going back to medieval times had put human victims in the foreground, but

Bonestell's paintings might have depicted (what was his usual subject) a distant astronomical event. Nonfiction writers offered little more. Usually they confined themselves to brief, hackneyed phrases about the dreadfulness of war and some abstract statistics about how many millions would be killed.[47]

When pollsters in the 1950s asked Americans what the world might look like after a nuclear attack, most people could scarcely come up with an image. Some talked vaguely about blasted landscapes as in the Second World War or worse. A small minority offered hopeful images of survival and recovery, while another minority spoke of utter doom. About a third gave purely emotional responses such as, "Oh my God it would be terrible, I can't imagine what it would be like."[48]

Although a few civil defense films of the early 1950s had offered more personal scenes, these were quickly outmoded by hydrogen bombs. The same obsolescence overtook early attempts at realistic fiction, such as the *Collier's* "War We Do Not Want" issue and Wylie's *Tomorrow!* After the IVY test, the stage was dominated by tales of mutant tribes and so forth. Nearly three decades passed without any technically accurate and widely seen portrayal of hydrogen bomb war.

There was one exception, *The War Game*, a television film produced for the British Broadcasting Corporation and released in 1966. The carefully researched scenes, such as a group of maimed children sunk in apathy or a bucket full of wedding rings stripped from the dead, seemed too grim to show the public, and *The War Game* was the first BBC film that the Corporation refused to broadcast. The film only played to relatively small audiences at colleges and the like. Most of the public, if they set aside fantasies, were left with the old, mostly impersonal photographs of bomb tests and Hiroshima.[49]

Several polls studied how people reacted to the idea of nuclear war. I will summarize the results in terms of the way eight representative Americans felt in the early 1960s. One of the eight was certain that nuclear war would mean the end of life on earth. A second representative citizen expected the end of civilization, or at any rate unbearably brutal conditions for the survivors. At the other extreme, one of the eight was confident that the United States could come through without much damage (further surveys showed that this person knew little about nuclear weapons). The remaining five, in the middle, felt that the United States had a chance of eventually rebuilding its society, but they expected great destruction first. The best-informed half of the public thought that their own towns would be annihilated or at best harmed by fallout, and felt their personal chances for survival were "poor" or "so-so." In Europe and Japan the range of opinions seems to have been similar, perhaps on average a bit more pessimistic.[50]

The most typical image was of the entire world becoming one boundless Hiroshima. A 1962 report to the American government by a panel of leading social scientists said that many citizens expected "physical destruction to be almost universal and the post-attack world to be a hopeless shambles, in which everything worth living for will be irretrievably lost." The panel noted, however, that opinions varied, with most people ignorant or confused about what might happen. As the woman who spoke of the Hiroshima photographs explained, the image was "not elaborated."[51]

In fact, specific conditions could be predicted to follow if a nuclear war were fought in the 1960s. Nations would be reduced to scattered islands of ramshackle industry in a sea of dejected, disease-ridden refugees. But aside from a few conscientious civil defense officials and *The War Game,* hardly anyone put the facts in such concrete terms; therefore even though the vague pessimism of the majority of people reflected the truth, the real after-the-bombs future did not become vivid enough to displace the fantasies that lay in the background. From the corners of their eyes the public continued to glimpse lurid visions of a sterilized planet, uncanny mutants, or a new Adam and Eve with backpacks.[52]

Things went differently in the Soviet Union, where the public never saw movies or novels depicting a future nuclear war. One or two science fiction writers did mention the possibility of postwar misery or lonely survivors, but they did not carry the images far. Some Western views filtered into the Central European satellites, as when *On the Beach* played in Poland, but only echoes of that got past the Russian border. Nevertheless, a 10-megaton capitalist bomb would produce much the same effects as a 10-megaton socialist one, and political leaders began to understand and explain that.

In 1963 Khrushchev himself said a nuclear war would flatten cities, kill populations, poison the atmosphere around the world, even exterminate all life in entire countries. This gradually became the official Soviet viewpoint, as expressed in public most forcefully by party ideologist Mikhail Suslov. Citing a calculation by Pauling that all but a few million Americans would perish in a war, Suslov said the same thing could happen elsewhere too, and "the question of the victory of socialism would no longer arise for entire peoples, inasmuch as they would have disappeared from the face of the earth." Soviet leaders took care never to say that Russia herself might perish, but sometimes the thought slipped into print. For example, in 1964 the President of India gave a speech in a rally at the Kremlin, saying that all humanity would die if nuclear weapons were used; the statement appeared in *Pravda*.[53]

These abstract words were reinforced from the late 1950s on by more visceral images in the omnipresent Soviet civil defense manuals and posters, which showed bleak pictures of bomb blasts and people huddling in shelters. Meanwhile traveling exhibits and other media gave many citizens of Communist countries a look at the Hiroshima photographs. The smashed landscapes and pitiful victims were meant to show American cruelty in the past, but a viewer might also think of the future.[54]

Meanwhile optimism remained the main official lesson in the compulsory civil defense courses, while the Soviet military, if only to keep up their own morale, insisted they would win any nuclear war. The Russian generals foresaw a sort of radioactive Second World War, with large areas of their nation scorched yet once again struggling to victory. After Khrushchev was deposed in 1964 this view became more common than exclamations about doom.[55]

Did the Russian public as a whole foresee destruction or victory? The question is meaningless, for no "public" in the Western sense existed, no citizenry aware of its opinions and freely expressing them. Ordinary Russians, feeling there was nothing they could do to influence policy, saw no more sense in thinking about nuclear war than in worrying about cancer. Nevertheless they had feelings, which can be reconstructed from the published Soviet literature and the reports of travelers and émigrés.

By the 1960s even the most confident generals agreed that nuclear war would be an unprecedented catastrophe. Civil defense instructors heard people saying, "When the atom bomb goes off, nothing can help." A scientist émigré from the Soviet Union, one of several I interviewed, told me that if nuclear war came, "Only cockroaches would be left alive." (She did not know that this was a cliché in the West.) Perhaps most characteristic of all was a joke that became known to most Soviet citizens by the early 1960s:

Q. What should you do in case of a nuclear attack?

A. Get a shovel and a sheet, and walk slowly . . . to the nearest cemetery.

Sometimes a second part was added:

Q. Why slowly?

A. You mustn't start a panic.

As a parody of the evacuations featured in Soviet civil defense instructions, this was the exact equivalent of a parody of duck-and-cover instructions that became known to most Americans by the early 1960s; in the event of attack, said the Americans, you should "get down on the floor, put your head between your knees . . . and kiss your ass goodbye."[56]

The average Russian image of nuclear war was close to the average American image, the world as one vast Hiroshima. Yet their reactions to the image were different. Total war was an abstraction to most Americans, but most Russian families had personal memories of a wretched life amid rubble at some point during the years between 1914 and 1945, and the memories were kept alive by a ceaseless barrage of patriotic films, stories, and ceremonies. Soviet citizens would imagine nuclear bombs less as bringing a Hiroshima than a Leningrad, a city that had survived only after suffering unspeakable horrors. They could believe that Russia would somehow come through, yet at the same time they could be desperately anxious to avoid a war, more realistically afraid of it than Americans with their mixture of abstractions and piquant fantasies.

Western Europeans and Japanese overlapped both sides. Like the Russians, they remembered real bombs and saw worse ones lurking right at their borders. Like the Americans, they had free access to published fantasies. Accordingly, it was in Western Europe and Japan, above all, that the most ghastly images of nuclear war would become connected with the most solid political realities.

13

The Politics of Survival

All the early active responses to the coming of nuclear weapons, from security hunts to civil defense, from the atomic scientists' movement to strategy analysis, had begun in small elite groups, and mostly within governments. The public's reaction was more passive, and through the 1950s can only be inferred from what people said to pollsters, wrote in letters, or went to see in movie theaters. Except for a few minor pacifist groups and the usual Communist propaganda, organized public effort regarding nuclear weapons was unknown outside Japan. But in 1957 the world's newspapers began to show a few photographs of protest demonstrations elsewhere, and by 1961 such photographs were appearing frequently. As the news about hydrogen bombs sank in and governments pursued Atoms for Peace as an answer, ordinary students and housewives were throwing themelves into a very different crusade. Like Atoms for Peace, this public movement had much to do with immediate political matters, which I will mention only briefly; my subject is the more enduring cultural, social, and psychological forces that also inspired the movement.[1]

THE MOVEMENT

The breakthrough came during November 1957 in response to the shock of Sputnik. The advent of ballistic missiles spurred liberal intellectuals and small pacifist groups to appeal to the public, and the response went beyond their hopes. When the writer J. B. Priestley published an article in *The New Statesman* calling on Britain to disarm, he got more than a thousand spontaneous letters of support; when a group led by Norman Cousins and others published an advertisement in the *New York Times,* they drew 2,500 responses. The writers began to organize a movement, but they did not guess what they were starting. The following Easter showed them.[2]

Some little-known pacifists had announced a protest walk from London to a nuclear bomb production plant fifty miles distant. They hoped for something like the demonstrations of a few dozen people that had been held here and there over the preceding decade, but when the column left London on Good Friday, 1958, it was two miles long. Cold wind and driving rain met the marchers during four days on the highway, but at the end 10,000 people stood shivering in silent vigil outside the barbed wire fence at Aldermaston.

Liberal and pacifist leaders meanwhile formed a Campaign for Nuclear Disarmament (CND) in Britain, a National Committee for a Sane Nuclear Policy (SANE) in the United States, and similar groups in other countries. Like the Japan Council against Atomic and Hydrogen Bombs, founded back in 1955, these organizations served as coordinating centers for swarms of smaller groups, all with little funds or staff. The movement lived on the donated services of hundreds of talented people, from artists to songwriters; on the long hours that thousands of volunteers gave to distributing petitions and arranging meetings; on the attendance of tens of thousands of protestors at demonstrations like the Aldermaston march, which became an annual event imitated around the world; and on the silent sympathy of millions.

The movement's message was embodied in a set of familiar images, bundled together and thrust in the public's face. A British pamphlet explained how bombs would kill people everywhere "in a lingering and horribly painful manner . . . just as they are still dying in Hiroshima." This was the movement's most characteristic picture, the world as Hiroshima. Many speakers and writers went on to offer science fiction images, grotesque mutants and all, as sober truth. For example, a physicist talking over loudspeakers to a hundred thousand people at a German rally predicted that only a few youths would struggle through to rebuild society, "perhaps beginning in the Stone Age." Many others said that nuclear war would exterminate the race. In short, like earlier groups only more single-mindedly, the movement spread powerful images with the deliberate aim of promoting nuclear fear. Few citizens escaped hearing the message—as a British leaflet put it, "Act Now or Perish!"[3]

The author of this leaflet was the movement's patron saint, a very old man with a shock of white hair and the grim look of a schoolmaster whose charges have been up to dangerous mischief: Bertrand Russell. When the philosopher preached the imminent end of all life on earth, he was drawing on a personal concern about death that was of long standing. All through his youth, frantic to escape an unbearable loneliness, he had considered suicide. In his most suicidal period Russell had written a story expressing his despair, telling of a scientist who invented an atomic gadget that could destroy the universe, and who found politicians so appalling that he decided to press the fatal button.

The feelings this story reflected may have begun when Russell was two years old, for his mother had died then, and he would have been unlike other humans if that had not left him with a desperate loneliness. When the news from Hiroshima arrived, Russell predicted world doom and fell into what he later called "a very much exaggerated nervous fear," which eased only in 1955 when he began to speak out publicly against bombs. He found that millions shared his feelings.[4]

Russell and his fellow campaigners focused not only on abstract future danger but on present-day bomb tests, until the anti-bomb movement became inseparable from the fallout controversy. A typical SANE advertisement hitched the two issues in tandem: "NUCLEAR BOMBS CAN DESTROY ALL LIFE IN WAR . . . NUCLEAR TESTS ARE ENDANGERING OUR HEALTH RIGHT NOW." This was the mingling of fears that left many of the public worrying chiefly about radioactive contamination and even displacing their fear of war onto strontium-90 in milk. However, leaders of the movement like Russell were convinced that the real contamination was the very existence of bombs.

Since the only way to do away with bombs was to attack the governments that fostered them, the movement was inherently political. An instructive case was in West Germany, the first strong campaign outside Japan. Germans had begun to stir when they realized that NATO's war plans rested on the idea, as a British Member of Parliament candidly put it, that "it is surely better, if bombs are to be dropped, to drop them only on the Continent, and not on Britain and on the United States and the U.S.S.R." The likely results were publicized by the NATO exercises where mock Soviet invasions were fended off with mock tactical weapons that killed millions of mock German civilians. The first loud protest came, as usual, from scientists. In April 1957 eighteen prominent West German physicists called on their government to renounce "explicitly and voluntarily the possession of atomic weapons of any kind."[5]

The left wing of the Social Democratic Party made that a political issue. When the German Parliament agreed in March 1958 to accept partial responsibility for holding NATO nuclear weapons, Social Democrats took furious exception. Hoping the issue would help them carry the next elections, the whole party joined with grassroots organizations in a fervent "Kampf dem Atomtod" (Campaign against Atomic Death). Countless speeches, pamphlets, posters, films, demonstrations, and strikes denounced nuclear weapons. The bitter confrontation tore at every German. "There seemed no basis for consensus," a political expert recalled; "it was impossible even to talk—you were either for or against and that was that!" In the end anti-Communism proved stronger and the Social Democrats lost the elections, but meanwhile their political ambitions had brought propaganda against bombs to the fore.[6]

Equally instructive was an opposite case, in France, where the government met little opposition as it developed a nuclear strike force. Although most of the French public too believed that nuclear war equaled national annihilation, they would acquiesce in any military decision taken by Charles de Gaulle and other admired national authorities. The French Socialist Party might have been expected to champion pacifism, but Socialists had led the government at times during the 1950s and had yielded to nationalism, allowing weapons designers to proceed with their work; therefore opposition to bombs was associated only with Communists. Besides, when the French thought of atoms they thought of Madame Curie, a national glory, and French filmmakers created no visions of radioactive monsters. Although anxieties about radiation and war persisted beneath the surface in France, as everywhere else, that was not enough: imagery can do nothing all by itself. Nuclear protest could grow strong only where cultural and political forces sustained it.[7]

Where the movement did flourish, it drew its strength from a new style of culture and politics. The CND, SANE, and similar organizations attracted not so much ordinary politicians and business leaders as people who took a stand outside the entire established order. Intellectual supporters like the British "angry young men" were writers already known for their alienation. A study of the American novelists who had written about nuclear weapons ever since 1946 found that their anxieties about the bombs were mingled with attacks on other institutions such as schools, churches, politics, business, and so on, all seen as dangerous failures. Less famous adherents came mainly from the middle classes, such as teachers and civil servants, but a sociologist who studied the CND found that these apparently solid bourgeois also suffered vague social discontents, questioning not only nuclear weapons but everything from the religious establishment to unfettered capitalism. A historian of the middle-class housewives who founded the Japanese movement likewise found them influenced by modern "cultural formlessness," with their movement partly a form of self-criticism. The Kampf dem Atomtod was similar, its leaders including scarcely any businessmen but many union leaders, writers, high school teachers, and unconventional churchmen. Alongside them marched openly disaffected Marxists and radicals.[8]

Most visible of all were the young. For the first time in modern history young people made up a large part of a mass campaign. To many students the marches were only an open-air diversion with guitars, part of the new youth culture. But this was an increasingly alienated, rebellious culture, and young people began to call nuclear bombs a prime example of everything they distrusted in adult society.

Another social category with an increasing distrust of authority—women—formed the backbone of the movement. Ever since the early 1950s, when the only protests had come from small church groups and the like, hard-working housewives had been invaluable organizers. Specifically female groups such as Women Strike for Peace sprang up around 1961. Perhaps one reason these tens of thousands of women took the lead in opposing weapons was that females in Western culture tended to suffer anxiety and to avoid risks more openly than males did. Some women had a different explanation, claiming that they had a special biological concern for issues like polluted milk and protection of children. "Men have always played irresponsibly with human life," they said, "while women have always protected it."[9]

This was less an overt feminism than a social viewpoint. Not only women but all the movement's followers wanted to stand up against unfeeling authorities and the entire aggressive culture that produced bombs. The sociologist who studied CND supporters found that they all "tended to translate the general anxiety arising from the dangers of a nuclear war into a full scale critique of contemporary society."[10]

Foremost of such social critics was Priestley, the writer whose compelling magazine article had helped found the CND. He confessed that he would prefer to be back in the sunlit Edwardian world of his youth, for he found modern society harsh, depersonalized, and entirely on the decline. People were spending ever more money on bombs, he complained, while giving novelists and playwrights (like himself) ever less attention, less prestige, and indeed less money. However, Priestley never said outright that his feeling of alienation had anything to do with his social role. He only said he despised the rigid technology that was taking command, and longed instead for "magic," the rich mystery of nature and feelings.

Reflecting in his old age, Priestley came up with a clue to his way of thinking. His early childhood had never felt secure, for his mother had died soon after he was born. "There were areas of dark bewilderment," he wrote. "Something was missing that should have been there," something that only magic might hope to recover. This is not to say that his mature character reflected only infantile thinking. When Frederic Soddy and Philip Wylie lost their mothers they grew up lonely and bitter, talking as if the world almost deserved to be blown up. From the same loss Priestley, and Bertrand Russell too, fought their way through pessimism to a sharpened perception of human failings and a courageous refusal to despair. They threw themselves into heroic struggle on behalf of life, taking their personal longing for a trustworthy world and transforming it into a call for a better world for everyone.[11]

In his writings and speeches for the CND, Priestley decried not only nuclear armament but every modern, mean-spirited form of conflict. He called on his countrymen to rise above partisan selfishness and start a moral "chain reaction" that could sweep the world; any noble cause would do, but renouncing the bombs would be best of all. Many other campaigners agreed that in fighting nuclear weapons they were fighting a deeper moral corruption. "The Aldermaston march," one churchman explained, "is necessary not chiefly to save our skins but to save our souls."[12]

Most explicit was a CND pamphlet describing a nuclear attack, not with the usual frightening tales of Soviet bombs on London, but in terms of British bombs on Moscow. After telling how millions of innocent Russian women and children would suffer horrific tortures, the writer asked: "Do you consent to this being done?"

Sermons, articles, and books debating the morality of nuclear weapons arrived in a flood, more in a few years around 1960 than in all the years before or since. Some preachers and authors, notably Methodists, argued against nuclear arms on traditional pacifist grounds, adding only that war was now more hateful than ever. Others, notably certain Roman Catholics, argued that the new weapons were uniquely evil because it was impossible ever to use them in a just and measured way. Many ignored theology and exclaimed that nuclear bombs were self-evidently blasphemous, or as a Kampf dem Atomtod pamphlet put it, "Atomic Weapons Are Sin."[13]

With vigils, fasts, pilgrimages, and sermons, the movement took on the air of a religious revival. After all, nuclear weapons posed questions about the world's fate such as only religion had once addressed. A British protestor, arrested at an illegal demonstration, wrote from prison: "Our position is not dissimilar to that of the early Christians. With us, as with them, the last things have arrived." Such feelings reached deep into the campaigners' personalities. Like certain of the Japanese victims who overcame their despair when they joined the first antibomb groups, so in other countries people who felt isolated and anxious, or who had even become desperate through the death of loved ones or thoughts of suicide, found a new life by throwing themselves into the moral struggle. The most committed people were a small minority in the movement, but by expressing the strongest emotions they attracted attention and helped set the tone.[14]

In its combination of personal, moral, and social goals the campaign came to resemble a millenarian movement—the kind that has turned up in many human societies stressed by change, from messianic "cargo cults," common in tribes confronted with Western civilization, to modern urban revolutionary sects. According to historians and social scientists who studied such movements, the followers would typically be convinced that their society was in the last throes of decay; that it could be transformed into a far better order; and that they could themselves be midwives of the world rebirth. Often a millenarian group simply projected onto society each member's hopes for personal transmutation. But some problems of an individual are in truth social in origin, and cannot be solved unless the distressed individual joins with like-minded people to reform society. Nuclear war was such a problem. Universal anxiety would never be relieved except by altering the entire world.[15]

The question posed by nuclear fear, then, raised a creative answer: world community. This was no longer a platitude but the realization that all humans truly depend upon one another. Fallout had begun the process, thanks to the sensitive instruments that could detect a handful of dust carried thousands of miles on the wind. Magazine drawings of ballistic missiles arcing around the globe drove home the point. The idea of world community became the movement's greatest strength not only morally but practically when groups of various nations joined in mutual support. The sense of fellowship extended still farther, as in an editorial cartoon showing animals looking at a nuclear blast and thinking, "Nobody bothered to consult us." Another cartoon showed babies labeled "future generations" peeking into a room where shadowy figures discussed a test ban; as the caption said, the babies were "Interested Parties." This compassion for all life, from the victims of Hiroshima to the citizens of Moscow, from irradiated fish in the sea to generations unborn, was the most profound answer that could ever be given to the nuclear question.[16]

ATTACKING THE WARRIORS

It was not enough for the movement to voice a general moral appeal; it had to oppose the specific advice of nuclear strategists. That meant entering the thermonuclear strategy debate with reasoned policies in mind. Many antibomb intellectuals did that, but from the outset they did much more. Their arguments shaded into a powerful personal attack, in the realm of imagery, against the men who controlled nuclear weapons.

The movement's rallying-cry was "unilateral disarmament," a refusal to own nuclear devices no matter what other nations did. Some admitted that this policy might mean yielding to Soviet conquest. The idea, simplified into "Better Red than Dead," was accepted by a substantial minority throughout Europe. After all, they argued, the Eastern Europeans seemed to prefer their life under Communism, however dismal, to nuclear annihilation. Scarcely anyone asked whether moving the Iron Curtain a few hundred miles westward would really make nuclear war less likely. Still less did people ask whether Communists could maintain stable rule over the entire planet. Few foresaw that nations led by Communists would have a penchant for attacking one another; that did not begin until 1968 when the Soviet Union and its allies invaded Czechoslovakia (followed by Soviet-Chinese border skirmishes, Chinese attacks on Vietnam, and so on). Nobody noticed a tremendous fact: it is democratically elected governments, alone of all types of government, that have never in history made war upon one another.[17]

Few supporters of the antibomb movement analyzed the problem of institutionalized aggression, let alone reflected that one might be both Red and dead. They simply hoped to be neither. The basic image of unilateral disarmament was sketched as early as 1954 in a play that later appeared on British television, Marghanita Laski's *The Offshore Island.* It showed a family living in a snug farmhouse, self-sufficient and content, when platoons of American and Russian soldiers arrived. Yanks and Reds each wanted to deny Britain to the other, so they callously agreed that the island must be wiped clean of life. The author's message was that everything would be all right if only the soldiers of both sides would go away and leave decent people alone. To achieve this, many decided the best course was pure moral action. Wouldn't the example of renouncing the blasphemous bombs shame other nations into laying down their own arms?

A few analysts added that hydrogen bombs could not be used anyway against Ghandian nonviolent actions, general strikes, or guerrilla warfare. If a nation gave up bombs but trained itself in less centralized forms of resistance to invasion, surely even the Russians would hesitate to swallow such a hedgehog?

A small number of people went farther still. Noting that nuclear debate was stuck in Alice-in-Wonderland contradictions, they said that was because people were afraid to examine their most basic assumptions about the state. Historically, national states were created chiefly for defense. But there was no defense against nuclear missiles; it was now governments themselves that threatened survival. The solution to the impasse must involve going back to subnational forms of association. Although few bomb opponents carried their thinking to this logical conclusion, in emotional terms many went from abhorrence of the bombs to abhorrence of modern governmental authority. After all, among everything else nuclear weapons stood for, they were a supreme expression of deadly state power.[18]

Loss of trust in the state and hopes for autonomous resistance turned into a call to action when Bertrand Russell and others organized illegal demonstrations. These reached a peak in the early 1960s when Polaris submarines began to use foreign bases; thousands of demonstrators came together in Scotland and up to a hundred thousand in Japan, the protestors attempting to throw their bodies in the way of the long gray ships. However, civil resistance works better when it expresses majority opinion than when it tries to change it. And when a few rebellious youths in each nation came to the protests to taunt police, they attracted the scorn of the press, weakening the standing of the rest of the movement.

Rational argument became less and less prominent in the controversy. One reason was that the means of persuasion available to political outsiders—demonstrations and hand-painted banners, appeals to passersby to sign a petition, occasional newspaper advertisements or thirty-second radio spots—did not allow much space for sophisticated exposition. Another reason was that many of the movement's supporters tended from the outset to reject abstract analysis.

On opposite sides of the debate stood opposite types of people. The ones most likely to use detailed logic and numbers were the men professionally involved in weapons decisions, a General LeMay who had to calculate how many bombers to assign each target, a Herman Kahn who based his recommendations on the distinction between a war that killed eighty million citizens and one that killed only ten million. Poles away from that mode of thinking stood people like Pat O'Connell, a housewife who was arrested again and again in the early 1960s for leading civil disobedience demonstrations. According to a writer who interviewed her, she saw things "simply in terms of human beings." People like O'Connell felt as a personal blow each individual story, each child who might die of leukemia from fallout, each maimed survivor of Hiroshima. Much later, in 1982, a survey confirmed that the people most active in opposing nuclear war were motivated not by technical statistics but by concrete personal images, charred bodies of people and the like.[19]

Literary personages said that everyone ought to approach things in this personal mode. Archibald MacLeish, for example, warned that civilization was in danger "when the fact is disassociated from the feel of the fact." It was nothing new for poets to say that society needed more of the kind of rich insight that poetry could offer. But now they insisted that lack of human feelings threatened universal doom.[20]

Nuclear strategists retorted that civilization was in danger when people relied on emotions and refused to bother with facts. Kahn complained that protestors were using impassioned slogans to hide from reality. He said nuclear weapons must be analyzed only with the coolest detachment, and demanded more support for studies by professionals (like himself). It began to sound as if the choice between building bombs and banning them was equivalent to the choice between rational logic and human feelings.[21]

Some protestors reversed the equation, declaring that they were the reasonable ones. They said that RAND's clever logic and calculations were at heart irrational, meaningless to decision makers trapped in the muddle of real politics or real war—a conclusion that was also reached by many military officers. Some protestors went on to insist that any talk of using nuclear bombs, for war or even for deterrence, was plain madness. CND titled its newsletter *Sanity,* while SANE's name spoke for itself.[22]

Who specifically was not sane? The RAND theorists in general and

Kahn in particular made the most obvious target. The portly analyst was first bewildered, then outraged, when he read the *Scientific American's* review of his book: "permeated with bloodthirsty irrationality . . . [and] insane lunacies." More precisely, *Saturday Review* warned that Kahn's "game of high military strategy has its own fascination, especially if one can train one's self to think in terms of nations and statistics, not human beings."[23]

A stereotype was at work, one that Priestley had elaborated back in 1955. In a notorious essay he had sketched the personality of "Sir Nuclear Fission," supposed to represent a breed of modern man. As a youth, said Priestley, Mr. Fission had isolated himself in abstract scientific studies, disregarding the claims of his senses and emotions until he lost his natural tenderness for life. Warped by scarcely recognized frustrations, "brilliant but unbalanced," even as he made scientific discoveries Dr. Fission became divorced from the real world. Because he lacked friends and pleasures, the eminent government adviser Sir Nuclear undervalued human life, perhaps even disliked it, and was all too ready for war. Priestley was not, of course, describing real scientist advisers, lively and often sentimental people whose careers were rich in personal interactions. He was describing the traditional mad scientist ready to blow up the world—a stereotype he had portrayed already in his 1938 thriller *The Doomsday Men.*[24]

There was at least one real nuclear authority, however, who seemed much like the stereotype. Since his student days Hyman Rickover had been unsocial and obsessed with knowledge, turning himself, according to *Life*, into "a tough intellectual." The magazine added that Rickover even looked like an incarnation of relentless thought, his body no more than a slender appendage "utterly controlled by the head, not permitted to engage in frivolities." The Admiral's public face was more grim and rigid even than General LeMay's.[25]

The same stereotype covered all Polaris submarine officers, men selected and trained by Rickover. Like the SAC airmen who had been publicized a decade earlier, nuclear submariners had to be impervious to sentiment if only because they literally held in their hands the keys to catastrophe. Magazine articles explained that the crews endured months locked in steel corridors, forbidden any contact with their wives; "after a while," a reporter wrote, "even their talk of sex stops." The public did not know that Rickover's men, like LeMay's, had morale problems and a high divorce rate. Nobody publicized the fact that in a typical year one out of every twenty-six missile submariners was referred to a psychiatrist and some had to be hospitalized for paranoid schizophrenia and other mental illnesses, a higher problem rate than in other branches of the Navy. The accepted image, rather, was of men who had made themselves into logical machines.[26]

The stereotype of nuclear officers took its purest form at the missile bases the United States built in the early 1960s. *Life* showed men in underground capsules, wearing white coveralls like laboratory technicians and telling the reporter, "We're like robots in a way." The only thing these men seemed to have in common with military heroes of the past was their suppression of sentiment in the name of duty.[27]

Nobody at the time gave these magazine articles a symbolic reading, and most people still claimed to admire military men. As the perceptive editor of the *Bulletin of the Atomic Scientists*, Eugene Rabinowitch, wrote in 1956, to ordinary citizens the word "war" still meant "a vast movement of armies, as in the two World Wars," while "security" suggested not deterrence but "a protective wall of steel and concrete, manned by soldiers." Into the early 1960s these traditional images remained familiar in magazine tales of staunch sergeants repelling enemy attack, in John Wayne movies, and in government-sponsored television documentaries about the armed forces, all highly popular in the United States and some other countries.[28]

A stronghold of tradition was the Soviet Union, where a ceaseless parade of Second World War novels and movies and televison programs, along with military training for all males beginning in early adolescence, helped the armed forces keep their image as men of the people. When Russians wrote about cold-hearted "atom maniacs" they were referring only to their enemies. In the early 1960s censorship slackened, and a few stories and movies appeared in which war was horrible and soldiers were less than noble; perhaps that showed how Russians really felt. The stories upset the Minister of Defense, who publicly lectured artists and writers about their duty to avoid pacifism. They all took his advice.[29]

Yet everywhere imagery was on the move in advance of overt opinion, for traditions were under pressure from new facts. Already by the end of the First World War, footsoldiers had looked more like victims than heroes, and by the end of the Second World War it was clear that they could not always protect their homes any more than themselves. That war, unlike most previous ones, had killed roughly as many civilians as soldiers; a limited nuclear war in Europe would kill ten times as many civilians as soldiers; an all-out global war could make the ratio a hundred to one. A physicist noted that tales of soldiers heroically interposing themselves between their loved ones and the desolation of war had become "fairy tales, and not nice ones at that."[30]

Meanwhile the military officer, once stereotyped as a romantic hero or, if not, then as a pigheaded tyrant or windbag, was gaining a grim complexity from his association with science. Links between officer and scientist were being forged in real places like the briefing rooms of RAND and the burgeoning weapons laboratories. Ever since the 1940s somewhat over a third of all the physical science and engineering researchers in the United States, and still more in the Soviet Union, had pursued military goals. The armed forces in turn had come to rely on technology more than on valor. As a Soviet general remarked about his missile troops, not only were rigorous training and technical perfection now essential, but "the commander who is at the same time an engineer has become a central figure." All this would have happened if there were no nuclear weapons, but the bombs did more than anything else to label military officers not only as intelligent professionals, but as robotic masters of technology.[31]

The officer's image was changing even in stories of past wars. The first worldwide bestseller in what would become a new genre was John Hersey's *The War Lover,* published in 1959 and soon made into a movie. Here the stalwart Second World War bomber pilot, hero of earlier tales like *Twelve O'Clock High!,* became a "hunk of machinery" who loved his airplane more than his girlfriend, loved still more to drop bombs with icy precision, and in the end rode his bomber down to suicidal death. Military toughness no longer seemed entirely admirable.[32]

RUNNING FOR SHELTER

While the movement for nuclear disarmament was getting under way, people who believed in tough-minded logic and military strength laid plans of their own. For advice on nuclear defense the Eisenhower administration called together a distinguished panel chaired by H. Rowan Gaither, a founder of RAND, and in November 1957 the panel submitted its secret report. It took Eisenhower aback. The Gaither Report warned that the Soviet Union was on its way to building so many missiles that, in the worst case, they would soon be able to destroy SAC with a surprise First Strike and rule the world. This was followed by 1958 military intelligence estimates that said the Soviet Union could soon have ten times as many missiles as the United States. (In fact the early 1960s would see just such an imbalance, but in the Americans' favor.)

The worst-case estimates quickly found their way into the press, and

Democrats who had worried about a "bomber gap" in 1956 now began to attack Eisenhower for allowing a "missile gap." The prospect of a real missile war began to preoccupy the government. This brought to the fore a particularly knotty problem, emphasized in the Gaither Report and every other analysis. For simple deterrence to work, it did not matter whether a nation expected to survive if the balancing act failed. But for a nation to fight a war in the old chess-playing military sense—missiles chiefly attacking missiles and some semblence of victory in the end—it was essential to have an effective civil defense program. That fact opened the way for a new assault of imagery on the public mind, striking closer to home than anything so far.[33]

American civil defense was in disarray. Whenever people like Teller called on the United States to spend billions of dollars on bomb shelters, others replied that by the time such shelters were finished the enemy would have enough weaponry to blast them open. Even Philip Wylie agreed with that, for the former civil defense advocate admitted that the coming of hydrogen bombs made him switch sides. "Where once I felt national apathy was dangerous," Wylie wrote, "I now feel it would be common sense."[34]

When Kennedy became President he revived the question. As a congressman and senator he had always spoken out for civil defense, and his presidential campaign had been filled with warnings of the missile gap. Now civil defense officials, starved for public respect and funds, hopefully bombarded the new President with memoranda. Imitating his campaign slogans, they said that a shelter program would "awaken the country" to defense needs and "show the world that the U.S. means business." The officials also noted that the Soviet Union's civil defense program was advancing swiftly; whether or not shelters would really work, the Soviets might become dangerously overconfident. Still more persuasive to Kennedy was an argument for shelters as "insurance" in case of war. A family that stayed beneath a few feet of dirt for a few weeks would greatly improve its likelihood of surviving fallout. How could anyone refuse a chance to save millions of lives?[35]

Kennedy came under special pressure from the governor of New York State, Nelson Rockefeller, whose long-standing interest in civil defense had sharpened when he befriended Teller. Rockefeller and other governors met with Kennedy in early May of 1961 to demand a shelter program. According to a secret summary of the meeting, Rockefeller hoped the program would "stiffen public willingness to support U.S. use of nuclear weapons if necessary." Other officials from mayors to congressmen joined the campaign, mounting a political challenge that the President, who saw Rockefeller as his chief rival for the 1964 election, dared not overlook. Besides, Kennedy was personally attracted to the John Wayne ideal of male vigor and military toughness. He was nevertheless unwilling to ask for the enormous multi-billion-dollar program that civil defense experts recommended. The one thing he could do easily was warn people to take care of themselves, and he began advising Americans to build shelters.[36]

The matter became urgent when Khrushchev threatened war over West Berlin and Kennedy made an equally belligerent reply. A nuclear scare built up, worse than any before, frightening the public in the United States, Western Europe, and the Soviet Union. It came to a climax in a tense speech the President gave over national radio and television in July 1961, implying that the world was on the brink of war. A draft of the speech written the day before had offered Americans detailed advice on taking shelter from fallout ("You will need water more than food," and so forth). This sounded too scary, however, and in the final version Kennedy only said that people should be ready to protect their families, and that he would ask Congress for funds to stock shelters with food, water, and first-aid kits—an indelicate mention of "sanitation facilities" was cut out at the last minute.[37]

Despite its cautious phrasing Kennedy's speech shocked the nation. The federal civil defense agency got more than 6,000 letters a day asking for information, more than it used to receive in an entire month. Civil defense was discussed in Rotary Club and PTA meetings, in school boards and churches, and for two months newspapers got more letters to the editor on shelters than on any other issue. In December a Kennedy aide canvassed the nation and reported that shelters had become the chief domestic concern, a fad verging on hysteria. Most citizens would have agreed with John Chancellor, who admitted on NBC's *Today* television program that he was "faced with a fear that I may find myself victimized by a nuclear attack." The reporter added that he was "ready to support a civil defense program that teaches me how to survive."[38]

The administration was slow to take up the challenge. Kennedy had announced he would send a civil defense booklet to every household in the nation, but his aides, taken aback by the uproar, kept delaying the publication for fear of setting off complete panic. Others hurried to fill the vacuum. Swimming pool contractors became experts on shelter construction, and manufacturers of everything from biscuits to portable radios discovered that they were already making survival equipment. Fly-by-night shysters and respected merchants advertised on television, encouraging the war anxiety. Banks set out free civil defense leaflets in their lobbies and offered people loans to build shelters; one bank, evidently not worried that war was imminent, advertised "up to 5 years to repay." [39]

The antibomb movement, seizing the issue as another opportunity to spread frightening images, vehemently attacked fallout shelters. Those would save only a few people, they said, who would anyway emerge into a world of rubble and would envy the dead. A controversy arose, one branch of the great thermonuclear strategy debate of the times. Shelter advocates rarely stressed the strategic value of civil defense in a limited chessboard war, usually arguing instead that shelters would help dissuade the enemy from launching the dreaded First Strike. Opponents replied that civil defense would give Americans false confidence, make the Soviets more nervous, and so render a First Strike more likely.[40]

The debate turned bitter when critics noted that there would never be enough completely safe shelter space for everyone. If fallout shelters did preserve anyone's lives, it would be the lives of farmers remote from cities and of suburbanites who could afford protection; tenement-dwellers would have worse luck. John Galbraith, in an acid letter to Kennedy commenting on a draft of the civil defense booklet, called it "a design for saving Republicans and sacrificing Democrats." More crudely, novels and films featured desperate fights at shelter doors—the old theme of a world reverting to savagery. As soon as the sirens sounded, a *Twilight Zone* television drama suggested, people would attack one another like "naked wild animals." In fact most shelter owners were willing to let in neighbors and even strangers, but about a quarter of the owners did say they would fight off intruders, and a few displayed rifles. Reporters exaggerated the issue, drawing peculiar statements from clergymen as to whether or not Christians had a duty to shoot someone who was trying to break into the family shelter. More pragmatic thinkers said that people who were locked out could block the shelter's ventilation shaft. A California entrepreneur responded with fake air vents to fool the neighbors. An aide warned Kennedy that the spirit of "do it yourself" that had inspired the home shelter program was slipping into a mood of "every man for himself." [41]

After all, what could be more American than the image of a father defending his homestead with a shotgun in a lawless world? Home shelters fitted nicely with the image of a new frontiersman who "could venture forth" after the bombs, as *Time* said, "to start ensuring his today and building for his tomorrow." This image of heroic survivors was encouraged by authorities who said that simple precautions could save most citizens. *Life* advised drinking hot tea for relief from radiation illness and printed a photograph of a teenage girl relaxing in a shelter, laughing and holding a bottle of Coca-Cola.[42]

A few critics noted sardonically that the much-publicized shelter somewhat resembled a womb, a safe place to await rebirth. Indeed the long wait in uncertainty, the cramped conditions, and the darkness sounded suspiciously like the traditional rite of passage into a new life, resembling the initiatory trials that young people in many primitive tribes were made to undergo in order to enter the adult world. Even if you were caught in the open when a bomb went off, civil defense instructions said, you should curl up with your head in your arms—a fetal crouch.

More commonly the shelter furor conveyed grimmer images, bringing back air raid fears with redoubled force. Most frightened of all were children, like a little girl who, upset when her parents decided not to build a shelter, picked out a closet to hide in when the bombs dropped. In many such cases the dark shelter beneath the ground represented victimization, separation, and pure death. Many agreed with the Soviet Minister of Defense when he told *Pravda* in 1962 that an atomic bomb shelter was "nothing but a coffin, a grave prepared in advance."[43]

To the majority, shelters did not clearly signify panic and savagery, nor heroic survival and rebirth, nor separation and the grave, nor anything they could picture at all. Most of the photographs in magazines and the sketches in American or Soviet civil defense booklets showed shelters as architecture, either empty or housing deliberately bland and impassive people. Except for a few antiwar stories that were not widely known, writers of fiction and nonfiction offered no lifelike pictures of people huddled for weeks underground, stench and all. The shelter debate drove home the idea of nuclear war as an indescribable catastrophe, while reinforcing murky associations with fantasies of victimization and survival, but it did little to bring the vague imagery into focus.

One definite result of the shelter debate was to center attention more than ever on radioactivity. By 1962 most Americans believed that if a bomb dropped on their city, fallout would kill or injure them; barely half feared the bomb's blast; fewer still feared death by fire. This was exactly backwards, for in fact a hydrogen bomb would do most of its damage with heat and next with blast. Fallout would be the greatest killer only in a limited attack that fell on remote military targets but generously avoided cities. Of course, it was natural for government officials and merchants to emphasize fallout, since that was the only thing that people stood any chance of protecting themselves against. But in the shelter furor, as in the concurrent debate over bomb tests, the great risk of fireballs covering every city took a back seat to anxiety about radioactive dust.[44]

Eventually the Berlin crisis cooled off. In January 1962 the Kennedy administration distributed its long-awaited booklet, setting out tens of millions of copies in post offices, but by then they had beaten the text into a featureless mush. Manufacturers of home shelters went out of business by the hundred, and talk of civil defense began to settle back toward a moderate level.[45]

The whole squalid controversy had been almost entirely confined to the United States. In the Soviet Union, shelter was a communal affair of mass evacuations; citizens were not told to build their own shelters but simply to sit through lectures. In the rest of the world (aside from Sweden and Switzerland, both steadily digging in), governments and citizens tended to think that any defense against hydrogen bombs was futile. Everywhere, mention of shelters did less to reassure people than to remind them of dangers that were rapidly increasing.

CUBAN CATHARSIS

"The very existence of mankind is in the balance," exclaimed the Secretary General of the United Nations. It was October 1962. Kennedy had come on television to announce that Soviet missiles were being emplaced in Cuba and to hint that if the missiles were not promptly removed he would attack them. In a secret message to Khrushchev, the President warned that the Soviets risked "catastrophic consequences to the whole world," and he told his advisers that he was not worried about the first step but about climbing the ladder to the fourth or fifth—"and we don't go to the sixth because there is no one around to do so." The Soviet Premier wrote back that he hoped Kennedy had not lost his "self-control," for the risk was "reciprocal extermination." He had not meant to provoke war, Khrushchev cried: "Only lunatics or suicides, who . . . want to destroy the whole world before they die, could do this." In the crisis the world's leaders abandoned the logic of strategic missile exchanges; instead they called up images of uncontrolled men blowing up everything.[46]

Among leaders and the public together, nuclear fear reached a higher peak during this crisis than at any time before or since. As Soviet ships approached the American blockade fleet, a considerable number of people from London to Tokyo thought they might not live to see another dawn. Young people in particular became deeply alarmed. The public was calm only in Moscow, where the press called for peace and did not mention until after the crisis was over that the squabble had something to do with nuclear missiles. In Washington shovels and sandbags were sold out at hardware stores, while Pentagon employees snatched up civil defense leaflets. In some cities food hoarding panics stripped supermarkets bare.

Those who had thought hardest about deterrence were especially worried. Undersecretary of State George Ball told his wife to turn their basement into a fallout shelter, and Herman Kahn carried a transistor radio everywhere he went. Two British antibomb campaigners hitchhiked to West Ireland to escape the coming fallout; some American campaigners fled to Australia; and Leo Szilard, most sophisticated of all, flew to Switzerland.[47]

Not everyone was so afraid. At Cornell University, where I was then a student, when the campus SANE group put a speaker on the steps of the student union building, members of the crowd shouted him down. Some of these students had been approached by the Minutemen, a national right-wing group that was stockpiling materials for guerrilla warfare in case of Communist invasion. When the exasperated speaker asked his hecklers, "Are you ready for nuclear war?" they roared back, "YES!" Similar clashes took place elsewhere, and in the crisis most people were ready to go wherever their leaders took them.

In the end the Soviets packed their missiles back home in exchange for an American promise to leave Cuba alone. Everyone except a few traditional militarists understood that a great nuclear war had been avoided only by difficult self-restraint. It had been true in crises throughout history that unless leaders exercised caution they were all to likely to trigger an avalanche; now hydrogen bombs had made the risk plain for any fool to see. Otherwise the bombs had been worse than useless. The solution had come through old-fashioned diplomatic accommodation on both sides. Nuclear warheads had only created the unnecessary crisis in the first place, and then carried it from a bothersome local confrontation right to the edge of catastrophe.

When the crisis ended most people turned their attention away as swiftly as a child who lifts up a rock, sees something slimy underneath, and drops the rock back. A survey of Americans in the wake of the Berlin crisis had found about half saying they had given some thought to building shelters, but after the Cuban crisis only a quarter could recall ever thinking about building shelters. In newspapers and magazines the space given to issues such as civil defense fell back to the pre-1961 level, and over the following years it kept falling. For example, the *New York Times* published more than 350 news items on civil defense in 1961, but only about 70 in 1963, 20 in 1966, and 4 in 1969. The *Readers' Guide* list of magazine articles showed the same collapse of interest in shelters.[48]

One reason for the collapse was a new spirit of accommodation among Soviet and American leaders. In August 1963 the two nations and Britain signed a treaty promising not to test nuclear weapons in the atmosphere. The treaty did nothing in practice to slow the nuclear arms race, for tests continued in underground shafts at a faster pace than ever. But out of sight was out of mind. There followed years of détente, with neither the crises nor the vehement official rhetoric of the 1950s. To symbolize the hopeful prospects when the Limited Test Ban Treaty was signed, the *Bulletin of the Atomic Scientists* set back the hand of its doomsday clock.

Governments promised to solve the problem for good. Since 1945 world leaders had sworn over and over that they were determined to bring about nuclear disarmament. Newspapers gave acres of front-page space to ingenious plans offered by one authority or another, along with complex negotiations on detecting hidden bomb tests, and controversies over inspecting countries for concealed weapons. All the speeches and negotiations constituted a major effort by governments to allay public anxiety. If this book were a history not of imagery but of public debate, close to half the chapters would have to be devoted to the nuclear arms talks.

How did the history of negotiations interact with nuclear imagery? One significant connection was that an obsession with control of secrets, already plain in 1945, dominated nuclear diplomacy. Objectively, the possibility that some fraction of weapons or bomb tests might go undetected was only one of many complex issues in arms control. But missiles hidden in mine shafts, bombs secretly exploded within huge underground caverns, endless talk of "inspection"—these were the questions that continually blocked progress. It is impossible to say whether the imagery of dread secrets brought an irrational element to diplomacy. But one result was plain: news about negotiations reminded the public again and again that nuclear energy was fundamentally connected with hidden things.

A more basic problem with the negotiations was noted already in 1947 by a Soviet delegate to the United Nations talks, Andrei Gromyko. One day an American delegate, Frederick Osborn, was walking down a corridor and happened to meet the Russian without his usual retinue of bodyguards. Osborn suggested that the two of them go off quietly to lunch and talk about disarmament, for they were both sincere people and could get to understand each other. Gromyko, exhausted from months of diplomatic dogfighting, leaned against the wall of the corridor. "He was quiet for quite a long while and looked at me," Osborn recalled. "Then he said, 'Mr. Osborn, you may be sincere, but governments never are.'"[49]

By the mid-1960s most of the world public had become equally cynical. Polls in several countries showed that a majority thought the world could avoid actual war but expected that little would come from arms talks. Of the many fictional works about nuclear war, scarcely any trusted diplomacy to work out a solution. Eventually the disarmament schemes that one side or the other proposed with loud fanfare did not even score propaganda points except among a gullible minority. Of course, true disarmament—an outbreak of peace—was not impossible; such a thing has happened between many nations in the past. Why were the talks so fruitless?[50]

David Lilienthal pointed out a main problem. In his early days at the AEC he had argued vigorously for international inspections, but in 1963, after Eisenhower and Kennedy had each proposed such schemes while the Soviets demanded still more utopian bomb bans, Lilienthal admitted that it was all "an escape from reality." Normal step-by-step diplomacy with limited aims in local situations could accomplish something, like the resolution of the Cuban crisis or the ban on open-air tests. But when leaders promised an age of perfect peace—and especially when they focused everything on the supremely sensitive issue of nuclear weapons—they only wound up in needless bitter quarrels. Lilienthal predicted that in the end nothing would come of the promises except disillusionment.[51]

He may have underestimated people's ability to keep hoping against hope for real peace, but his prediction that interest would falter came true. The space that the *New York Times* gave to arms control had been high in the first years after Hiroshima, fell very low during the Korean war, and rose again beginning in the mid-1950s; during the Kennedy years disarmament talks grabbed many front-page headlines. But once the Limited Test Ban Treaty was signed, the space the *Times* gave to arms control abruptly dropped to less than a third of the peak level.

Over the next two decades important progress was made with a few limited agreements that encouraged détente and even slightly reduced the number of weapons built. As Lilienthal and others had pointed out, it was by avoiding attempts at complete solutions that progress might be made. These were important events in history, but not in imagery. Bewildered by the technical complexities and discouraged by the imperfections of the treaties, the public found nothing there to change its views on nuclear weapons.[52]

When bomb tests went underground the nuclear disarmament movement's tactic of concentrating on fallout lost its point, while the growth of détente robbed the outcries about war of their urgency. The Kampf dem Atomtod in West Germany was the first to pass its peak, for already in 1959 the Social Democrats saw that disarmament would not win a majority at the polls and began to realign their defense policy. The CND peaked later as a political force, but at most a third of the British public supported them, and by 1961 the Labour Party had decided not to champion unilateral disarmament. Meanwhile, as the adolescent and radical fringe clashed with police and took up a variety of left-wing causes, the British movement's prestige and cohesion declined. SANE followed a similar pattern: a failure to go beyond slogans to a political program that a majority of citizens would vote for; increasing dominance by radical young people; internal schisms over the role of Communist adherents; and finally, after the Limited Test Ban Treaty, abandonment of the nuclear issue in favor of agitation for black civil rights, peace in Vietnam, and other causes. The Japan Council against Atomic and Hydrogen Bombs declined in the same way, eventually splitting into rival factions.[53]

Along with the popular protests, technical debate over nuclear strategy also began to fade into silence, as if there were no more to say. The worldwide collapse of interest in nuclear war showed up in every indicator I have studied, including bibliographies of magazines, indexes of newspaper articles, catalogs of nonfiction books and novels, and lists of films. From their peak around the time of the Cuban crisis, all these measures plunged by the late 1960s to a quarter or less of their former numbers. Even comic books with "Atom" in the title faded from the newsstands.[54]

REASONS FOR SILENCE

What were the reasons for this astonishing event, the only well-documented case in history when most of the world's citizens suddenly stopped paying attention to facts that continued to threaten their very survival?

The coming of détente and the apparent success of arms control and deterrence satisfied many people on a rational level. Yet reason could also point out that the bombs remained, more numerous each year. Close observers believed the precipitous drop of interest reflected other, less rational factors. Some said the Cuban crisis had served as a catharsis. But what did that explanation mean?

The public had been bombarded again and again by war scares and by hydrogen bomb stories, by warnings against fallout and by fantasies about monsters, by antibomb outcries and by shelter debates, with the Cuban shock only the last in a long series. Bomb protestors and civil defense advocates alike had always said they wanted to make the problems of nuclear war seem real to people; basement shelters and missile emplacements had done that. Yet awareness of peril did not spur action. Scarcely one in eight Americans actually took precautions during the Berlin and Cuban crises, and only about one in fifty built any sort of fallout shelter. Even RAND analysts and civil defense officials rarely built shelters for their own families.[55]

A panel of social scientists that Kennedy quietly assembled in 1962 inspected the psychological consequences of such decisions, and concluded that citizens faced "cognitive dissonance." This referred to a theory that was popular among academic psychologists, and that moreover had common sense in it. The theory began with the observation that most people held a confusion of half-formed beliefs that might not logically agree with the way they acted. If they were forced to pay attention to the contradiction, to the dissonance, people would strive to adjust either how they acted or what they believed (unless they could thrust away the matter altogether).

In nuclear affairs, ever since 1945 many people had admitted to deep confusion. But it was hard to tell what they were doing to rcconcilc inconsistent beliefs and actions; the survey results were as fragmented and confused as public opinion itself. Even the minority who decided to build shelters, when pollsters hunted them down, turned out to retain self-contradictory beliefs. The overall pattern, however, was compatible with cognitive dissonance theory.[56]

A small minority of Americans made a deliberate decision not to protect themselves, and these people held the corresponding conviction that nuclear war would be so ghastly that shelter was pointless: they had brought action and belief into accord. Others thought survival was possible, and built shelters. Most Americans, however, thought that civil defense would be of some use yet did nothing about it. The panel of social scientists noted that the worst dilemma was faced by people in areas liable to be wiped out by bombs—and polls found a majority of the population believed that they lived in target areas. The only way people in such areas could survive the bombs would be to uproot themselves and go live elsewhere, but very few families had done that. A survey found that during the Cuban crisis a third of Americans had discussed with their families what to do in case of war, but only one-twentieth recalled giving thought to fleeing their homes. Thus the majority could keep their beliefs compatible with their actions only if they insisted there was little chance of a nuclear war actually happening soon.[57]

Clever surveys could discover the mental process at work. One poll found that a third of Americans said nuclear war was likely to come, but when the questioners pressed on to ask more specifically, "How do you think things are going to turn out?" only 7 percent flatly predicted disaster. Similarly, French polls in the 1950s and 1960s found up to a third of the population saying that there was "great danger" of another world war, yet when another poll asked them point-blank whether they expected to see a nuclear war in their lifetimes, only 8 percent said yes. Would the real tomorrow, the one you actually prepared for, reveal a limitless Hiroshima, or would it hold familiar scenes of everyday life? The question had been posed ever more insistently, beginning in 1945 as a problem for the misty future and becoming by October 1962 a demand about what the next morning might reveal. Most people chose to live with the image of peace.[58]

When not forced to make a decision, people could follow a course of taking no action while still worrying vaguely about danger. Ever since the early 1950s roughly a third of the world public admitted strong fear of nuclear war, with most of the rest at least somewhat worried, and the numbers did not change markedly once the Cuban crisis was over. In fact, into the 1980s responses on polls about nuclear war showed little change, shifting by scarcely 10 percent one way or the other. In the few years after 1962 when published attention to nuclear war dropped to a quarter or even a tenth of its previous level, this was not because of any great change in the public's beliefs and concerns. People still admitted their nuclear fear if asked about it, but they no longer brought it up spontaneously. Brushing the issue aside was the easiest way of all to ward off cognitive dissonance.[59]

The poll results had some other curious features. From 1945 on, sociologists could not find their usual correlations. Some people seemed desperately afraid of war and others completely uninterested, but the difference had little to do with social class. Well-educated people tended to be more aware of the dangers of nuclear war, but that did not make them more anxious than others. A study that addressed the matter directly found that the amount of knowledge a person had about fallout had no relation at all to the amount of anxiety the person felt; anxiety did not come from a special knowledge nor from a special ignorance of danger, but from somewhere else. On the average, women were somewhat more worried about the bombs than men were, and poor or uneducated people somewhat more worried than wealthy or educated ones, but studies have found those groups tending to express more insecurity on almost any issue. The differences were greatest not between groups but between individuals within a given group. Attitudes and images had taken root far below normal social categories, in levels of the mind that simple questions were ill suited to probe.[60]

One research team constructed an "index of general anxiety" and found that this did agree closely with fears of nuclear war. In other words, people who expressed worry about nuclear bombs were people who expressed worry in general, and that was that. The bombs, like death itself, had become part of the general background of living.[61]

What difference did it make to have this appalling danger as a normal part of daily life? Sensitive observers believed that it seriously disturbed young people at least. Well after the Cuban crisis a poll found 40 percent of adolescents admitting a "great deal" of anxiety about war, more than twice the rate found in older groups. A survey that said nothing about bombs, but only asked schoolchildren to talk about the world ten years ahead, found over two-thirds of the children mentioning war, often in terms of somber helplessness. In 1965 a song lamenting that we were on the "Eve of Destruction" became the first song on a political issue to become a number-one popular tune in the United States; the nineteen-year-old writer explained that he felt war like "a cloud hanging over me all the time."[62]

There were effects more subtle than fear. As early as 1946 the British Council of Churches had warned that bombs capable of annihilating the future might bring people to live in the present, and that already many youths, feeling "helpless in the grip of forces quite beyond their control," were withdrawing from action. By the 1960s, observers from Teller to Dr. Benjamin Spock of SANE were reporting talks with young people who said it was pointless to study or save up money when the world might end tomorrow. In 1982 a psychiatrist, summarizing decades of studies, said that the nuclear problem had left many young people with "a sense of powerlessness and cynical resignation."[63]

But among the young, as among their elders, nuclear fear was expressed less and less openly. A college teacher in the 1960s asked his students every semester about such feelings, and noticed the shift. Those who had entered adolescence before the bombing of Hiroshima frankly admitted anxiety, but the next generation did not. The teacher felt that all the students were nevertheless acutely aware of danger, and he concluded that the danger had become so frightening that they denied it.[64]

Denial: one of the few mental defense mechanisms whose existence is accepted even by severe critics of psychological theory. It is a familiar fact that many people turn away, in particular, from the possibility that they or their loved ones may die. Such a response can become embedded in a culture, and indeed during the mid-twentieth century the tendency to look away from mortality (for example, by closing coffins at funerals) was observed to increase. Not only death in general but also immediate threats were often pushed aside, as by children in Second World War air raids who rigidly insisted that their homes were perfectly safe.[65]

Since the early 1950s acute observers had noted that many citizens were refusing to face the issue of nuclear war, and by the mid-1960s the defense mechanism was ubiquitous. It almost made sense to close one's eyes, the way sensible children cover their faces in a horror movie. As a young adult said in 1965, "If we lived in fear of the bomb we couldn't function."[66]

Not all denial was under such conscious control. Doctors found that many patients, when told their disease was incurable, simply forgot the intolerable fact; psychiatrists knew patients who were so disturbed by certain things that they literally became blind rather than face them. Nuclear war images were all the more likely to be repressed in this way because the imagery was mixed up with fantasies originally inspired by fears and desires that many people hid even from themselves. Projected onto the bombs, the imagery still proved unbearable.

To test denial directly, in the mid-1950s a psychologist showed people a set of drawings ambiguously related to nuclear war and asked them to make up stories around the drawings. The more explicit and potentially frightening the pictures were, the *less* some people would allow the idea of war into the stories they told. More ominous still, those who failed to see war in the pictures made up stories that were nevertheless gloomy, and also relatively brief, lacking a central subject, and with little effective problem solving. This was the phenomenon that psychologist Robert Lifton and others later called "numbing." Somebody who denied an idea would also thrust away everything reminiscent of the idea; refusal to feel, to think, or to contemplate action could spread outward indefinitely. One of the great unanswered questions of our age is how far denial of nuclear dangers has promoted such numbing.[67]

The experts themselves were not immune. A few psychologists and social scientists remarked that there were scandalously few detailed studies of reactions to the idea of nuclear war, and after the early 1960s even this trickle of scholarly papers stopped. One psychologist noted that while working on the subject she "experienced great anxiety about what I found and a continuing desire to stop work on it." Another psychologist reported he felt "overwhelmed and frightened," had disturbing dreams, and wanted to quit.[68]

Herman Kahn noticed denial even among officials whose jobs required them to consider nuclear war. He found many of them refusing to discuss deterrence because they felt that war with hydrogen bombs was, in a word, unthinkable. Nowhere was the tendency so surprising as among American military men. A spate of theorizing about tactical use of nuclear bombs had filled the U.S. Army's *Military Review* in the five years up to 1960, but this dwindled to fewer than half as many articles during the entire twenty years that followed.[69]

Denial was supported by more elementary psychological processes, particularly one called habituation. Animals exposed to a loud noise that is never followed by physical harm will eventually behave as if they do not hear the noise. The principle was applied with notable success to cure people of phobias, and one of the classic papers explaining this therapy, published in 1963, is of special interest.

A salesman was afflicted with morbid fear of nuclear war. His terror had become so acute that he avoided the radio or anything else that could remind him of the international news, until he lost his job and spent most of the day hiding with covers over his head. He claimed that his real fear was of losing his wife, on whom he was childishly dependent—which is in accord with suggestions I gave earlier about thoughts of world doom—but his psychologist chose to treat the phobia directly. The patient was told to put himself in a relaxed and pleasant state of mind and then imagine glancing briefly at a newspaper. After many repetitions he could do this without anxiety. He then progressed to the radio and so forth, until he was cured.[70]

How many people in the early 1960s were similarly "cured" of nuclear fear? There can be no sure answer to such a question, but I believe that what happened in an extreme case to one patient also happened in a milder way to the entire populace, all exposed to war scares and other stimuli that were never followed by actual harm. Unfortunately, as in the story of the boy who cried wolf, growing accustomed to outcries was not the same as getting rid of the wolf.

Reinforcing the move to apathy was another elementary process, one later named "learned helplessness." Psychologists found that a caged dog given electric shocks at random, with no way to avoid them, would eventually cease trying to save itself. Later, put in a different cage where a normal dog could learn to avoid shocks, the creature that was taught to feel helpless would only lie down and whimper. The corresponding human state was fatalistic apathy, which could be experimentally induced simply by presenting a person with a series of insoluble problems. Bertrand Russell complained of this on a larger scale in 1964, saying that the waning of the CND and the growing silence about bombs did not mean people were truly indifferent to the prospect of catastrophe. He said they were overtaken by a "sense of helplessness . . . most people feel utterly paralysed by the vast impersonal machinery of war and state power." Turning to other matters was in fact a sensible approach for those who believed they could do nothing to bring about the transformations of world society that alone could solve the problem of nuclear war.[71]

These were not just matters of personal psychology but also of social forces, of propaganda in the most general sense of the term. Propaganda does not always aim to excite particular opinions and actions; most societies have a ceaseless background propaganda pushing toward an *absence* of opinion, a silent acquiescence. This may be done deliberately, or it may happen simply because it is the nature of propaganda to work to suppress diversity of thought and reflection itself. For example, the perennial publicity about arms control talks and about new types of weapons tended to persuade people that national authorities were working hard to solve the nuclear problem. It seemed that only the authorities could understand the intricate diplomatic and technological details; there was no role for ordinary citizens. This was not learned helplessness, but taught helplessness.[72]

Whether people rationally decided that nuclear war was not an urgent matter because deterrence and détente appeared to be successful; or avoided cognitive dissonance by choosing to believe that war was unlikely at the same time as they decided not to flee their city; or found the images so terrifying and disturbing that they consciously or unconsciously denied them; or became accustomed to warnings and no longer felt fear at all; or felt so helpless that they turned to other matters; or accepted the authorities' frequently expressed belief that they should be trusted to handle the problem; or, most likely of all, submitted to many of these influences together, the result was the same. The images of nuclear war remained a mixture of the fantastic and the abstract, never developed into anything closer to realistic and conceivable actions. Anyone who tried to think about the issues found nothing to say that had not been worked through exhaustively by RAND or the CND or those in between. Like the denizens of an enchanted Sleeping Beauty's castle, each image and each concept kept its silent place while nearly two decades passed. It was a stagnation of military, political, and moral thought without modern precedent.

Yet important and subtle changes in thinking were under way. More and more people were becoming suspicious of the control that authorities held over nuclear energy, and some began to wonder about controls within technological society in general. This particular question, the most fundamental one, did not fall into silence after the missiles left Cuba.

part four

SUSPECT TECHNOLOGY

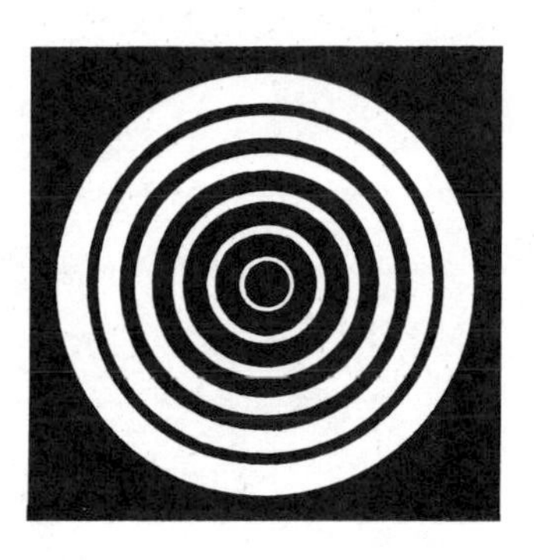

1956–1986

14

Fail/Safe

What exactly had doomed the human race to slow death? According to Fred Astaire, playing a drunken physicist in *On the Beach,* it was "a handful of vacuum tubes and transistors—probably faulty." In the book that inspired the film, however, the end of the world had begun with nations deliberately attacking each other. Between 1957 when the book appeared and the 1959 film there had come a change of emphasis, not only in this story but throughout public thinking. The threat of catastrophe was beginning to seem less a question of international politics than of technology.

During the decade after *On the Beach* appeared, as complaints about weapons went silent, complaints about another nuclear technology—civilian reactors—began to grow loud. In practical terms, accidental war and industrial accidents could scarcely have been farther apart. In terms of imagery, they could scarcely have been closer. In each case the central concern was, as Fred Astaire said, technical systems proliferating until "we couldn't control them."

UNWANTED EXPLOSIONS: BOMBS

Public concern about catastrophic technological error had grown up along with nuclear weapons. Before 1945 scarcely anyone had imagined a war set off by a technical accident; if writers spoke of inadvertent war they meant a breakdown in the fragile structure of diplomacy, with civilian and military leaders rather than electronic tubes going awry. Beginning in 1946 atomic scientists warned that automated war systems could increase this danger: what if a general should panic and push the buttons too hastily? The idea became part of the background of informed thinking. Yet during the next decade little time was spent worrying about such accidents, for in fact push-button war was still a matter for the future.

Aside from questions of war, people were as fascinated as ever with the picture of an accidental nuclear blast triggered by some reckless scientist. For example, *The Mouse That Roared,* which amused millions as a magazine serial and book in 1955 and as a movie four years later, featured a "Q bomb" the size of a football but powerful enough to destroy a continent. The inventor, a typical unworldly scientist, had thoughtlessly rigged a hair-trigger firing mechanism. Toward the movie's end, as the characters chased each other while clumsily juggling the bomb, a vast explosion suddenly burst across the screen. "This is not the end of our picture," an announcer told the quivering audience. "However, something like this could happen at any moment!" Most people saw that as a fact of life.[1]

Could the problem extend beyond a localized accident to touch off world war? Military realities began to make the idea plausible. In deep secrecy, American warplanes were regularly flying across Soviet territory to take photographs and test defenses. The flights annoyed Khrushchev, and in 1957 he hinted discreetly about practices that might set off an accidental war—the first official notice that such a disaster had become possible. The warning took on new meaning when Sputnik went up that October, bringing push-button rocket attack out of the realm of science fiction. The old idea of a scientific accident destroying the world began to creep into serious military discussions.

Real failures took a hand, for during 1958 the Strategic Air Command's new B-47 bombers suffered seven crashes and other accidents. The most spectacular came in March when a nuclear bomb fell from a B-47 over South Carolina. The chemical explosives that were part of the warhead went off, injuring five people and scattering about radioactive material. Radio Moscow took a leap to a greater problem, complaining that a similar mishap could panic SAC into attacking Russia.[2]

At the end of the 1950s SAC did seem to be on a hair trigger. When "bomber gap" and "missile gap" theorists calculated that America's bombers could all be blasted on the ground like sitting ducks, SAC made it a rule to put a small fraction of its fleet into the air at the first suspicious signal. In practice such signals included an unidentified group of airplanes, a radar error, even a meteor shower. The issue came into the open in April 1958 when the president of the United Press Service described the system in a news story. Introducing the public to some engineers' jargon, he explained that "Fail Safe" rules would always turn the bombers home if anything went wrong.

The reality was that American nuclear weapons, at first guarded jealously under the AEC's civilian control, had gradually come into the hands of military field commanders, until by the end of the 1950s certain commanders and even some individual pilots had the ability to drop bombs on their own. Those facts were secret, but as often happened with crucial matters in the United States, plain hints reached the public. After 1961 the Kennedy administration installed safeguard devices and also halted the flights over the Soviet Union, but the fear had already been aroused. Further bomb accidents, such as one that scattered radioactive material around Palomares, Spain, in 1966, continued to darken the public image of "the atomic enterprise generally," as an Atomic Industrial Forum spokesman lamented.[3]

Thousands of voices were now warning against inadvertent apocalypse, but without reviving much overt anxiety among the general public. Surveys in various nations found that most people expected a war to begin not by accident, but either in a local conflict that expanded out of control or in a deliberate sneak attack. The public was showing common sense, for indeed nearly all past wars had begun in one of these two ways. If nuclear war ever came, it would most likely be in some crisis where the leaders of both sides would decide they had less to lose by fighting than by yielding.[4]

Yet few could forget the faulty transistor or the careless fool. However implausible, these images of fatal innocence had a peculiar fascination, for they masked more intimate thoughts that were far from innocent.

The mask was partly lifted by two professors, Eugene Burdick and Harvey Wheeler. Their novel *Fail-Safe* was serialized in the *Saturday Evening Post* in October 1962, and it chilled the readers (with some help from the Cuban missile crisis that same month). The book, selling over two million copies, was the only nuclear novel ever to make the list of the top ten best sellers for a given year. The story began in 1967 with the failure of a minor electronic device causing a coded signal that sent an American bomber group into Russia. One pilot fought his way through to Moscow and obliterated it; the American President, to forestall Soviet vengeance, had SAC blow up New York as atonement.[5]

Air Force spokesmen complained that the story was all wrong, for bombers would not set off for a target without a meticulously checked voice command. The authors of *Fail-Safe* brushed that aside. The novel's lesson was not about such technicalities but about human fallibility. Burdick and Wheeler pointed out that the chance of an accident, however unlikely, was never exactly zero, and the consequences of failure in nuclear control would be so dreadful that, from their viewpoint, even the most implausible chance must be regarded exactly as if it were certain to happen eventually. The only solution was to dismantle the entire nuclear war apparatus. Critics attacked this viewpoint, blind to any distinction between the extremely implausible and the inevitable, as "intellectually scandalous." But the *Fail-Safe* viewpoint—shrugging aside realistic technical and political questions in order to focus on the most dreadful risk—would remain at the core of nuclear debates.[6]

Ideas about risk fed upon more visceral feelings, as the producer Max Youngstein recognized. A member of SANE, he wanted to further the disarmament cause, and his movie *Fail Safe* put millions into a cold sweat. Both film and novel succeeded in giving people what Wheeler called a "vocabulary for discussing their deep secret horror." The vocabulary said that technology, ever "less personal," had gotten "out of hand." While the story had no robots running amok, it did have humans who would have been at home in a robot story. For example, the film had a strategist who coldly advised a mass attack when he calculated that war was inescapable, and the book also had a scientist who ignored his wife and his own physical needs to devote himself to fail-safe gadgetry. Most of all it was military officers who were "automated." The pilot of the bomber heading for Moscow refused to heed radio commands to return, since his orders told him that voices could be enemy ruses; the film put the pilot's wife at the microphone, tossing her head in a frenzy as she begged her husband to turn back, but he stolidly switched her off. This officer placing his task ahead of wife and feelings was straight out of the SAC movies, but he no longer looked reassuring.[7]

The real danger, then, came not from automated machines but from automated men. When the President ordered New York destroyed he knew his own wife was in the city, and the Air Force general who carried out the order had left his wife there too. Dropping a hydrogen bomb on their wives was surely the last word on the problem of authorities' marriages! Yet there could still be one more step, and the general took it by killing himself.

Even more penetrating and influential imagery came out of a 1958 British novel that few people read, Peter George's *Two Hours to Doom*. This author too wanted to take a stand against hydrogen bombs. Perhaps he also drew on more personal emotions; a few years later, shortly after finishing a novel in which bombs reduced humanity to warring savages, he killed himself. The plot in *Two Hours to Doom* openly addressed something akin to suicide, for here it was not electronic breakdown that launched bombers into the Soviet Union but frank mental breakdown, a psychopathic general.[8]

Stanley Kubrick found the idea plausible. The veteran film director had thrown himself headlong into studies of nuclear war, collecting shelves of books, talking with strategists, pondering how to send the world a message. He happened upon Peter George's little-known thriller and found the plot he had been seeking. The result was the most important of all nuclear films, *Dr. Strangelove.*

It was no use when aggrieved officials insisted that now, in 1964, no general any longer had the means to send bombers into Russia without a code word from the President. Logical explanations fell apart when faced with the movie's hilarious, shocking action. At last the Man Who Started the War showed his face, and it turned out to belong to a SAC officer. But the film's message was only beginning when General Jack D. Ripper, convinced the Communists were poisoning his "precious bodily fluids," privately arranged to attack Russia. Kubrick satirized not only right-wing fanatics but also the well-meaning liberal President, turning the whole great apparatus of authority and technology into a haywire clown's contraption. Audiences giggled, for example, when an officer who tried to warn of world disaster by calling up the White House on a pay telephone found that he didn't have any coins. Men and machinery, each perversely flawed, spiraled into total breakdown.[9]

The central figure was Dr. Strangelove himself, a scientist-strategist in a motorized wheelchair, scornful of human feelings. As a critic noted, he "comes to us by courtesy of a Universal horror movie. In his person, the Mad Doctor and the State Scientist merge." Even Strangelove's crippled, black-gloved right hand was a sly reference to the similar feature of a robot-making scientist in a classic 1926 film, *Metropolis.* That film in turn had lifted its imagery from the Frankenstein story and the paraphernalia of medieval sorcerers. At last the secret that had been hinted at for decades was proclaimed aloud. The psychological problem of nuclear war was precisely what the traditional mad scientist had always stood for: the problem of human authority losing its self-control.[10]

This was not entirely fantasy. The public was beginning to sense that nuclear deterrence strategy built its logic squarely on the lunacy of its threats, on First Strikes and Doomsday Machines. This suggested that the lessons of mad-scientist myths could be brought to bear on the real world, the world of politics. For a few years that did seem to be happening; nuclear stereotypes influenced the way the public saw living politicians.

Nuclear weapons had been an issue in every American election since 1952, when Eisenhower had gained votes as a military expert who could master the new peril. In the test ban controversy of 1956 and the missile gap debate of 1960 it was again the more military-oriented candidate who won the presidency. But the mood was different by 1964 when Barry Goldwater became the Republican candidate. President Lyndon Johnson, with advertisements and speeches about the horrors of nuclear war, played on fears that his opponent was, as *Time* gently suggested, "nuke-happy." The Republican's slogan was "In Your Heart You Know He's Right," but Goldwater's casual remarks about dropping bombs left many people suspicious that, as one Democrat put it, "In Your Heart You Know He Might."[11]

Such a politician could look much like an overweening military officer (and Goldwater really was a general in the Air Force reserves). During the election campaign an antibomb film was in production, a story of a Navy captain who made so many demands on his crew that one of them broke down and set off a nuclear warhead, symbolic of accidental war; the actor who played the rigid captain modeled himself on film clips of Goldwater. A more popular book and movie, *Seven Days in May,* which featured a general who grasped for political power to block a peace treaty, referred in passing to the real anti-Communist politician and retired General Edwin Walker. This merging of images, politician with hair-trigger militarist, added to Goldwater's electoral defeat. It similarly entered the 1968 election when General LeMay, running for vice-president on George Wallace's third-party ticket, drove away more voters than he attracted.[12]

From then on every serious political candidate, in the United States and everywhere else, scrupulously avoided the military label. While the public no longer entirely separated its stereotypes of mad scientists and nuclear officers, politicians managed to escape the spreading taint. After the 1960s few people made a connection between Dr. Strangelove and his ultimate employers.

The stories of accidental war resulted, finally, only in further catharsis. Like the Cuban crisis but on another plane, they could help people to bring up their secret anxieties and to partly resolve the dissonance of incompatible beliefs and actions—if only by deciding to set the whole business aside. That particular resolution was easier to reach because these novels and films drew attention away from real politics, focusing instead on cracked transistors and cracked minds. What was the point in fretting over remote possibilities lying outside the citizen's purview or even beyond human ken? There was an edge of truth in *Dr. Strangelove's* arch subtitle: *How I Learned to Stop Worrying and Love the Bomb.* The 1958–1965 spate of films, novels, and magazine articles about accidental war brought down the curtain on the long series of important nuclear war fiction and nonfiction published since 1945. Debate over accidental war was the final burst of serious argument before attention turned elsewhere.[13]

A faint echo lingered, for there was one way nuclear accidents could plausibly connect with the continuing history of international events. In 1960 France had exploded its first atomic device, and China followed suit in 1964. Everyone understood that more nations with bombs meant more chances for disaster from a faulty transistor, an insane authority, or (more realistically) an escalating local conflict. Reporters mingled warnings against accidents with warnings against the spread of weapons technology, and foresighted policymakers got into complex negotiations about restricting the export of uranium and so forth. They called the problem "proliferation"—a biological term, as if bombs were prone to reproduce all by themselves like giant ants.[14]

In 1968 a Non-Proliferation Treaty was signed, empowering at last a corps of inspectors. But the International Atomic Energy Agency only inspected civilian reactors, and of all routes to building a bomb, civilian reactors offered one of the most expensive and the hardest to keep hidden; a determined nation might well choose other paths. While much expert attention went into hypothetical connections between bombs and the rising nuclear power industry, the real question of proliferation lay elsewhere, simply in whether or not more nations would decide to build bombs.

To everyone's surprise, very few nations felt that such a choice would be to their advantage over the long run. Public debate over proliferation therefore grew desultory, and there were no extended fictional treatments of the subject. When the idea did appear, it was as a minor device of rhetoric or plot. For example, Nevil Shute started his war in the book version of *On the Beach* with the dictators of Egypt and Albania capriciously bombing their neighbors; his choice of nations was clearly arbitrary, without meaning. The public found the threat remote when the likes of Egypt and Albania did not in fact own atomic bombs.

Instead of following the difficult negotiations over proliferation, public imagery wandered off in a direction of its own, back to tales of fanatics building or stealing bombs. Genuine national leaders were forgotten while stereotype scientist-criminals brandished infernal devices. That tradition was carried on, for example, in 1962 by the first of the immensely popular James Bond movies, *Dr. No,* in which the agent faced an evil genius (complete with black gloves) on his nuclear-powered island stronghold. Unlike other nuclear fiction, such stories did not disappear from the paperback book racks and movie theaters after the mid-1960s but remained roughly constant in numbers over the next decades. Time and again James Bond or his imitators saved the world from secret criminals who wielded hydrogen bombs. The movies won the widest audience of any tales relating to nuclear energy.

If there was any social message or propaganda here, it was that ordinary citizens had no role but could only wait helplessly, trusting in the hero with his special abilities and technological gadgets. However, the most important messages were even more simple. As one film critic remarked, scriptwriters had domesticated the hydrogen bomb, transforming it into "just another technological toy" in the perennial spy fantasy. Nuclear energy had become a theatrical prop in stories that combined violent adventure, mystery, and casual sexual encounters. But this was a peculiarly well-fitting prop. Integral to the spy fantasy were linkages among aggression, secrecy, and sex. Nuclear bombs already had such associations, so bringing them into spy tales helped to grab an audience. Meanwhile the tales reinforced the associations clustering around bombs. Talk about nuclear war had gone as far as it could go in a direct line; it was reverting to more primitive and hidden levels.[15]

UNWANTED EXPLOSIONS: REACTORS

While people spoke less and less about controlling nuclear weapons, some began to notice the problem of controlling nuclear reactors. Such talk had been muted during the early years when the barrage of Atoms for Peace publicity made worries seem unreasonable. An image of dreadful risk nevertheless appeared—one originally created out of public view by the very men who were responsible for building reactors.

Most nuclear authorities had always felt confident that sound engineering could prevent disasters, yet they had insisted since the Manhattan Project days that reactors should be handled with a care going beyond anything known in other industries. Unmindful of how many thousands of lives were routinely sacrificed in ordinary mines and factories, atomic scientists and other nuclear industry leaders insisted, with a perverse pride, that reactors were "by long odds the most dangerous manufacturing process in which men have ever engaged." That idea gradually became dominant in the thinking of reactor builders and eventually of everyone else.[16]

As usual it was American scientists who opened up discussion. A first disclosure came in 1953 when Edward Teller gave a public speech to a group of nuclear experts. Going well beyond what other authorities had said in public, he exclaimed that "a runaway reactor can be relatively more dangerous than an atomic bomb." But his warning attracted little notice from the general press.[17]

More publicity came at the 1955 Geneva Atoms for Peace congress. The most important paper was again by Teller, as coauthor with two other scientists. They discussed the radioactive fission fragments that would build up inevitably in any reactor, isotopes they believed to be a million times more toxic, gram for gram, than ordinary industrial chemicals (this was before the hazards of certain chemical processes were recognized). The authors warned that running a reactor was like "conducting both explosive and virulent poison production under the same roof."[18]

Further thoughts about reactor hazards became public in 1956 when the Joint Committee on Atomic Energy held hearings on industrial radiation and when the National Academy of Sciences' report on fallout, stressing that bomb tests were not as bad as civilian sources of radiation, mentioned nuclear power. Most reporters ignored these hints from the experts, but a few journalists mentioned the matter, not to deny that Atoms for Peace was a wonderful idea but to call for caution during its development. Characteristic was the subtitle to a *Popular Science* magazine article: "Experts bare the weird hazards of nuclear-electric stations and tell of the measures being taken to protect you."[19]

A few real critics stood up. Already in 1947 when the French Atomic Energy Commission had set out to build a prototype reactor in the countryside south of Paris, local farmers had circulated protest petitions, fearing that radiation might damage their crops or that the device might blow up. And in 1956 when the United States offered to build a reactor in Brussels, the city authorities politely refused to permit it anywhere near the king's palace. Such isolated incidents had little importance, but they reminded the industry that the seeming public confidence in Atoms for Peace hid persistent unspoken anxieties. People were familiar with science fiction stories about fission plants causing the most horrid accidents.[20]

Sensing that they were on thin ice, a few of the nuclear industry's public relations experts warned that it was poor tactics to promise that accidents were impossible. If anything ever did happen, nuclear officials would never be trusted again. As Teller explained, "in any great development fatal accidents occur sooner or later," so the public ought to hear the blunt truth beforehand. Many agreed that a single serious accident, the kind that happened now and then in almost any industry, could be a "psychological disaster" that would impede nuclear progress. Technical calculations had given engineers reason to advance with caution, and now they had a second reason: not only the public itself but the public image of nuclear energy must be kept from harm.[21]

Another force impelling the experts to caution came from more inward anxieties. I have spoken a great deal about associations among ideas, but I have said little about associations made unconsciously. The visible and consciously acknowledged imagery would suffice to explain most of people's behavior in nuclear affairs. However, at least one group could be expected to put nuclear fears literally out of mind, namely the nuclear workers themselves. People in any trade naturally tend to set aside thoughts of risk rather than worry all day, but such refused anxieties could still influence their behavior.

Psychologists who studied nuclear industry workers in later years found such influences. An American who interviewed reactor operators in the 1970s did find them all insisting they felt safe, but when pressed to say how they responded to some strange event at the control panel they confessed to surges of fear, as if they secretly felt they lived on the edge of disaster. A study of European nuclear industry workers in the mid-1960s likewise found them outwardly calm but revealing through certain classic signs (anxious responses on the Rorschach test, a high rate of psychosomatic illnesses such as ulcers or impotence) that they harbored unconfessed fears. In another European study a psychologist found superficially confident nuclear workers taking refuge in silence and remarks about secrecy; as she probed further, the interviews wandered into talk of bombs, monstrous progeny, marital difficulties, children's questions about sex. Apparently nuclear fear, with all its associations, was not excluded from the reactor control rooms.

No such professional study was made of top experts, but from anecdotal evidence I believe they had much the same feelings as everyone else in the industry and among the public.[22]

Nuclear leaders created formal institutions to perpetuate their concern about reactor dangers. The first and most important was set up by the U.S. Atomic Energy Commission in 1947, following the advice of their General Advisory Committee of scientists. The new Reactor Safeguard Committee was only an ad hoc group of a half-dozen experts serving as part-time advisers, but within a few years they would force the entire nuclear industry to think as they did, and eventually their way of thinking would impress itself upon the public at large. Teller was the committee's chair for its first six years, the formative years; dedicated to a particular view of the safety problem, he would dominate all decisions.[23]

In the spectrum of opinion about safety, ranging from nonchalant overconfidence to apprehensions of disaster, the Reactor Safeguard Committee inclined toward disaster. That accorded with the chairman's personal concern with every possible sort of nuclear doom. More important, the committee members knew that official responsibility for averting mass death lay on their heads. Besides, since they held as firmly as anyone to the hope that nuclear power would do excellent things, they worried that the careless attitude of certain reactor builders might cause accidents that would retard the industry's growth.

The Safeguard Committee took its job so seriously that before its first year of work was done it was at odds with many of the AEC's other experts, including the entire General Advisory Committee. Teller's group believed that the only way to protect the public for certain would be to keep large nuclear facilities far from populated areas; when scientists planned reactors at Brookhaven and Oak Ridge, the committee frustrated them by ruling that these neighborhoods were too populous. The Reactor Safeguard Committee was sometimes called, even to Teller's face, the Reactor Prevention Committee.

The cautious approach met a better reception at the Argonne National Laboratory outside Chicago, where the AEC was concentrating reactor design work under the leadership of Walter Zinn, a frank-spoken scientist who knew reactors like the back of his hand. Conservative about safety, Zinn suggested that all large experimental reactors be isolated on a thoroughly remote reservation. The AEC came up with a suitably desolate area in Idaho, and eventually dozens of pilot reactors were built there, scattered across wastes of sand and sagebrush. Among these reactors was a small one originally planned for Argonne; Teller's committee, more conservative even than Zinn, had forced him to move it there.[24]

A reactor design group at General Electric in upstate New York came up with an alternative. They were not eager to commute to a desert thousands of miles from their laboratory, and what was more important, the GE staff saw that commercial nuclear power would be costly if reactors had to be located many hundreds of miles from the cities they served. The staff therefore proposed to enclose their reactor within a tremendous steel shell, so that if an accident did happen the radioactive substances would be safely contained. This idea the Safeguard Committee accepted.

The GE architects now raised a fateful question. Exactly how thick should they make the containment shell? Up till then everyone's attention had gone into preventing accidents, and nobody had tried to calculate the details of what an accident might look like. But the architects needed precise numbers. The reactor designers therefore had to come up with a realistic catastrophe, in fact the very worst that could plausibly be imagined. The container would then be built to withstand even that. It was a new idea to plan such massive precautions for such a remote possibility. "We basically went through the whole nuclear controversy," one of the GE staff recalled. "We ourselves played the devil's advocate."[25]

Soon the Reactor Safeguard Committee itself began to insist that any group who wanted to build a reactor must describe the worst failure—what came to be called the Maximum Credible Accident. The committee members would review that and try to conceive an even more dreadful failure. Engineers would calculate in detail what the worst case would look like and then design something to contain it. Meanwhile the committee would use the calculations to decide how far from populated areas the reactor must be located, just in case even the containment failed. In short, Teller's committee made all thinking about reactor safety revolve around the Maximum Credible Accident. This only confirmed the viewpoint, held ever since 1943 when Compton's team had calculated how many miles to put between the Hanford piles, that reactor safety meant staving off the most unspeakable calamity.

That was not how engineers in other fields thought about safety. Other industries gave primary attention to the more modest accidents that from time to time actually killed people; they groped toward reliable practices through experience with hundreds of small-scale tragedies. The unspoken assumption was that great accidents would stem from the same causes as little ones, so that once you understood and prevented every ordinary failure, there was little more to worry about. The nuclear industry did not and could not adopt this approach. In the 1950s engineers were developing reactors so swiftly that before experience could reveal the drawbacks of one design they were building the next, quite different and several times larger. Furthermore, the 1950s reactors suffered such a small number of serious accidents that they gave few examples of real problems.

The Maximum Credible Accident, then, was logically similar to the First Strike of nuclear strategists: something spun out of pure thought rather than out of the record of experience, yet considered more painstakingly than any actual event because of the horror attached to it. This was the logic of the U.S. Air Force's "worst case" guesses about Soviet capabilities, guesses that led Teller and others during these same years to campaign for ever more weapons to remedy supposed bomber or missile gaps. It was the same viewpoint popularized in *Fail-Safe,* the argument that when nuclear catastrophe was at stake the most implausible failure must be treated as if it were inevitable—and never mind the details of how it might come about. All such worst cases were based less on engineering than on fear of human failure and loss of control. Indeed, the Reactor Safeguard Committee was particularly worried that some vicious or unstable person might set off a reactor accident on purpose, much as SANE worried that such a person might start a war.

Concern about accidents, maximum and otherwise, became institutionally embedded in reactor laboratories. The first formal organization outside of Teller's committee came in 1952 when Zinn set up his own Reactor Safety Review Committee to oversee work in Argonne and Idaho, that is, to oversee roughly half of all reactor research in the world. Zinn's committee devoted much time to imaginary maximum accidents, but they also uncovered real cases where overconfident workers had created hazards. The group spent long days and weekends interviewing engineers and reactor crews and going over every piece of every reactor with a fine-tooth comb. Enjoying Zinn's unqualified support, the committee could and did get careless people fired.[26]

Brookhaven and Oak Ridge, the other AEC laboratories working on reactors, later set up their own independent teams to review the design of each new reactor, as well as separate committees to watch over the safety of operating plants. The scientists created these institutions not in response to outside prompting but out of their own concern. General Electric, Westinghouse, and other private firms likewise spontaneously set up in-house safety committees that reviewed reactor designs and reported their findings straight to top management. All these groups could look at things much more closely than the AEC's Reactor Safeguard Committee, scrutinizing every line of blueprints, grilling engineers, analyzing actual minor mishaps in exhaustive detail, making inspection trips to working reactors, coming down hard on failures to obey safety rules, and forcing everyone to think about possible disasters.

In the end these attitudes led to engineering designs. For example, American engineers spent countless hours calculating a numerical coefficient that described how this or that reactor design would behave if the temperature changed. In Fermi's original pile the chain reaction would automatically slow down if the temperature rose. That would not hold for all designs, however; in some cases as the reactor got hotter the reaction could go faster, heating the reactor still more, running away into an explosion—not like an atomic bomb, of course, but like a TNT bomb salted with intensely radioactive materials, which would be bad enough. Zinn and his colleagues devoted themselves to designing reactors in which the numbers made that physically impossible.

Most other nations followed the American pattern. For example, leaders of the British program, from contacts with Teller and from their own calculations, agreed that there was a chance of fatal accidents that could set back nuclear power for years. When a cabinet minister complained that since the nuclear safety record was far superior to that of many other industries, the program ought to cut its costly safety budget, British scientists replied with scorn. Not only were ethical issues involved, but they felt they would never win the public's trust, nor be able to recruit radiation-shy technicians, unless they continued to show an outstanding record. The French and Germans too, heeding warnings from Teller and others, set up painstaking controls under independent committees. Nearly everyone began to place their reactors away from populous areas and build containment shells around them.[27]

The exception was the Soviet Union and its satellites. Authorities there publicly insisted that there would never be a reactor calamity, and they probably believed their own publicity. Besides, to criticize a technology that was controlled by the state was tantamount to criticizing the state itself. Soviet nuclear experts did not even insist on automatic stability: the type of reactor they built most abundantly, moderated with graphite, was a type in which the chain reaction could run away if the temperature rose.

This reactor design was not an arbitrary choice, for the laws of physics have a role in history. One result of these laws is that a graphite reactor is particularly suitable to produce plutonium for bombs; moreover, for this purpose the reactor should be refueled frequently. Soviet civilian reactors were directly descended from ones built to produce military plutonium, and they used many of the same design shortcuts. It was simply cheaper not to alter the design in order to provide automatic self-regulation. Moreover, Soviet reactors built before the 1980s—such as the one that later became famous at Chernobyl—had on top nothing more than a screen through which refueling took place, a screen literally full of holes. Doing without a full containment shell not only made refueling easy but was economical in itself. In short, for Soviet reactor designers, safety was less important than building civilian reactors that could produce military plutonium if desired, and building them cheaply.

The Soviet program was also marred by sloppiness in daily affairs. This was part of a general disregard of the dangers of technology, an attitude that meanwhile produced accidents in other industries, and disease from chemical pollutants, at a higher level than in most other nations. The reactor program, spurred by its leaders to advance with breakneck haste, seems to have suffered more than one serious accident in which workers and perhaps members of the public were injured by radiation. However, only vague hints of this came through the walls of secrecy.[28]

Not even the most dedicated effort could make a flawless reactor. The American program had a number of minor mishaps during the 1950s. In 1952 and 1958 Canadian reactors caught fire, the workers narrowly escaping injury, and in 1959 and 1961 came the world's first fatal accidents as reactor operators died in Yugoslavia and in Idaho. Each accident caused a stir, drawing newspaper comments on the problem of reactor safety. But these were all experimental reactors, where sensible people had expected problems. Since in each case there was no sign that anybody outside the building was harmed, and since officials redoubled their assurances that there was no danger of anything worse, the public showed little concern.

The response was different when a reactor released an important amount of radioactive material into the open. To get plutonium for bombs the British had constructed large reactors at Windscale in northern England, looming above verdant meadows where sheep and cows grazed. In October 1957 some of the uranium fuel rods caught fire, and they kept on burning with a cherry-red glow while the authorities made soothing announcements. For a day or two newspapers around the world spoke of heroic firemen, but on the third day black headlines exploded: traces of radioactive iodine had been discovered on the pastures for dozens of miles around Windscale, and officials were confiscating milk from the area. The press furiously criticized nuclear officials while frightful rumors spread. Even before the accident, farmers near Windscale had worried that radiation might be sterilizing their cows, and now people over half of England shared such worries.[29]

The accident may have slightly increased the risk of cancer in the region and thus resulted in perhaps a dozen deaths over the next several decades, but at the time nobody was visibly injured. Most people believed that the affair ended without harm, and the press turned to other matters. Yet if the outcry was brief, it had been strong enough to show that the public remained extremely wary of any escape of radioactive substances.

ADVERTISING THE MAXIMUM ACCIDENT

The first case of sustained public opposition to a reactor came in the United States during 1956, and it revealed much about the roots of the opposition. The central issue was political—a feature of the controversy that would later be half hidden under layers of seemingly nonpolitical imagery. Another fact that was later obscured was that the negative imagery and evidence that supported the opposition had originated within the nuclear community itself, where they served quite different purposes. More revealing still, these negative factors first emerged into public view just where positive imagery was greatest, evoked by the enthusiasm for what was the most utopian of all reactor types.

Around 1945 Manhattan Project scientists had became excited by the fact that uranium-238, which made up the bulk of the natural uranium in reactor fuel, was transmuted within a reactor into plutonium that could itself be a reactor fuel. Enthusiasts soon began to speak as if the process were not simply a way to convert an unwanted isotope into a useful one, but a philosophers' stone producing endless new gold. A Soviet scientist said that it sounded like burning coal in a stove and getting back more coal. Leo Szilard, an ardent promoter of the process, dubbed it "breeding," as if it captured the limitless reproductive powers of life. The Russians frankly termed the process "reproduction," and Americans said the uranium-238 was a "fertile" material.

As a Westinghouse pamphlet later put it, "breeder" reactors promised "what might be called perpetual youth."[30]

The breeder attracted some of the most audacious minds in industry, and that meant Walker Cisler. The sober-faced executive with a social crusader's soul became interested in the device first as an engineer, second as chief of the Detroit Edison Company and founder of the Atomic Industrial Forum, and last but not least as the creator of a new private consortium whose goal was to break the government monopoly on nuclear power. Cisler wanted to prevent the government from strangling free enterprise and freedom itself; at the same time he expected that the breeder reactor would add wealth to his company, his nation, and eventually the whole world. Since Lewis Strauss and others on the AEC shared a similar vision, Cisler struck an agreement with the agency for help in building a prototype breeder. It would go up at Lagoona Beach, Michigan, twenty miles outside Detroit.

The "fast" breeder Cisler planned would use a dense flux of swift-moving neutrons, and would therefore be more sensitive to mistakes than almost any other type of reactor. If by mischance the fuel rods warped or melted, there was a chance of a runaway chain reaction heating the device so fast that it would explode roughly like an ordinary TNT bomb. Such a thing actually happened in 1955 when a small experimental breeder in the Idaho desert was deliberately tested beyond its limits.

The problem worried the Advisory Committee on Reactor Safeguards (as the Reactor Safeguard Committee was now named) when they studied the Lagoona Beach project in June 1956. They were not concerned about any identifiable flaw in the reactor design so much as about the AEC's failure to fund enough breeder safety research. As usual, their discussion revolved around the worst case. What if the fuel somehow melted, and what if that somehow caused an explosion, and what if that somehow broke the containment shell, and what if the wind were just then blowing toward Detroit? After days of debate the committee advised that such a breeder should not be permitted until more research was done.[31]

The Atomic Energy Commission rejected the Safeguards Committee's advice. Strauss felt sure that by the time the reactor was completed, the research needed to keep it safe would be in hand. Since the Safeguards Committee's report was confidential, Cisler's group might now have gone ahead without interference, but for one man. Commissioner Thomas Murray, concerned about safety and disgusted with Strauss's high-handed ways, decided to block the project even if he had to throw a monkey wrench into the AEC's machinery. He revealed the Safeguards Committee's misgivings to the chairman of the Joint Committee on Atomic Energy, Senator Clinton Anderson.

The Democratic senator, an enemy of Strauss since the Dixon-Yates controversy, saw Lagoona Beach as the worst move yet in Republican plans to give nuclear power away to private business. He set out to pry the Safeguards Committee's report away from the AEC and bring it into public view, and meanwhile stirred up opposition in Michigan. In particular Anderson got in touch with Walter Reuther of the United Auto Workers, who was happy to take up cudgels against the Eisenhower administration in an election year. Joining with other unions in the Detroit area, the UAW took the breeder to court.[32]

The unions drew on the AEC's own ideas about Maximum Credible Accidents. They tracked down reports prepared as part of the safety review process and publicized calculations that the worst breeder failure might kill thousands. The plaintiffs fired away at Cisler and the AEC in a legal battle that lasted four years, although eventually every court ruled in the AEC's favor.

The affair disgusted Strauss. In private he told Eisenhower that the opponents were out to sabotage the Atoms for Peace program, while in public he said that "Communist propaganda" must not destroy the "God-given opportunity" to use atoms under free enterprise. He was supported by people like a Republican congressman who felt that "alarmists" were conspiring to seize control of atomic power as "a socialistic monopoly." In fact, the unions and other breeder opponents strongly favored nuclear power in general and were fighting in Congress for a speeded-up program to build government reactors, a true atomic TVA; they distrusted nuclear power only in private hands. The breeder opponents' motives were summarized by a Michigan senator who told Cisler that the chief concern was safety, of course, but that he also feared the private consortium would grab "monopolistic control" of this grand development.[33]

In the Lagoona Beach fight, opposition to a reactor had as much to do with political fears as with nuclear ones. Reuther could exclaim that the breeder "might convert itself into a small-scale atomic bomb," and his lawyers could talk about the risk of "fallout" upon people's homes, but they were still enthusiastic about nuclear power in principle, and their warnings brought little response. If newspapers mentioned the fight at all, the typical headline was "Physicist Calls A-plant Safe." The real fear on each side was that the other would gain strength by seizing a monopoly over what promised to be a centrally important new industry. At Lagoona Beach this question of control within society became associated with the technical safety issue. In particular, the Maximum Credible Accident had broken out of the obscurity of engineering reports to become an actor on the political stage.[34]

The move from concern over social controls to concern over dreadful technical failures was reinforced by Congress in its most ambitious attempt to ease public worries about reactors, the Price-Anderson Act of 1957. Some key provisions of the act came straight out of the Lagoona Beach fight. Senator Anderson insisted the AEC must hold public hearings before giving any license to build a reactor, for he expected that the process would make the Commission more responsive and ultimately would strengthen public confidence in nuclear power. The results were precisely the opposite. Over the next decade many hearings were held, each one conducted by the AEC only after its staff had fought out a reactor's safety questions with the designers and decided to give approval. With the engineers' doubts resolved out of sight, most hearings were manifestly nothing but a legal charade. The hearings therefore reinforced feelings that the AEC was arrogant and secretive. Moreover, the AEC used every available legal and bureaucratic maneuver to brush off anyone who questioned its decisions, converting modest skeptics into embittered enemies.[35]

Another provision of the Price-Anderson Act did still more to distress rather than soothe the public. Cisler and other nuclear spokesmen had been complaining to Congress that reactors, although insured for higher sums than any other industrial installation, did not have enough underwriting. Insurance experts and even the Atomic Industrial Forum told Congress once again that the "catastrophe potential" of reactors was "more serious than anything that is now known in any other industry," and some companies said they might flatly refuse to develop nuclear power for fear an accident would bankrupt them. Only the government itself could underwrite reactors. The Joint Committee, determined to push Atoms for Peace with all speed as part of national grand strategy, requested half a billion dollars of protection, and as usual Congress passed with little debate what the Committee requested. The press scarcely noticed, for reactor catastrophes seemed a remote problem. By comparison, floods in New England and California had recently killed over a hundred people, prompting Congress to pass a five-billion-dollar subsidized flood insurance program; next to that, the Price-Anderson Act was small change.[36]

But an ominous image lay buried within the new law. Evidently the idea of a single accident doing half a billion dollars' worth of damage had to be taken seriously—and for no industry but nuclear power. Congress had put an official stamp of plausibility on the Maximum Credible Accident.

Meanwhile the AEC inadvertently made the idea more concrete. At first the Joint Committee had not been convinced that the nuclear industry needed a special indemnity, and if it did, they wanted dollar numbers. The AEC therefore asked its Brookhaven laboratory to prepare a report, which it said should "resemble studies of the 'Maximum Credible Accident' now generally made for each reactor in obtaining approval from the Advisory Committee on Reactor Safeguards." The Brookhaven study group accordingly took a worst case: what if half the contents of a reactor were turned to powder and hurled into the sky? It seemed extremely unlikely that this would happen with a typical American reactor (and no technical sequence of events was suggested), but nobody could prove it was impossible. If it did happen, the Brookhaven group added, the radioactive powder might somehow all escape the containment, and perhaps at a time when the wind was blowing straight toward a nearby city, but not so fast as to disperse the cloud before it got there.[37]

The Brookhaven study group had gotten into what Szilard back in 1944 had called the "question-answer game." The AEC's question was, What is the worst possible event, regardless of other considerations? The scientists obliged with a maximum catastrophe, a cautionary tale. They were willing to encourage Congress to set a high level for insurance on reactors, and meanwhile they hoped to rub peoples' noses in the potential problems so that the nuclear industry would take full precautions. The scientists felt sure that once everyone was taught to take risks seriously, safety problems would be worked out. At the close of 1956 the Brookhaven group privately circulated preliminary conclusions estimating that damage from an accident could cost up to to one and a half billion dollars.

The most important critique of the draft report came from Teller, who warned that it "understates the actual dangers." He could imagine still worse circumstances, for example that just at the moment when a radioactive cloud from an accident reached a city, rain would fall and sweep the powder to the ground. As a result of such criticism the final Brookhaven report, designated WASH-740, estimated that damage from a reactor accident could amount to as much as seven billion dollars, while immediate deaths—not even counting later cancers—could exceed 3,000. Of course that was still not the upper limit, for anyone could dream up worse if increasingly implausible situations (what if a tornado struck, sucked up the reactor's contents, and distributed them evenly over a city?). Some experts felt that WASH-740 had already gone over the border into the incredible. But the AEC had an answer to its question—a new set of authoritative "facts" about reactors. The idea of a Maximum Credible Accident, from its beginnings as the science fiction tale of an experiment gone awry, had been condensed into numbers.[38]

Nobody expected the dry WASH-740 report to have much impact on public opinion, but opponents of the Lagoona Beach and subsequent reactors made much of its story of deaths by the thousand. One of the authors of WASH-740 later remarked angrily that he and his colleagues had expected their words to be read like a normal laboratory report for the technical content; "We did not anticipate that selected ideas would be excerpted and used to support preconceived ideas." The public took up mainly the part of the scientists' thoughts that matched familiar images of doom. Other AEC studies in later years had similar effects. For example, a 1965 calculation found that in a worst case, temporary agricultural restrictions such as the confiscation of milk near Windscale, as well as a slight increase in cancer rates, might affect an area the size of the state of Pennsylvania. Opponents used those calculations to imply the kind of picture that *Astounding Science-Fiction* had offered in 1940—a vast territory left as barren as the moon.[39]

The old undercurrent of anxiety was working itself free from the countervailing confidence that experts would keep everything safe. From the late 1950s on, in the United States, Britain, Germany, Japan, or wherever someone proposed to build a reactor, vague thoughts of disaster afflicted the neighborhood residents. There were a few cases of overt protest at the grass roots. For example, in February 1963 when the Consolidated Edison Company of New York proposed to build a reactor at Ravenswood, literally in the shadows of the towers of Manhattan, the local Community Council held a public meeting in a church. Some 250 residents, anxious and hostile, gave a Con Edison representative a hard time.

Reinforcement arrived from an unexpected quarter. After David Lilienthal had retired as chairman of the AEC he had grown increasingly dismayed at the actions of his Republican successors, and in the privacy of his diary he scoffed at the "zealots" of Atoms for Peace with their "orgasms of promises." Finally, in early 1963 he made his criticism public in a series of lectures. The Joint Committee on Atomic Energy, annoyed by Lilienthal's apostasy, called him to testify and peppered him with angry questions, but Lilienthal had faced hostile committees before, and far from retreating he said more than he had planned. Asked about Ravenswood, he declared, "I myself would not want to live around one of these plants." The fundamental problem, he said, was that engineers "are human beings who are fallible."[40]

The words became a rallying cry. In the summer a small group of ordinary New Yorkers formed an ad hoc Committee Against Nuclear Power Plant in New York City, "CAN POP in NYC," and began sidewalk demonstrations; a typical placard read, "Why Take a Chance?" All this criticism was based on ideas close to the *Fail-Safe* viewpoint on weapons that was under debate at just that time. Con Edison officials replied much as the military officials were doing, insisting that ingeniously designed safeguards made an accident impossible. The AEC was less confident, for although the agency's staff found nothing specifically hazardous in the proposed design, they had always refused on principle to permit large reactors right next to a city, and they were not ready to make an exception for Ravenswood. Eventually Con Edison withdrew its proposal, defeated as much by the experts' misgivings as by grass-roots protest.[41]

Meanwhile in November 1963 Lilienthal met his former colleagues on their home territory, a meeting of the American Nuclear Society where industry people gathered to trade notes and make speeches praising the atom's future. He warned that unless the nuclear community became more open to criticism they might let loose a "wave of disillusion" against all of science and technology. The audience booed him, and in the lobby officials buttonholed reporters to warn that Lilienthal was in the pay of the coal interests. But as he noted, the louder the industry tried to shout down its critics, the more the public would wonder. The maneuvering among elites that had characterized nuclear history up to and including the Lagoona Beach lawsuit was now being submerged in a larger question that had begun to interest people in general, the *Fail-Safe* question: Could anyone, however expert, be trusted to control nuclear energy?[42]

15

Reactor Poisons and Promises

In 1958 the Pacific Gas & Electric Company decided to build a power station north of San Francisco at Bodega Bay. This was pristine coastline with immense vistas, ocean waves surging against a headland, fields of grass bending in waves of their own under the wind. In 1961 PG&E announced that the proposed plant would use a nuclear reactor, but even before then local residents were organizing opposition, complaining that any kind of power plant would "ruin the scenic value of Bodega" and "deflate real estate values."[1]

Similar arguments had been coming up elsewhere, especially in crowded Britain where any isolated spot was bound to have value for someone. For example, sportsmen opposed one reactor because it might damage nearby duck breeding grounds and because the access road would cut across hallowed fox-hunting fields. At Bodega too, as one official put it, the conflict was between an industrial plant and "unspoiled natural beauty." Centuries-old worries about the ravages of industry were beginning to focus on reactors.[2]

Aesthetic scruples had never done much to stop industry, and they were not going to stop PG&E; nuclear anxiety was more effective. Deliberately invoking the anxieties of the bomb test debates then under way, critics warned that an earthquake in the nearby San Andreas Fault might split the plant open and bring "fallout." Opponents went to Bodega and released batches of balloons marked "Strontium-90," and the

Consumers Cooperative of Berkeley, an important purchaser of milk in the region, sent a letter to dairymen warning that contamination was threatening them both from bomb tests and from the proposed reactor. More and more people were drawn into the battle.[3]

Meanwhile geologists discovered that a small fault ran right through the site. The AEC staff, never convinced that the reactor would be safe, now vetoed it, and in 1964 PG&E gave up. As at Ravenswood, the reactor was defeated not so much by local opposition nor by specific problems with its technical design as by general misgivings among experts. But the details of this battle pointed to a new image: with reactors as with bombs, people were beginning to fear not just that the things would explode, but that they would horribly contaminate the entire natural order.

At the same time nuclear publicists were devising more detailed visions than ever of the clean and well-ordered White City that reactors would bring. As the 1960s proceeded these contrasting images, the polluted world and the redeemed one, each came to seem possible for a nearby rather than a distant future. Gradually the opposition between the two visions became a matter of overt politics.

POLLUTION FROM REACTORS

The idea of contamination from civilian nuclear industry, like most nuclear ideas, started among a minority of experts at an early date and only gradually reached the public. Since 1945 Herman Muller and a few others had warned against careless leakage of radioactive industrial by-products. Inside the Reactor Safeguard Committee a sanitary engineer, Abel Wollman, fought for better handling of the AEC's radioactive wastes; dissatisfied with the response, he warned outside public health experts about the potential problem. The critics first attracted public attention in 1955 when they brought up the subject at the Atoms for Peace conference in Geneva, and in the next few years they got into the newspapers now and then, as when a National Academy of Sciences panel declared that radioactive wastes were more hazardous than any other industrial material and must be handled with extreme care. The press and the public gave the matter only passing attention, preferring to leave nuclear sanitary engineering to officials. Officials left it to nuclear experts, and most nuclear experts left it alone. Wastes were messy, unglamorous stuff, and even the expert critics agreed that there were many feasible ways of disposal. The AEC and its counterparts in other nations studied the matter at leisure, meanwhile finding temporary expedients for storing unwanted radioactive materials.[4]

But as the bomb test controversy mounted, more and more people noticed a similarity between reactor by-products and fallout. For example, in January 1957 *McCall's* magazine ran a banner on its cover, slashed across a picture of a baby, declaring that "RADIOACTIVITY is Poisoning Your Children." The featured article began by exclaiming that fallout from bombs could kill most of the human race, but gave about the same amount of attention to the way the nuclear industry's wastes "polluted the earth." The first book to criticize civilian nuclear energy at length, published in France in 1956, declared that the people who put hundreds of kilograms of isotopes into the atmosphere with bomb tests could not be trusted with reactor wastes either. The first book in English critical of the industry, Robert and Leona Rienow's *Our New Life with the Atom* of 1959, also devoted about half its length to bombs and mixed talk of fallout with talk of reactor by-products. The Rienows were veteran conservationists who had cut their teeth in the traditional type of battle over scenic woodlands, then had gone on to jeremiads against resource depletion on a national scale, and now warned against contamination of the globe; they called isotopes a sort of poisonous dirt that could doom Mother Earth. Intemperate and ahead of their time, such writings attracted little notice, but they showed how ideas of pollution could be transferred from military to civilian materials.[5]

A 1959 letter that a couple in Florida sent their congressman displayed the ideas all intermingled. Worried by a local report of frogs with extra eyes, they demanded a ban on bomb tests. A few lines later they demanded a ban also on civilian nuclear work. Turning back to bombs, they spoke of the end of life on earth; turning again to industry, they said that nuclear wastes threatened future generations. In the same vein, a British film about a monster lizard engendered by radioactivity put newsreel clips of bomb blasts alongside talk of nuclear industry wastes. "After all, 'radioactivity' is a frightening word," remarked a newspaper editor in 1959, indicating how much things had changed since the start of the century.[6]

The editor was addressing a political storm that had arisen over the AEC's practice of dumping drums of slightly radioactive materials into the ocean. That made only an insignificant addition to the huge amount of natural radioactivity already spread thinly through the world's waters, so experts had not objected, and in mid-1959 the AEC had routinely announced new offshore dump sites. Unexpected shouts of protest came from fishermen and beach property owners. Local citizens' groups caught the ear of the national press, while sophisticated critics warned that isotopes might be concentrated in food chains until fish became contaminated, just like the tuna after bomb tests. Such worries cropped up around the world. For example, fishermen and housewives on the Riviera fought French government plans to dump nuclear wastes in the Mediterranean. Even if the dumping was in fact technically safe, explained the mayor of Antibes, it would be "disastrous to citizens' morale and the tourist economy." The French, along with the AEC and most other nuclear agencies, decided to leave wastes in temporary storage tanks on remote land sites, which somehow seemed to threaten people less.[7]

Outcries over waste had some unmentioned resonances. In reality the world's most dangerous pollutant by far was not nuclear waste but human excrement, a carrier of disease that killed tens of millions of people every year; during the 1960s the greatest waste problem even in the United States was uncontrolled sewage. When some critics called radioactive wastes "sewage" they were suggesting thoughts of excrement. The nuclear industry itself described its wastes as products of the "back end" of nuclear fuel processing, a transparent metaphor, and some industry workers described radioactively contaminated objects as "crapped up." A French psychologist who ran association tests on the word "wastes" found that it evoked disgust and sometimes direct thoughts of feces. This association was hard to discuss, since many cultures suppressed such matters so strongly that the common words for human wastes were used mainly in anger.[8]

Excrement was often associated with hostility, and therefore, curiously enough, with bombs. Clinical psychologists noticed that small children thought of their feces not only as poisonous but sometimes as explosive, and fantasized using the stuff to blow things up. Second World War bomber crews occasionally remarked that they were symbolically defecating on the enemy, and some flew warplanes like the "Privy Donna," decorated with a painting of a flying outhouse depositing bombs. I would not go so far as Robert Oppenheimer, who one day despondently told Szilard that their weapons were of no use for any good purpose: "The atomic bomb is shit." I am only suggesting that on every level of human thought, radioactive wastes—in association with weapons—were seen as filthy insults against the proper order of things.[9]

Not only primal associations but real problems made the issue important. Deaths numbering perhaps in the hundreds, an entire countryside contaminated and abandoned—in 1958 newspapers reported hints that something like that had happened in the Soviet Union. Some said the disaster involved fallout from bomb tests, but nuclear authorities scoffed at the notion and the matter was quickly forgotten. Yet there really had been a disaster, not at a bomb test site but in the Ural Mountains at the Soviet equivalent of Hanford. Twenty years later some facts emerged. In a way never disclosed but evidently involving the carelessness typical of Soviet nuclear work, tons of radioactive materials from plutonium processing had spread across hundreds of square miles. Although the problem involved wastes rather than a failed reactor, the result was about as bad as a typical Maximum Credible Accident calculated by the AEC. The minority of nuclear experts who had insisted that radioactive wastes could do serious harm, if badly enough mishandled, were right.

Yet it was almost impossible to see the genuine, straightforward technical problems of nuclear industry amid the images of catastrophe and horrid pollution. A few people kept the horror in the foreground of their thinking. The great majority of citizens, however, kept waste problems in the background alongside science-fiction exaggerations, all obscured behind the dazzling visions of Atoms for Peace.[10]

THE PUBLIC LOSES INTEREST

After feeling a flick of the whip of opposition, the nuclear community redoubled its efforts to soothe the public. During the five-year period 1963–1967, American agencies and corporations made available over twice as many films about reactors and three times as many about nuclear safety and the environment as they had offered during the five years preceding. In the 1960s roughly forty million people attended screenings of AEC films and four times as many saw them on television, while in other countries too a great many people saw films or visited promotional exhibits. Only now the result, as intended, was less to excite the public about nuclear power than to calm them.[11]

The new productions often diverged from the 1950s publicity by toning down utopian promises, focusing on ordinary electricity production rather than medical and agricultural fantasies, giving more space than ever to prosaic reassurances. A color film available on free loan under the joint sponsorship of the AEC and the Atomic Industrial Forum had a representative title, "Atomic Power Today: Service with Safety." Similarly, the author of a popular children's book on atoms inserted a few pages about safety between the first edition in 1961 and the third edition in 1966, just to say that experts had everything well in hand.[12]

The occasional criticism of reactors grew fainter, perhaps less because of public relations work than because the forces behind the opposition had slackened. After the ban on atmospheric bomb tests, with scientists and street demonstrators no longer lashing out against explosions and radioactive dust from weapons, people were less often reminded of such problems from reactors. Moreover, in the United States the proponents of private nuclear power won their battle against government ownership, so the political dispute was put in cold storage and both sides got on with promoting reactors. Meanwhile the conflict between the Joint Committee on Atomic Energy and Strauss's AEC, a conflict that had occasioned much congressional criticism of the agency, ended when the Joint Committee proved stronger. Deferred to by the executive branch after Strauss retired in 1958, and also deferred to as nuclear experts by their fellow legislators, the committee members came to dictate American nuclear policy without official dissent. Lilienthal remarked in disgust that these congressmen had abandoned their watchdog role to become high priests of the magic atomic secrets, deflecting normal democratic questioning.[13]

Public interest in the nuclear industry waned still more because it became clear that an atomic industrial revolution was not coming soon. The economics of electricity from reactors began to look shaky, for new technologies such as offshore drilling and strip mining were holding down the costs of oil and coal, while nuclear power plants ran afoul of costly technical surprises. Around 1960 almost every nation cut back its goals for the amount of electricity to be produced from reactors in the next decade, settling for a half or a quarter of the amount in the first ambitious plans. A British reporter wrote that the romantic days of nuclear energy were over: a young scientist could no longer get a date with a girl just by telling her he worked with atoms, and senior physicists no longer got invitations from tycoons for a lobster and champagne dinner.[14]

Among periodical articles from various countries, listed in the *Internationale Bibliographie der Zeitschriftenliteratur,* the total space given to civilian (as well as military) nuclear energy reached a peak around 1960 and then steadily dropped. By the early 1970s these matters took up less than a quarter of the space they had formerly taken. A study of the Paris newspaper *France-Soir* also found that the space given to nuclear questions, about half military and half civilian, after growing swiftly in the 1950s, dropped sharply between 1962 and 1972.[15]

A more precise picture emerges from the titles listed in the *Readers' Guide.* Magazine articles on peaceful atoms shot up to a peak in the second half of the 1950s and then died away, a passing fad. By the second half of the 1960s, articles about civilian nuclear energy were so few that they took up about the same fraction of all magazine space that radium had occupied during the early decades of the century. This rise and fall was parallel to the wave of interest in hydrogen bombs, fallout, nuclear strategy, and so forth, which also reached a peak in the latter 1950s; however, interest in weapons was kept high through 1964 by bomb tests, war scares, and the proliferation and fail-safe debates, so the early 1960s were the one period when military atoms got far more attention than peaceful ones. If bombs held the public's interest longer, they lost it more abruptly, and by the latter 1960s weaponry had fallen to about the same modest fraction of magazine space as civilian uses.[16]

Citizens at a few localities in the United States, supported by a handful of hardy critics who kept the fight alive on the national level, kept on opposing certain proposed reactors. But the nuclear industry, having learned to be wary, tried to choose reactor sites where the neighbors would feel that tax advantages and new jobs outweighed any risks. Besides, whatever battles did take place no longer had the sense of novelty or urgency that could make something newsworthy.

Even accidents aroused little interest. In 1966, while the "Fermi-I" breeder reactor that Cisler had built at Lagoona Beach was going through its shakedown tests, a piece of metal came loose and partly blocked the flow of coolant. The result was a melting of fuel roughly comparable to the Maximum Credible Accident that the designers had prepared the reactor to withstand, and no radioactive material escaped. Critics promptly pointed out that if somehow there had been a far more extensive blockage, and if every safeguard had then failed and somehow there were an explosion, and if the containment shell had also somehow failed, and if a breeze had come in just the wrong direction, Detroit could have suffered grievously. That was an old story, and the press scarcely noticed. As far as polls can show, a majority of people around the world were as willing to accept nuclear power, and a minority as suspicious of it, in 1966 as they had been in 1956 when the Lagoona Beach fight began, or for that matter in 1946.[17]

Underneath the overt opinions, however, there had been important changes during the late 1950s and early 1960s. The nuclear industry had become plausibly associated with drastic error, mass death, and dreadful pollution. As the visions of wondrous progress receded into the haze of a distant future, these anxious images remained in place. At the peak of the Atoms for Peace campaign, around 1956, fully half of the articles on civilian uses in the *Readers' Guide* had optimistic titles, with most of the rest displaying no particular tendency. Only four years later, less than a quarter of the titles sounded positive, with most of the rest still neutral. Meanwhile articles with negative or fearful titles, and about as many more with titles that noted questions about nuclear risks but did not take sides on the issue, were climbing in proportion to the rest, as gradually and regularly as the advance of a tide. In the mid-1960s the total number of articles that showed doubt was still small, since the press was giving little attention of any kind to nuclear affairs. But as a proportion of all articles on reactors, the anxious titles were beginning to catch up with the cheerful ones. Back in 1957 a World Health Organization study group had remarked that educated people, faced with official confidence about reactor safety, hesitated to voice their inner worries lest they appear ridiculous. Five years later it had become respectable, if not yet popular, to voice such worries.[18]

Nuclear officials could remain confident as long as most of the public showed little interest in the industry. In February 1969 a leader of the Joint Committee gave a pep talk to nuclear engineers at Oak Ridge. He admitted that there were "still some critics who like to get their names in the paper by making muckraking charges or fabricating half-truths," but he said that because the nuclear power industry had taken enough care to forestall any serious accident, it had "largely solved the public acceptance dilemma it faced just a few years ago." The public, he believed, was learning to ignore the occasional cries of misinformed critics.[19]

THE NUPLEX VERSUS THE CHINA SYNDROME

It was time to move ahead boldly, and the Oak Ridge staff meant to be in the lead. In their "City of the Atom" they already lived contentedly alongside reactors and other nuclear installations, scattered among the rolling hills of Tennessee. This was TVA country, with a local tradition of visions of high technology blended with rural self-reliance. Oak Ridge itself had not quite managed to marry the White City with Arcadia, for the buildings seemed out of place, like a machine set down in a cow pasture, and electrification of the backwoods had somehow not eliminated all poverty. Scientists felt challenged to work still harder toward a utopian community. Oak Ridge had a technical word for it: the nuplex.

A nuplex, a nuclear complex, would be a town centered on reactors. Confident about safety, engineers figured the reactors ought to be right in the middle so that their output of heat could serve factories. The nuplexes, virtually independent of fuel supplies, could be dropped into jungle or tundra. For some people this was not just a technical idea but a move toward social rebirth. Ungainly cities, those "great clots of humanity," would wither away leaving only centers for thought and art, while industry would be handled by reactor complexes buried safe and invisible beneath meadows. Oak Ridge leader Alvin Weinberg proudly explained that this was precisely the dream of a world set free that he had learned from H. G. Wells.[20]

Oak Ridge drew up detailed plans for a "nuclear-powered agro-industrial complex," with reactors located in a desert to pump up underground water or remove the salt from seawater. Supported by Senator Anderson and many others, the idea turned into serious plans. In 1965 President Johnson announced that if Israel and Egypt promised not to build nuclear weapons, the United States would give them reactors to pour forth electricity and pure water, while the Palestinian refugees would be bought off with fertile new irrigated lands. The world's knottiest political problem, threat of bomb proliferation and all, would be solved by the power of the atom!

The nuplex, a monumental cluster of white hemispheres with clean-cut personnel and miles of pipe, was the most specific plan ever devised for a futuristic town. And it was technically feasible, as the Soviets set out to prove in practice by siting reactors where they could help to heat towns. But the image had at least one drawback. In Oak Ridge's colored drawings a nuplex was neatly demarcated from the surrounding countryside, as if the designers remembered that the original nuclear complexes such as Oak Ridge itself had at first been literally enclosed within fences. As Lilienthal later noted, the social corollary of the nuplex was "an invisible cadre of experts" doing as they thought best in isolation from the larger community. Such a nuclear "priesthood," as Weinberg approvingly called it, would appeal to experts like Weinberg himself more than to ordinary citizens.[21]

The main promoter of nuplex imagery during the 1960s was Glenn Seaborg, who served as chairman of the AEC for a decade. Seaborg had begun as one of the bright science students at the University of California in Berkeley, youths with flat Midwest accents who aspired to climb the arduous paths of research. He made his name directing the Manhattan Project's effort to produce plutonium, yet he had none of the bravura that people might expect in a top scientist-administrator. Tall and rangy, austere in his personal life, cautious and orderly in his thinking but with a romantic streak when his mind turned to the future, and dedicated to improving the world through science, he seemed to embody the values of H. G. Wells's virtuous technocrats. Seaborg had not begun his career hoping to win money or power or leisure, things a scientist could scarcely expect; like many other scientists he longed especially for the respect of his peers and a good name among future generations. Since his career was based on his work with plutonium, it was only natural for him to hope the metal would prove supremely valuable. Close observers suspected that this was what gave him his enthusiasm for the plutonium breeder reactor.

Seaborg reported to President Kennedy that only the breeder could "really solve the problem of adequate energy supplies for future generations," and he gave speeches predicting that the nation would soon have a "plutonium economy." The element he had helped to discover would become the basis of world prosperity, perhaps even replacing gold as the monetary standard; alchemical gold was nothing next to the "large-scale alchemy" of reactors. Moreover, Seaborg and his colleagues expected economic trouble unless they had breeders in hand by the 1980s to stretch out the supply of uranium fuel. For the number of uranium reactors was swiftly rising.[22]

The breakthrough had come in 1964 when General Electric announced that it would build a reactor in New Jersey under a contract that guaranteed to deliver electricity more cheaply than a coal-burner. Soon GE and Westinghouse were furiously bidding against each other for orders while utilities jumped on the bandwagon. In fact the new reactors ran into every obstacle that could be imagined for an untried technology, and when the dust cleared it appeared that between them GE and Westinghouse had lost a billion dollars. But the cost seemed worthwhile to make nuclear power a permanent fixture of the economy. Western Europe, Japan, the Soviet Union, and some less developed nations followed suit, returning to the ambitious plans they had laid aside a few years earlier.[23]

Behind the scenes at the AEC the reactor boom gave headaches to the staff, who had to answer a flood of applications for reactor licenses. The proposed new plants would each generate close to a thousand megawatts of electrical power, but so far nobody had any experience in operating a reactor larger than a few hundred megawatts. Worse, the big new plants would each contain an unprecedented load of radioactive materials. Some safety experts grew so concerned that they left the mainstream of optimism.

After 1965 the Safeguards Committee began to write confidential warnings about something worse than the usual Maximum Credible Accident. If most of the fuel in one of the new reactors ever melted, there was a chance that it could burn its way down through the floor of the containment building. If the molten mass hit groundwater and set off a steam explosion, the containment shell might be no more reliable a line of defense than the Maginot Line. One engineer dubbed the problem "the China Syndrome." In fact the molten fuel would probably descend only a few dozen meters toward China, but everyone adopted the phrase. Like President Truman's exclamation that atomic bombs could blow a hole clean through the earth, like the deep shafts in mad scientist movies, the words caught a feeling of uncanny dread.[24]

The best defense against the China Syndrome would be to prevent a reactor's fuel from ever melting. If something stopped the flow of coolant there ought to be a system to pour a torrent of fresh water into the reactor's core, an "emergency core cooling system." Safety researchers began to focus on that idea, as usual giving the most attention to the worst conceivable case. The AEC defined that as a total break of the largest pipe leading into the reactor, as if some giant were to wrench the pipe in two. It was hard to imagine how such a failure could happen, but it was closely studied as the most severe test any reactor might face.

Engineers were challenged to calculate what would happen as steam rushed away from a white-hot set of fuel rods while cold water flooded in. The work offered government safety researchers fascinating insights into the nature of violent fluid behavior, and also offered them (as reactor manufacturers wryly noted) years of funding or even an entire career. The problem turned out to be so challenging that it could not be solved by calculation but only through lengthy tests. By the mid-1970s it had been proved that emergency core cooling systems would work, but throughout the preceding decade nobody could be quite sure.

Not many noticed a worse flaw in the research: it was closer to fantasy than to anything known in real life. As a few experienced engineers pointed out, and as events would later show, actual accidents were most likely to begin not with a huge pipe wrenched in two but in more routine ways—a hidden flaw in the design of one of the thousands of minor components, combined with instrumentation that gave a poor picture of what was going on in the reactor, combined with mistakes by thoughtless or misinformed workers. But that sort of combination was less calculable than a maximum failure, less liable to cause utter catastrophe, and, in short, less interesting. Few people studied minor design flaws, instrumentation, or workers' mistakes. Although the AEC did require reports on every little failure, the reactor operators naturally described all these as innocuous. The reports about small things that went wrong piled up, unnoticed, while studies of spectacular imaginary disasters forged ahead.[25]

There was one official at AEC headquarters who fought against the preoccupation with calamity. Milton Shaw, head of the division that developed new reactors, said that studying emergency cooling was like studying how an aircraft disintegrated during an accident; he thought the money would be better spent making sure that accidents never happened in the first place. Shaw's opinion counted, for he was in charge not only of developing and promoting new reactors but also of overseeing research on their safety. In 1965 he began to reshape this work with an iron hand.[26]

Shaw was a Rickover man who had joined the AEC after working on naval reactors, and like the Admiral he was a tough, tireless administrator. His favorite word was "disciplined": more disciplined organization and more disciplined control over manufacturers, just as in the Navy—that was the route to safety. The AEC laboratories were strangers to such an approach, for they had been purposely set up with a loose structure to attract top scientists, the sort of people who chose their own tasks. Shaw slapped down such attitudes. Soon research assignments were dictated from Washington, and people who did not toe the line might find their jobs yanked away. Shaw became not only the hardest working but also the most despised administrator in the history of the AEC.[27]

The internal conflict came to a head over safety questions. Shaw felt that light water reactors had come of age and could be freed from government tutelage; his heart was in the next leap forward, the breeder. Seaborg, the other commissioners, and the Joint Committee on Atomic Energy all agreed. During the latter 1960s, just when the Safeguards Committee and many researchers at AEC laboratories were becoming concerned about the China Syndrome, Shaw and his superiors directed effort away from ordinary reactor safety and toward breeders. Only the Safeguards Committee, with increasing exasperation, tried to oppose the trend. But when final decisions were made the members of the Safeguards Committee went along, for they too wanted the nuclear industry to grow, if not at such a headlong pace.

In one way Shaw's strategy succeeded. From the early 1970s on, every least valve and switch on every American reactor was forced to meet rigorous tests. If a pipe failed there would be records to trace back who had inspected it and even where the steel had come from. People in the industry hated the truckloads of paperwork and the skyrocketing costs that Shaw's methods imposed, but in the end they had to obey.

Such controls had kept submarines reasonably safe, and they could work in other places too. For example, French nuclear authorities hired former naval officers with their tradition of discipline, trained a highly professional corps of reactor operators, and concentrated their attention more on actual experience with small-scale failures than on imaginary catastrophes; the result was a fine safety record. But by the 1970s France, like Rickover's Navy, had settled on one standardized type of reactor (based on a Westinghouse design), one builder (a combine called Framatome), and one user (the nationalized utility, Electricité de France). American civilian industry, with its bewildering variety of reactors—each unlike the next, designed by different manufacturers, built by a variety of contractors, and run by all sorts of utilities—did not fit so easily into the harness.[28]

A strict yet flexible and realistic control over American reactor safety had been achieved in the 1950s by the independent safety review committees that had appeared spontaneously in each laboratory, but in the late 1960s these began to weaken. Engineers who had to meet Shaw's demands begrudged taking further months to work through the problems again with an in-house safety committee. And if the local committee had its own ideas about safety methods, these were swept aside by the AEC's rigid specifications. One by one the laboratory safety committees either withered away altogether, or became mainly concerned with previewing designs so the AEC staff would find no embarrassing flaws. The primary concern of engineers was no longer how they would look in front of a jury of their colleagues who reported to the organization's bosses; now they had to measure their work against shelves of written standards issued from AEC headquarters.[29]

Most fateful of all, nuclear experts had become divided. Some researchers at Oak Ridge and other laboratories, and a majority of the Safeguards Committee, kept discovering further possibilities for accidents, and begged for more funds so they could study whether the problems were trivial or deadly. When Shaw gave such research little money, these experts began to feel that the AEC leadership had made an alliance with industrialists to deliberately suppress every awkward safety question. These frustrated feelings, if they ever came into the open and mixed with the public's latent anxieties about all things nuclear, would make an explosive combination. The spark needed to set off the explosion would arrive from an unexected quarter: some startling new ideas about nuclear war.

16

The Debate Explodes

Thousands of warheads would slant down from the direction of the North Star. When they hit the upper atmosphere they would burn like meteors, drawing lines of fire pointing to their targets. Deep in caverns at the end of those lines the defenders, forewarned by Argus-eyed satellites, would activate their computers and await the outcome; the next acts would be too swift for human reflexes. Electronically guided rockets would dart upward like meteors in reverse to thwart the attack, nuclear fireballs by the thousand would illuminate the air, and then . . . and then what? Nobody knew how it might end, this warfare of the 1970s.

The potential threat brought an actual attack of imagery, a renewed eruption of hallucinatory visions across the landscapes of the mind. Missiles were not the only disruption. By the 1970s many people began to long for a more human scale in every technology, for a world that was nearer to the cozy villages of children's books and farther from the incandescent puzzles of computer screens.

When technologies came into question, nuclear energy would stand at the head of most critics' lists—and by nuclear energy they would mean reactors. Yet public criticism of the civilian nuclear industry became important only after other problems had stirred the public. The first radically new technology to run into widespread and effective opposition was not the reactor, but the antiballistic missile.[1]

THE FIGHT AGAINST ANTIMISSILES

In the late 1960s a new game began in the rooms where military strategists assembled. Between 1960 and 1966 the United States had quadrupled the number of vehicles that could reliably drop a bomb on Soviet territory, leveling off at more than two thousand B-52 bombers, Polaris and Minuteman missiles, and the like. Half a dozen years behind came the Soviet reply, giant rockets whose numbers doubled every two years. Meanwhile engineers on both sides were designing multiple independently targetable reentry vehicles, MIRVs, each of which could divide up to destroy several enemy bases. Of course, a few missile silos would survive such a First Strike and launch their multiple warheads in retaliation. But now American and Soviet engineers claimed they could build a rocket to shoot these down, an antiballistic missile or ABM. A decade earlier RAND and other groups had developed theories about using missiles to attack enemy missiles like pieces in a game of chess; now these theories were crystallizing into hardware.[2]

Poring over the photographs sent down by spy satellites, in 1966 American analysts discovered primitive ABM bases under construction around Moscow. Perhaps the Soviets believed that with multiple warheads and ABMs, they could launch a First Strike and win? This time it was the Republicans who were out of power, and they cried that an "ABM gap" threatened the nation, until President Johnson announced that the United States too would defend itself with amazing inventions.

Some knowledgeable scientists were dismayed. They suspected that to build such defenses might frighten the Russians into attacking while they still had a chance, or might even delude Americans into thinking they could safely start a war themselves. Besides, technical analysis strongly indicated that no defense system, faced with a barrage of thousands of warheads mixed together with decoys, could offer salvation: missiles would always get through. Physicists who had done little since the atomic scientists' movement of 1946 began to organize a new campaign. They were middle-aged now, university professors with high prestige and inside knowledge of weapons, yet their lobbying within the government had failed against other experts who thought that ABMs were worth a try. The opponents began to publish magazine articles, but at first these did little to arouse the public. It took the United States Army to do that.

In November 1967 the Army announced it was choosing sites for

ABM bases near Boston, Chicago, Pittsburgh, Seattle, and six other major cities. The announcement hit a nerve. Suddenly in the outskirts of each city there was a particular location singled out as a bull's-eye for Soviet attack. That neighborhood would also be a home for hydrogen bombs on the tip of each ABM rocket, and might one explode by accident? "H-Bombs in the Back Yard" was how one physicist put it, and the phrase became a slogan. By early 1969 local groups ranging from Audubon Society chapters to town boards were straining to shove the fearful things away. In the chosen Seattle and Chicago suburbs, resistance was so fierce that the Army decided to shift to other locations, but that only added two more towns to the opposition. People's feelings were caught by a newspaper headline: "Anywhere Except Near Us."[3]

The scientists lifted these feelings from a local to a national issue. A few hundred dedicated physicists crisscrossing the country gave speeches, debated generals on television, distributed bumper stickers, and marched on the White House, raising the first strong outcry of nuclear anxiety that the world had seen since bomb tests went underground. A college student remarked that he had tucked the threat of nuclear war out of consciousness for half a decade, but the ABM debate made it all "real to me once again." A SANE newspaper advertisement, reproduced as a poster that showed up on community bulletin boards and college dormitory walls, featured a caricature of generals slavering with goggle-eyed glee over a model ABM rocket. "They're mad," said the text. "They're absolutely mad . . ."[4]

SANE labeled these generals "The People Who Brought You Vietnam"; the stereotype of crazed nuclear warriors was linking up with larger issues. Many people around the world had begun to turn against warfare in former colonial regions, and at the same time against the economic damage done by the arms race and against "militarism" in general. Earlier debates had concentrated on nuclear bombs while scarcely ever addressing the specific social factors that underlay all decisions about armaments. But when critics in the late 1960s looked anew at nuclear war, they examined the bombs within a larger context.

The first precise questions had been posed in 1967 by no less a personage than the American Secretary of Defense, Robert McNamara. Disgusted by the pressures that were forcing him to build an ABM system, he declared in a speech to news publishers that a "mad momentum" was pushing the nation to arm itself above any sensible level. The warning was heeded by many, for example by a veteran atomic physicist and critic of the AEC, Ralph Lapp. In a 1968 book he went beyond the customary technical study of nuclear strategy to look at the congressmen and industrialists who promoted new weapons. Were their views, he asked, "colored by a vested interest in the arms race?" Soon many were explaining that weapons were indeed built because of interests vested in cherished ideologies, in political ambitions, or in cold cash. A few authors even considered how these interests combined with the imagery of fear, reporting on the extensive public relations efforts that encouraged military spending by playing up foreign threats, and suggesting how national leadership groups had burdened themselves with mortal suspicions that were only loosely connected to the actual behavior of other nations. For a few years after 1968 a crowd of books and articles about the "military-industrial complex" rushed into print, urging citizens to strike down the ABM and other weapons at their economic, social, and ideological roots.[5]

The battle was short-lived. In March 1969 the new Nixon administration announced that it would not put H-bombs in the cities' back yards but would only defend remote missile bases, and in 1972 the ABM was essentially abandoned by mutual agreement with the Soviet Union—the first true victory for arms control. As nuclear weapons once again receded from view, and as the Americans retreated from Vietnam, study of the military-industrial complex faded away as rapidly as it had arisen.

The brief ABM debate had only agitated a small fraction of the world public, but it showed that nuclear fear had not disappeared. All that was needed to bring opposition roaring forth was an issue where some neighborhoods were singled out to face a threat and where they felt they had a chance to push it away. Meanwhile the debate reminded citizens that government decisions about nuclear technology could be scientifically suspect, and politically vulnerable besides.

SOUNDING THE RADIATION ALARM

The Atomic Energy Commission and the Joint Committee on Atomic Energy were only peripherally involved with the ABM, but before that battle faded it gave birth to another which would destroy both institutions. The new battle, like the fallout controversy of a decade earlier, was about radioactive substances. But this time the associated images of mass death would gather around civilian industry.

The opening gun was an article in *Esquire* magazine by the physicist Ernest Sternglass. Sternglass had known about the problems of radiation since childhood, for his father, a doctor, had often used X-rays. In 1947 the physicist's infant son became desperately ill, and he worried, according to his later recollection, that he had passed on a defective gene originating in his father's exposure to X-rays. It turned out that the baby had Tay-Sachs disease, a fatal genetic flaw of ancient lineage, yet Sternglass never felt certain that radiation had not caused the tragedy. Meanwhile he was inventing idiosyncratic theories about elementary particles; when his colleagues rejected his ideas he made a career for himself as an expert on radiation effects. Sternglass became worried about radiation from bombs, and from time to time he spoke out. For example, in early 1968 when there was a minor flurry of debate over whether the United States should use nuclear weapons in Vietnam, he publicly warned that bombs would cause widespread illness.[6]

Sternglass became famous with his *Esquire* article, titled "The Death of All Children." His argument, which he repeated in other magazines, in newspaper interviews, and in appearances on major radio and television shows, began with the fact that infant mortality had not declined recently in the American South as in other regions. He said this happened because the South lay in the path of fallout from the Nevada bomb tests. Extrapolating his results to the whole world, he figured that one out of every three infants who had died in the 1960s had been killed by bomb test fallout.

Sternglass concluded that in a war, even if ABMs destroyed every missile with nuclear explosions high in the atmosphere, the ABMs' own fallout would doom every baby born for decades in both America and Russia. In short, the ABM was a "doomsday machine." He found the prospect of racial suicide almost hopeful, for if using weapons would mean the end of humanity, then no soldier would dare to use them. He insisted that he had proved this; his statistics would help save the world from war.

For people anxious to abolish bombs, it was almost a disappointment when Sternglass's calculations turned out to be worthless. It was not even true that southern states were downwind from bomb tests. The reason infants in the South died disproportionately was the region's "socioeconomic conditions," that is, poverty. No other scientist who studied the matter found any value in Sternglass's fallout theory, and in later years not even the most passionate opponents of radioactivity would invoke it.

While the experts were dismissing Sternglass, ordinary citizens who read his statements or saw him on television could not be so sure. He was a physics professor, just like others who were arguing plausibly against the ABM. He was also a good performer, making his points clearly and persuasively. When he and an AEC expert each had a few minutes to present opposing views on a television show, it was impossible to tell who was right. Most effective of all was the way Sternglass shifted ground. When his bomb test study was thoroughly discredited, he came up with statistics about infant mortality in the vicinity of nuclear power plants.[7]

Others had already begun to worry about the fact that all nuclear reactors released slight traces of radioactive materials into the neighboring air and waters. For example, the Minnesota state government was squabbling with the AEC over how much radioactive material should be allowed to escape from a nuclear reactor upstream from Minneapolis. Into these debates strode Sternglass, insisting that radiation from existing reactors had already killed babies by the thousand. Worried public health experts scrutinized his figures and discovered that they were full of holes. The regions where Sternglass pointed to high death rates had typically received a far more serious dose of poverty than of radiation. However, by the time one set of statistics was refuted the professor had a new set which he found even more persuasive. Not one other scientist was ever convinced by his figures, but there was always some newspaper reporter who would publish Sternglass's latest horrific calculations. The game went on all through the 1970s.[8]

Although Sternglass was distressed that reactors, as he believed, slaughtered babies, his main motive was his original one of discrediting bombs. "The military is behind the entire nuclear reactor program," he declared. He believed that evil officials would do anything to make people think that radioactivity from reactors was safe, since otherwise nobody would tolerate nuclear weapons either. The situation was so desperate, he hinted, that his enemies might kill him.[9]

Nuclear experts were disgusted by Sternglass's attacks. They believed that reactors were going to be much cheaper than other sources of energy and therefore would reduce poverty, the real slayer of babies. AEC staff tried to learn in advance where Sternglass was heading, and worked with occasional success to persuade magazine publishers and television stations that he did not deserve a public stage. The AEC also encouraged its own scientists to refute him, and that turned out to be a mistake.

Nowhere in the AEC's domain was enthusiasm for nuclear energy higher than at the Livermore Laboratory. Fenced in among the hills of central California, the laboratory was Teller's creation, built to rival Los Alamos in designing weapons. During the fallout controversy Livermore had set up a health research program under John Gofman, a respected radiation medicine expert with a fringe of graying beard and a friendly smile, who insisted from the start that he would never kowtow to the AEC. Despising government authority in general, and Teller's enthusiasm for explosions in particular, he remained apart from the Livermore spirit. Gofman later explained that he found all officials nauseating, little better than murderers; "They can only think how to obliterate, control, and use each other."[10]

When the AEC sent Sternglass's *Esquire* article around its laboratories, Gofman asked another Livermore scientist, Arthur Tamplin, to look into the matter. Tamplin, an opponent of the ABM, later said he had found himself in a dilemma once he understood that Sternglass's argument against the weapon did not hold water. He solved his problem by recalling the reasoning that Pauling had used during the fallout battles: stray radiation might well have killed some thousands of babies even if that could not be proved. With Gofman's backing he insisted on publishing this fact. Like any bureaucracy, the AEC disliked doubters within its own ranks, and the pair found themselves increasingly unwelcome at Livermore.

The conflict drew Gofman and Tamplin's attention to claims that reactors were absolutely safe. That was false, of course, for however slight the amount of radioactive material a reactor released, there was at least a possibility that it would cause somebody harm. Disdaining to back away from a fight with authorities, the pair toured the nation, giving talks and interviews to warn against materials routinely emitted by reactors. Meanwhile they wrote a book titled *"Population Control" through Nuclear Pollution.*[11]

Gofman and Tamplin built upon the worries that Sternglass had originally provoked. For example, all three were given generous time in a 1971 television documentary broadcast in Los Angeles. Viewers were left with the thought that, as narrator Jack Lemmon put it, "nuclear power is not only dirty and undependable . . . it's about as safe as a closetful of cobras." The AEC and California electric utilities, outraged, bombarded the television station with rebuttals until the station broadcast an editorial saying that it thought nuclear power was a good thing. That could scarcely erase the first impression.[12]

While the AEC was losing the battle of images it also fared poorly in the scientific dispute. Gofman and Tamplin scotched the notion, still upheld by some of the nuclear industry's friends, that there was a threshold below which radiation was proven to cause no damage. However, they did not emphasize the fact that nobody could prove what effect, if any, tiny traces of radiation had on health; instead they suggested that the effects were almost certainly serious. Gofman and Tamplin's most effective argument was modeled on Pauling's large numbers: they said repeatedly that if everyone in the United States got a dose at the maximum level the AEC would permit reactors to emit, there could be 32,000 excess cases of cancer every year. Nuclear industry spokesmen replied that the calculation was not only purely speculative but grossly unfair, since most members of the public got only a minute fraction of the maximum permitted dose. Very well, said Gofman and Tamplin, then why not set the rules so that public exposure would remain low when the reactor industry expanded?

In 1971 the AEC reluctantly bowed to this logic, imposing far stricter regulations on the amount of radioactive material that a reactor could legally release. Now even Gofman and Tamplin's calculations showed less damage to the public from reactors than from various ordinary industrial hazards. Reactor builders grudgingly agreed to install the expensive pollution control equipment, if only to buy freedom from public outcries. The radiation controversy faded away. But it left behind a new image—reactors insidiously slaying people at a distance.

Real events reinforced the idea that radioactivity might contaminate people not just near reactors but anywhere. The first visible problem came at uranium mines in the American West, poorly regulated by the responsible state agencies and thick with radioactive dust. In 1967 the Secretary of Labor announced that about a hundred miners had perished needlessly of lung cancer, with hundreds more probably doomed. Over the next few years the federal government imposed tight standards on uranium mines, much as it began clamping down during the same years on deadly practices in coal mines and other industries. Meanwhile the uranium miners' deaths had demonstrated that nuclear work was not somehow set above other industries; it really could bring a health disaster.[13]

Another lesson came from the tailings of uranium mines and mills, waste heaps that emitted radon gas and other radioactive substances. Some of the substances found their way into the air and rivers of Colorado and New Mexico; more outrageous still, tailings had been trucked away and used for construction, so that thousands of houses stood on mildly radioactive sand. In 1971 the Colorado Board of Health warned that some families were exposed to an extra risk of cancer. This was no visible epidemic such as the uranium miners suffered, but it was, for the first time, a problem clearly endangering people outside the nuclear industry.[14]

Even less visible and less specific problems disturbed people even more. A particularly worrisome revelation of the early 1970s came from Hanford, where intensely radioactive liquids from the atomic bomb factories were temporarily stored in underground tanks among the barren hills. Since the late 1950s some tanks had leaked, releasing hundreds of thousands of gallons of wastes into the sandy soil. Nobody around Hanford had been exposed to excess radiation, but the AEC had kept silent about the leaks for over a decade, so when the news came out, official reassurances were unconvincing. Was any place safe?[15]

Researchers who had been studying reactor by-products in a leisurely way had concluded that the problem could be solved. It was only necessary to mix wastes with cement or glass or whatever to make a rock-hard substance, seal that in thick canisters for extra protection from ground water, and bury the canisters. After all, in some parts of the world natural radioactivity in rocks, buried near the surface with no precautions, had lain for billions of years without straying out of the rock bed. With criticism building up over the Hanford leaks and a number of similar cases, the AEC announced in 1971 that it would solve the problem for good. It would bury wastes in an underground salt dome, since such rock formations were perfectly dry and self-sealing.

With his usual confidence Milton Shaw told the Joint Committee that the chosen site in Kansas was "equal or superior" to all others in the nation. Then Kansas officials showed that the region was riddled with bore holes into which salt miners had injected water; the AEC had hit upon a remarkable rarity, a wet and leaky salt dome. Faced with an uproar among geologists and the Kansas public, the AEC abandoned its plans. Now there was no place to put wastes permanently, and little confidence that the government could be trusted to find a place. Further government efforts over the next decade led only to paralysis, for the experts' reassurances of perfect safety met a fearful and determined cry: "Not in my back yard!"[16]

According to polls in the United States and Europe, in 1960 few people had worried about radioactive wastes, and even at the start of the 1970s few would bring up the issue. But by the mid-1970s a majority saw the wastes as a major problem, and many worried about it more than any other nuclear hazard. Articles on wastes listed in the *Readers' Guide* climbed steadily in numbers from the 1950s through the 1970s, while other nuclear industry issues came and went. A survey found that almost everyone, whether they were for nuclear power or against it, would rather live near a reactor than near a nuclear waste disposal site. Many felt that nuclear wastes were somehow unique, more dreadful than other industrial dangers.[17]

The most persistent nuclear disputes of the 1970s would center on such by-products, particularly in France and Germany, where bitter opposition arose against installations for reprocessing used nuclear fuel. No other nuclear issue brought out such a wide spectrum of passions. At the highest level were moral objections. What gave anyone the right to contaminate the earth, to endanger innocent animals, to make money at the expense of future generations? At the other extreme the imagery of "wastes" continued to remind some people of infantile anal concerns. American and Japanese critics likened building reactors without a waste repository to building houses without toilets. A French tract called nuclear wastes "a heap of excrement," and when public relations experts in Germany sent an "information bus" into villages near a proposed nuclear reprocessing plant, protestors trimmed the bus with toilet paper and smeared it with dung.[18]

Some believed, as a worried citizen told poll-takers, that reactors threatened a "nuclear infection eventually destroying life on this fair planet." Such a view ignored the fact that reactor wastes were remarkably compact, indeed the only type of industrial waste with a volume small enough to bury all in one locality with scant risk to people in other regions. Even experts who from the start had criticized the nuclear industry for carelessness generally agreed that the risks were relatively minor and local. But their calculations could not outmatch the image of nuclear objects buried in the earth—what people had associated for decades with wicked violations and dreadful dooms. The imagery deployed by critics said that "the silent bomb" (as one writer called it) posed just the same threat as nuclear weapons: contamination of all the world.[19]

Characteristic was a 1972 British television movie, *Doomwatch,* about radioactive waste dumped in the ocean near a coastal village. Fish grew to a great size while the fisherfolk became shambling monsters with a penchant for insane violence. As if to demonstrate the links with tradition, a cheap horror film about oversized ants resulting from nuclear wastes was produced by the same man who had made one of the 1950s films about monstrous insects engendered by bombs.[20]

From the early 1970s on, whenever a nuclear reactor was proposed in the United States a local group of housewives and students, reinforced by a few professional people, would spring up to oppose it. Opponents usually began by reciting Gofman and Tamplin's arguments against the faint radioactivity that regulations allowed reactors to release, but behind that was a persistent suspicion that no amount of regulation would keep reactors from violently spewing their entire contents over the countryside. Experts had scoffed at such beliefs, but in 1971 the opponents suddenly found support among scientists themselves.[21]

The Union of Concerned Scientists was a group of students, scientists, and others, centered on the universities of Boston, who had come together in 1969 during the ABM battle. According to a historian who studied the group, their "primary organizational goal" was to effect "an ideological change in society." For they doubted that science and technology would lead to the betterment or even the survival of civilization unless they were controlled in a more democratic way. The UCS leaders moved on to other issues as the armament battles died away, and their membership declined. Drawn into a controversy over a nuclear reactor site not far from Boston, they cast about for information and came across the China Syndrome.

Despite five years of prodding from the Advisory Committee on Reactor Safeguards, Milton Shaw had funded AEC researchers to make only limited efforts to study emergency systems for cooling a reactor's core. The problem was no secret, but neither was it publicized. Then in July 1971 a UCS report thrust the question into newspapers and television news broadcasts. Daniel Ford, a graduate student, and other UCS members had been talking with scientists from the Idaho reactor laboratory and Oak Ridge and were surprised to find that some AEC experts were unsure that the emergency core cooling system would work. Secret meetings were held late at night; unmarked envelopes arrived at the UCS office with copies of internal AEC safety reports that Shaw had refused to take seriously. All this the UCS publicized vigorously. For the next half-decade the group would focus its efforts on reactor safety, along the way gaining a sharp increase in membership and funding.[22]

Hoping to answer the new public criticism, the AEC opened hearings on emergency core cooling systems. The hearings lasted far longer than planned, giving Shaw's enemies a fine platform. The critics showed how he and other nuclear officials had squelched the doubts not only of outsiders but even of some of their own experts. When Shaw took the witness stand he only made things worse, rejecting criticism with scorn, looking the prototype of a pigheaded official, reminding people of all their suspicions about a powerful and arrogant "atomic establishment."[23]

Meanwhile the UCS and its allies showed that it was not yet proven that a reactor could be prevented from ejecting radioactive materials if it happened to lose all its cooling water at once. Whether that particular, spectacular accident would ever happen in a typical American reactor, and whether research should concentrate on that rather than on more mundane routes to failure, was less discussed. The effect of the emergency cooling controversy, combined with the stories of Sternglass, Gofman, and Tamplin about the perils of radioactivity, and reinforced by the news of real accidents with wastes, was to make everyone from student opponents to nuclear engineers, from television commentators to AEC officials, more preoccupied than ever with the most ghastly conceivable possibilities.

REACTORS: A SURROGATE FOR BOMBS?

Some familiar fears were displayed in a 1977 television movie, *Red Alert.* That had been the title of the American edition of the novel that foreshadowed *Dr. Strangelove* and *Fail Safe;* like those films, the television production was built around forbidding mechanisms, for example a control center where anxious experts studied a computerized wall map. The story was again one of nuclear doom threatened by electronics gone out of control and by a rigid official who himself acted like "a robot." However, the threat in the 1977 production was not nuclear war but the simultaneous explosion of every reactor in the United States, and the official was no general but a reactor expert. The film showed how the image of reactors could borrow wholesale from the image of bombs—and even displace the weapons as a focus of public anxieties.[24]

Not only Sternglass, Gofman, Tamplin, and the UCS, but many others who started with worries about military authorities, began in the 1970s to turn the same thoughts against reactor builders. For example, one of the most effective activists was Helen Caldicott, a lively and tireless pediatrician. As a youth in Australia she had been deeply impressed by *On the Beach,* and when she grew up she helped lead a campaign against French bomb tests in the Pacific. When she moved to the United States and found nobody there interested in bombs any longer, she began to fight reactors. Entire groups followed a similar evolution. For example, one of the most effective groups was in St. Louis, inspired by the biologist Barry Commoner. They had begun work in 1957 with scientific studies of fallout, but in 1971 as the

Scientists' Institute for Public Information they brought a lawsuit against the AEC's breeder reactor program. Their critique was summarized in the first methodically argued and widely read antireactor book, *The Careless Atom,* published in 1969 by a Commoner protégé, Sheldon Novick. Most other organizations that opposed reactors similarly drew much of their strength from individuals and groups who had first fought the AEC over weapons. In the coffeehouses where students gathered in the early 1970s, bulletin board posters opposing the Vietnam war and the ABM were covered over with notices of meetings to protest reactors.[25]

This development was partly due to a merging of images. Polls found a third of the world public still convinced that a reactor could blow up exactly like an atomic bomb, while many more thought that reactors offered at least a vague danger of "nuclear explosion." Even among better-informed people, the image of a poisoned territory following a China Syndrome accident was not very different from the bleak science-fiction landscapes of postwar forbidden zones. And for almost everyone, radioactive reactor products seemed as alive with unspoken horror as if they were pieces of a bomb.[26]

There was a core of truth in this last thought, for reactors do make plutonium. In the mid-1970s the world press took up more vigorously than before the idea that somebody might steal a few kilograms of the metal from civilian industry and craft an atomic bomb. For example, a 1975 American public televison documentary showed that an undergraduate student could design some sort of fission explosive. A closer look showed that the requirements for success were so stringent that his design, or any other plutonium device that could be constructed without a cadre of top experts, would produce a blast closer to a large ordinary bomb than to a Hiroshima catastrophe. But the news reports sounded as if the decades of fiction about entire cities blown up by terrorists could come true at any moment.[27]

Far more realistic was the risk that a nation might steal from itself, secretly diverting plutonium from a reactor to make bombs. Beginning around 1970 there were authoritative hints that India and Israel were doing exactly that, and in 1974 India dismayed the world with an atomic explosion. Although both nations were drawing plutonium not from civilian industry but from "research" reactors they had built with explosives distinctly in mind, critics warned that the nuclear power industry could tempt a nation into making weapons. Most experts on both sides of the nuclear debate had agreed ever since 1945 that this was the most worrisome of all liabilities of civilian nuclear industry, and in the mid-1970s high government circles gave it more attention than any other nuclear hazard, getting into elaborate arguments over the sale of reactor materials to Brazil and so forth. The restrictions only made some nations more determined than ever to build a nuclear industry free of foreign supervision.[28]

The press took only passing notice of these arguments. A survey of American magazine and newspaper articles of the mid-1970s found that only 8 percent of all items on nuclear power focused mainly on weapon proliferation, and a minuscule 3 percent on theft or sabotage. Similarly, when poll-takers asked the public in America and Europe to explain their worries about nuclear power, the replies centered on health, safety, and wastes; unless asked about it specifically, scarcely anyone brought up the fact that reactors could be used to make the stuff of bombs.[29]

It was remarkable how the world public continued to avoid thinking about nuclear weapons. Nothing could have been more spectacular than images of missile warfare, the rockets blasting up from their hiding places, the warheads falling like shooting stars, yet the imagery remained undeveloped. Aside from one or two almost unknown exceptions, movies and television and popular fiction simply did not show what missile warfare would look like. As for nonfiction, in the 1970s for the first time since 1945 civilian atoms dominated military ones in the *Readers' Guide* titles, and by two to one. The same effect could be seen in *France-Soir,* in the *New York Times,* and elsewhere.[30]

As the 1970s wore on there erupted passionate protests about the chance that some reactor might explode despite every effort of its designers, but scarcely any protests against the chance that thousands of devices built to explode might fulfill their purpose. In various nations emotions flared over shipments of civilian reactor fuel along highways, while few mentioned the far more extensive travels of warheads. No agency dared to bury industrial fission products permanently, but masses of the same isotopes were casually implanted underground in every bomb test. Anyone reading the newspapers might have supposed that as soon as governments crafted plutonium into a bomb it ceased to be of interest.

This disturbed the few people who continued to struggle against weapons. Lilienthal, for example, critical of reactors but far more concerned about war, worried that citizens had given up hope of controlling the arms race. Since nuclear reactors did seem politically vulnerable, he said, people attacked them as "a surrogate for bombs."[31]

Lilienthal was saying that fear and hostility had been displaced from weapons onto civilian nuclear power. All the elements for classic displacement were indeed present. There was a persistent anxiety about nuclear war. There was an inability to dispel the anxiety in the only genuine way, by getting rid of bombs. Finally, there was a target toward which the frustrated feelings could redirect themselves, and all the more easily because of the many associations between bombs and reactors. If you had spent your life in a room with a grimacing Russian who kept a flamethrower pointed at your head, you might well feel upset when somebody struck a match.

Displacement could be conscious, as with a person who told me, "I can't do anything about bombs, but I can do something about reactors." Such displacement was not simply the pursuit of a mistaken object. A decade earlier during the fallout debate, when many people had seemed more concerned about milk than about missiles, leaders of the movement had explained that they were attacking weapons obliquely, using an argument that came close to home. The ABM controversy had similarly brought the arms race into people's back yards. Now reactor critics spoke of threats to daily life as intimate as fallout, as visible as a neighborhood ABM base. It was one way to attack the entire "atomic establishment," military and civilian together.[32]

Among the general public, polls showed some striking similarities between approaches to reactor catastrophe and to war. People believed the two possibilities were about equally likely. Probing for social mechanisms, surveys around the world found that with reactor fears, as with war fears, the most remarkable result was no result: only minor differences were found between the standard social groups, a sign that all varieties of nuclear fear had deep roots in individual personalities. A closer look showed that the same groups who worried a bit more on average about nuclear war—the young, the lower social and economic classes, and especially women, or in short, the relatively powerless—also worried more about reactors. Perhaps both varieties of anxiety reflected a more comprehensive concern that extended beyond reactors—and even beyond bombs.[33]

ENVIRONMENTALISTS STEP IN

Negative attitudes toward science and technology were cropping up all over, from scholarly publications to student demonstrations. Polls showed that the number of Americans who felt "a great deal" of confidence in science declined from over half in 1966 to about a third in

1973, while large majorities around the world felt that in some ways scientific progress was too swift. A main reason for the misgivings about science, according to an earlier poll that had studied the matter in detail, was "unspoken fear of atomic war." In the 1970s atomic bombs remained the one technology that most people frankly distrusted.

But public confidence in every form of authority was also falling, with the military, business, and government all increasingly distrusted. Public support for pure scientific research remained high in comparison with support for nuclear power; what was chiefly failing was faith in the officials who ran such an industry. Nuclear affairs were caught up in a larger social tide.[34]

While reactors faced growing opposition, so did many other industries. Until the mid-1960s only a few writers had voiced disgust with science and industrial growth. But it was becoming plain that new technologies such as offshore oil drilling and strip-mining of coal could harm broad regions, and increasingly people began to protest. Even in the Soviet Union a few writers published stories about robots threatening the world, while many more wrote mournful elegies for the simple village life of olden times; others wrote reports on the rampant pollution of rivers and forests in the West, or even nearer home. By the mid-1970s talk of environmental harm was so widespread that Soviet authorities began to censor it, while in other industrial nations protest climbed without restraint.[35]

Groups like the Sierra Club, which traditionally had devoted themselves to local affairs such as preserving redwood groves, began to question industry on a national scale. They were joined by new and more militant groups such as Friends of the Earth. Such organizations burgeoned in the late 1960s and early 1970s, accompanied by tremendous press attention that reflected an equally great rise in public concern. Governments responded with strict laws and new bureaucracies to enforce them. In 1970–1971 alone, major environmental agencies or programs were founded in Britain, France, Japan, the United States, and the Soviet Union.

A historian of the movement, Walter Rosenbaum, found the new mood had three main components—ones with a curiously familiar ring. First was a sense of dire crisis, a fear that humanity was headed for self-imposed catastrophe. For example, a 1970 book titled *The Doomsday Book: Can the World Survive?* opened with the usual quotes from the Revelation of St. John, then went on to exclaim that chemicals, oil pollution, radioactive substances, and much else could bring a crash in the world's population. Other nonfiction works along with novels and movies warned the public against new ice ages, stupendous earthquakes provoked by injection of liquid wastes, and dozens of other agents ready to poison, choke, or otherwise destroy civilization. Predictably, many of the fictional apocalypses led through social disintegration to a new and better world.[36]

A second theme of the environmental movement that Rosenbaum noted had less to do with traditional fantasies. It was a concern with whole systems, with the ways in which pollution could affect ecological relationships around the globe, and the ways in which our entire modern culture and economic system encouraged pollution. A third theme was disillusion with these modern values and authorities, especially those that supported technology.

There were many reasons for the rise of the environmental movement, but the basic themes—dismay with technological authorities and systems that seemed about to doom the entire world—had first been thrust upon the public by hydrogen bombs. Had nuclear fallout taught anxiety about all technology? Barry Commoner, most prominent of the 1960s ecologist critics, said so: "I learned about the environment from the United States Atomic Energy Commission in 1953." The publication that his group began during the fallout debate, *Nuclear Information,* turned into *Environment* magazine, one of the new movement's main vehicles. But popular environmentalism had really started with a book attacking pesticides, *Silent Spring,* begun by Rachel Carson in 1958. She wrote privately that in earlier years, despite scientific evidence about harmful chemicals, she had clung to the faith that "much of Nature was forever beyond the tampering reach of man . . . the clouds and the rain and the winds were God's." It was radioactive fallout, she said, that had killed this faith. Her book, published in 1962, opened with a fable of a town dying from a white chemical powder that snowed down from the sky, a powder that Carson likened to fallout. Many other environmentalist writings likewise used images taken over directly from nuclear fears. For example, one successful children's book closely resembled John Wyndham's *Re-Birth,* with industrial pollution replacing bombs as the cause of humanity's downfall and transmutation.[37]

Nuclear fear was by no means the only form of imagery, let alone the only social force, behind the environmental movement. But nuclear fear took a special place. It raised emotions earlier and on a more visceral level than any other issue. And it served as a banner that could rally everyone around. In return, environmentalism would give a solid base for the opposition to reactors, lending organization, seasoned leadership, masses of followers, and a set of ideals.

As early as 1965 a poll showed that the great majority of environmentalists, unlike most other Americans, opposed nuclear power plants. A study of National Intervenors, a coalition that battled the AEC over emergency core cooling systems, found that two-thirds of the leaders had been in environmental groups. These leaders were usually mature, middle-class liberals, the sort of people who had distrusted the AEC ever since the campaign against fallout. Many of them had begun with no specific dislike of civilian nuclear power, and during the 1960s some environmentalists had even accepted reactors as less polluting than coal. At first, when such people opposed a reactor at a particular site such as Bodega Bay, they did so for reasons that would have led them to oppose any industrial plant there. But raw nuclear fear was adding a further driving force.[38]

The fiercest nuclear controversy of 1971 showed the mingling of anxieties. At issue were plans to test an ABM hydrogen bomb warhead in a shaft a mile deep at Amchitka Island, Alaska. The Sierra Club and other environmental groups joined with SANE and other disarmament groups (and John Gofman too) in a furor of opposition. Warnings about militarism mingled with warnings about contaminated fish, if not still worse calamities. It was almost suprising when the test did no visible harm.[39]

A process of conversion spread through the environmental movement. An example was the experience of David Pesonen, a leader of the Bodega Bay opponents, who at first had only opposed the reactor as an industrial object desecrating the shore. One evening as he was reflecting on the fight he underwent a remarkable experience. "It was a beautiful evening, a touch of fog," Pesonen recalled. "I had a feeling of the enormousness of what we were fighting; that it was antilife." It struck him that nuclear power was "the ultimate brutality, short of nuclear weapons." He became a vigorous campaigner against all reactors.[40]

More typical was the conversion of Ralph Nader. As a crusader against the hidden evils of industry and government, in 1970 he had been less concerned about nuclear power than about air pollution from oil and coal. But Gofman and Tamplin's warnings caught his attention, and so did the emergency core cooling system hearings. He became convinced that in nuclear power the public faced another official "coverup" of hazards, and in 1974 he convened a national conference that brought together more than a thousand reactor opponents. A reporter said the assembled militants, wearing everything from backpacks to business suits, "had the grim flavor and messianic fervor of the movement to end the war in Vietnam." Nader institutionalized this spirit the following year in Critical Mass, a coalition of hundreds of local and national groups.[41]

The most powerful organization to join the fight was the Sierra Club. The Club had traditionally been wary of all power plants including hydroelectric dams, but among the local chapters a particular hostility to nuclear power was rising. The issue was hotly debated within the Club, and in January 1974, compelled by a spontaneous wave of feeling at the grass roots, the directors took an official stand against reactors.[42]

Other environmental causes were faltering. Polls in the mid-1970s found that the public had become less worried about things like strip-mining and industrial pollution, if only because the new laws really were doing some good, and the press turned to newer fads. Following this general trend, newspapers began to carry fewer articles on routine reactor emissions and similar nuclear environmental accidents. But meanwhile the emergency core cooling system battle had come along, raising a wave of articles about reactor meltdowns; and then new radioactive waste problems appeared; and then home-made atomic bombs and proliferation each had their season. It scarcely mattered that the nuclear industry was at last learning, and better than most industries, to take extreme care with its wastes. Articles critical of nuclear power kept piling up, looming ever larger in proportion to other technological concerns.[43]

Nuclear opposition absorbed back into itself much of the driving force that nuclear fear had lent to environmentalism; along the way the force was transferred to reactors. The public might be growing used to alarms about oil spills and pesticides, as it had grown used to bomb threats. But now nuclear reactors represented another rapidly increasing anxiety, one that addressed the goals of all industrial society.

17

Energy Choices

The stakes in the nuclear power controversy were abruptly raised in October 1973, when oil shortages and a huge leap in fuel prices stunned the world. The economic consequences were so severe that some people began to take a closer look at Frederick Soddy's old claim that all wealth was based on energy. A few decades earlier most advanced nations had been nearly self-sufficient in energy; now nearly all of them relied heavily on imported oil. When the flow faltered, governments laid frantic plans to safeguard national prosperity and security. These plans included ambitious long-range programs that would build hundreds of reactors and make them a linchpin of the economy. For the first time such matters became a central issue; a great energy debate erupted, as complex as the nuclear weapons debate of 1956–1964.

Up to 1973 the reactor controversy had mainly involved on one side nuclear officials and on the other side small local groups backed by a few full-time opponents, arguing technical questions at government safety hearings and the like. Certainly the technical questions were important, and I will glance at them in this chapter, for over the long run decisions could not ignore physical reality. But over the short run of a decade or two the debate would pivot on other issues, involving engineering less than society and politics.

ALTERNATIVE ENERGY SOURCES

Technical, social, and political issues were all tied together in the central question of the great energy debate: if we did not power civilization with reactors, what would we use instead?

The first few antinuclear critics back in the 1950s had looked to high technology for an alternative. They said there was no need to bear the risks of uranium fission because, as many Atoms for Peace enthusiasts also claimed, "the atomic golden age" would come by the 1980s through hydrogen fusion. Once scientists found a way to bottle up the energy within hydrogen bombs, the true atomic golden age would begin. But in fact nobody knew how to build a fusion power plant. Each decade some experts predicted a breakthrough in the next, while other experts said it would take much longer. The most common informed view from the mid-1960s on was that fusion power would not begin to play a significant role in the economy until the early decades of the twenty-first century—and perhaps never. The public's hopes persisted, and during the 1970s almost the only optimistic magazine articles classified under "Atomic" in the *Readers' Guide* were about fusion.[1]

One reason fusion sounded attractive was an association with solar powers. Journalists explained that fusion was what made the sun shine, so mastering the process would be like "catching the sun," as in ancient worldwide myths about mastering a solar god. The crudest form of such symbolism could be found in certain schizophrenics who declared they would personally control or even become the sun to give the world new life. In Western tradition, where the sun had been a mystical symbol for divine illumination and an alchemical symbol for gold, the future Golden Age had long been identified with a perfected "solar" age. For example, one of the most influential Renaissance utopias was a "City of the Sun" centered around a solar temple.[2]

If reproducing this divine energy in fusion devices was not possible, the next most attractive idea would be to use sunlight itself. Nineteenth-century dreams of driving a civilization with solar power persisted, as seen for example in a 1940 *Astounding Science-Fiction* tale by Robert Heinlein: a scientist fought industrialists who wanted to suppress his discovery of solar electric-power panels. Heinlein took it for granted that his readers would understand energy from the sun to mean a decentralized civilization with riches and freedom for all. Some reactor critics spoke in equally glowing terms of the beneficent powers of sunlight; the old image of good radiation was not dead after all.[3]

Large majorities of people, particularly the younger and more highly educated ones, whether or not they favored reactors, as an alternative had far more enthusiasm for the sun than for any other energy source. Solar and related technologies, according to a prominent reactor critic, were "ideally suited for rural villagers and urban poor alike," and so forth. Only a few historians noticed that the promises of prosperity and freedom made for the smiling sun had been made two decades earlier on behalf of nuclear reactors and fusion, and before that on behalf of hydroelectric power, and before that on behalf of coal and steam.[4]

The sun did turn out to be surprisingly useful for heating houses, but that was not the question when the sun was set against reactors; the question was how to make electricity. Almost every means proposed for turning solar energy into electricity used schemes that a few scientists and engineers had been working on for decades. In the 1970s many more set to work, yet they still failed to produce solar electricity economically in wholesale amounts. Estimates on when that would happen varied, and every year some experts spoke of imminent breakthroughs, but the informed consensus wound up the same as for fusion: solar energy would not begin providing truly important quantities of electricity until the early decades of the twenty-first century—and perhaps never. As the history of reactors showed, no matter how good an industry appeared on paper it could not be built up to full scale in less than a generation, and then it might come up with nasty problems. Sunlight and its derivatives such as wind are extremely dilute forms of power, so any significant production would have to cover thousands of square miles of land and use staggering quantities of factory products and potentially hazardous metals.[5]

Environmentalists came up with another proposal for the near term: conservation. In the abstract the idea had always appealed to practically everyone; a famous nineteenth-century German chemist had built an entire moral system on the categorical commandment, "Thou Shalt Not Waste Energy!" The idea made more sense than ever when fuel prices rose. For example, a major "energy resource" turned out to lie in giving houses better insulation, and another in giving automobiles better fuel economy. Yet conservation had limits. After 1973 the advanced nations, beset with rising fuel costs and economic stagnation, did reduce their total energy use, yet their use of electricity slowly rose; electricity seemed to be more indispensable than other forms of power. In any case the population in many regions was rapidly climbing. When people in a city came home at night and turned on their lights, something would have to furnish the electricity.[6]

For most of the "atomic age" that something had been oil, but the oil feast was coming to an end. By the mid-1970s informed people realized that half of all the oil that could easily be taken from the ground in the United States was already gone, and the world's halfway point would be reached during the next generation. Anyone who doubted that there was a problem only had to notice that the most promising new fields were in Alaska and Siberia amid crippling ice or in the storm-swept Gulf of Mexico and the North Sea, where hundreds of men died at one stroke as their rigs foundered; the world was beginning to glimpse the bottom of the oil barrel. Natural gas might eventually be able to take oil's place, but that remained entirely uncertain.[7]

For most utilities responsible for lighting a city, the choice came down to a nuclear reactor or coal. Perhaps, as some argued, governments and utilities could have gotten around that choice by aggressively pursuing conservation in the short term and promoting solar energy as well as natural gas, tar sands, and so on for the long term. But in practice, when a utility decided against building a reactor, a coal plant usually took its place. The use of coal by American utilities increased 50 percent between 1975 and 1982 as oil burners were converted and new coal power plants were built, while similar sharp rises took place in other nations. Reality was posing a blunt question: Which would the public rather draw its electricity from, a coal plant or a nuclear reactor?[8]

That choice was not central to the world's economic difficulties, which revolved less around electricity than around liquid fuels such as gasoline. Even for electricity itself, by 1980 nuclear power provided only about a tenth of the world's supply. Despite the anguished cries on all sides, then, the nuclear industry had no immediate prospect of either saving the world economy or subverting it. In the worst case, making the wrong choice between uranium and coal would probably not change a nation's total economic product by more than a few percentage points in the next few decades. Whatever all the shouting was about, it was not about reactor economics.

REAL REACTOR RISKS

Just how good or bad were reactors—and compared to what? That risks should be weighed against benefits was an idea developed by generations of lawyers, and was also common sense. During the early decades of the nuclear industry the expected benefits of cheap and clean electricity had attracted everyone. But regarding risks, critics predicted tens of thousands of deaths while the industry promised total safety, so there seemed to be no reliable way to compare risks with benefits.

During the early debates neither side knew the number of deaths in the 1958 Urals disaster and among American uranium miners—not ten thousand and not zero, but perhaps a few hundred in each case. Neither side noticed the frequency of tragedies on that scale with other technologies, for example the 1959 collapse of a dam in France that left four hundred dead. Few imagined that nuclear problems might be comparable with the problems of less magical industries. In the mid-1960s a handful of nuclear scientists did wonder what a Maximum Credible Accident might look like at a chemical factory or a hydroelectric dam, but at the time nobody took the trouble to study that question seriously.[9]

As the antinuclear movement grew strong, experts began paying closer attention to comparative risks. The first comments were sweeping ones, as in a letter that a former chairman of the Advisory Committee on Reactor Safeguards wrote to a magazine in 1969. He noted that "human beings are crushed, dismembered, fried, poisoned, infected, or otherwise done away with at a steady high rate by such agencies as firearms, automobiles, aircraft . . . and even lightning. Such mayhem seems simply to be accepted. I do not find any such stories of steady human attrition connected with nuclear power reactors, and indeed there have been no such stories."[10]

That sort of broad comparison was relevant from the standpoint of engineering, a profession dedicated to the best use of limited resources. Engineers pointed out, for example, that hundreds of millions of dollars were being spent on improvements in reactor pollution controls that would save a few lives at most, yet only a few million extra dollars given to a city's Health Department could save hundreds of lives by attacking infant mortality among the poor. Nothing could make an engineer's blood boil like resources wasted on an inefficient way to achieve a given result. It was true that if people wanted to have a fit planet, they should look in this holistic way at all the problems and benefits of every technology together. Yet a comparison between things as unlike as reactor construction and a city budget was not going to get far. Talk of choices was more useful when the alternatives were real and closely comparable.

There was one area in which nuclear power could be and indeed had to be compared with a familiar technology. Coal too had its perils. Well known in mining and industrial regions, the hazards became obvious to everyone after a week in 1952 when a "killer smog" of coal smoke settled over London and left some four thousand dead. Might a switch from coal to uranium save more lives than it endangered? A pioneering attempt at an answer was published in the trade journal *Nucleonics* in 1959. The author compared risks sketched by the WASH-740 group with known fatalities among coal miners and so forth, and suggested that in terms of risk, reactors might already be preferable to coal. The idea provoked angry replies from engineers who felt such calculations were irresponsible, for nothing should divert people from making reactors as safe as humanly possible.[11]

The rise of the antireactor movement made the comparison with coal more urgent, and beginning in the early 1970s a few people worked out the relative risks and benefits with increasing precision. It was the birth of a new type of analysis, soon applied to many other industries; nuclear questions had become the paradigm for detailed analysis of technology. By the end of the 1970s the key facts regarding coal and uranium were clear. Without getting into the details I will give an example of some conclusions, to show to what extent the fear of nuclear reactors relied upon facts and to what extent it did not.[12]

Analysis suggested that the nuclear industry's chief health risk, not only in popular belief but in fact, was radioactive wastes. Here experts on both sides of the debate agreed that the worst problem came from the mining and milling of many thousands of tons of uranium ore, which could release a far greater amount of radioactive material into the environment than was ever likely to escape after the fuel was put in compact form. What was the comparison with coal? That would be clear to anyone who visited areas where uranium was mined and then visited coal country. Thousands of square miles of barren, misshapen land, thousands of miles of creeks where the water ran yellow and stank—those were the marks of coal mining. The clearest case of gross damage was the 1972 collapse of a dam holding coal-mining wastes at Buffalo Creek, West Virginia, where the flood killed 117 people. In contrast, uranium-bearing areas were scarcely altered from their original state.

What about more insidious effects across still larger regions? Here the nuclear industry's most severe problem turned out to be radon gas that seeped from excavated uranium ore, perhaps to cause lung cancers. Radon-producing uranium was not found only in concentrated ore, however, for minute traces were everywhere in rocks and soil. American coal mining brought about 20,000 tons of uranium to the surface each year, mixed with dirt and a billion tons of coal. Taking into account the regulations that required uranium mine tailings to be covered over, but not coal residues, the harm to health through radon from the two sources of energy turned out to be on more or less the same level.[13]

With coal, unlike uranium, the worst problem was not radon. One material found in coal was arsenic, and there was good evidence that traces of arsenic could bring cancer and genetic damage. As with plutonium, experts debated whether there was a threshold below which no damage would be done. But for both metals the conservative assumption held that even a single atom might cause a cancer or mutation. Unlike plutonium, however, arsenic did not have a half-life of some twenty thousand years: it was a stable element and would persist in the environment forever. Similar problems held for mercury and lead, also found in coal and not easily removed by pollution control devices. And whatever the devices did remove would be dumped in the open as part of the billions of tons of sludge and slag that piled up each year.[14]

These metals were not the worst problem of coal. Scientists had never studied most chemical compounds with anything remotely like the care that they applied to radioactive isotopes, but in the 1970s some began to study the fly ash that went up smokestacks and wound up either in the air or in the sludge piles produced by pollution controls. It turned out that these complex chemicals too might damage genes or cause cancer. It was no coincidence that such chemicals posed health hazards exactly like those from radioactive isotopes. Radiation does not bring harm through some uncanny magic, but by breaking apart molecules within the body, that is, by causing chemical changes. That is why cancers and mutations caused by radiation are virtually indistinguishable from the same ills caused by chemicals.

Complex chemicals were still not the worst problem of coal. The limited knowledge available by the 1980s suggested that the gravest health hazard came from simple smog chemicals. Their effects, like those of radioactive isotopes, were too diffuse for scientific proof, and some estimates pointed to only hundreds of illnesses. But equally plausible estimates suggested that where coal was heavily used, as in the central and northeastern United States and much of Europe, it could be causing tens of thousands of premature deaths each year from cancer and lung disease.

Smog chemicals also brought acid rain. Already in 1978 a conservative estimate by a blue-ribbon panel, not enemies of coal, concluded that damage to crops, erosion of building surfaces, and the like were costing the United States a billion dollars a year, and that cost was rapidly rising. Reliable reports told of dying lakes and sick forests all across eastern North America and Europe. The panel's hair-raising (and unpublicized) conclusion was that at some time in the future the effects of acid rain might become "irreversible, as well as intolerable." Pollution controls, if the laws could ever be passed and enforced, would alleviate the problem but not solve it.[15]

In sum, the legacy of pollution that a normal nuclear reactor left to future generations would be some acres of isolated, paved-over mine tailings and a few thousand tons of compact radioactive materials buried with all the safeguards human ingenuity could offer. If coal were used instead, future generations would be left with mountains by the mile containing noxious and very long-lived wastes on top of the ground, and probably wholesale environmental havoc from residues poured directly into the air. Most scientists who studied the facts with care concluded that coal was already killing people, not only through wastes but in mining and transportation accidents and so forth, on a scale that nuclear power was unlikely ever to match in the same categories.[16]

None of this meant that burning coal was unconscionable. The advanced nations reduced the total of their air pollution in the 1970s, and if the political will remained they could continue both to increase the use of coal and decrease pollution for several decades. Besides, even the worst industrial contamination added only a small fraction to the total of human sickness and death, while poverty was killing wholesale; the most industrialized countries had by far the healthiest populations. My only point is that, by every sensible measure I can find, as a health and environmental issue coal deserved many times more attention and care than nuclear power.

There was one category, however, where the comparison was harder to make, and it bothered some people most of all: the maximum accident. Nobody could estimate the chances of a catastrophe for which there had never been a single example, but the AEC tried, organizing a lengthy study under the engineering professor Norman Rasmussen. The Rasmussen report, completed in 1974, was designed to reassure the public. Its heavily publicized conclusions said that a citizen was more likely to be killed by a meteor than by a reactor accident. The report immediately came under fire not only from antinuclear groups but also from independent scientists who showed that the conclusions were riddled with methodological problems and in some ways far too optimistic. Pronuclear scientists replied that in other ways the study was too cautious. The final consensus was that the whole subject was a swamp of uncertainty. Yet buried in the details of Rasmussen's study were some clear and trustworthy facts, overlooked by both sides.[17]

The study group had begun with the fact that every industry suffers hundreds of minor mishaps each year. A valve would get stuck, an instrument would give a false reading, an operator would make a mistake. By itself one such failure was trivial. But what if two or three of them happened one on top of another? Such coincidences were much less likely than a single failure, of course, but from time to time a valve would stick, an instrument fail, and an operator make a mistake all at once, especially since one problem might provoke the next. Given that there were hundreds of individual failures, such a cluster of failures might be expected to happen a few times a year, and experience with reactors confirmed this. Such incidents were still not enough to melt down a reactor's fuel, but even more elaborate chains of failure could also be expected, if more seldom. That sort of chain could lead to partial melting of the reactor's fuel, as had happened in 1957 at Windscale and in 1966 at Lagoona Beach.

These possibilities were little noted. Nuclear engineers and their critics continued to focus single-mindedly through the 1970s on the spectacular events of the China Syndrome. Meanwhile the hundreds of little mishaps each year and the occasional double or triple coincidences kept on happening. In 1979 came an elaborate chain of failures, any one of which would have been harmless by itself but which all came together to cause a medium-sized accident: partial melting of the fuel in the Three Mile Island reactor near Middletown, Pennsylvania. Although nobody was injured, the reactor was ruined and the public for many miles around was alarmed almost to the point of panic.

Later, calculating from the detailed analyses buried in the Rasmussen Report, engineers found that some such accident might have been expected somewhere in the world, with fair probability, before the end of the 1970s. Even back in 1957 members of the WASH-740 study group had predicted that a medium-sized accident could happen within their lifetimes. The exact types of failures that came together to wreck the Three Mile Island reactor had been occurring at one or another reactor for years without happening all at the same time, but reports on those failures had been overlooked amid the storms of debate over maximum disasters and the blizzard of regulatory paperwork. Taken aback by the Three Mile Island accident, American officials and engineers at last began to pay less attention to imaginary vast catastrophes and more to the real problems of reactor safety.[18]

The accident could have been worse. A runaway heating explosion was not possible in this type of reactor, but another one or two coincidental failures might have melted the rest of the fuel; there remained the containment shell, but even more coincidences might somehow have breached it, releasing radioactivity to threaten the public. The fact that coincidences had to keep piling up before the worst could happen brought a crucial, poorly understood type of safety.

Just as hundreds of minor individual failures had been observed for each troublesome incident, and hundreds of those for each serious accident like Windscale or Three Mile Island, so among reactors with containment shells there should be many such cases of partly melted fuel for each great disaster where a still longer chain of failures let a radioactive cloud break into the open. Before such a cloud dispersed it might kill a few hundred people in the vicinity, reaching the level of commonplace disasters like dam failures. Finally—and here the calculation became entirely reliable, for it depended only on weather statistics—there would be hundreds of these ordinary disasters for each one in which the wind was exactly wrong and a radioactive cloud settled on a city to slay tens of thousands. In short, for typical American and Western European reactors, a great many accidents like Three Mile Island could be expected before the first dreadful catastrophe. Long before then the owners would either make reactors more reliable or abandon them, if only because each ruined reactor would cost them billions of dollars.

The problems were made still plainer by a far worse accident, the 1986 explosion of a reactor at Chernobyl in the Soviet Union. Once again the immediate fault lay in a series of small, almost routine errors. This time they were magnified out of hand by the Soviet design that made runaway heating possible, and that provided only radiation shielding around the sides with no cap strong enough to contain such an explosion. It was what many had warned against, a Maximum Credible Accident without containment. Only the last line of defense held, the weather, for most of the cloud drifted upward to disperse in the atmosphere. The old predictions for a reactor catastrophe were demonstrated when agricultural restrictions and stark fear disrupted lives across an entire continent. Yet the immediate death toll was only thirty or so, chiefly reactor workers and fire fighters. What caused most of the disruption was government and individual precautions to handle the possible effects of the radioactivity dispersed on the winds. Experts calculated that the long-term damage from that would be at roughly the same level as the damage from bomb test fallout two decades earlier: a loss of life, spread over decades, somewhere between a few dozens and hundreds of thousands, depending on the still unknown effects of low-level radiation. (Some experts meanwhile found new evidence supporting the old theory that such weak radiation stimulates *improved* health, but their comments did not reach the public.)[19]

Soviet authorities struggled to correct the mismanagement that had caused the disaster. But there was no way to be sure there would not be at least a few more such accidents somewhere in the world, killing dozens and leaving millions uncertain about their health, before everyone learned the lesson and either handled reactors more safely or gave them up. This prospect was far from the shortcut to an ideal future that Atoms for Peace had promised—yet it was not the global doom that some opponents feared. The anxieties and precautions that followed reactor accidents could temporarily disrupt entire nations and inflict serious economic blows, but in this section I am dealing for once not with human responses but with the objective physical consequences of the technology. In those terms, a long series of Chernobyl-like reactor explosions would resemble nothing worse than a combination of normal coal mining disasters plus the wholesale emission of damaging chemicals that routinely results from coal burning.

For coal, the probable catastrophe is more certain and on an immeasurably greater scale. Burning fossil fuel increases the amount of carbon dioxide gas in the earth's atmosphere. This heats up the planet and at some point will change climates and even melt the polar icecaps, with consequences almost too large to contemplate. The process is inexorable so long as coal burning continues, but the harm will probably begin in our children's generation, not ours. Most people have shrugged at the problem.

There were striking possibilities of more immediate disaster from non-nuclear power production. Beginning in the 1970s a few engineers, provoked by the antinuclear critics, pointed out that oil storage tank fires and natural gas explosions occasionally killed people by the hundred, and discovered that a Maximum Credible Accident with such things could slay thousands. Hydroelectric power dam catastrophes were so likely that they actually happened, as when a 1963 flood in Italy killed 2,500 people. There were dams in the United States whose failure could kill a quarter of a million outright. Those who carefully studied possible disasters found that reactors, even in the Soviet Union, were not necessarily more dangerous than many forms of industry that the public accepted.[20]

The public heard little about comparative risks. When Seaborg and his colleagues remarked that a typical coal-fired plant routinely emitted more radioactive material into the air than a nuclear reactor, and when other experts estimated that using coal instead of uranium might cost hundreds of lives each year for each power plant, they spoke with restraint. Most scientists and engineers, along with a majority of the public, did not despise any form of civilian technology, having observed that over the long run technology saved many times more lives than it cost; whoever favored reactors usually favored coal as well. Government officials felt no urge to insist that their nations' use of coal was poisoning forests for thousands of miles downwind, and utilities did not rush to advertise that their hydroelectric dams could slay thousands as readily as any reactor.

Environmentalists were even less inclined to publicize comparative risks. Many who fought against reactors fought with equal conviction (if less success) against coal plants, finding no need for comparisons when they condemned both alike. Others claimed that coal was "far cleaner" than nuclear power. Some widely read authors imagined that the problem of coal wastes could be solved by removing substances from smoke and dumping them in sludge heaps; others ignorantly assumed that radioactive isotopes were the only chemicals that had long lifetimes or could cause genetic damage; still others simply took it for granted that nuclear things were more horrible than any alternative. Committed opponents of nuclear power had little incentive to study, still less to advertise, comparisons that favored reactors. Besides, they had more urgent concerns.[21]

"IT'S POLITICAL"

In 1978 a French sociologist arranged a private meeting between a group of antinuclear militants and André Gauvenet. The latter was an elderly gentleman, trained in engineering, who had worked his way up within the French Atomic Energy Commission to become its head of nuclear safety. Calm and refined, intensely civilized, Gauvenet had at his fingertips a rebuttal to every technical criticism his opponents raised. The militants quickly tired of the argument. Particularly irritated was a leader of the local Friends of the Earth, a young man who had left his family and studies to commit himself to life in a commune and antinuclear campaigning. He raised his voice to insist that the real problem was not in technical safety games—"It's political."[22]

The confrontation between the silver-haired authority and the frustrated militant took place on many levels. The two were poles apart not only in their views on reactors but in their deepest social and political attitudes. The nuclear debate had exposed a rift that ran straight through the core of modern society.

The young militant, like most other nuclear protest leaders in the 1970s, owed many of his views to the worldwide countercultural movement that had burst forth a decade earlier. This movement had many causes, of course, ranging from the advent of television to the emergence of a large population of educated youth. But of special importance among these causes was a matter particularly close to reactors: the shadow of nuclear war.

As one protestor put it, recalling civil defense drills, "In many ways, the styles and explosions of the 1960s were born in those dank, subterranean high-school corridors near the boiler room where we decided that our elders were indeed unreliable." A psychological survey of young people in the mid-1960s confirmed that in their thoughts of imminent nuclear bombing, reality was reinforcing adolescent fantasies about inadequate and destructive adults. Most people who observed the young radicals agreed with the 1962 manifesto of the Students for a Democratic Society, which said they were "guided by the sense that we may be the last generation." Some eminent social commentators said that fear of nuclear war was not just one force, but the primary force, behind the rebellion.[23]

In the political sphere the first signs of a youth rebellion had come with the movement against fallout. The most revealing case was in Germany, where the short-lived Kampf dem Atomtod of 1957 had given birth to a Campaign for Disarmament (not just, as in Britain, a Campaign for *Nuclear* Disarmament). Unlike other groups of the period, the German movement held together after bomb tests went underground, and during the late 1960s they could mobilize more than a hundred thousand people in demonstrations. These were no longer against nuclear weapons. The increasingly young, increasingly radical protestors struck out against militarism in general, government police powers, the American presence in Vietnam, and the "fascist" aspects of modern capitalism. By 1970 the Campaign for Disarmament was fragmented, but it had been a school where rebels learned the tactics, the moral approach, and the outsider politics they used in their many subsequent pamphlets and demonstrations.[24]

In other countries too, the antibomb movement of the early 1960s was a main source of both practical lessons and experienced leaders for the later, more general protests. During the mid-1970s, when the countercultural movement concentrated a large part of its attention on nuclear energy, it was not being diverted but was returning to basic concerns. Even the early bomb protesters had addressed issues beyond war, and now the general disaffection with modern technology and the officials who controlled it was becoming plain.

Nuclear officials were reluctant to recognize that they faced something more than technical criticism. At first they believed, as Seaborg put it, that reactor critics "have just not considered the facts." For decades Seaborg and his colleagues had seen themselves as realistic experts and responsible citizens, bothered by uninformed nobodies. But the opponents of the 1970s were different from the people who had worried over nuclear energy back in the 1950s; on average they were now younger, better educated, and from a higher social level. As these opponents grew more visible the nuclear industry began to call them spoiled children or wild radicals, while some hinted that the antinuclear movement was manipulated by Moscow in a bid to undermine the Western economy. The pronuclear view was summed up by one man's description of an antinuclear demonstration: "attended by a number of hippy characters professing no knowledge of nuclear energy." But by the mid-1970s a few thoughtful people had noticed that the matter ran deeper. There seemed to be an entire type of person opposing reactors, a type with strong values of its own.[25]

The antinuclear people agreed with that. They called themselves democratic, warm, spontaneous, frank, and free; they called their enemies an anonymous, hierarchical, and secretive establishment of cramped bureaucrats who cared little for humans. The debate had gone beyond technical issues of safety or economics to more intimate issues.

The social questions were laid out most persuasively in a 1976 article in *Foreign Affairs.* This was an intellectuals' journal, and no cartoonist could have invented a better picture of an intellectual than the face of the article's author, Amory Lovins. A broad forehead, a little chin, and mild eyes behind severe dark-rimmed glasses bespoke a man who lived by knowledge and thought. Indeed Lovins's main occupations in the 1970s were studying, writing, and speaking, except during the summers when he went backpacking. It had been hikes in Wales that brought him together with local environmentalists and pointed him to a goal in life, for he had left his college studies with no particular career in mind. Lovins delved into the general problems of industrial society. Through the force of his thinking and the clarity of his words, he soon came to be called the nuclear industry's "Public Enemy No. 1."

Lovins was the foremost of many who were going beyond environmentalism to a fundamental critique of modern society. The critique had been pioneered by earlier intellectuals such as Lewis Mumford and, in particular, the economist E. F. Schumacher, whose 1973 book *Small Is Beautiful* argued for restricting industries to a scale that could be controlled by an individual or a small organization. Schumacher envisioned a network of diverse, modestly sized, quasi-autonomous groups, each using only technologies that were readily understood, each giving close attention to human and spiritual values. It sounded roughly like fifteenth-century Europe, with windmills grinding out electricity as well as grain. People like Schumacher scoffed at the TVA dream of using central government organization to promote a modern Arcadia; they found hope for Arcadian harmony, economic prosperity, and even world peace only through complete decentralization from the start.[26]

In his 1976 article Lovins brought this vision into collision with nuclear reactors. He pointed out that reactors necessarily meant centralized power systems, which by their very nature were inflexible, hard to understand, unresponsive to ordinary people, inequitable, and vulnerable to disruption. He said that even if nuclear power were clean, safe, and economical, "it would still be unattractive because of the political implications of the kind of energy economy it would lock us into."[27]

Beneath this political issue lay more basic questions of personal ideology. The career of a German journalist, Robert Jungk, offered a good example of how this worked. In the mid-1950s he had gone to Hiroshima to interview bomb victims; while talking with them he was struck by a realization that he too was a "survivor," for in his childhood his family had escaped the Nazi death camps through emigration. Inspired by the Japanese victims who found personal rebirth in working for peace, Jungk dedicated his life to fighting against both nuclear bombs and fascism. He supported the Kampf dem Atomtod and other opposition groups. Earlier he had written a well-known book, *Brighter Than a Thousand Suns,* which infuriated Manhattan Project veterans because it distorted evidence to imply that they were morally corrupt by comparison with the German scientists who failed to make an atomic bomb. Jungk kept his commitment to black-and-white morality when he moved on to opposing civilian reactors. In *The New Tyranny,* internationally the most widely read antinuclear book of the late 1970s, Jungk described the fight as an apocalyptic battle: the nuclear power industry was driving us into a robotic slave society, an empire of death more evil even than Hitler's.[28]

This industry, according to Jungk and many others, imposed a pattern of decision making by remote and heedless authorities. Some opponents had come to this view through frustrating experience, for nuclear officials had often patronized or ignored any nonexpert who worried about safety. However, people like Jungk were worrying about surviving deadly authorities long before they began to worry about reactors. They believed that citizens were not just ignored but victimized. Jungk described gruesome accidents that had befallen nuclear workers, including injuries that could have happened in any industry, as evidence of the peril from all callous experts. He and others also told how nuclear authorities—worried about possible theft of plutonium, and stirred by public threats of sabotage along with a few actual bombings of their offices and installations—ordered surveillance of opponents. Were they not preparing the way for a garrison state? Once again we were at a crossroads. The choice between nuclear and solar power, one writer declared, was "a choice between corporate rule and social democracy, between suicide and survival."[29]

THE REACTOR WARS

The social ideas and passions built into the nuclear debate became most apparent in Western Europe. Opposition to reactors had been almost unknown there before 1970 and remained tentative until 1974, when governments announced that they would meet the oil crisis with hundreds of reactors. Now energy policy became a matter of concern far beyond the limited circles that had been handling it. In various nations a new set of groups with their own goals burst into action.

Like their American counterparts, most of the European groups that pioneered the campaign against nuclear power could trace their origins to indigenous struggles against weapons or environmental pollution. But often the immediate spark that ignited opposition came from the United States. Europeans translated and repeated much of the American polemics, beginning with the Sternglass, Gofman, and Tamplin warnings. For example, a landmark in the growth of the Swedish movement was a well-publicized 1973 conference on nuclear safety that featured speeches by Tamplin and two other American critics. Similarly, a British debate in 1973 over what type of reactor to adopt got bogged down in arguments imported from the AEC's emergency core cooling system hearings. If Europeans borrowed ideas and tactics from the

United States, they would soon return the favor, bringing a new level of political sophistication.[30]

The European reactor opponents proved especially effective in mobilizing support at the local level. Their greatest early success was in the French and German Rhineland, where they won over citizens from wine growers to schoolteachers, from housewives to town officials, all marching behind a new slogan: "Better Active Today than Radioactive Tomorrow!" A climax came in 1975 at Wyhl on the German side of the Rhine when 20,000 people advanced on the site of a proposed reactor and tore up the fence surrounding it. Despite police resistance people swarmed onto the site, setting up barricades and putting up tents in the forest. Farmers from the surrounding countryside brought in food as the protestors built a communal kitchen and settled in.

Over the next months students and other activists from all over Western Europe came to the encampment at Wyhl, mingling with local citizens and organizing study groups at a "people's school." The discussions covered not only reactor technology and ecology but also the police state, the right to resist, and similar political topics. Here and at demonstrations elsewhere, the legacy of the student countercultural movement of the 1960s was in evidence—the flowers and songs, the rejection of bourgeois society, the spontaneous informal organization almost like a rural commune. But communard attitudes mingled with the solid conservatism of German farmers who only feared that a nearby installation might spoil their vineyards. The battle moved into lawcourts, and the Wyhl reactor was never built.[31]

Subsequent years saw further attempts to occupy nuclear industry sites in Europe and, by imitation, in the United States. Government authorities had learned a lesson, however, and were determined to prevent another Wyhl. The climactic battle came in France, where the government had put its greatest hopes in nuclear industry. France had to import nearly all its oil; most of its hydroelectric sites had already been developed; its cheap coal was giving out, with the peak of production already passed in the 1950s. Yet French leaders still expected to secure national independence and prosperity, thanks to nuclear energy.

If the Wyhl campsite was a living model of Arcadian hopes, its counterpart was visible in the offices of Electricité de France and the French Atomic Energy Commission. These were located in prosperous sections of Paris amid quiet boulevards where strollers could linger at an occasional café or marble fountain—boulevards deliberately modeled on the nineteenth-century vision of the White City. The offices themselves were in perfect order, with polished glass, a guard in uniform near the door, silent elevators, efficient secretaries. Of a piece with this social order was the ingenious device going up at Malville in the Rhone valley. Because France had only a modest amount of uranium as of all other fuels, the government had determined to build a complete breeder reactor economy, staking the nation's long-term future on a prototype at Malville. The device was named "Super-Phoenix" after the mythical bird reborn in its ashes, another of the countless symbols of transmutation.

July 31, 1977, was a rainy day at Malville. Some 50,000 protestors advanced on the fortified reactor site, slogging ahead in columns, silent in the dismal weather. The antinuclear movement with its environmentalist and pacifist background was by nature nonviolent, but by nature it also lacked strict organization, and the demonstrators were a haphazard assembly ranging from local villagers to anarchists. Small bands came with iron clubs, helmets, red or black flags, and Molotov cocktails. The police met the marchers with truncheons, high-pressure water hoses, and tear gas grenades. When the cries and confusion ended there were over a hundred injured, and one demonstrator was dead.[32]

Construction of the Super-Phoenix and dozens of other French reactors went ahead on schedule. The opponents could not win in the field, and they could not win in the French lawcourts either. Unlike the American and German systems for licensing reactors—mazes of hearings in which a determined group of opponents could raise endless obstructions against a builder—French licensing was a straightforward matter of working things out among government experts. Moreover, the French governing system did not leave as many openings for opponents as did the American and German systems with their numerous rival local and national authorities, any of which might be turned into a pulpit or a stumbling block. French reactors would go ahead so long as a bare majority of the nation's citizens continued to trust their leaders.[33]

The importance of the political constitution was likewise evident in Japan. No nation felt more anxious about radiation, yet minority opposition to reactors could make little headway so long as the majority trusted government leaders; and these leaders, like the French, felt an urgent need to free their nation from its abject dependence on foreign fuel. Minority misgivings had still less chance to make a difference in the Soviet Union and some other nondemocratic countries, where nuclear programs went ahead with only technical impediments. Public attitudes were never more than the raw material of politics; actual outcomes depended on political systems themselves.

In some nations the opponents' demands were effective. The American movement was the first to achieve a major political goal: the destruction of the Atomic Energy Commission. Promoting and regulating a technology at the same time, although common among government agencies, seemed less and less acceptable for nuclear energy, and in 1974 Congress tried to satisfy critics by splitting up the AEC. One part became the hapless Nuclear Regulatory Commission while the rest, demoralized, eventually joined the new Department of Energy. Meanwhile Congress stripped the Joint Committee on Atomic Energy of its unique powers. Nuclear energy had lost its strongest institutional supports.

More direct political gains followed. A 1980 referendum left Sweden with a nuclear program scheduled to end; a 1978 referendum in Austria stopped reactors there forthwith. Referenda against nuclear power failed in seven U.S. states during 1976 but succeeded in three out of five states where they were held in 1980. A few other states passed laws that barred new reactors until a way to dispose of wastes permanently was demonstrated. Similar results were achieved in other democracies such as Germany, Italy, the Netherlands, and wherever else political and judicial systems gave a determined minority a veto. The mechanics of the process were most visible in Germany, where a whole new political party, the Greens, rose up with opposition to nuclear power the keystone of their program. The Greens had some local electoral triumphs, but their chief success lay in making the major German political parties, scrambling to regain votes, hesitant about reactors. Similar pressures worked around the world; by the early 1980s far fewer reactors were on order than had been planned a decade earlier.[34]

That was not only for political reasons. Largely because of the oil price rise, the world economy stagnated while interest rates soared, undercutting every kind of massive industrial project. Reactor construction was further wounded by managers and workers who performed poorly, partly because the technology of giant reactors was new and untried, and partly because a ceaseless stream of new government safety rules preoccupied everyone. These and other factors raised costs far above the cost of reactors finished in the 1960s; some utilities wound up in deep financial trouble. Billions of dollars were lost particularly in the United States, with its confused patchwork of management groups, unions, and government authorities. The big losers were utilities that, having predicted a swiftly growing economy, had launched ambitious plans despite limited experience in nuclear work. One example was the TVA, whose reactor program, designed to scatter power plants across the countryside in the best utopian style, wound up miserably with little accomplished at great cost. On the other hand, there were some American utilities, and foreign ones such as Electricité de France, that used experienced personnel to build reactors of proven design, kept down costs, found buyers for the electricity, and were highly pleased with the results. In economic terms nuclear power could be either a bust or a boon, depending on circumstances.[35]

Thus if utility executives tried to choose among coal, uranium, and other alternatives strictly in terms of money, they would still be subject to pressures from imagery. Planning an electrical system meant peering thirty or forty years into the future, yet nobody could predict even a few years ahead what would happen to fuel costs, interest rates, or demand for electricity. Estimates about reactor economics necessarily depended less on facts than on images of the future. Therefore the estimates varied wildly from one time to another and from one expert to another. By the mid-1980s American utilities had shied away from nuclear power, only completing reactors already begun or even abandoning them halfway. The French were meanwhile completing an electrical system that drew more than two-thirds of its power from reactors—the cheapest electricity in Europe, some of which they exported to neighboring countries at a tidy profit. Most other nations hesitated in between, planning a gradual and wary increase in their use of nuclear energy.

One thing was certain: the decision would depend on guesses about what requirements might be imposed in future by the national government and the local public, requirements in the form of pollution and safety regulations, mandatory engineering changes, and legal or extralegal harassment. Therefore, even economic projections depended on the attitudes of everyone who held a piece of power, from national officials to patchworks of local groups. Eventually, if all technologies got a fair trial, experience might show which was cheapest and safest. Until then the great energy debate would turn largely on public images—and on the little-known forces that promoted images.

18

Civilization or Liberation?

Forces long obscure came into the open when proponents and opponents of reactors stood up to debate face to face, startling onlookers with their vehemence. Many nuclear opponents frankly displayed their anger and anxiety, and if advocates of reactors showed a calmer rationality, on that side too anyone listening closely could detect anxiety and anger. With growing exasperation each side accused the other of flagrant nonsense. In formal debates nobody seemed able to refute anyone else's statements; arguments flew past each other almost without touching. Observers began to notice that the two sides held widely divergent assumptions about not only reactors but many other things. Sometimes the debaters spoke what amounted to different languages, using different words and expressing themselves in different modes.[1]

Exactly what things were separated by this great division in modern society, and what forces caused it? For an answer I will look at three related features. First and most evident was a division in attitudes toward authority, which closely corresponded to opposed political ideologies. The second feature was a divergence in attitudes toward the overall relationship between technological civilization and the natural world; these attitudes often turned out to be connected to people's specific roles in society. Finally, both the political and the cultural divisions frequently corresponded to divergent personal styles of thinking, with one side tending more to logical analysis and the other to intuition. This last divergence did much to shape the battle of imagery itself, with people on opposite sides addressing the nuclear issue in strikingly different ways.

THE LOGIC OF AUTHORITY AND ITS ENEMIES

By the mid-1970s many nuclear opponents were saying that their fight was not just against the reactor industry but against all modern hierarchies and their technologies. The lines of battle were mapped by polls that found a seeming paradox. People who identified themselves as "conservatives," supporters of the established order, were more likely to accept new technology than were "liberals" who said they favored alterations in society. This was not really a paradox, for scientific and technological change had become so embedded in modern civilization that to uphold such change was to uphold the status quo, that is, the existing social authorities and ways of thinking. More specific surveys in Canada and Massachusetts found, among many factors, only one that clearly predicted whether a person would be for or against reactors: it depended on whether or not the person had confidence in basic social institutions.[2]

Of course the movement included a wide range of viewpoints. Many antinuclear groups were so allergic to authority that they came close to paralysis in their interminable democratic consensus-making; other groups left all decisions in the hands of a few charismatic leaders; still other reactor opponents, far from wanting less government, demanded much stricter bureaucratic regulation of industry. All, however, distrusted the currently established authorities.[3]

A few activists hoped the movement would serve as "a mechanism for raising the political consciousness" of people who had originally feared only for the environment. Reactor dangers would teach them the necessity of a "revolutionary movement" against capitalism. Ordinary opponents of reactors did not share such a radical political critique, yet the critique revealed much about the thoughts that helped sustain the movement.[4]

Perhaps the sharpest analysis was made by feminists who saw nuclear energy, both military and civilian, as an outrageous form of male domination. A few women dragged into the open the sly messages long hidden in journalistic talk about atoms, accepting the phrases as disgracefully true. Mastery of nature, they said, was only a variant of male domination over women. Some were offended to see "the earth raped by the insertion of missiles." Others claimed that a reactor spilling radiation was equivalent to incestuous assault, pollution inflicted by father figures: "The patriarchy . . . has created the rapist-energy of radiation-ejaculating nuclear power." Most of the antinuclear movement would not follow the analysis that far, but they all responded to images of victimization at the hands of authorities.[5]

Many of the symbolic associations of the nuclear battles could fit neatly into an ancient structure. According to the analysis I gave earlier in this book, using the evil scientist's ambiguous monster as an example, a widespread feature of modern culture is a bipolar structure of symbols that might be written in shorthand as *authority : victim.* On one side stands the scientist with his dangerous devices, or the cruel parent, or the domineering male (especially a government, industry, or military official), or the entire generalized threat from science and technology. On the other side stands the guinea pig, the rejected child, the enslaved worker, the dominated woman, the individual crushed by modern society.

The evil atomic scientist stereotype, which had already partly invaded the public's images of strategy theorists, Air Force officers, and other government authorities, now invaded the image of private industry too. The heartless and calculating businessman, another centuries-old stereotype, had originally had little to do with science, but he fitted much the same psychological pattern as the cold-blooded scientist. With modern industry becoming increasingly tied up with science, a few stories began to show brilliant technologists as masters over industrial serfs and even as owners of a runaway atomic power plant. That was the role played, for example, by Ming the Merciless in the 1936 Flash Gordon serial and by Dr. No in the 1962 James Bond movie. The combined stereotype took a more overt meaning in the 1970s as opponents accused industrialists of using perverse technology to pollute nature.

The emotional force became obvious after 1974 when the antinuclear movement announced its first martyr, Karen Silkwood. She had suffered radioactive contamination from the processing plant in Oklahoma where she worked, then died in a peculiar automobile accident. Because she had been a union activist digging for secrets about mishandled plutonium, many were convinced that Silkwood had been murdered by order of some anonymous official. Much that was written about her portrayed her as the complete victim: a woman, a suppressed proletarian, and even a childlike innocent, who learned the evil secrets of her masters, was polluted by unwholesome technology, and then was struck down by malign authorities.[6]

The most characteristic images came in a Time-Life television movie modeled on the Silkwood legend, *The Plutonium Incident*, shown in 1980 and later. When the naive young heroine stumbled across wrongdoing in a fuel processing plant, officials falsely accused her of smuggling plutonium out of the plant in her vagina, deliberately contaminated her, and finally killed her with a burst of radiation. Along the way the movie showed the heroine lying terrified in her bed as callous "radiation security" guards broke into her house, and twice showed her forced down struggling and crying on a bed while technicians inserted a "womb monitor." Weird murder and outright symbolic rape at the hands of the nuclear industry—facts were of no importance in such a production, for the imagery held a greater power.[7]

As a perceptive critic wrote of the 1983 movie *Silkwood*, another major production using the same themes, the true subject was contamination, not just of one woman but "the contamination of America by a heedless, enveloping capitalism." In another widely seen antireactor movie, *The China Syndrome* of 1979, the heartless male authority, possessor of exotic and perilous nuclear forces, was shown explicitly as the chairman of the board of an electric utility.[8]

Yet the dramatic core of these films and many other antireactor productions included something more subtle. Much of the tension came from a conflict between the smooth and reasonable assurances of officials and the protagonist's gut feeling that something was dreadfully wrong. Here the division between authority and its victims stood alongside another division, one that could be summarized as *logic : feelings*.

Social commentators had long recognized the distinction. The success of scientists relied upon rational systems from which everything personal or emotional had to be excluded; at an opposite extreme worked artists and poets, preoccupied with individual human feelings. The difference between occupational styles probably reflected native differences in personality and styles of thinking, but it also held social meaning. Not only scientists but engineers, industrialists, and government officials were increasingly called mechanical and unfeeling—as if the orderly social processes they administered were precisely analogous to the orderly structure of scientific thought. Contrasted with them were groups traditionally described as closer to emotion and instinct, more attuned to spontaneity and intuition: women, children, primitive peoples, peasants, manual laborers. These were precisely the groups excluded from power, and especially from power over technology.[9]

Identification of the authority : subject dichotomy with the logic :

feelings dichotomy did not always make sense. Peasant women could be as rational as anyone, while good scientists followed their intuition. A model logician like Socrates might take no part in either technology or social authority; engineers rarely had much to say about the uses of their work; the people who in fact held power usually cared little for either engineering or strict logic. However, in modern civilization many took it for granted that authority and rationality were inseparable, and together were opposed to intuitive human feelings.

The countercultural rebellion and many in the antinuclear movement took the side of intuition. Starting with attacks on established institutions, they went on to criticize the laboriously established structure of science and verified knowledge itself. Nuclear reactors had long since been offered as the quintessence of modern science, and now that could be another reason to denounce them.

The lines of division were not always clear. Some reactor opponents kept a faith in the power of science to fulfill every wish: it was not the laws of nature that hindered a quick solution to energy problems, but only the greed of corporations. Many who wanted a less hierarchical society were friends of particular technologies, devoting much effort to windmills or solar heating. They would welcome clever devices provided those could somehow be developed with the participation and consent of every citizen. Opponents of nuclear reactors often deployed reasoned arguments backed up by an impressive array of facts, and their ranks included a number of experts. For example, a 1975 French petition questioning nuclear power was signed by 4,000 scientists and engineers. Because of such publicity many believed that a majority of scientists and engineers rejected nuclear energy; the reactor opponents were satisfied that logic was on their side. Was this really, as its enemies charged, an antirational movement?

A French sociologist sympathetic to the reactor opposition studied some of the activists, including a few of the petition-signing scientists, and uneasily concluded that they never saw themselves as champions of rationality. Although they marshaled technical arguments as a debating tactic, at heart they were profoundly skeptical of expertise. The antinuclear scientists seemed especially concerned about the bureaucratic organization of their laboratories. Other committed opponents disavowed modern science altogether in favor of intuition or a frank belief in magic.[10]

Distrusting technical arguments was not the same as being irrational. Most people on both sides of the controversy were fully rational, in the sense that their views were commonsense deductions from what they believed to be true. However, people might believe false things. For example, studies found that the average citizen severely misperceived the frequency of accidents, both nuclear and other kinds. Compared with the statistics of actual events, most people tended to overestimate greatly the likelihood of various sorts of spectacular disasters, while underrating the risk from commonplace causes of harm such as heart attacks. This fitted with the normal urge to believe that things one does every day, such as overeating, are not really dangerous. The misperception also agreed with what citizens saw in the press, where ten people dying together in an unexpected event could mobilize more attention than ten thousand dying separately. In short, images of catastrophe could be strong enought to make "common sense" differ from the conclusions reached by experts.

Beyond that, many people were simply not interested in mortality rates and all the rest of the rational apparatus for calculating risks. They found the mathematical probability of death less important than whether an industry seemed likely to bring everyday harm or shocking catastrophe, whether it seemed to lie within their control or not, whether it looked familiar or alien. To people who valued intuition and human feelings above logic, accident rates were just not the point. What really mattered was a question of values.[11]

When it came to assigning values, the perceived structural division between logical officials and humanistic opponents could be read in two opposite ways, each with its own imagery. Some people saw the antinuclear groups as brave coalitions of potential victims, while officials were dangerous men with ice water for blood. In editorial cartoons the callous and deceptive nuclear official became a frequent stereotype. Meanwhile other people identified officials with benevolent forces of reason, opposed by an ignorant and rebellious mob. For example, during the oil crisis a church near the Hanford nuclear reservation declared that opposition to nuclear power was immoral, for reactors were "our nation's best hope to avoid the darker world of civil chaos, unemployment and hunger." A West German minister suggested that lack of nuclear energy could lead to an economic collapse that would engender a new Hitler.

Reactor builders tended to keep their feelings to themselves, but in private conversation they revealed a growing bitterness and sense of isolation. Indeed, both sides sometimes displayed the paranoid style common in millenarian groups: each felt besieged by evil forces that conspired to impede the coming of peace and plenty. The most ardent debaters on each side tended to see themselves as members of a community engaged not just in a political dispute but in a mission to drive back death and transform the world.[12]

NATURE VERSUS CULTURE

The difference between the two extremes came down to a question of just what sort of world transformation they desired.

Of all the dichotomies people have used to organize their thinking in debates over the future of society, one of the most powerful has been the bipolar structure *nature : culture.* Anthropologists, sociologists, and historians of ideas have uncovered an endless variety of themes that people have fitted into this overall pattern. The most primitive form was found in tribes who saw the world as a jungle surrounding a village of cultural artifacts and social rules. In Western thought this grew into polarities such as countryside versus city or wilderness versus civilization, polarities embedded everywhere from folk tales to social philosophy.

The contrast between wilderness and civilization implied a contrast between wild and controlled things. In personal terms the side of nature was often associated with intimacy rather than formality, with instinctive, "natural" impulses rather than self-control and planning, and, in brief, with feelings as opposed to logic. The next seemingly straightforward link followed the traditional association of logic with authority. Wasn't it true that civilization stood for hierarchical control while the natural state was unrestrained liberty? Weren't technical authorities the backbone of civilization while women, primitive peoples, peasants, and the like lived close to nature?

In fact none of these associations was necessarily true. Anthropologists described tribes that did not see women as closer to nature than men, tribes that would laugh at the idea of a Mother Nature. Even in Western thought, a comparison of anthills and parliaments could lead to doubt about whether it was "nature" or "culture" that was really closer to mechanistic control. Given the right form of government, the advance of technological civilization had in fact fostered not subjection but liberation from arbitrary authority. Nevertheless, in Western tradition the pattern of association was so pervasive that most people took it for granted. Nature was to culture as wilderness was to civilization, wild to self-controlled, victim to authority, feelings to logic, and female to male, not to mention liberty to order, freedom to secu-

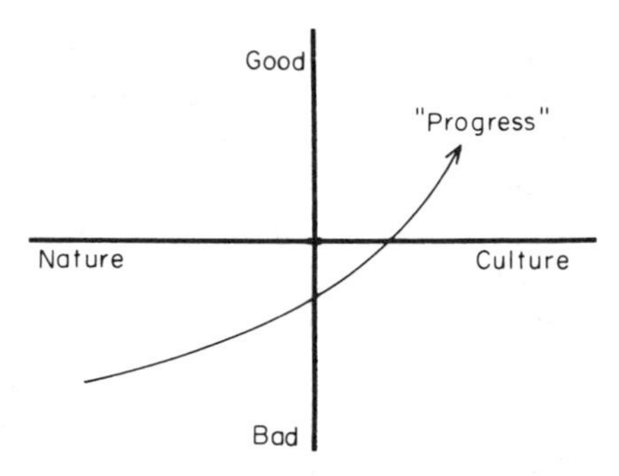

Figure 1. Scheme of ideas, common in modern thinking, distinguishing nature from culture, or similarly wilderness from civilization, organic from mechanical, feelings from logic, and so forth. Either extreme could be seen as good, bad, or a mixture. In the lower left of the diagram we would find the old view of wilderness as a thorny forest full of wolves and demons; in the upper left, an unspoiled fruitful Arcadia; in the upper right, an orderly utopian White City; and in the lower right, a robotic slave state. The traditional arrow of "progress" or economic growth is shown as an ideology of replacing lawless wilderness with beneficent civilization. Opposition to this view of "progress" was at the core of antinuclear ideology.

rity, Dionysian to Apollonian, organic to mechanical, "soft" to "hard," and so forth indefinitely.[13]

In many societies the pattern of associations carried a scale of values. Usually nature and culture were seen as equally capable of good or ill. Good arose when human activity came into harmony with the natural environment and each worked appropriately upon the other, an ideal depicted in thousands of Western and Oriental paintings. However, another way of thinking was also present from the start in many societies: distrust of the wild. Western civilization in particular had always felt itself surrounded by threatening wastes, seeing the deserts of the Near East and the dense forests of Europe as savage and pagan regions, the haunt of ravening beasts and outlaws. Man's duty was to tame and civilize, to transform the useless disorder of desert and forest into an organized landscape. By the nineteenth century this idea of "progress" was universally identified with the growth of science, industry, and social order (see Figure 1).

Such values could be useful to certain groups. The rule of colonial officials over primitive natives, of men over women, and of industrial elites over workers could all be justified by an assignment of values along the nature-culture scale. Moreover, linking civilization with faith in reason helped to justify the power accumulated by rationalized bureaucratic organizations. That was part of another long-term trend:

beginning with the rise of the modern state during the Renaissance and the invention of the business corporation soon after, Western civilization had increasingly pinned its hopes on the prodigious abilities of large, rationally structured organizations.

Values could also be assigned in the opposite sense. All through history some people pointed out that unrestrained "progress" could do harm, and in the nineteenth century intellectuals began to directly praise untouched wilderness along with spontaneity and freedom. By the late twentieth century such critics of progress had become numerous, for the balance between nature and culture had reversed. Where wilderness had once surrounded and menaced our frail villages, now the world-city surrounded and menaced whatever unspoiled nature remained. In parallel with this was the increasing distrust of industrial and state bureaucracies; even the most structured form of both, the Communist system, no longer attracted many people.

The antinuclear movement was thus taking its stand on powerful ideas that were more and more widely accepted. A French survey addressed the question directly. It found that among many ideas held by the public, the ones that correlated most closely with opposition to nuclear power were the beliefs, first, that economic growth brought more harm than good, and second, that the advance of technology harmed personal relationships. American surveys similarly found that roughly two out of three people in the general public who opposed reactors also opposed all economic growth; two out of three who favored reactors favored growth.[14]

Nobody designed a poll covering the whole range of attitudes I have listed, but there were a few suggestive studies. For example, an American survey found that, as compared with pronuclear people, antinuclear ones gave less emphasis to such values of civilization as "national security," "family security," and "a comfortable life," while giving more emphasis to values such as "inner harmony," "equality," and "a world of beauty." In sum, I suspect that most opponents of reactors could have been distinguished from advocates with a test that measured how people assigned values to the entire structure of nature : culture, feelings : logic, liberty : order, and all the rest.[15]

By the mid-1970s the controversy had become, for the most committed, an explicit battle of ideologies. An ideology serves many purposes, but in a narrow sense it is a set of beliefs and values that a group accepts when it adds legitimacy to their political goals. In this sense, experts and officials would normally be expected to support the values of rationality and the corresponding ideology that decisions should be made by experts and officials—that is, by themselves. The fact that nuclear reactors would always require a large, well-paid corps of engineers and administrators was scarcely a drawback in their eyes. Whether or not their decisions were really rational, they would insist that logic was their ally. A similar elementary analysis can be applied to the opponents of reactors and of large-scale technology in general.

If power was taken away from hierarchies, then decisions would still have to be made—but by whom? With remarkable unanimity, leaders of the antinuclear movement spoke up for local initiative, small "affinity groups," interdependent networks, the populace at large. In such a situation the power to make choices for society, including technological choices, would rest with those skilled in the communication of ideas and feelings. Journalists and novelists; television producers and movie makers; artists and actors; schoolteachers and professors; lawyers; ministers; community activists and organizers: in a society without centralized authority, these were the professionals who would ultimately say what technologies were acceptable.

These groups made up a "New Class," according to some of their critics, although it was neither exactly new nor exactly a class. Sociologists agreed that the New Class, such as it was, emerged from the modern revolution in manufacturing technologies and the explosive growth of education and communications. The fastest-growing professions were ones remote from the struggle of industrial production, resolutely independent professions that manipulated ideas and symbols. This was far from the traditional sort of class distinction; a television film director and the vice-president of a manufacturing corporation might live side by side, having the same income and sending their children to the same schools, yet stand poles apart in their values.

In fact film directors and other American media leaders, unlike industrialists, wanted changes in the social order. Such was the finding of a survey by a team of social scientists under Stanley Rothman. Whereas business leaders felt the press had too much power and business too little, media leaders felt the reverse; they would give relatively more power to intellectuals, consumer groups, blacks, and the like—and above all to media people like themselves.[16]

There was evidence that New Class professions were unusually numerous in the rank and file of the antinuclear movement. Such people were certainly the most visible, here as in many other movements that opposed authorities. And by the very nature of their skills they were the most influential people in such movements.[17]

Going straight to the point, the sociologist Stephen Cotgrove separated out two viewpoints in British polls. On one side he set people who distrusted science and technology, saw environmental problems as extremely serious, and were against industrial growth; on the other side he set people who felt the opposite way. Those who favored economic growth and so forth typically turned out to hold jobs close to industrial production: engineers, industrial scientists, and managers. Their environment-minded opponents were usually remote from production and even from the market economy: research scientists and other academics, teachers, clergy, social workers, and artists. Similarly, a poll among influential Americans found none so pronuclear as professional engineers and none so antinuclear as university social scientists. At one extreme stood the technologist who had devoted thirty years to his company's hardware; at the other Amory Lovins, the writer and activist with no fixed salary.[18]

Industrial officials remained firmly in power, yet the intellectuals were gaining leverage. Earlier writers such as Henry Adams, Raymond Fosdick, and J. B. Priestley had bemoaned the helplessness of humanists like themselves in the face of the juggernaut of bureaucracy and technology. Their counterparts of the 1970s felt strong enough to fight the machine head-on. They were not making an explicit bid for power over other people; it was enough simply to demand autonomy. The idea appealed also to farmers, independent tradesmen, students, and housewives, all of whom were numerous among signers of antinuclear petitions. Such people felt that once the technocrats were discredited, once decisions were made in community networks, their ideals would have a better chance to guide the world.

Beyond this opposition to hierarchical authority, a more general allegiance to the side of "nature" was the touchstone of the movement's sometimes surprising politics. Since the antinuclear movement was popular among leftist radicals, some observers thought it strange that institutions of the established Left such as labor unions and the European Communist parties remained largely pronuclear (or at least their leadership did). The explanation was that union and Communist leaders were firmly on the side of both rationalized organization and economic growth. Meanwhile housewives could march alongside famous poets, disaffected students bent on transforming society could make common cause with conservative peasants who sought only to preserve their crops, because the banner of "nature" united them all.

Similar if less powerful coalitions formed on other issues. The history of attitudes toward nuclear energy is interesting not just in itself but as a sensitive indicator of something far greater: attitudes toward all rational knowledge, technical progress, and organized decision making. The nuclear debates made plain how the trend toward confidence in all these things, a confidence that had been growing for a millennium, turned back around the middle of the twentieth century—whether temporarily or permanently could not yet be guessed. In view of the central place that technology and rationalized bureaucracy hold in modern civilization, this reversal has grave implications. The mechanisms revealed in the nuclear debates are therefore worth still closer inspection.

MODES OF EXPRESSION

The deep division between the two sides in the nuclear debate affected how they expressed themselves, and these differing modes of expression pointed to some of the most obscure roots of the controversy. Meanwhile the different modes helped to determine how each side influenced the public at large.

A political scientist who studied 1973 congressional hearings on reactor safety found that reactor advocates spent four-fifths of their time talking about technical facts or administrative expertise, while the opponents spoke of such things in less than one-fifth of their testimony, devoting the rest to warnings against careless officials. Outside such formal settings, the nuclear industry filled volumes with numbers and charts, marshaling painstaking evidence, while their opponents communicated with stirring phrases and slogans, cartoons and songs. Of course there existed antinuclear calculations and even a few pronuclear poems, but few people spent much time reading such things. A reactor opponent like Helen Caldicott, in the same vein as speakers in the earlier campaign against fallout, would tell of watching a particular child die painfully of cancer; meanwhile a reactor advocate would deploy statistics of illness in entire populations, waving away emotional talk about this or that individual case.

The opponents thought it a virtue to feel emotions. The most widely read polemic of the movement's early years, the 1969 book *Perils of the Peaceful Atom,* had a foreword titled "In Defense of Fear," in which the authors said they hoped that "some of our distress is communicated to the reader." In the most widely read polemic of a decade later,

The New Tyranny, Jungk said he wrote "in fear and anger" and wanted such outrage to replace calm discussion.[19]

Antinuclear works deliberately used the most emotive imagery. The word "nuke," which originally meant nuclear bomb, was applied to reactors in such a way as to make the two devices seem equivalent. Cartoons about nuclear wastes showed two-headed babies or giant rats. *Perils of the Peaceful Atom* warned of puny mortals defying God, and said the AEC had breathed life into a "robot" which had "risen from the table and was marching on the nation's cities"; *The New Tyranny,* more precise, called nuclear power a "Frankenstein's monster" that threatened doom to all mankind. Particularly popular was the greatest of all antecedents of the evil scientist, Doctor Faust. Alvin Weinberg and a few other nuclear scientists had compared themselves almost proudly with the ever-striving magus, but their opponents seized the idea that nuclear power was a "Faustian bargain" with the devil. In novels and cartoons the archfiend himself appeared, embracing reactors. A cheap 1978 horror movie about an experimental reactor that threatened to end the world featured a dream sequence in which the device turned into the Beast of the Apocalypse. Some opponents exclaimed that they were engaged in a genuine apocalyptic battle against evil and death; in street demonstrations they wore skull masks, carried coffins, and set up mock graves.[20]

Advocates of reactors also attempted to mobilize symbols. For example, major propaganda campaigns were launched by government ministries in Germany and France in the mid-1970s, and a few years later in the United States a Committee on Energy Awareness mounted a large industry-funded advertising effort. These were intended chiefly as rational "information" campaigns, but raw facts would be most convincing to those people who had technical training and aptitude. For everyone else the main message would lie in what the array of information itself, regardless of its actual content, stood for: confidence that matters were safe with the experts who understood what it was all about.

The potential results of using information as a symbol were revealed by a study of the public relations work undertaken, in the years preceding the Three Mile Island accident, by the utility that owned the reactor. A steady stream of news releases couched in technical jargon that only an engineer could understand, combined with flat statements that all was well, had indeed left most citizens in the region feeling ignorant but reassured. However, when the accident came it shattered trust in all statements by nuclear authorities. That trust had already been weak for decades, undermined by events from the Oppenheimer affair to the ABM controversy and above all by the fallout battles. To more and more people, a show of information by nuclear officials no longer conveyed any confidence at all.

Pronuclear advertisements often aimed more directly at the public's attitudes, appealing to patriotism and hopes for progress. But this was done mostly in abstract words, as when an American advertisement stated that nuclear power, by forestalling war over foreign oil, would give "Energy Independence and World Peace." Some publicists used more concrete images. One example was a 1969 AEC film, *Nuclear Power and the Environment,* which included a rural scene of girls fishing by a river near a white reactor building. Just such images of an electric power plant fitting harmoniously into the landscape, a machine in a meadow, had been used in advertisements since the 1920s; indeed, nuclear publicity generally tended to follow the established modes of industrial propaganda. Thus other films, advertisements, and exhibits extolled progress by showing impeccable mechanisms (such as gleaming uranium fuel rods) or fully-equipped kitchens enjoying the benefits of (nuclear) electricity. The only alternative, according to a cartoon in an Electricité de France advertisement, was the short and brutish life of a caveman. A veteran of the Joint Committee on Atomic Energy advised the nuclear industry to use such imagery, attacking opponents as people who wanted "to make you cold, hungry and shivering in the dark."[21]

The advice was seldom taken. Colorful invective did not come easily to senior officials in government agencies and industry, and still less to nuclear scientists and engineers. A psychologist who studied European nuclear personnel in the late 1960s noted that, preoccupied with the engineering of reactors, they were indifferent to psychology and human relations. They wished the public's worries would just go away, or at least that they could be dispelled by publishing accurate technical information. The whole aim of the engineer and scientist was to replace the shifting sands of personal emotion with the solid ground of rules and logic. A related ideal, dispassionate decision making, gave legitimacy to the power held by officials in industry and government. All these occupations would be expected to attract people with the corresponding personal tendencies.

Subtle psychological tests confirmed this. A study of people with no technical training found that most of them tended to decide various issues on the basis of feelings, that is, on the basis of personal values and sentiments about what seemed best in human terms; in contrast, professionals involved with radiation preferred to make decisions by thinking the problem through in a logical, analytical way. It was no wonder that the great majority of pronuclear publications concentrated more on explanations and economics than on mythopoeic language.[22]

The antinuclear mode of expression also reflected both the movement's social character and the personal characteristics of its members. Like the antifallout campaigners two decades earlier, reactor opponents stood on the outside and could usually make themselves heard only through posters, demonstrations, and other means poorly suited to rational discourse. Besides, outsiders trying to change the course of policy had more reason to use value-laden statements than did officials who could silently take for granted the established values of industrial progress. Within the antinuclear groups themselves, relying as they did on volunteers and direct-mail fund raising, passionate cries of horror would do more than statistics to keep up spirits and to attract new members and money. Finally, since the antinuclear movement was attacking hierarchy, it particularly attracted poets, artists, actors, and the like, who naturally expressed themselves in their own professional and personal styles.[23]

The differing modes of expression had very different impacts on the debate. There were roughly as many pronuclear as antinuclear people among the educated public, yet the ratio was different in the popular press. Most of the articles explaining that nuclear industry was safe and beneficial were printed in the recondite technical journals of physics, health physics, and engineering. On the paperback racks in bookstores, most of the books about nuclear power were openly hostile.[24]

One important example was a work by John Fuller, a television playwright and journalist. He was best known for books that pretended to prove the existence of psychic healers, ghosts, or flying saucers, books that intrigued the public while disgusting the few who investigated Fuller's use of evidence. Around 1975 he set out to write a novel about a nuclear reactor meltdown, then decided to try nonfiction instead and published a book about the 1966 accident in the Fermi-I breeder reactor at Lagoona Beach. The title was *We Almost Lost Detroit.* In reality that particular accident had been physically incapable of causing a great disaster, but here, as in his books on flying saucers and psychic healing, Fuller selected a particular set of facts, ignored the rest, and wove a gripping tale which convinced even seasoned book reviewers that (as the paperback cover put it) a "Hiroshima" had nearly struck Detroit.[25]

Four years later the American Nuclear Society published *Fermi-I:*

New Age for Nuclear Power. The title was scarcely more accurate than Fuller's, but what lay behind the cover was another matter. With numbers, diagrams, and lengthy engineering descriptions, a team of experts attempted to set down accurately every technical fact about the reactor. Unlike Fuller, the senior men who assembled this book did not like to discuss exciting political fights, personalities, and scarcely possible catastrophes, and wrote instead about hardware. The journalist's book was widely read; the engineers' book was never heard of outside the little group that published it. Nonfiction was simply more readable when it was antinuclear.[26]

That was even more true for fiction. Roughly a dozen ephemeral thrillers about nuclear reactors were published in the 1970s, and their titles reveal their viewpoint: *The Accident, Meltdown, The Nuclear Catastrophe,* and so forth. Typical plots used the idea that Robert Heinlein had invented for "Blowups Happen" in 1940: an advanced reactor, officials stiffly determined to keep it going, an individual striving to avert doom. None of these books won a large audience, but that came in 1979 when *The China Syndrome,* finely crafted to make reactors and their owners look ready to kill at the slightest provocation, brought thrills of terror to millions of moviegoers.[27]

There were no equivalent thrillers in support of nuclear power. Some writers were pronuclear, but they wrote no more than a few paragraphs about the matter in novels on other subjects. After all, who could attract a big audience with a story about routine life at an industrial plant? Such a story would be far closer to the truth than, for example, the claim made in the movie *Silkwood* that a single flawed fuel rod could slay two million people, but without such claims there would be little excitement. It was a simple matter, this bias toward colorful talk of danger, but it gravely influenced the image of all technology.

Another, more subtle bias affected not just the media's approach to issues but the very issues that were raised. An analysis of coverage of the nuclear industry found that American news publications and television gave little attention to technical problems of design and operation, the problems that preoccupied engineers and others within the industry. The press concentrated instead on human problems such as management, regulation, and politics. These were not just two different approaches but two different realms of thought. In the world of technology the problems would eventually give way to neat solutions; in the world shown by the press the problems were endlessly ambiguous and contentious.

The mechanism was explained with disarming frankness by Mike

Wallace of CBS television. In 1976 his popular *60 Minutes* program ran a segment on the nuclear controversy. The opportunity was the resignation of Robert Pollard, an engineer on the staff of the Nuclear Regulatory Commission, in protest against the NRC's overconfidence about safety. The television crew, having heard about Pollard's decision in advance from the Union of Concerned Scientists, interviewed him sympathetically in a pleasant atmosphere. For the other side of the story the crew tricked the chairman of the NRC into an unprepared interview and sprang the news of the resignation on him, confusing him with hostile questions until he looked the stereotype of an evasive official. In both interviews the actual technical problems raised by Pollard got scant attention. Wallace later told critics that he had no apology for these methods: the confrontation was "very dramatic" and made his show "interesting."[28]

One of the most respected scientists in the field of medical radiation, Lauriston Taylor, reported that since the 1950s editors had asked him point-blank to give alarming statements, telling him there would be no audience for calming explanations. He indignantly complained that whenever a radioactivity story came along, the media would eagerly take up the "wilfully deceptive" statements of a half-dozen or so antinuclear scientists, always the same ones, people whose reputations with their colleagues had long since vanished; "collectively they account for more news lines than the hundreds of reliable professionals accepted by their peers." This complaint was confirmed by a study of newspapers combined with polls. People like Sternglass had a very poor credibility rating among radiation medicine experts yet were widely cited, while people like Taylor who were admired by academic scientists appeared in the media seldom or never.

Of course it was normal for the press to play up the unusual—there was no news in an expert supporting nuclear power—and journalists could also be expected to play up attacks and human emotions that would sell publications. But beyond the sensationalism that inevitably attends a free press, nuclear advocates began to accuse journalists of frank antinuclear bias.[29]

Detailed studies of American newspapers, magazines, and television, and briefer surveys of the European press, found that in the 1970s antinuclear presentations came to outweigh pronuclear ones by two to one or more. No other technology was so despised in the press. Yet polls in the same period found the public at large slightly favoring reactors. The press, then, was not simply mimicking the public's views. Neither was it reflecting the views of government officials and business leaders, who were largely pronuclear. Still less was the press reflecting the views of scientists and engineers—for such professionals, even those who had no connection with the industry, supported nuclear power by the strongest majority of all. When the press gave the few antinuclear experts somewhat more space than the many pronuclear ones, the implied picture of the scientific community was far from accurate reporting.[30]

Direct surveys in the United States found that not only the articles, but the journalists who wrote them, were far more antinuclear on average than the groups they supposedly drew upon for their information. Even the new generation of science writers, while defending science as a whole, were not such believers as William Laurence's generation had been, and other journalists were still more skeptical. At the extreme were television journalists, openly antinuclear. It was not surprising, then, that studies found that television had an outright antinuclear slant, more consistent than in any other medium. Such press coverage had consequences. For example, after a heavily biased 1985 ABC television special which suggested that the nuclear power industry could end life on earth, a survey of the viewers found a significant and lasting shift against nuclear power. In sum, there was good evidence that the reactor wars reflected a profound division in modern society, and that this division influenced the public debate.[31]

THE PUBLIC'S IMAGE OF NUCLEAR POWER

In the last several chapters I have focused on articulate and active minorities. What did everyone else think? In most nations, and at most times since polls on nuclear power began in 1945, it would have made sense to divide the public into five groups of roughly equal size. The first group were strong advocates, confident in the ability of experts to keep reactors safe and impressed by the benefits of economic growth. Directly denying all that and about equally numerous were the convinced opponents, at first silent but increasingly vocal. In between were those who leaned one way or the other: a group who favored reactors even while harboring some misgivings about their safety, and a group who frankly feared reactors but were not convinced they should be banned. Both these middle groups would usually go along with whatever the authorities decided. In polls and referenda from the 1940s through the 1970s, as governments pushed their reactor programs, these groups usually joined with convinced advocates to give a solid majority in favor of nuclear power. But whenever authorities themselves were divided the middle groups hesitated, giving a majority attitude of wait-and-see. The fifth group consisted of people with no position, and often no interest, in the whole question.[32]

The antinuclear campaigns had some effect. Polls in Germany and France showed that around 1975, when the issue of nuclear safety began coming into the headlines, support for reactors fell off from a dominating majority to a bare majority. Support among Americans similarly declined, especially after news of the Three Mile Island accident, dropping below a majority. These were modest shifts rather than major swings, for negative attitudes, like positive ones, had been present everywhere over many decades. They could even be found in Communist nations: well before the Chernobyl accident Soviet scientists warned against radioactive contamination; Soviet officials complained about their citizens' anxieties; in Shanghai workers reportedly raised problems for authorities who planned to build a reactor. Emigrés from Communist countries told me in the early 1980s that misgivings about reactors were much the same there as in the United States, although less effective.[33]

If views on reactors changed only modestly from one region of the world to another and even from one decade to another, then what did make for differences in opinion? Among the leaders of each side it was clear that pronuclear feelings were strong among business people, scientists, engineers, and government officials, while opposition usually came from the liberal wing of the New Class. But among the ordinary people who made their voices heard only in opinion polls or at the voting booth, advocates and opponents of reactors (as of nuclear weapons) were spread quite evenly through the traditional social groups.

Even education made little difference. Nuclear experts often suggested that the antinuclear movement was based on ignorance of facts, while the opponents too felt they would win support if only they could teach people the truth. Neither side was correct, for studies showed that the way people felt about nuclear power was mostly independent of how much they knew about it. (In fact most people had only rudimentary knowledge, mixed with various bits of misinformation.) It was a phenomenon already seen two decades earlier in the fallout debates: when a person took a nuclear stance it was not from some special knowledge or lack of it, but as part of a total approach to society.[34]

A close look showed some groups with more misgivings than others about various technologies including nuclear power, although the differences were not great. The median opponent of reactors turned out to be a bit younger and worse paid than the median supporter. This was compatible with the fact that younger and poorer people had less control over technology and more reason to question whether authorities were doing what was best for them. Another group with relatively little power was women, and they too were a bit less inclined than men to take risks on various technological and environmental issues. With nuclear power, however, the difference became striking: sex was the one area where a strong divergence turned up in every poll, in every country, at every time. For example, a survey conducted near several American reactor sites asked people to complete the sentence: "When I think of the nuclear power plant, I feel ———." Twice as many men as women came up with talk of progress and economic benefits; twice as many women as men spoke anxiously about dangers.

Why did no other major group oppose reactors anywhere near as much as women did? Careful analysis of polls suggested that the phenomenon was connected with a tendency, when any technology was mentioned, for women to think in terms of safety, the environment, and their children, whereas men would think more in terms of the benefits and perils of scientific progress. But this did not answer the question of why women were especially likely to associate nuclear energy with an assault on the natural world. On other issues too, women were found to be less friendly than men to scientific matters, but only marginally, the differences being a matter of a few percent. Why did women diverge from men far more on reactors than on any other technology?

The only explanation I can find is that to many women nuclear energy stood for the worst in technology, because it had become most specifically associated with aggressive masculine imagery: weapons, mysteriously powerful machines, domination of nature, contamination verging on rape. On no other technological issue was the sexual imagery so thoroughly developed and, from a woman's standpoint, so viscerally disturbing. More generally, when women were more apt than men to connect technology with harm to the environment, in principle the cause must have been some sort of associations related to one's sex. I will not speculate further on this, for it is an area of popular imagery that, like the connection between masculine culture and war, has not been systematically studied to the extent that its importance warrants.[35]

Whatever the effects of nuclear imagery, images remained at the center of public thought, for all the old myths kept their vigor into the

1980s. Many people in Eastern Europe, in South America, and even in the United States continued to visit spas that discreetly advertised the healing powers of their radioactive waters. Meanwhile antinuclear writers told of weird sea creatures found near nuclear facilities (unfortunately no specimens of these marvels were preserved). Most commonly the creatures were said to be abnormally enlarged, if not quite so much as in the 1950s monster films. Novels and popular movies continued to display every sort of transmutation imagery, and anyone curious to see fantasies of rays with uncanny powers needed only to watch a few hours of American children's television on a Saturday morning.

Radiation and monster stories were largely an American phenomenon, exported along with movies, but sometimes the imagery sprang up on its own in Western Europe, while many Japanese childrens' comic books featured devastating rays. Not even the Soviet Union was exempt. The best-known book by the most popular Russian science fiction team was *Roadside Picnic,* also made into a Russian art film, *Stalker.* Both works centered on a mysterious technology that produced wondrous and horrid magic, threatening nature and human values; the film prominently featured a nuclear physicist, electric power technology, an atomic bomb, and a mutant child.[36]

As ever, the strongest fantasies dealt with human transformations. Nobody was so successful here as Stan Lee, creator of the characters that brought Marvel Comics to sales of over a hundred million copies a year, and not only among children. Lee's first popular creations were the Fantastic Four, whose powers were induced by cosmic radiation in 1961; the next year came Spider-Man, bitten by a radioactive spider; meanwhile, with Dr. Frankenstein and Dr. Jekyll explicitly in mind, Lee created the mighty Hulk from a physicist struck by nuclear rays. Lee's characters remained immensely popular into the 1980s and inspired a host of imitators, such as "Doctor Solar, Man of the Atom . . . A scientist caught in the fury of an atomic accident . . . changed into a being with fantastic powers!"[37]

Nuclear energy remained the leading manufacturer not only of superheroes but of monsters. For example, in a 1981 comic book Superman and Spider-Man teamed up to quash a monster—a nuclear industry worker transformed by exposure to an experimental isotope. With scarcely a pause the heroic pair then quashed a super nuclear reactor, built by a villain to control the world, which threatened to reduce the planet to ashes. Children who missed the complex associations among monster, reactor, and evil scientist could learn the lesson on a television cartoon where "the atomic robot, Mr. Atom," swore to "make the human race my servants."[38]

The outcome of the whole history of images was reflected in a word association test conducted in the mid-1970s by a French psychologist. Asked to respond to the word "atom," many people gave only banal descriptions, as if reluctant to bring up emotions. Of those who did give emotion-laden responses, only a small minority included such hopeful words as "progress" or "future." The rest were fearful: "Hiroshima," "disaster," "death." Some people came up with what the psychologist called "personal preoccupations centered around the problem of origins," for example the word "father."[39]

Other interviewers found that the pronuclear and antinuclear public together shared a distinct image of reactors. The white buildings behind a fence stood for industry and science, an immense and mysterious force that was barely contained. Following the Three Mile Island accident, the plant's gigantic, curiously shaped, and symmetrical cooling towers became the most common pictorial representation of a reactor. Scarcely anyone remarked that some non-nuclear industrial plants used the same kind of cooling towers, for that did not matter; like reactors themselves, the towers had become a symbol for all modern technology. Asked about nuclear plants, people tended to mention on the one hand knowledge and prosperity, on the other hand danger, mad scientists, and pacts with the devil. Neighbors of one reactor believed that it had vast subterranean installations, like the laboratory of the comic-book genius. In short, whether one saw a reactor as good or evil, or more likely as a mixture of both, it stood as a full symbolic representation of attempts by the forces of order to master nature.[40]

That symbolism can explain better than anything else why nuclear reactors were singled out, far more than any other technology, for utopian hopes on the one hand, fear and hostility on the other. In terms of hierarchical control by a technical elite, reactors were not obviously worse than, for example, the telecommunications industry; in terms of prospective hazards to our daily lives, reactors were not obviously worse than, for example, the chemicals industry. What reactors did offer that nothing else could match was a unitary image tying together everything involved in the battle between "nature" and "culture."

Reactor accidents provided a litmus test of how the public in general, and the press in particular, saw reactors. The 1986 Chernobyl accident set off an explosion of fear as some newspapers claimed an immediate death toll in the thousands or more and warned of radioactive fallout. Throughout Europe, even thousands of miles from the

Ukraine, people doubted the safety of their food, mothers worried about whether they should let their children play outside, and temporary residents pulled up stakes to move farther away. In all this the press and public were making the best they could of scanty and confused information. Soviet officials were not only reticent but were themselves largely in the dark, while some Western governments took the opportunity to exclaim about the horrible Communist failure.

Meanwhile the radioactive dust came down chiefly wherever rain happened to fall, creating local hot spots; the result was wildly divergent reports about how much radioactivity was really at hand. Of course, even when fallout was accurately measured the long-term hazards of such faint radiation would remain unknown. Many European governments chose to play it safe by imposing restrictions on foodstuffs, but the rules varied so much from place to place that the public only became more anxious. As with bomb test fallout a quarter-century earlier, it was impossible to sort out real problems from imaginary ones. The only clear conclusion was that where radioactivity was involved, the public could never be calm.

More revealing than the confused facts was the accident's imagery. Seeking a comparison with events outside the nuclear industry, the press most often settled on a 1984 disaster at Bhopal, India, where a chemical cloud had escaped from a pesticide plant. That cloud had killed outright not a few dozen people but over two thousand; the long-term damage to the health of another ten thousand or so was not hypothetical but visible; yet to the press and most of the public, the Chernobyl accident seemed more serious. That was largely because reactor worries centered on faint but widespread radioactivity, whereas in Bhopal the hypothetical long-term effects of dispersed chemicals for a wide population were hardly mentioned. As an emblem of contamination, the radioactive atom remained supreme. It was not pesticides but nuclear power that *Newsweek* announced to be "a bargain with the Devil."[41]

Hardly any reporter described the most likely, but prosaic, long-term health consequence at Chernobyl: an excess of thyroid abnormalities among nearby residents (this was the chief problem plaguing the natives of Rongelap Atoll who had inhaled radioactive iodine from the BRAVO hydrogen bomb test). Much more attention in the Western press went to the plight of the Lapps in the far North, where government authorities ruled that radioactive fallout had made their reindeer unfit to eat. As innocent natives in a pristine environment, the pastoral Lapps and their animals could well symbolize victims of the modern civilization that reactors represented. In fact around the world the fragile tundra was already endangered—the most persistent threat to its ecology came from "arctic haze" caused by coal smoke and other pollutants—but this was not mentioned. The lesson of Chernobyl, according even to some Soviet specialists, was a problem seldom suggested with other industries: "We can no longer control technology."

The imagery that people invoked to give meaning to the Chernobyl disaster did not stop there, for many went on to link it with nuclear war, not omitting references to apocalypse. To the *New Yorker* the accident was "all that is given to us to know of the end of the world." Even Soviet premier Mikhail Gorbachev called the accident "another sound of the tocsin, another grim warning" against the nuclear arms race. A Russian documentary on Chernobyl exclaimed that in a war, "the destruction of Soviet reactors would destroy all life in Europe, North Africa and parts of Asia." That was untrue, but a good warning to potential attackers. Other Russians pointed out to one another the Bible passage where at the end of the world "part of the waters became wormwood, and many men died of the waters," noting that the Ukrainian word for wormwood was *chernobyl.* According to a reporter, for Russians the disaster had become something "almost supernatural." A reactor was not just a reactor.[42]

The Three Mile Island accident was still more revealing, for scholars studied the reaction of the press and public in detail. They found that the press could not be blamed for raising fears, for these had already been alive in the people the reporters interviewed, including nervous staff members of the Nuclear Regulatory Commission itself. For example, the press made much of a "hydrogen bubble," described as a mysterious new danger. In fact the hydrogen problem had been known since the explosion that almost killed Heisenberg in 1942, and reactor designers had prepared for it; later analysis showed that hydrogen could not have done much damage at Three Mile Island. Yet during the accident journalists had been accurate in reporting that some officials, especially those far from the reactor, were keenly worried about hydrogen. The most important process had been one of distorted communication, for at each stage as information passed from reactor operators to officials, from officials to reporters, and from reporters to the public, attention concentrated on the most frightening possibilities.

Alongside this mechanism was frank sensationalism. Journalists sought out the most worried people for interviews, while on national television Walter Cronkite philosophized about Frankenstein and man's "tampering with natural forces." The world press made numerous references to monsters, robots, and the Devil, as well as to Hiroshima, nuclear war, and the end of humanity. Many observers felt the press had gone overboard. David Lilienthal found the treatment of the accident so "shocking and indecent" that, at the age of 80, the veteran nuclear critic threw himself into writing a book to warn that reactors must not be rejected out of hand.[43]

Did the press treat Three Mile Island differently from other industrial problems? The plainest difference was not in the nature but the strength of the coverage. For a solid week Three Mile Island dominated evening television news and crowded the front pages of newspapers, not only in the United States but around the world. Residents of the area were so upset by the stories that about half moved away temporarily; some, calling themselves "survivors," eventually suffered psychological difficulties much like victims of real massacres. When the time came for lawsuits, this psychological harm was one of the strongest claims that people made. Such problems were probably caused less by the nature of the press coverage than by its very existence, for studies have indicated that even a neutral article could make readers anxious simply by raising questions about health and safety that people otherwise would not have thought about. Here was where the antinuclear movement had its greatest effect: less by injecting images into the media (although there was plenty of that) than by preparing the ground so that the press would cover Three Mile Island with an intensity far beyond any previous industrial accident.[44]

Eight months after the accident, while arguments were growing over the feasibility of evacuating people from the neighborhood of reactors, authorities in Ontario had to rush the evacuation of a quarter of a million people when a tank car leaked deadly chlorine gas. The near disaster got only a few columns in the newspapers, quickly replaced by further talk about hypothetical reactor accidents. Around the same time a dam collapsed in India, killing well over a thousand people. Outside India it passed almost unnoticed in the press.

Still more instructive, a year after the Three Mile Island accident a chemical waste dump in New Jersey, next to New York Harbor, caught fire and sent a tower of noxious smoke high into the air. Schools on nearby Staten Island sent children home, and residents waited behind closed windows as chemical fog drifted past. If the weather had been different (a gentle breeze settling poisons directly upon populated areas, a Maximum Credible Accident) millions of people might have been endangered. Overall, the waste dump fire came many times closer to a great catastrophe than did the accident at Three Mile Island. The fire made headlines for a day or two and was then forgotten. A presidential commission was convened to investigate the Three Mile Island accident but not the New Jersey one; teams of sociologists and doctors searched for anxiety and cancers among residents near Three Mile Island but not on Staten Island; Ernest Sternglass discovered a harvest of dead babies, quickly proved to be spurious, in Pennsylvania but not in New Jersey; demonstrations brought tens of thousands into the streets in Europe and a hundred thousand in Washington, D.C., to protest reactors but not the far more ubiquitous and far less regulated chemical waste dumps. This was not a bias of the press alone. Nor was it an accurate response to the complex problems of the times. This was nuclear fear at work, single-minded and insatiable, stretching from ordinary citizens to the highest authorities.[45]

Nevertheless the Three Mile Island accident only temporarily revived interest in a topic that was going out of fashion; during the early 1980s the entire movement against reactors grew torpid. In my counts of titles in the *Readers' Guide,* by 1984 the number of articles dealing with civilian nuclear energy had dropped to the level of a decade earlier, which was half the peak value; a new peak following Chernobyl proved equally transitory. This falling off of attention may have come partly because not many new reactors were being ordered (and none at all in the United States), so there was less to fight about. Moreover, the press and public could not sustain a fierce interest in any issue for more than a few years. The decline of interest did not come because feeling against reactors had mellowed: the proportion of *Readers' Guide* titles showing doubt or open hostility toward reactors continued to climb. In 1950 barely one-tenth of the titles had suggested that there was anything to worry about from the industry, but by 1986 the fraction had reached nine-tenths. Wherever something new in nuclear energy was proposed, the opposition would be there, quietly waiting.

In summary, I can identify four main themes in the web of associations that influenced the way people thought about nuclear power. At the origin were (1) the technical realities of reactors, both the economic opportunities and the hazards, as seen by scientists and transmitted to the public. From these realities particular "facts," such as the hypothetical effects of widespread low-level radiation from a Maximum Credible Accident, were selected and stressed. That was largely because of (2) nuclear energy's social and political associations, including especially ideas involving modern civilization and authority. These associations explain what happened when reactors became a condensed symbol for all modern industrial society. They do not explain why nuclear power was singled out for this role. That was largely a result of (3) the old myths about pollution, cosmic secrets, mad scientists, and apocalypse that were historically associated with atomic power and radiation, indestructible myths with deep psychological resonances. A final association was (4) the threat of nuclear war, never for a moment forgotten.

19

The War Fear Revival: An Unfinished Chapter

In the late 1970s interest in nuclear weapons revived, with some people demanding that far more of them be built, others far less. Even more than earlier waves of attention, this resembled a passing fad. Yet it reaffirmed that the world public never forgot any of the apocalyptic imagery of nuclear war, and it showed new ways in which the imagery could affect history.

The renewal of organized effort against the bombs began in Japan, where the disarmament movement of the 1950s had also begun. Survivors of Hiroshima and Nagasaki, refusing to tolerate the world's acquiescence to an endless arms race, redoubled their efforts to spread their message. Their most effective tool was the film taken by daring Japanese photographers after the bombings and then confiscated by American authorities. Returned to the Japanese in 1968, the films were made into a documentary, showing doctors attempting to treat horrific wounds and vistas of streets paved by skulls. Excerpts appeared on television in a number of nations during the 1970s, while franker extracts played to audiences in colleges and churches. The audiences were stunned, scarcely able to speak after watching ten or fifteen minutes of film; some viewers fainted.[1]

In the universities of Cambridge, Massachusetts, and a few other centers of dissenting thought, the disarmament advocates who had persisted almost unnoticed since the early 1960s began to find support among the opponents of nuclear power. Helen Caldicott concluded a polemic against reactors by warning, "When compared to the threat of nuclear war, the nuclear power controversy shrinks to paltry dimensions." Around the end of the 1970s Caldicott and numerous other long-time reactor opponents such as the Union of Concerned Scientists and the Friends of the Earth, along with newer ones such as Physicians for Social Responsibility and the Green party in West Germany, turned their main attention to armaments. Like Caldicott, many of these people were moving back to their original concerns—striking evidence that anxiety about reactors had included a displaced component of nuclear bomb anxiety. Now this returned to its real target.[2]

The antinuclear people were running hard to keep up with the public and the press as debate over bombs awoke from its enchanted sleep. One sign of the awakening was a steady rise in the number of magazine articles dealing with nuclear weapons, as seen in the *Readers' Guide* index of titles beginning around 1976. At first this seemed to be only a renewed debate about terrorists or petty dictators with atomic bombs. But novels on the subject also began to appear, and some of these were not the hack works of the preceding decade but thrillers by masters of the genre, each selling more than the one before. *Goodbye California* in 1978 had a mad terrorist threatening to destroy the state with bombs; *The Fifth Horseman* in 1980 had a mad dictator threatening New York City; *The Parsifal Mosaic,* the number one best-seller of 1982, had a mad statesman threatening to trigger world war. The implicit message of all these stories was, as *The Parsifal Mosaic* put it, that "this world can be set on fire by a single brilliant mind."[3]

Following the mad-bomber thrillers came articles and books that dealt with nuclear weapons more realistically. By 1982 the number of American magazine articles dealing with nuclear weapons had shot up to a level attained only once before, in 1964. Meanwhile paperback racks in the United States and Europe, which a few years earlier had been loaded with books on reactor dangers, filled up with sensational fiction and sober factual warnings about war. The most important was Jonathan Schell's *The Fate of the Earth,* the first nonfiction book about nuclear war to become a best-seller since Hersey's *Hiroshima* of 1946. Hundreds of other books in various languages joined it in a remarkable burst of publishing.[4]

Why did anxiety about nuclear war come into the open again around 1980? Some of the works (including this one) were begun in the years when the problem seemed forgotten, and we hoped to reopen debate. There may be a cycle to such things, as thoughts that have been suppressed for decades are faced again at last. But there were more direct causes. Citizens usually responded rapidly to new facts about weapons, and there were dreadful new facts.

The United States had rapidly multiplied the number of MIRV warheads that it could hurl at an enemy, until the number leveled off around 1977 at roughly 7,000 nuclear bombs on intercontinental missiles along with 2,000 carried by bombers. The Soviet Union, as usual a half-decade behind its adversary, went through a similar process of missile warhead multiplication, passing the 7,000 mark around 1982. The number of shorter-range weapons also increased steadily; according to one estimate, by the mid-1980s the Americans had stockpiled about 37,000 warheads in total and the Soviets 17,000 more.[5]

Some people began to take notice of all this in 1977 during the first substantial weapons controversy since the ABM fight of a decade earlier. The U.S. Congress was debating whether to allow production of "clean" neutron bombs. Most Americans continued to ignore such issues; the debate raged chiefly in Europe, the most likely place for a neutron bomb to explode. Next, when the Soviet Union began emplacing a new type of medium-range missile and NATO decided to reply with new missiles of its own, passionate protests erupted in West Germany and neighboring nations. They noted that these ballistic missiles were only a beginning, for soon would come thousands of elusive cruise missiles and other novel weapons.

At first Americans scoffed at the Europeans' fears. As late as 1979, a senator said he got more letters about the plight of the Alaskan timber wolf than about the SALT II arms control treaty that was before the Senate. But when the United States hesitated to ratify this treaty, most people realized that diplomacy was not even coming close to solving the arms race. Then the Soviet Union redoubled tensions by invading Afghanistan.

Now as in the 1950s, those who wanted an increased military budget joined their opponents in warning about nuclear peril. A group of conservative Americans formed a "Committee on the Present Danger" which launched a publicity effort to revive fear of a Russian attack. Of course there were economic interests that worked to hold down military spending (groups with other uses for the money), but powerful interests supported it; the Committee had abundant links, such as interlocking boards of directors, with firms that lived off military contracts. The committee was well funded and dedicated to its work, and many observers felt that it strongly influenced American opinion. The campaign culminated in 1982 when President Reagan, who was closely associated with the Committee, announced that the Soviet Union had "a definite margin of superiority . . . there is . . . a window of vulnerability." This was as false as the bomber and missile gaps of earlier decades, but many Americans believed the President. Accelerating a trend begun under President Carter, the Reagan administration worked to raise military spending by at least 50 percent. It fitted well with their effort to divert as much money as possible from federal social programs to private industry.[6]

Opposition spread rapidly. With the environmental movement quiet, the battles for equal civil rights and against reactors at least partly won, and other problems such as poverty commanding little public sympathy, the energies of social critics turned to the now urgent matter of new weapons. Many of the activists had won their spurs in other movements, especially the fight against reactors, and retained those movements' general distrust of modern society and authority. But unlike the reactor opposition and the 1960s antibomb campaigns, this movement did not rely almost entirely on relatively peripheral people such as housewives, students and intellectuals. It drew equally on respected elites—physicians, mayors, elder statesmen, even some generals and admirals. A minority of liberal clerics had helped lead the 1960s campaigns, but now the full weight of important church organizations in Western Europe and the United States, even including the American Catholic bishops in a strongly argued report, turned against the weapons buildup. Moreover, the new movement did not focus on the supposed horrors of bomb tests and radioactive fallout but looked closer to the heart of the problem, the arms race itself.

In the fall of 1981, in ten major northern European cities, crowds of 100,000 or more demonstrated against deployment of a new type of missile. In 1982 there was an enormous demonstration in New York City together with a series of resolutions and referenda across the United States on a Nuclear Freeze, that is, a verifiable halt to nuclear weapons deployment. Both the demonstration and the referenda mobilized Americans on a scale well beyond any other such outcry in the nation's history.[7]

The leap in the attention paid to nuclear war was not a reflection of gross changes in public beliefs. Polls around the world found that fear of imminent nuclear war, which had subsided somewhat in the 1960s, rose in the late 1970s and early 1980s, returning to the level of the 1950s—but these were shifts of no more than 10 percent of the population. Most people had known all along that they were in danger, whether or not they chose to talk about it.[8]

A psychiatrist who brought up nuclear war problems with acquaintances in the early 1980s found results exactly like those reported in the early 1950s: "The people interviewed shrugged their shoulders, continued their activities, remained uninvolved, and the dialogue ceased." Around the same time a science fiction editor issued an invitation for new after-the-bombs stories, and got tales full of elves. "Our writers, the communicators of our collective unconscious," she explained, found it too hurtful "to grapple with the subject matter directly." Most insightful was a 1981 poll that gave Americans a choice for describing how they felt about future war. Now as always since 1945, some professed little concern and others admitted serious fear. But half the population agreed with the statement, "While I am concerned about the chances of a nuclear war, I try to put it out of my mind."[9]

When adults asked students what they thought of their future, they found the same disturbed responses as in the 1950s and early 1960s: strong fears of nuclear war (greater than in the late 1960s and early 1970s), tending to lead to denial. "We are the last generation," a young man told me—and went back to his pursuit of a career. The one clear difference was that this generation had heard far less than the children of 1945–1965 about weapons. Many American high school students in the 1980s did not even know that the United States had once dropped atomic bombs on Japan.[10]

Attitudes shifted farther in the Soviet Union. Around 1966 Soviet leaders had adopted the escalation ladder thesis: war could and must be kept below the level of all-out destruction. Beginning in the early 1970s Leonid Brezhnev and some of his Politburo colleagues publicly insisted that nuclear war would be unacceptable now that bomb stockpiles were "capable of blowing up the entire planet." Such apocalyptic talk had seldom been heard in Russia since Khrushchev's heyday, when it had been less widespread and less carefully considered. Authorities may have been actively promoting worries about nuclear war in order to win support, since fear of foreign attack was still a main reason for acceptance of a strong government, in the Soviet Union as elsewhere.[11]

In the early 1980s the average Russian lived in increasing dread. To be sure, some Soviet leaders felt they could survive a nuclear war; indeed, they had built a costly system of special bomb shelters for themselves (so had the American political leadership). Ordinary citizens were far less protected and less confident. The belief that Soviet society would survive, as it had survived the Second World War, was still ritually invoked in military circles, but it was now largely discredited among the public. A group of American psychiatrists who managed to poll some Soviet adolescents found them heavily concerned with the threat of global doom. Only 3 percent said that they and their families would survive a war, compared with 16 percent of a comparable sample of American students. A 1986 Russian film, *Letters from a Dead Man,* showed the nation destroyed in a war triggered by an error in their own computer; the few survivors lived underground beneath a dimly lit wilderness of debris and corpses. The Soviet civil defense program tried harder than ever, but its posters and textbooks, seen everywhere, offered precisely the same uncanny picture: a desert of rubble through which somber figures moved, the few survivors shrouded in rubberized protective clothing and wearing gas masks, like a spectral procession commemorating the triumph of death.[12]

Public imagery outside the Communist countries changed in a more complex way. In the early 1980s, for the first time, television specials thrust into the public's face a realistic and detailed picture of missile warfare. One example was a 1981 CBS television news special that included a brief but disturbingly lifelike depiction of a missile exploding over Omaha, Nebraska. Nothing like that had ever appeared in a major television production, and this was only the first of several shows seen around the non-Communist world. In 1983 came ABC's "The Day After," the most famous nuclear television show ever, giving hundreds of millions of people their first look at what a real nuclear war might be like.[13]

Not all the images were realistic. At one extreme was a British general's best-selling fictional book called *The Third World War,* which ended with one hydrogen bomb dropped upon Minsk, whereupon white flags sprang up all over the Soviet Union. At the other extreme was an NBC television thriller "World War III," about how Soviet and American leaders stumbled into conflict; their decision to fight was directly followed by the closing credits, as if the beginning of nuclear war would equal the end of the world's time. Between these extremes were the usual gaudy stories, for example *Damnation Alley,* an after-the-bombs tale first published in 1967 in a science fiction magazine, then expanded into a book and in 1977 into a movie. It could have been written in the 1950s, with its brutalized survivors battling swarms of giant insects, disastrously altered weather, and one another. Especially traditional was the movie's ending, where the hero and heroine fought through the storms to reach a land of blue skies and "a new life." Fantasies of heroic individualism also inspired detailed "survivalist" manuals, paperback novels, and even a magazine, as well as a minor industry selling equipment with particular attention to guns. Meanwhile yet another widely sold juvenile book rehashed the theme of magical postwar rebirth, complete with a triumph over mechanistic authority by psychic mutants in a golden village.[14]

The most carefully researched books and films likewise carried an element of fantasy. From congressional studies to "The Day After," they showed war as a matter of a few thousand bombs all used within a few days. In fact, most past wars had gone on far longer than they had been expected to last. Hardly anyone seemed aware of this, or realized that once a nuclear war began it would as likely as not continue for months or years, with tens of thousands of additional warheads hunting down the survivors. Even civil defense enthusiasts, preoccupied with insisting that nations could handily survive the first salvo, seldom considered what such survival might imply about later events. While civil defense officials selectively overlooked possible features of nuclear war, their opponents gave all their attention to preventing war in the first place, which also begged the question. When experts failed to face squarely the real after-the-bombs world, the public could hardly be expected to grasp what nuclear war actually meant.[15]

Schell's *Fate of the Earth* troubled people because it was the first widely read work to show clearly what could happen if *all* the available bombs were dropped. Little expertise was needed to understand that. If the Soviets simply exploded their 7,000 missile warheads at equally spaced intervals in the sky across the United States, then about half of the area of the nation would be within eight miles of an explosion. Within that area the heat from an average bomb (one megaton) would be strong enough, at minimum, to blister exposed human skin and set fire to newspapers and dry leaves. Of course the blasts would not really go off over desert and forest, but would erase every inhabited place larger than a village. Meanwhile Europe and the Soviet Union, their populations more concentrated in urban areas, would suffer a worse fate. Perhaps the Southern Hemisphere would escape, and peasants in Bolivia might scarcely notice the war. Yet there was no saying whether, with so many thousands of spare warheads, once the mad slaughter began someone might not destroy "neutral" regions to deny an opponent their resources.[16]

People in every nation knew enough of that to be increasingly concerned about plain survival. The best guess was that a fair fraction of the human race would survive even 50,000 bombs, but was that certain? Of course, the chance of extinction had already become a commonplace in the 1950s with talk of bombs knocking the earth off its axis or the like. Only gradually did this become plausible. In an unpublicized 1956 study for the Atomic Energy Commission, the United States Weather Bureau noted a "possibility that a massive nuclear exchange might usher in a new 'ice age'" by throwing dust into the stratosphere. In the early 1960s a few scientists publicized this threat, along with the suggestion that nuclear war could bring plagues of insects and so forth, perhaps exterminating all life higher than cockroaches. In 1974 a new threat was added, the hypothesis that nuclear war might temporarily thin the protective blanket of ozone in the earth's atmosphere. However, most experts saw little evidence that such events could actually end the human race.[17]

In 1982, after talk of nuclear war became widespread again, a few scientists realized there could be another effect: the smoke from thousands of burning cities and forests might darken the skies for months. In 1983 a large group of scientists launched a sophisticated publicity campaign warning against this "nuclear winter." The idea of a Great Winter, an end of the world in ice, was a vision millennia old, but it now began to look as possible as missile warfare itself. Experts felt that the chance of outright race extinction was still small, but now many of them suspected that the chance did exist—growing year by year as warheads multiplied.[18]

People have become almost incapable of responding to such facts. It is not only that the facts are too dreadful to face, and not only that they are buried amid a junk-heap of fantasies, but also that they seem an old tale, nothing urgent. Even the widely discussed television movie "The Day After" had little impact, according to a survey of viewers. In all the noisy controversies and writings and films of the early 1980s, I have seen not one idea about military or political or diplomatic policies that was not available twenty years earlier as the great nuclear strategy debate came to its exhausted end.[19]

However, beginning in the mid-1970s a few people took a closer look at some of the central ideas of the strategy paradox and made those ideas more explicit. Analysts agreed that once both sides had a thousand or so missiles, possession of greater numbers or more modern types could scarcely make a nation come out ahead, in any meaningful way, in a nuclear war: not winning the war but preventing it was the only rational goal. But while more weapons became redundant, the ways people perceived them remained crucial. In the game of chicken it matters little what sort of car you are driving headlong toward your opponent; what matters is what is in your opponent's mind, and in your own. And most people were still under the illusion that the nation with the biggest and most modern nuclear armament was somehow the most powerful. That belief could affect how leaders would act in a crisis, and therefore the level of armament was important after all. The nuclear strategy paradox—as Churchill had put it, "the worse things get the better"—was revealed to be not a military but a psychological conundrum, true exactly so long as it was not recognized to be false.

This "perception theory" was accepted, tacitly or explicitly, by many American military and political leaders. They were willing to act according to the illusions of the Soviet leadership and the American public, or even to foster those illusions, in hopes of weakening resolve on the one side and strengthening it on the other. They argued, for example, that there was urgent need to deploy modernized missiles in Europe, insisting that the missiles' presence would reassure the NATO allies and give the enemy pause; they neglected to add that the deployment would have negligible practical influence on the military situation. In every such case the drive was to make armament "worse" as a physical destroyer in order to make it "better" as a mental deterrent. By the 1980s it was clear to all careful thinkers that nuclear policy had less to do with the physical weapons than with the images they aroused.[20]

As for imagery itself, there were some effective reminders of the old themes but only one striking new image, the nuclear winter. Pictures of a smoke-shrouded planet were propagated deliberately to dramatize the evidence that at some numerical level the stockpile of nuclear warheads would become—perhaps already was—the Doomsday Machine. Yet the idea had no evident effect on official strategic doctrine. After all, in terms of imagery the nuclear winter looked no worse than the clouded chaos shown in *Damnation Alley* and some earlier movies and books. The Doomsday Machine, otherwise called Mutual Assured Destruction or MAD, had long since been accepted as the keystone of the illogical paradox of deterrence; new evidence only seemed to confirm the old themes.[21]

The most revealing use of traditional themes came from President Reagan in a March 1983 speech. He called on the nation to build a strategic defense system that would render nuclear weapons "impotent and obsolete." Like Kennedy's bomb shelter speech and many other defense announcements, the motive seems to have been largely political. The Secretary of State later said the speech came from a desire for "a big public relations splash." More precisely, the President's science adviser and other insiders said that when the Nuclear Freeze movement showed that the public was not willing to keep endlessly spending money for nuclear weapons as deterrents, Reagan decided to change the rules of the game.[22]

The speech took the government's official advisers by surprise; the impetus behind strategic defense had come from outside lobbying groups, supported by wealthy right-wingers and spearheaded by Edward Teller, rather than from a consensus among experts. In fact scarcely any experts thought it likely that a cost-effective system to defend the populace could be built in the foreseeable future. In a poll of members of the National Academy of Sciences, an overwhelming majority denied the possibility. At many of the nation's leading universities majorities of physicists signed petitions declaring the whole project bankrupt; an impartial panel of experts appointed by the American Physical Society made a close study of the energy-beam devices that were central to strategic defense plans, and announced that such weapons would not be feasible for decades if ever. A few brave government experts resigned in protest, while most of their colleagues secretly scoffed at the President's promises. A former Secretary of Defense publicly said the only result of the scheme would be to "make lots of money for lots of groups." In private an eminent military laser specialist was still more blunt: referring to the Depression-era program of government funding for jobs, he called the new plan "a WPA for the rich." Seldom had experts so consistently opposed a technical plan proposed by a democratic government.

Yet important segments of the press supported the plan; a majority of citizens in the United States (although not in Europe) believed it could work; Soviet political leaders, chronically anxious about the mysterious potential of American technology, seemed likewise impressed; and Congress appropriated billions of dollars. The opponents had been far less effective than in the antiballistic missile controversy of the late 1960s. At that time they could battle specific weapons, scheduled for specific neighborhoods; now they were hitting out against nothing tangible but only a vision of some future device. It was a striking demonstration that imagery could shape history in spite of what scientists and engineers held to be true.[23]

For imagery was at the core of "Star Wars"—as various observers, spontaneously and independently, named the program within hours of Reagan's speech. They were referring to the space battles in the most famous of all Hollywood death-ray spectacles, but earlier precedents were also plain. As I showed in Chapter 3, all-seeing and all-destroying rays had fascinated people for millennia. The numerous movies, television shows, toys, and other productions of the "Star Wars" genre further appealed to the public by linking rays with spaceflight, for cosmic flight was another theme of ancient power (it went all the way back to the magical ascents of shamans). The theme had come into modern times not only undiminished but strengthened by many forces; for example, NASA and the Air Force had been glad to aid selected spaceflight movies. Among the young scientists at Livermore Laboratory whose work led directly to Reagan's speech, there was high interest in both spaceflight and the death rays of science fiction.

Reinforcing these primitive themes was the enduring modern belief that scientists could do anything if they really tried. When Little Orphan Annie was kidnapped in 1986 by villains hunting strategic defense secrets, it was a replay of her kidnapping forty years earlier for the secret of Eonite, the impenetrable defense against atomic bombs. The fact that no such defense was physically possible, a main point in the atomic scientists' publicity campaign of the 1940s, had been a great blow against naive faith in science and technology. In promoting the Star Wars program Reagan was apparently working to restore this trust by affirming that technology, even weapons technology, could be inherently moral and humane. His supporters took for granted what critics promptly denied—that the novel weapons they meant to build would necessarily be benign.[24]

Of course the image of scientific ray weapons, however potent, could do nothing by itself. It served rather as a harness of a particular shape, in which a particular set of historical, social, and psychological forces could readily join together to pull the nation in a particular direction. That is what I mean when I speak of the power of imagery.

If the administration's strategic defense program was built on images, to a large extent that was deliberate. Perception theory strategists who knew that a missile defense system could not work still wanted to build one, for they thought the Star Wars imagery itself would instill caution in the enemy, and at the same time would make their own side more willing to fight. One might not even have to actually build the system; the claim that one was designing such a thing became, all by itself, an important tool in diplomacy and domestic politics. If nothing else, such a claim could handle the criticism of existing weapons by diverting it into a sterile debate over weapons that were not yet even on the drawing boards.

Whether or not that was intended, it happened. The Nuclear Freeze movement in the United States began to flag. After 1983 public attention to nuclear weapons, as measured by attendance at demonstrations and by the number of American magazine article titles, fell with remarkable speed (see Figure 2). Meanwhile articles on missile defense rose rapidly; by 1986 they exceeded the combined total of articles on nuclear winter, bomb testing, and other negative aspects of nuclear weapons. The decline in articles directly concerned with bombs was due not only to the Star Wars diversion but to something that had also sapped the antibomb movement of the 1960s: opposition began to seem futile. The government, with the acquiescence of a majority of American citizens, remained determined to build more weapons. The European movement similarly withered when it became clear that its drive to block new missiles was a failure.

Meanwhile tensions eased as a new Soviet leadership pleaded for arms control and peace, while Congress slowed the American arms buildup; now as in the 1960s, politicians were hastening to reassure the public by moving toward détente. The antibomb movement was perhaps not so futile after all. For whatever reason, polls in the Western nations found fear of imminent war receding. Mechanisms of reassurance and denial were still at work, and the news media could not concentrate their attention on a particular issue for more than a few years. Yet spy thrillers about nuclear war threats remained common on the paperback racks, and polls found that latent public concern over the bombs remained high, as always.

Was the war fear revival of the late 1970s and early 1980s no more than a pointless repeat of the temporary surge of the late 1950s and early 1960s? The evidence is not yet all in hand, but I think that something new arose. In addition to whatever permanent political effects there may have been, the real meaning of many of the old nuclear ideas and images became better understood. For example, widespread use of Robert Lifton's term "numbing" pointed to a recognition of the psychological mechanisms of denial, learned helplessness, and so forth. The listing of scholarly research in *Psychological Abstracts* showed that the effect of nuclear war threats on people's thinking, a subject almost entirely neglected prior to the late 1970s, was now attracting numerous careful studies. Deepened insight also emerged in a burst of excellent poems, paintings, short stories, and novels that began where earlier artistic works on bombs had stopped.

Popular movies too were more subtle and complex, and more conscious of unconscious motives, than any nuclear movie of the 1950s. For example, the widely seen *Wargames* made the theme of the dangerous bad boy explicit. Here nuclear catastrophe was almost set loose by an irresponsible scientist, a childish computer—and a rebellious boy. In this and other productions the Men Who Started the War began

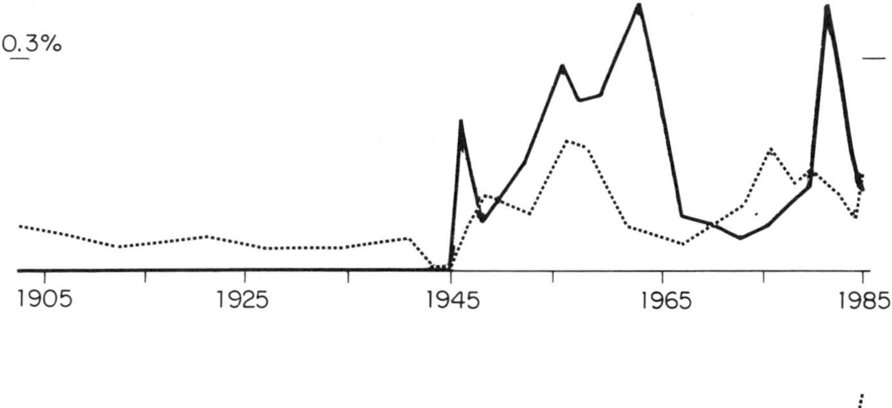

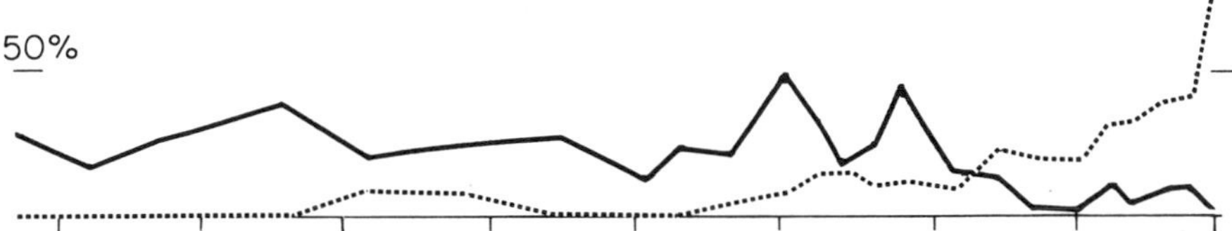

Figure 2. My counts of articles listed in the *Readers' Guide to Periodical Literature* under headings such as "Atomic," "Nuclear," and "Radioactivity," in selected volumes spaced from 1900 through 1986. Minor wiggles in the curves (10 percent change or less) may reflect my uncertainty on how titles should be categorized.
Top: Nuclear articles as a fraction of the total number of articles listed on all subjects in a given volume. *Solid line:* Military uses. The curve shows there were almost no articles before 1945, then a brief post-Hiroshima burst of interest. With the coming of hydrogen bombs and test fallout, attention revived. It fell off slightly at the end of the 1950s but reached a maximum in the early 1960s with war scares and talk of accidental war. Attention then fell dramatically until the late 1970s. A burst of renewed interest peaked around 1982 at the time of the Catholic bishops' statement on nuclear arms. *Dashed line:* Civilian uses. A surge of post-Hiroshima interest was followed by a second surge fueled by Atoms for Peace publicity. After a period of disinterest the antinuclear debate brought attention to a peak with referenda in 1976; interest then declined except for revivals following the Three Mile Island and Chernobyl accidents.
Bottom: Titles on civilian uses of nuclear energy with clearly positive (hopeful) or negative (fearful) implications, as fractions of all titles on civilian uses. *Solid line:* Positive titles. The optimism, present throughout the early decades of the century (especially for medical uses), climbed higher with Atoms for Peace and again in the 1960s with talk of peaceful explosives and desalting seawater, then declined. After the mid-1970s almost the only positive titles dealt with fusion energy. *Dashed line:* Negative titles. These made a temporary appearance in the late 1920s with tales of hazards in industry and from quack medicines. Anxiety began rising again in the 1950s, paused during the period of disinterest, and rose steadily thereafter; by the 1980s more than over a third of all articles had titles that implied the nuclear industry was harmful. Adding to these negative titles other titles that showed concern but did not take a stance (weighing risks against benefits) gives another curve, not shown: the fraction of titles implying that problems existed was under 10 percent through 1952 and then began a fairly steady climb, reaching 90 percent in 1986.

to step into view. Now they were not evil madmen, but people wrenched by familiar personal and social strains. Meanwhile after-the-bombs genre films, reaching a peak in the popular *Mad Max* series, connected sadistic violence with technology; the savage battles of bomb survivors were no longer seen as the opposite of modern civilization, but as an exaggeration of trends already present in society.[25]

Attitudes and images have a slow, ponderous way of pressing upon the systems of thought and politics that finally determine events. I have argued that the early waves of nuclear fear had much to do with the radical rethinking of modern society that swept the world a little later. I expect that the wave of the early 1980s will have its own surprising effects. This history is not finished.

part five

THE SEARCH FOR RENEWAL

20

The Modern Arcanum

Another name for the philosophers' stone was the Arcanum, the great secret. Alchemists surrounded this central mystery with lesser symbols—naked kings coupling with queens, green lions devouring the sun, a thousand pictures whose meanings were concealed from the profane. In the twentieth century nuclear energy became the modern Arcanum, no less surrounded with images that most people could not entirely understand. In this part of the book I will survey the whole postwar period to see how nuclear energy was addressed, not in speeches, newspapers, popular magazines, and movies, but in the more subtle symbolism of our greatest writers and other artists. At the same time I will not ignore the public at large, who invented their own mysteriously compelling symbols.

DESPAIR AND DENIAL

In 1947 the poet W. H. Auden characterized postwar times as the "Age of Anxiety." A bleak mood had fallen upon the artistic and literary elites. Novelists, painters, social critics, theologians, all spoke of an anguished hollowness. This was not a matter only for intellectuals; psychological tests repeated over decades suggested that since 1940 raw anxiety had strengthened among the general population too. But it was chiefly artists and intellectuals who expressed the new despondency clearly and related it to social trends.

Scholars who looked into the problem said that covert anxiety had been growing steadily among thoughtful people throughout the twentieth century and had only become more overt. The diminished role of human will in the impersonal universe of Newton, Darwin, and Freud, the failure first of religious faith and then of humanistic values, the disintegration of traditional social relationships—a hundred influences had created an existential dilemma. But many sophisticated thinkers connected the problem most particularly with nuclear bombs. Hardly anyone claimed that bombs were the only source of modern anxiety, but intellectuals worked with symbols, and what could be a better symbol of our problem? No discussion about contemporary anguish and futility seemed complete unless it mentioned the bombs. Typically such mentions were brief and stereotyped, but they came at a key point, as if a phrase or two said it all. Like reactors only more thoroughly, the bombs and nuclear energy in general served as a condensed symbol for the worst of modernity.[1]

Nuclear weapons gave the twentieth century's nihilism a dismal solidity. Immediately upon hearing the news from Hiroshima, sensitive thinkers had realized that doomsday—an idea that until then had seemed like a religious or science-fiction myth, something outside worldly time—would become as real a part of the possible future as tomorrow's breakfast. Worse, where time had once either stretched linearly forward without end or else circled back though apocalypse to a Golden Age, now the future might just lead to nothing. In the 1950s J. B. Priestley and others noted that characters in the newest novels seemed flat and withdrawn, "as if already they were on the other side of the monstrous bombs threatening us." In 1960 the literary analyst Charles Glicksberg wrote that all through modern literature ran a "monomaniacal emphasis on the inevitability of death," an emphasis that the bombs intensified. "Overwhelmed by a feeling of his own insignificance in a world ruled by malign atomic energy, the individual concludes that life is . . . an ephemeral and idiotic episode."[2]

Others from the 1950s on noted that death itself lost meaning when bombs could rob victims of their dignity and individuality, making people ridiculous grains of dust in a mass extinction. What use was anything when a crazy accident could wipe out not only oneself but one's progeny, the memory of one's lifework, indeed everything human, perhaps all the natural world? This became a recurrent theme in poetry and other art forms: the bombs had changed our ideas about death, and therefore about life too, placing at the base of things an unspeakable void. A psychologist remarked that even music and dance now "express the despair of a humanity that no longer believes in its own future."[3]

Such despair was akin to denial. People looking in the direction of modern war might perceive a great void simply because they dared not open their eyes. For example, Kurt Vonnegut, whose best novels dealt with nihilism, personally struggled with the classic denial mechanism. As a captured soldier he had been in Dresden on the night in 1945 when incendiary bombs made the city a desert of ash; afterward he simply could not remember details of the experience. He kept telling friends that his next book would be about Dresden, but for decades he could not write it. His innocent faith in science and humanity had gone, and not just at Dresden (according to his later recollection) but still more when he heard the news from Hiroshima. In his novels of the 1950s and 1960s Vonnegut kept groping for images of meaninglessness and annihilated civilization without ever quite facing bombs directly.[4]

Other writers had similar difficulties. Particularly frank was John Braine, a British novelist, who admitted that he could not write a book except when he forgot about nuclear weapons. As for writing about the hydrogen bomb itself, he said, "It frightens me too much . . . whenever I think about the H-bomb, I can only give way to despair, which is to say that I can only stop thinking of it." Filmmakers were equally baffled. In 1962 a magazine surveyed the world's greatest directors of artistic films, asking them how they would treat hydrogen bombs in a movie. Ingmar Bergman replied: "I find it impossible to make a picture 'about the Bomb and its effect on modern man.'" Louis Malle: "For nothing in the world would I make a film about the H-bomb." Elia Kazan: "It's too much for me." Michelangelo Antonioni: "I do not have the vaguest idea." John Cassavetes: "Frankly, the very thought of atomic warfare . . . threatens to throw me into a panic."[5]

Those who managed to get into the subject tended to produce works of despair. Pat Frank, for example, confessed that he forced himself to write nuclear war novels from a sense of duty and failed to do the subject justice. A writer like himself, he said, "stares at his typewriter and the typewriter stares back and says, 'This thing is bigger than both of us.'" Each of Frank's three novels on nuclear war was more pessimistic than the last. The same held true for Philip Wylie and Upton Sinclair, who both wrote several bomb tales. Characteristic of sophisticated nuclear fiction was a novel that won critical respect and millions of readers around the world, Mordecai Roshwald's *Level 7* of 1960. Its characters were depersonalized soldiers, locked deep in sterile tunnels under the command of unseen authorities. They inadvertently began a war that destroyed all life on the surface of the earth, then perished themselves one by one, obedient and unthinking, until no life remained. Such works found only numb emptiness in the modern situation.[6]

From 1945 through the 1970s a sense of anxiety and disintegration overtook not only writing but the other arts from music to painting, although it was impossible to say how much that had to do specifically with nuclear weapons. What was one to make, for example, of Jasper Johns's famous "targets," anonymous human fragments placed behind painted bull's-eyes? Or of Jean Tinguely's notorious contraption of motors, noise, and smoke, which in March 1960 performed its sole function—self-destruction—before an elite audience at New York's Museum of Modern Art? A few critics suggested that such artistic obscurity and pessimism were partly a response to the arms race. But aside from occasional polemical drawings and writings with the usual political messages, most painters, musicians, playwrights, and so forth made no direct reference to the bombs, speaking rather of the internal paradoxes of art.[7]

In the few cases where artists dealt directly with the bombs, they had even less success than authors. Representative of these few pictures was Philip Evergood's 1946 painting *Renunciation*. In the background an atomic bomb cloud appeared as a huge, evil brain; in the foreground a group of apes turned away, renouncing human intelligence in disgust. Apparently the only alternative to denial was plain despair.

Some thought that nuclear weapons were crudely handled or ignored because they were beyond the reach of the arts, but that proved untrue. Beginning in 1980 as war fears reawakened, hundreds of young artists began to treat nuclear weapons in paintings and sculptures of high quality. Some revived old images of transformation: a vine-covered ruined city, a Christ crucified on a nuclear bomb cloud. Others found new ways to create moving statements, for example using the bomb shelter's grave and womb symbolism. Although many of these works remained, as one of the artists confessed, "just elaborate despair," to work that out explicitly was itself an important step, rarely taken before 1980.[8]

Although the bombs were hard to face head-on, they could be approached indirectly through irony. A combination of social criticism with pessimism and absurdity in art could be traced back to the time of the First World War, but it was not until the 1960s that black humor became a major movement. One leading work in the genre was *Cat's Cradle,* in which Vonnegut joked about an atomic scientist whose new invention accidentally and ridiculously destroyed the world. A later landmark novel owing much to the black humor movement was *Gravity's Rainbow,* in which Thomas Pynchon played paranoid games involving corporations, ballistic missiles, a secret invention, and the end of the world. Most influential of all was Joseph Heller's *Catch-22,* the end of the trajectory that Second World War bombing novels had traced through *Twelve O'Clock High* and *The War Lover.* Beneath jeers about military insanity Heller told a mordant story of a man who, with classic denial, refused to face the fact of death. "I wrote it during the Korean War and aimed it for the one after that," Heller explained. "The Cold War is what I was truly talking about." Nuclear concerns were still more obvious in the most successful black humor movie, *Dr. Strangelove.* And the one great painting done in the ironic mode, James Rosenquist's *F-111* of 1965, placed handsome consumer goods atop a warplane and an atomic bomb cloud. In sum, while no doubt many conditions fed the new taste for combining criticism of modern society with thoughts of death and absurdity, by far the most obvious influence was nuclear fear.[9]

Macabre irony was a step upward from plain denial or despair, for the black humor works implicitly compared our current state with more rational and kindly possibilities. But these works held out little hope that those possibilities could be reached, offering no advice except to grin bitterly and scrabble for personal escape. It seemed that, as Vonnegut insisted, "there is nothing intelligent to say about a massacre."[10]

HELP FROM HEAVEN?

While some people who looked at nuclear energy saw dark emptiness, others saw a dazzling light. The dismal viewpoint was common among artistic and literary intellectuals who felt that the gospel of scientific progress had proved hollow and that God was long since dead. Meanwhile the majority of citizens, giving up on neither progress nor God, held a view of nuclear energy that was poles apart from nihilism. Some came peculiarly close to an attitude of religious awe.

"Atomic power! Atomic power! It was given by the mighty hand of God!" Country and western singer Fred Kirby sang that on the radio in

1945, making such a hit that he started a fad for atomic lyrics. Many agreed that the new power was a gift from beyond the mortal sphere. From 1945 on, scientists, statesmen, and ministers all spoke in terms of transcendental forces and the sacred duties of stewardship. As President Lyndon Johnson put it in 1967, atomic power could help us fulfill God's command in the Book of Genesis to "replenish the earth, and subdue it," or else "to fulfill the prophecy of Armageddon." Many understood such words not as trite metaphors but as explicit religious statements.[11]

Within days after the bombing of Hiroshima, watchful intellectuals warned that feelings of awe, insignificance, and hope for salvation, which properly belonged to holy things, were being directed instead toward atomic energy—what top scientists called "the basic power source of the universe." The word "atomic" itself could stand, as in a 1951 gospel song, for the power to "cure the sick and destroy their evil . . . power known by God alone." Articulate groups scorned such ideas, so as time passed the ideas came to be expressed through inarticulate symbols.[12]

Nuclear energy was frequently represented by a blaze of light, for example radiating from a bomb blast or from a nucleus. Thus the "inner space" ride at Disneyland climaxed in an impressive vision of the nucleus as a mysterious glowing sphere. A real nucleus emits no visible light, but with their usual skill Disney's staff had found an emotive symbol. Observers at bomb tests, exclaiming over the apocalyptic fireball, appealed to the same tradition. It was no new tradition: flaming balls or disks, solar spheres, and radiating halos had stood for divine power in pictures everywhere from ancient Egypt to China. As the literary scholar Gerald Ringer pointed out in 1966, fireballs were also a traditional symbol for the goal of alchemists and mystics—Godlike wholeness, the Arcanum itself. A luminous sphere was bound to evoke holy awe, suggesting Godhead in its aspects not only of limitless power but still more of the "numinous," to use the theologians' term for a tremendous otherworldly presence. "The fireball of the Bomb," Ringer wrote, "is probably the prime symbol of the numinous in the world today."[13]

Like any form of divine energy, nuclear energy could be seen not only as solar but as the chthonic opposite. Writers never tired of repeating how Oppenheimer at the Trinity test saw the bomb's cloud as God in His aspect of Death the destroyer—which was only an elegant version of the cartoonists' drawings of a bomb as war-god. According to some psychologists, by the mid-1950s for many people "The Bomb" had become something even more: a prime symbolic representation, an apotheosis, of universal death. Antinuclear activists of the 1970s were simply borrowing the association when they linked not just bombs but reactors with effigies of the Devil and Death.

For people of a religious inclination, it was more than a primitive myth to connect nuclear energy with divine energy. Atoms were after all created by God, and like everything in creation they should be approached with holy faith. This could lead to the inverse of despair—a pathology of hope—for the connection between bombs and apocalyptic power could take on a precise Christian meaning.[14]

A gospel song of 1950 warned that on the Day of Judgment, Jesus would "hit like an atom bomb." Some took that literally. Preachers and religious tracts in the 1950s and 1960s (most but not all of them American) said that the Second Coming of Christ would be heralded by nuclear missiles, fulfilling in plain fact the biblical prophecy of falling stars, scorching heat, rivers of blood, and so forth. After all, the chaos of war and a Final Battle were central to Western apocalyptic tradition. Now as throughout history, the preachers and their flocks hoped to be among the remnant of the faithful who would be saved in the Last Judgment. Some millenarian sects moved to remote areas or built fallout shelters, just to make sure.[15]

This was no minor tendency. In the United States the most popular nonfiction book of the 1970s, reportedly selling some fifteen million copies (about twice as many as *Catch-22* and at least three times as many as any other book I have mentioned except the Bible) was Hal Lindsey's *The Late Great Planet Earth*. It predicted a nuclear Armageddon, miracles and all. The book spun off a movie and four sequels, inspiring many other paperback books as well as radio sermons and so on, all prophesying nuclear Doomsday. In the 1980s an estimated five to ten million Americans, including President Reagan, believed there was a good chance that the Armageddon of the Bible was imminent in nuclear form. In other nations the belief, while less openly expressed, was not unknown.[16]

Numerous other secular causes for the Apocalypse had been proposed, but nuclear bombs fitted here, as with so many other ancient themes, best of all. In this case the tightest bond attaching old ideas to physics was the transcendental fire of the bombs. The fire envisioned by medieval preachers had served three overlapping purposes: it separated the good parts of humanity from the bad, a trial by fire; it purified the virtuous with its divine force, as gold was refined in the alchemical furnace; and it punished the wicked. Nuclear war and its aftermath, as described in after-the-bombs tales, could serve well as a secular replacement for the traditional afterlife—a time of testing, purification, and punishment. However, that subtle thought was rarely expressed; what became widely heard was a cruder and more explicitly religious version.

The most convinced millenarians emphasized the miraculous "rapture": at the sound of a trumpet the minority of good Christians would be bodily lifted from the earth, "caught up together . . . in the clouds, to meet the Lord in the air" (1 Thessalonians 4:17). One religious writer noted that a surprisingly large number of people believed "the rapture will provide an escape hatch for Christians." As a Texas preacher put it, "My church, my people, you're not gonna be there when the bomb starts falling!"[17]

Or would God spare not just good Christians but all of us? A substantial minority of people responding to polls about nuclear war said they took refuge in a faith that God would never permit the world to be destroyed. In the first decade after the bombing of Hiroshima, some magazine stories and movies imagined a direct message from God warning humanity against self-destruction. The thought came up again in a popular 1981 novel in which war was miraculously averted after the Pope received a revelation that doom was imminent.[18]

Stories of miraculous intervention were not just pathetic escapism, for they came close to the statements of sophisticated Christian thinkers. There could be no escape from nuclear war, nor from any other form of sin and death, except through the gift of God's grace. Thus Pope John Paul II declared that in our times Christ's prayer, "Deliver us from evil!" could mean, "Do not allow us to kill! . . . Defend us from war!"[19]

Although theologically impeccable, such prayers were admittedly less than a complete answer to the problems of nuclear energy, even for people of a religious temperament. And overt religious messages had little meaning for the rest of the world's population. In a secular age, symbols of salvation would be more impressive if they were wrapped in an aura of science.

OBJECTS IN THE SKIES

In 1951 Arthur C. Clarke published a story about a colony on the moon where the human race survived after nuclear war had left the planet a phosphorescent ruin. Survival of a remnant rocketed into space became a familiar idea, and not only in science fiction. During the 1970s a variety of people suggested that colonies in orbit could answer the threat of war by giving humanity a common goal, a source of prosperity, and if necessary a celestial bomb shelter. We didn't need God to lift us from danger. Teller and others campaigned more specifically for orbiting weapons platforms, perhaps under international ownership, to shoot down all missiles—H. G. Wells's airmen in outer space. The Star Wars program included the same theme with Americans cast as the virtuous airmen. Through all this talk ran the idea of salvation by way of technology in the heavens.[20]

A sociologist who surveyed science fiction pulp magazine stories found that after 1945, much more than before, they tended to solve problems less by the hero's prowess than by intervention from outer space. For example, it became a cliché that a threat from alien invaders would make the nations forget their differences and unite. An alternative solution appeared in Clarke's famous 1953 novel, *Childhood's End,* which transformed the aliens themselves into virtuous airmen, hovering in great ships over the world's cities and forbidding war.[21]

Critics said that Clarke was dressing up Christian symbolism in scientific clothing, for the novel ended in a terrestrial apocalypse and a reborn race rising into the sky, all described with barely concealed religiosity. Many were attracted by this approach to the world's problems. Other examples were an award-winning children's book of 1978 where a glorious being descended from the heavens to help save the world from nuclear doom, and the popular 1951 film *The Day the Earth Stood Still,* where the messenger from the stars demanding peace was a patent Christ figure.[22]

This last messenger had arrived in a glowing, silvery disk propelled by atomic energy: a "flying saucer." Since 1947 an avalanche of reports about things in the skies, first in the United States and then around the world, had caught intense public attention. Some observers called it an "atomic psychosis" brought on by fear of the bombs, and the idea is worth a close look.

Flying saucers were no trivial phenomenon. American sightings of Unidentified Flying Objects (UFOs), as the things were later named, continued at a rate of hundreds each year, rising to much higher peaks in 1952, 1957, 1966, 1973, and 1978. The press gave the matter only limited attention once the novelty wore off, but extremely successful movies featured wondrous visitors from the skies, and polls found that about half the population believed that UFOs were somehow real. By the 1970s more than one-tenth of Americans said they had personally seen a UFO. This was another belief indifferent to boundaries of class and education; the sightings could come from sober businessmen or engineers as much as anyone else. For example, in 1973 the governor of Georgia, Jimmy Carter, reported having once seen a large, brilliant, unidentified object hovering in the sky.[23]

What the future President had seen was the planet Venus in an unusually bright phase. In fact anyone who answered the telephone at an astronomical observatory learned to check the location of Venus when someone reported a strange object moving and flashing. I am not addressing here the vanishingly small number of reports that proved difficult to explain; I am interested in the fact that millions of people looked at planets, weather balloons, and other ordinary things and thought they were seeing something uncanny. There has never been a better example of the mass projection of cultural images onto external stimuli.

The projected material, especially in the first decade of sightings, included much about nuclear energy. This was commonly suggested as the UFOs' mysterious power source, and rumors circulated that radiation monitors had gone wild when the objects were nearby. The AEC had to issue a public statement that the objects did not result from bomb tests. Americans wrote their congressmen about secret Russian aircraft sent to reconnoiter, planning perhaps to return with atomic bombs. That worry increased when people noticed that sightings were reported especially often near AEC installations.[24]

In the 1950s many people began to see the objects less as a threat than a hope. If the unknown pilots were not Russians then they were alien beings, no doubt far advanced not only in technology but in wisdom—the Martian engineers. Perhaps nuclear bomb tests had drawn their attention to our planet and they had come to observe what was happening, or even to save us from war.

The only deep psychological study of the sightings, by Carl Jung, suggested they had much to do with the Cold War. For example, in a dream reported to the psychologist, a woman was walking down an avenue in Paris when she heard an air-raid siren howl. She tried to hide but could find no shelter, and saw in terror that the avenue was empty: the typical nuclear war dream of separation, known to many in the 1950s. But then, in the sky, "instead of the expected bombers I saw a sort of Flying Saucer." Her next dream transformed the saucer into a godlike eye in the heavens. Jung suggested that in the anxiety caused by the division of the world into two hostile camps, people would like to pray for a divine answer, but many found such a miracle implausible and looked instead for salvation through superhuman technology.[25]

Such an answer was openly promised in tracts and personal proselytizing that inspired the growth of some 150 American cult groups. Men and women reported that beings from outer space had taught them to surmount fear of death and to move the world toward an advanced and peaceful utopia such as other planets already enjoyed. Many of the reports resembled paranoid transmutational fantasies. Movies and widely read books conveyed a preoccupation with secrecy and with superior beings at once menacing and benign, who sometimes immobilized and wrought bodily changes upon their favorite victims, and who conveyed redemptive messages of cosmic importance. A straightforward case was that of a schizophrenic man who believed, into the 1980s, that flying saucers would come to rescue him when the world was destroyed in nuclear war. Less apocalyptic, but in the same vein, was a Walt Disney movie in which benevolent saucer beings thwarted an evil scientist's nuclear reactor catastrophe.[26]

The UFO theme plainly addressed anxiety about all modern technology. By the 1970s some men and women were reporting messages from space that called for reforms to prevent not only nuclear war but also environmental perils, including nuclear reactors. Meanwhile leaders of the UFO movement bitterly attacked the scientific "establishment" that found their messages worthless. One reporter who studied the matter concluded that "UFOlogy as a powerful social movement is fundamentally a reaction against science and reason." But it was even more than that.[27]

The UFO phenomenon was a rare historical event: the emergence of a major popular symbol. A host of seemingly ordinary people had worked a grand creative act, inventing a new representation for the dangers and hopes of personal and social transmutation, a new myth peculiarly appropriate to the modern age. This myth had originated in close connection with nuclear fear, and carried an implicit response. Awesome modern technology, said the UFO stories, must be accompanied by a full-scale transmutation of civilization and everyone within it.

MUSHROOM AND MANDALA

Themes of transmutation and the sacred showed an astonishing affinity for nuclear energy, turning up in all sorts of curious places, giving nuclear imagery an inner force that few recognized. There were examples less transparent than flying saucers, and more important. One of these was the most impressive of all nuclear symbols, found in the great majority of nuclear films, books, and pamphlets, and at first sight seeming entirely straightforward: a towering white cloud. In the late 1970s the psychologist Michael Carey found that photographs of these clouds had made an unforgettable impression on nearly everyone he interviewed, imparting a vision of overwhelming and numinous power. As *Pravda* remarked, the "mushroom-shaped cloud" seemed to hang "suspended over the future of mankind."[28]

Why mushroom-shaped? Other words could have been chosen. Observers of the first Trinity test wrote of the "multi-colored surging cloud," the "giant column," the "chimney-shaped column," the "dome-shaped" column, the "parasol," the "great funnel," the "geyser," the "convoluting brain," and even the "raspberry." A Japanese witness of the Hiroshima explosion described a "pillar of black smoke shaped like a parachute," and a Bikini test in 1946 was aptly described as a "cauliflower" cloud. Yet at Bikini a reporter also spoke of "the mushroom, now the common symbol of the atomic age," and the term or its equivalent in other languages became almost synonymous with nuclear bombs. Already at the Trinity test more observers had mentioned mushrooms than anything else. Somehow it was hard to imagine people feeling the same awe for a "cauliflower cloud."[29]

The traditional symbolism of fungi was studied in the 1950s by a respected scholar, R. G. Wasson. He noted that in Western culture mushrooms were usually associated with dank, dark places, rot and poison, and, in short, death. But if that was all people had in mind at bomb tests, they could have used the English term for a specifically poisonous fungus, namely "toadstool." Wasson pointed out that mushrooms were also associated with food, and therefore with life. The mysteriously swift growth of these organisms was part of their folklore, and indeed descriptions of the Trinity test spoke of the cloud not only as a static shape but equally often as something that "mushroomed" up. Mushrooms could also shelter life, as in children's book illustrations where they arched protectively like an umbrella over small creatures. The mushroom, whether an atomic bomb cloud or simply a knob growing on a rotting log, could represent life opposing death—perhaps even life arising from within death, that is, transmutation?[30]

Mushrooms had traditional associations with thunderbolts, witches, and fairies, or in short with magical power. Wasson suggested that these associations were connected with a particular mushroom, the fly amanita, well known among folklorists for its poisonous and magical associations, and used as a hallucinogenic drug by Siberian shamans to help them into ecstatic trances in quest of cures and rebirth. He found evidence that precisely this mushroom was also the divine transformational food featured in ancient Hindu texts, the same place where the earliest hints about alchemical transmutation had appeared. "Magic mushrooms" of an unidentified species also figured prominently in the lore of medieval Chinese Taoism, again connected with religion and the elixir of immortality. Beyond these and a few other scattered hints, nobody has traced the trail. I will only note that at the time of the Trinity test mushrooms were still associated with magic in children's books, and the pictures usually showed them as red with white spots: the fly amanita. For example, in the 1930s classic *The Story of Babar,* the climax came when young Babar replaced the old king—such an act was among the major symbols of transformation in world folklore and alchemical writings—after the old king died of eating a mushroom, red with white spots.[31]

In 1955 Wasson discovered isolated Mexican villages where another mushroom, the white psilocybin, was traditionally used to induce religious trances. The mushroom eater, Wasson reported, experienced "the very fission of his soul." Other people followed his path, until the drug culture of the 1960s adopted the "magic mushroom" as a vehicle of mystical transports that they expected would lead toward a rebirth of individuals and all society.[32]

The atomic bomb mushroom likewise could be used optimistically. A 1955 AEC film declared that "the towering cloud of the atomic age is a symbol of strength . . . for freedom-loving peoples." Occasionally the umbrella metaphor was invoked. A 1981 perfume advertisement pictured a couple embracing within a vaguely phallic atomic cloud representing the aroma's sexual power. Most striking of all, a 1947 magazine article about medical benefits of nuclear energy was illustrated with a montage of a smiling man rising from a wheelchair, lifted as it were by a superimposed golden atomic cloud.[33]

The mushroom cloud was a folk symbol, created by nobody in particular for reasons that nobody explained. Surprisingly few artists, prior to the new insights of the 1980s, used the image in sophisticated works. The single outstanding exception was Henry Moore's 1966 sculpture "Nuclear Energy," a bronze tower of enigmatic curves. One thing clear at a glance was that, as Moore wrote, "The upper part is very much connected with the mushroom cloud of an atomic explosion, but also it has the shape and eye sockets of a skull." This combination of death's head with atomic cloud was already a visual cliché, and the sculptor would have made only a crude political statement if he had stopped there. But a viewer who came closer and studied the monument's subtly decorated surfaces and interior spaces could grasp Moore's further comment: "I had reminiscences . . . of the inside of a church or cathedral." In short, the theme of this sculpture was a tension between bodily death and awesome eternal force, what Moore called "contained power." That had been one meaning of the mushroom for a very long time; now it was a meaning of the mushroom cloud.[34]

Contained transcendental power was more clearly manifested in a variation on the mushroom cloud image. General Electric's 1952 animated film *A Is for Atom* opened with a bomb explosion, then showed the cloud transforming into a faceless giant with muscular arms crossed, an ominous warrior or benevolent helper. The 1957 Walt Disney movie *Our Friend the Atom* was still more explicit with its tale of a fisherman who uncorked a genie, murderously enraged by aeons of imprisonment. The image was an old one; already in 1908 Anatole France had compared the cloud rising from a city blasted by an atomic device to the cloud from the Arabian Nights fisherman's bottle. By 1963 the symbol was so familiar that a reporter could ask President Kennedy about bomb test negotiations in a sort of shorthand: "Where is the genie, sir? Is it out of the bottle or in the bottle?"[35]

The atomic genie was a cousin of the golem. With his woeful cramped imprisonment and rebirth into the world, his irresistible rage that could be converted to benevolent power, the genie had always excelled as a symbol of the peril and promise of psychological transmutation. But here as with the mushroom cloud, scarcely anyone used this symbol in a work of fine art or literature on nuclear energy. Did these popular symbols really have little significance, being mere arbitrary signs only accidentally connected with transmutation? The answer may be found by looking into the most universal of all symbols of nuclear energy.

At one point in *A Is for Atom* the genie faded into a representation of an atom itself. The picture was utterly conventional—a little ball ringed by ellipses, standing for a nucleus with electrons whirling around it. This ringed atom was by far the most common of all pictorial images found in nuclear energy presentations from the 1920s on. Anything so hackneyed merits close scrutiny.

Early in the century Niels Bohr and other physicists had found that certain features of an atom could be handled mathematically by imagining electrons to circle the nucleus, rather like planets orbiting the sun. However, physicists abandoned that preliminary model in the late 1920s when they found that electrons do not in fact orbit, but are found in a pattern of probable locations spread around the nucleus. In 1931 the correct representation of an atom was pictured in the *Physical Review,* looking more like a fuzzy snowflake than a solar system. None of this had anything to do with nuclear energy anyway, for only the central nucleus gives energy, and the electrons that dominate the design are irrelevant to that. In physical fact, a ringed atom was scarcely more accurate than a genie as a representation of nuclear energy.[36]

Nevertheless, journalists and even scientists persisted in using the solar-system picture. Cut off from real physics, the ringed atom evolved in a remarkable way. Bohr's original model had not been symmetrical; the elliptical orbits were skewed off to one side. Yet scientists and popularizers began to draw symmetrical pictures, and most often with a fourfold symmetry, which had little scientific justification. After 1945 representations of the atom were almost always symmetrical, and some designers went on to enclose the whole thing in a circle. The definitive version was promulgated by the Atomic Energy Commission on its official seal, commissioned from an artist in the Post Office Department. After laboring on the project for months he submitted a number of designs, and the Commission chose a ringed atom with four ellipses around a central nucleus, the whole enclosed in a circle. "I consulted some of our scientists," the artist later recalled, and "they told me that the design . . . did not mean anything. It is only symbolic."[37]

Any scholar familiar with symbology could identify in an instant a design characterized by a central object, symmetry (especially fourfold symmetry), and perhaps a surrounding circle. It is a mandala. This class of patterns is the most important of all human symbols.

Jung pointed out the importance of mandalas after he found such patterns in the recondite alchemical texts he was studying and independently in some of his patients' dreams. Subsequent work by many scholars found mandalas in the earliest scribblings of children around the world; in plans for sacred cities from Middle America to China; in therapeutic drawings of schizophrenics, including tribesmen who had never held a pencil before; in prehistoric solar diagrams pecked into rocks; in high religious art, such as the rose windows of cathedrals; in drugged psychedelic visions—almost anywhere one looked. The mandala proved to be a universal human archetype.[38]

Whether in the drawings of children or psychotics or alchemists or mushroom eaters, the mandala typically appeared in a context of spiritual and psychological questing, as a signpost toward the mysterious goal. The most thorough students of mandalas, Indo-Tibetan sages, used them as maps of a mystical state embracing everything from sexual conjunction to the entire cosmos. As Jung put it, mandalas stood for the sacred and potent union of every opposite, male and female and all the rest—that is, for the philosphers' stone. In sum, the ringed atom, most ubiquitous of all symbols of nuclear energy, followed a pattern universally associated with transmutation.[39]

The ringed atom diagram as most commonly used seemed apart from any sacred context, standing for nuclear energy or for science in general, with their inner mystery and their power to transform the world. But the diagram was also used directly to show personal transformation. Most striking were two minor comic book superheroes, The Atom, appearing in the 1960s, and Firestorm the Nuclear Man, originated in the 1980s; to show one or the other hero metamorphosing into his magical form, the illustrators drew ellipses around him as if he were the nucleus in an atom diagram. Such ellipses could even displace traditional mandalas in a strictly religious context. I have seen the ringed atom as an abstract title-page design in a 1969 textbook for Catholic schoolchildren, *The Story of Salvation,* and I have seen it superimposed on a drawing of a meditating man in a 1987 poster for a storefront spiritual-exercise school.[40]

Fireball, flying saucer, mushroom cloud, genie, ringed atom, in their various ways all said the same thing. If one knew how to read them, these pictures were posted across the public image of nuclear energy like a banner: "CAUTION—HIGH SYMBOLIC POWER!" Nuclear energy had become a full symbolic representation for the entire bundle of themes involving personal, social, and cosmic destruction and rebirth. By symbolic representation I mean a structured set of associations, known to everyone and not easily altered, condensing into a single thing a large number of meanings all loaded with emotional power. Indeed, over the years the meanings grew endlessly more numerous and rich, for like other great symbolic representations this one was open-ended. Nuclear energy had become precisely the new Arcanum, permanently connected with the most terrible, fascinating, and sacred of all human themes.[41]

21

Artistic Transmutations

How could we deal with the overwhelming power of nuclear energy? The symbols developed anonymously in popular thought scarcely answered that question, pointing only toward hazy and grandiose total transmutations. Fortunately a few of our best writers, filmmakers, and other artists did not follow their colleagues in turning from nuclear energy in denial or despair, but explored its implications in major works of art. These works pointed toward a more modest and more realistic kind of spiritual rebirth. They spoke in terms of immediate human themes that I have had little occasion to mention so far—the failure of responsibility and love, or their triumph.

Against the nihilism adopted by many intellectuals, others held up traditional values, reasserting religious or humanist ethics. Many said the bombs required a revival of moral courage, an acceptance of personal responsibility, a determination to replace hatred with love. This was sensible, yet incomplete. Writings that carried the message were frequent in the first decade or two after 1945 but then faded away. For the moral message ran into difficulties that were only gradually understood.

THE INTERIOR HOLOCAUST

The problem of nuclear weapons was that some humans sometimes wished to cause harm; this was hardly a new problem, but it was one that in the twentieth century found far broader scope than before. The old answers needed modification to deal with the new scale of danger. For example, one major artistic production that preached a solution, written before the Second World War, was extensively revised in response to the news from Hiroshima: Bertolt Brecht's *Galileo.* In the prewar version the playwright saw Galileo as a flawed hero, guilty only because he failed to persist as a champion of truth. In the next version the scientist became still more of an intellectual traitor because, isolating himself from social concerns, he permitted his discoveries to be abused by evil authorities.

It became common to fasten the problem upon scientists. Another widely seen play, Heinar Kipphardt's *In the Matter of J. Robert Oppenheimer,* directly convicted a physicist of betraying society by heedlessly turning over his knowledge to authorities. It was the old warning against Faust's sin of prideful power divorced from moral responsibility. Other serious fictional works portrayed atomic scientists as still more warped, for example unable to approach women normally, as in the mad scientist movies. But putting blame on scientists, as if the forces that drove them were alien to normal people, evaded the main problem. In fact the majority of citizens were more eager than most scientists to see their nation possess great weapons.[1]

Other writers did address human responsibility in general. A striking example was Philip Wylie's best-known story, "The Answer," published in 1955. Winged angels were discovered lying on the ground near American and Soviet bomb tests, slain by the blasts, still bearing tablets inscribed with a message. National authorities locked away the tablets in deep secrecy along with the message—"Love One Another." In a similar vein was the 1955 animated film where animals watched humanity exterminate itself, after which an elderly mouse said that people should have simply obeyed the biblical command, "Love Thy Neighbor." The advice was no doubt correct, but repeating it did not seem likely to solve matters.

Unheeded moral speeches were likewise central to the only explicitly bomb-related work of high literary quality to achieve a great popular success: *Lord of the Flies.* William Golding directly connected his 1954 novel to nuclear war, for it was while fleeing atomic bombardment that his schoolboys were marooned on an island, and Golding put in reminders of adult warfare at key points as the boys descended into murderous savagery. He concluded that although some of us might preach the ethical restraints of civilization, primal evil was always liable to burst these bonds.[2]

Such works of the early nuclear age did not quite despair. For all their bleakness they upheld morality as a conceivable solution, and they did not find human evil wholly beyond remedy. Galileo or Oppenheimer (geniuses unlike ourselves) might be tempted to subtle error; some of Golding's schoolboys (not his protagonists) might torture one another; but was it really plausible that normal adults would deliberately incinerate whole cities of innocents?

There was a word to answer that question—"Hiroshima." Yet that event could be gradually thrust aside as unique and outdated, irrelevant to the current situation. Perhaps within the closed circle of nuclear imagery this balancing of half-hearted moral preaching against half-hearted despair could have continued indefinitely. But from outside the circle people heard, like a distant shriek, a word still harder to dismiss—"Auschwitz."

Images of nuclear energy, like all modern thought, felt a tidal pull from the near extermination of the Jews of Europe. As the decades passed it became common for intellectuals to name Hiroshima and Auschwitz in the same sentence. Even the term most often used for the Nazi genocide, "the Holocaust," was also widely used to describe a future nuclear war, as if the two things were akin. That association could add something essential to the concept of nuclear war.

It was not only that the slaughter at Auschwitz was ten times greater than at Hiroshima, worse even than at Leningrad, more slaughter than in any other single location in history. The real lesson of the Nazi camps was that intelligent and educated adults, looking their innocent victims in the face, could organize torments that until then had been literally inconceivable. Jung rightly called this outbreak of collective sadism an "unmitigated psychic disaster" with far worse repercussions than the destruction of Hiroshima. And the Nazi atrocity was only one example of the twentieth century's horrors, from the 1915 slaughter of Armenians in Turkey to Stalin's concentration camps to the Cambodian "holocaust" of the late 1970s. As people gradually understood such events they recognized that moral preaching was sometimes as meaningless as a glass of water in a firestorm.[3]

Understanding these events required a long struggle. Of all the great atrocities only the one perpetrated by the Nazis inspired a major body of literature, and a main point in that literature was that the evil universe of Auschwitz could scarcely be comprehended even by those who had suffered within it, still less by outsiders. Most people had tried from the start to ignore the Nazi death camps, and it was not until the late 1970s (around the time that open fear of nuclear war revived) that film and television in the United States and Europe made a serious attempt to deal with the camps. Until then by far the most widely seen work on Nazi genocide was the book, play, and movie *Anne Frank: The Diary of a Young Girl,* a tale that stopped short at the gates of Auschwitz. In a biting critique, the psychologist and death camp survivor Bruno Bettelheim noted that this story turned the murder of a family into a sentimental tale about basically kind people. The Franks had not tried to escape or fight back but hid themselves in a garret, eventually dying helplessly because they could not quite believe in the monstrous evil that threatened them. They had clung to a conviction that, as the movie sweetly suggested in Anne's last words to the audience, everyone was "really good at heart."[4]

Such sentimental blindness also persisted in writings by survivors of Hiroshima and Nagasaki. Well known in Japan and to some extent elsewhere, the survivors' testimony shared with the concentration camp memoirs an unutterable vision of desolation. But the Japanese did not achieve Bettelheim's bitter insight, for they had not looked into their torturer's eyes and seen a human being like themselves. Japanese writers either self-righteously blamed everything on wicked Americans, or else expressed quiet sorrow at the tragic workings of fate. Few asked about universal human impulses. The Japanese counterpart to *Anne Frank* in popularity was *Thousand Cranes,* another true story of a young girl who had died, in this case of leukemia presumed to be caused by the Hiroshima bomb. During her illness she and her young friends set out to fold a thousand paper cranes, symbols of long life. The superstitious attempt at survival, and the storybook appeal to the idealism of children, were as pathetically inadequate as Anne Frank's hiding place.

Such works implied not only that all problems could be met with simple good will, but that this good will was at hand—at least among one's own people. The Nazis would have agreed. They had justified their death camps as an unfortunate necessity in a grand struggle for renewal; once the "Final Solution" eliminated Jewish evil, everyone else would enter a glorious new life. However, after 1945 people looked on Nazis as cardboard villains rather than as the ordinary humans that most of them in fact were. Hardly any work on nuclear warfare paid attention to Auschwitz. Few people considered how, in a wrongly organized society, a person's normal will to dominate and harm could be entangled even with a crusade for rebirth. Thinking about that came too close to thinking about one's own situation and one's own desires.[5]

A few psychologists who studied nuclear fear warned that the world public, like the Frank family, was shutting its eyes to the human forces that could destroy it. People who tried to understand our situation hesitated when they found that the investigation led them on a frightening journey into themselves. Failure to control the murderous hostility of others was terrifying enough, but what of our own desires? This was of course the bond joining the mad scientist with his monster, a theme that hundreds of works associated with "control" of nuclear energy. Yet most works preferred to project those feelings outward, assigning them to atomic scientists, Cold War generals, or Russians. Very few works associated the theme with universal human circumstances.

A suitable symbol stood right at hand, for explosions were a traditional symbol of inner destructiveness. In the 1930s Melanie Klein noticed that small children sometimes took literally the idea of "bursting with rage." Other psychologists described such cases as "Mary," a schizophrenic child who feared that she might explode and annihilate herself by losing self-control; at the same time Mary feared "being bombed" as punishment for her evil impulses, such as murderous feelings toward her neglectful mother. Another example was a paranoid woman, emotionally neglected by her parents, whose usual listless stupor sometimes gave way to violent outbursts, and who believed she "had an atom bomb inside her." Such damaged people indicated how suppressed rage could create imaginary bombs or real ones.[6]

The explosion symbol could fit especially well with nuclear energy, a truly great force locked within all things, liable to be released in wild abandon. In 1946 the cliché expert Mr. Arbuthnot hinted at the connection. "The atomic energy in your thumbnail," he told his interviewer, "could, if unleashed, destroy a city." Upon which the interviewer exclaimed, "For God's sake, stop, Mr. Arbuthnot! You make me feel like a menace to world security in dire need of control by international authority." The only thing to add would be what Karel Čapek's fictional atomic explosives expert learned back in 1924: "If it had not been in you, it would not have been in your invention."[7]

The explosion of inner forces was a symbol uniquely appropriate to nuclear energy. It is the only deep symbol I know of that was *not* developed in either popular or artistic works about bombs, prior to the new art of the 1980s. This failure suggests how difficult it was for people to connect nuclear war with the sort of internal forces, from which none of us is exempt, that built Auschwitz.

REBIRTH FROM DESPAIR

Some works of art relating to nuclear war did address the question of human evil, the universe where Auschwitz could exist, without losing hope. Although few, they were among the most respected and powerful works of the age. One reason was that they did not offer simplistic commandments to "Love One Another" but spoke of redemption through individual love, achieved not by inborn good will but through painful effort. A second reason for the power of most of these works was that they showed redemption in terms of transmutation imagery.

The first major after-the-bombs novel to address human bestiality was Aldous Huxley's *Ape and Essence* of 1948, which showed love between the main characters hampered not only by repressive tribal authorities but by the savagery within each individual soul. The same theme appeared still more clearly in 1980 in another fine novel, Russell Hoban's *Riddley Walker.* Young Riddley too had to survive as a victim in a brutalized postwar world, and at the same time he had to work out his angry and loving relationship to dangerous father figures—which could easily remind readers of their relationship with nuclear authorities. A closely related subsequent after-the-bombs novel, Denis Johnson's *Fiskadoro,* also took up a boy's relations with father figures alongside broader problems of the loss of reason in the individual and in a degenerated society.

A few sentences can say little about such themes, which entire novels were barely able to unfold, but I can address the symbolism. The end of Huxley's story was deliberately symbolic: his lovers escaped the repressive tribe to journey toward a hopeful new community, while the author rose to lyrical praise of a sacred marriage of flesh and spirit and the birth of a higher soul. Riddley Walker was equally preoccupied with rebirth of a spirit torn asunder by the conflict between scientific "cleverness" and intuition. In his pilgrimage Riddley encountered such symbols as a mutant telepathic community, a mandala laid out across the ruins of England, and even an overweening technologist's personal explosion. The boy in *Fiskadoro* got caught up in still more explicit symbols, a primitive rite of passage and an apocalyptic religious movement. Such novels showed that many of the transmutation themes associated with nuclear energy could be woven together with serious artistic insight.[8]

Although such works seemed at first to be about world catastrophe, they really centered on individual rebirth through the hard-won ability to love. That was particularly clear in Ingmar Bergman's great film *The Seventh Seal.* In this tale of a fourteenth-century knight amid the desolation of the Black Plague, Bergman addressed lifelong concerns about innate evil, despair, and death. The main actor said that the film, made in 1956, also reflected thoughts of the new hydrogen bombs, and when Bergman himself was asked whether *The Seventh Seal* was supposed to forebode nuclear catastrophe, he replied, "That's why I made it." However, the film's concluding vision of life beyond death, with the knight sacrificing himself so that one family could survive and begin anew, addressed the apocalypse as only one aspect of the timeless theme of personal redemption.[9]

The move from exterior social catastrophe to interior spiritual problem was structural in one of the best-known art works with a nuclear theme, Alain Resnais's *Hiroshima mon Amour.* The movie opened with grisly reenactments of the city's desolation, a shock to audiences in 1959. Scene by scene the film moved away from that until the conclusion seemed like an ordinary story of a sexual liaison amid the bright lights of the rebuilt city. The plot centered on a woman who, during the war in France, had been locked in a dark cellar and descended into madness. Eventually she emerged from that symbolic grave, but it was only when she saw how her lover from Hiroshima had survived that she could put the past behind her and truly live again.[10]

A psychologist reported patients reacting to the movie with dreams in which there would be an atomic explosion, probably symbolizing "complete dissociation, equivalent to madness," unless the lovers joined in "the union of inner opposites." But in transmutation imagery the way to such a saving union led through chaos, the anguish of temporary dissolution. That passage was the subject of the film, which strongly hinted that destruction was essential for spiritual progress. Three times the woman said to her lover, who forced her to face her memories, "You destroy me. You are good for me."[11]

This transmutational theme of hope arising in the heart of destruction found its most complete and paradoxical expression in *Dr. Strangelove.* One aspect was the movie's perverse sexuality; all its warriors blatantly got more manly pleasure from weapons than from women. This was not transmutation but its exact opposite, a rejection of sacred marriage, a failure to emerge from the core of aggressive destruction. Such inversion was perfectly expressed by Dr. Strangelove himself when he slid forth in his wheelchair to bring the film a happy ending. He told the men in the War Room that they could survive the doomsday machine by burrowing underground, with ten nubile females for every male. But it was not really the thought of a reborn world nor even of sexual domination that attracted this evil scientist; the phrase that rolled most deliciously off his tongue was that in the deep shafts, "animals could be bred and *slaughtered*." Alone among characters in nuclear movies, Strangelove was portrayed as a sometime Nazi. In his person Auschwitz came forward to explain what underlay all nuclear weaponry—the "strange love" for doing harm. As the world exploded he sprang from his wheelchair, reborn.

Strangelove's unholy rebirth would have made a fit ending, but Kubrick felt a need for more. One after another white mushroom clouds bloomed on the screen in ethereal beauty, accompanied by a woman crooning a Second World War song, "I know we'll meet again, some sunny day." Combining bedroom promises with hints of resurrection, the sequence gave a seductive beauty to apocalypse. Hardened critics found themselves unaccountably moved to tears at this ending, and audiences were perplexed and disturbed. When the film concluded not with charred corpses but with talk of unrestrained sex and slaughter in the womb of the earth, then moved off into mushroom heaven, it was showing nuclear war as desirable. No doubt Kubrick meant this to be ironic and instructive. But the imagery of transmutation had a power beyond irony, and it took over the conclusion.

All these excellent novels and films did help people to explore the motives that have supported nuclear weapons, and most of them pointed out the path to personal redemption: love working a laborious way through our evil and despair. However, these works went astray when they drew artistic strength from the analogy between personal and social transmutation. That analogy might help audiences to consider their spiritual progress as individuals. But comparing the personal passage through destruction with an entire society's travails was a false analogy; therefore it could never show citizens how to solve their collective peril.

A few other nuclear stories, poems, and books, in some ways the best of all, avoided this trap by concentrating on individual psychology. They explored attitudes toward nuclear weapons and death, using the familiar war anxieties of our society as a setting, and showed how love itself could contribute to insane violence unless it was unselfish love. These works are too subtle to analyze in a few paragraphs; I will only note some titles here: Akira Kurosawa's 1955 film *I Live in Fear,* Timothy O'Brien's closely related novel *The Nuclear Age,* and Robert Butler's *Countrymen of Bones.* Except for the film, which was scarcely noticed or understood in its time, these were works of the 1980s.

They did not point toward any solution. For all their valuable analysis of personal motivation and the modern situation, they might as well have drawn the pessimistic conclusion that Vonnegut reached in 1980s novels involving nuclear death and the end of humanity. Humans were destined to think their way into disaster, he said, thanks to "our overelaborate nervous circuitry." To go beyond that and find hope without invoking false transmutational analogies, a writer would somehow have to restore what was left out, something beyond individual psychology and the general mood of the times—an exploration of the influence of particular social structures, and the possibilities for changing them.[12]

TOWARD THE FOUR-GATED CITY

Art on nuclear themes rarely gave a social analysis as thorough as its psychological analysis. A few fine early novels, notably *Level 7* with its military hierarchy and *Lord of the Flies* with its tribal savagery, did explore the social roots of organized violence. Moreoever, these novels implied a solution: a society structured to respect individuals, with all their rights and feelings. But these and similar works were allegories that had no plain connection to our actual social organization. Not until the late 1960s did some people look in detail into the relations between nuclear war and the structures of modern civilization, and their attention soon shifted to debates over the environment and nuclear reactors. Would the reactor debate and the other vigorous questioning have something new to offer?

Reactors were seldom addressed in the arts, but the 1970s did bring a few poems by significant postmodernist authors. Like most poets, they believed that people should use more feelings and less abstract logic—an antinuclear stance. For example, Denise Levertov became sensitive to nuclear threats when she joined demonstrations against fallout in the late 1950s. She moved on to support the North Vietnamese against the United States, wrote political poems about unfeeling American military killers, and in 1970 said that the peace movement must become a "revolutionary movement" against the mechanical "profit system." By the late 1970s she was writing poems about her fear of the end of the world in nuclear war and meanwhile opposing reactors.

Responding to a controversy over a reactor near her home in Maine, Levertov wrote "The Split Mind," a poem about a state governor whose way of thinking made him a sinister authority. With his granddaughter the governor was warm and caring, but in his official role he contemptuously tossed aside doubts about radioactive wastes, which Levertov suggested would one day kill the child. Where the governor saw the reactor as a "planned facility," the poet saw it as a deathly "simulacrum of his will to power." In like fashion other poets, along with a few painters, drew art out of the antinuclear movement's ideal of spontaneous intuition opposed to mechanical authority.[13]

Nuclear power advocates produced nothing like that; they poured their skills and their souls into another art form, reactors themselves. Wherever such a device was built as intended, clean-shaped and ingenious, powerful and efficient, it could be a statement as impressive as any poem.

In all works involving reactors, the cluster of symbolic polarities of the reactor controversy, nature versus culture and feelings versus logic and all the rest, became connected with transmutation, the cluster of themes that dominated works on nuclear war. The most obvious concern shared by these two clusters involved authority. That concern had the potential for putting problems of social structure, the central question in reactor debates, together with problems of human aggression, the central question in nuclear war. However, most writers on nuclear issues spoke only vaguely about how the patterns of civilization might be encouraging and organizing destructive forces. Many settled for depicting an authority as a mad scientist, which linked reactor and war themes but sidestepped real social questions.

These were addressed through another connection between the nature-culture polarities and transmutational imagery: the choice of what sort of transformed society to seek. The alternatives became clear during the reactor controversy as antinuclear thinkers rejected the vision of a rationally organized White City in favor of Arcadian individualism. That vision was in accord with their goals for personal renewal, the private aspect of transmutation; here they aimed for the victory of human feelings over mechanical logic.

The combination of ideas was laid forth in full by Doris Lessing, the finest author of any who wrote time and again about technological catastrophe. In her childhood during the 1920s, on a remote farm in Africa, she had already met imagery of atomic apocalypse; in one of the few early memories she reported, her father sat on his porch looking at the stars and speculated that we might blow up our planet, but other planets would survive. As she grew up Lessing was temporarily attracted to another kind of apocalypse, the primitive Marxist dream of bloody revolution leading to a glorious new society. When she moved to London she retained leftist sympathies and became a literary star of the Campaign for Nuclear Disarmament.[14]

In the late 1960s Lessing summed up her thinking in a highly regarded book, *The Four-Gated City.* Beginning as an ordinary study of relations between parents and children, the novel proceeded through scenes of nuclear disarmament marches and concluded with a vision of a future world irradiated and ruined. "I think it's a true prophecy," Lessing told an interviewer. She expected civilization to fall as a result of all the problems of modernity—its corrupted values, its hypertrophied intellect, and in particular its nuclear technology. That made her feel, she said, "as if the Bomb had gone off inside myself." It was "as if the structure of the mind is being battered from inside. Some terrible new thing is happening."[15]

Lessing's metaphor of an interior bomb might recall the schizophrenic child "Mary," and her writings as a whole might recall the thoughts of catastrophe that preoccupied other writers—Mary Shelley, Frederick Soddy, Jack London, Philip Wylie, J. B. Priestley, Bertrand Russell, Helen Caldicott, and others—all of whom, like "Mary," had a background where maternal care was abnormally lacking. It is no surprise, then, to find that the struggle against a cold, annihilating mother, and the corresponding feelings of anger and desolation, were central in Lessing's own semi-autobiographical novels. This cannot be dismissed as mere mental disturbance. Everyone has had thoughts of violence and abandonment, but only a few have felt the pain so urgently that they dug it out and pondered its larger meanings.

The Four-Gated City used an impressive metaphor to relate inner and outer catastrophe. On the walls of a room, that common symbol of the mind, a man obsessively posted up news items in a schizophrenic jumble, clippings about nuclear bombs, radioactive and chemical pollution, riots and oppression. Lessing was saying that our civilization is driving us mad, while our madness in turn is breaking apart our civilization. Unlike most writers who addressed nuclear issues, she did not just use disaster as a dramatic backdrop for social criticism or as a metaphor of inner trials, but described the illnesses of the individual and of society as two sides of a single problem.

In 1957 she wrote that she was "haunted by the image of an idiot hand, pressing down a great black lever" to start nuclear war. She understood that the hand could be anyone's. "Exhausted with the pressure of living," each of us might say, "'Oh for God's sake, press the button, turn down the switch, we've all had enough.'" Such evil was nothing exceptional, but a reflection of blind forces that modern civilization stimulated within each person.[16]

The solution for Lessing's characters was passage through the chaos of madness to rebirth, a process the author herself, disturbed by dreams of world destruction, had once undergone. Her goal was plainly revealed by a vision she had around 1963 when she took a hallucinogenic substance. "I was both giving birth and being given birth to," she recalled. "Who was the mother, who was the baby? I was both but neither." It does not trivialize this vision to note that the schizophrenic child "Mary" also one day acted out birth as both mother and baby, and from that point began to cure herself. This was answering the threat of annihilation with one's own inner vitality. Renewal is never without perils: the idea came dangerously close to Frankenstein's fantasy of generating life like a god without aid from a partner. But the vision could also symbolize transmutation in its highest sense.[17]

Lessing, a dedicated student of esoteric religion and alchemy, consciously employed the concepts and symbols of transmutation. For example, the title of *The Four-Gated City* referred to an image long associated with the perfected soul, the utopian community laid out in the form of a mandala. Lessing added nuclear transmutation symbolism, saying that the chaotic stress of radioactivity would bring mutations, perhaps monsters but perhaps wondrous superhumans. At the novel's end, when the world was all but annihilated by pollution from innumerable accidents involving industry and weapons, a gentle race of telepathic children did appear.[18]

In later works Lessing developed more fully her idea: our own present times are already the transmutational night. We are not to await a final collapse and a new Adam and Eve, for we are right now under disintegrating pressures that can drive us to create prototypes of a better society. With the old civilization falling she saw new towns arise, not wounding the earth like our cities but seeming to grow from it, symmetrical and graceful as flowers. The new people, knowing to the full both nature and one another, would be radiant and powerful beings in intuitive communion—and for some of us such a life can begin today. Lessing was offering something valuable and rare, a vision of a welcome future that we can work toward.

It was still an incomplete vision. A convinced mystic resolutely ignorant of elementary science, Lessing foresaw the millennium only in terms of spontaneous communities in a wilderness, sustained by magical love that brushed aside logic and technology. Her hoped-for social transmutation would be a shift from culture to nature in every respect. From the individual standpoint too, Lessing saw modern civilization only as a threat, and her protagonists typically found personal rebirth by abandoning cities and reason itself. Many other writers and artists likewise held that the goal of personal transmutation was an escape into intuitive oneness with nature.[19]

An opposing vision aimed to achieve individual and social rebirth not by avoiding civilization but by embracing it. Technology inspired no great author, but in numerous minor writings and speeches one or another engineer or administrator would voice the hope that as civilization advanced, each member of society would reach a better spiritual life. Liberated from abject superstition by scientific knowledge, brought to a deeper sympathy with foreign peoples by electronic communications, saved from the degradation and violence that so often come with material poverty, each of us would become a better person.

The idea that the betterment of all society would come through technology remained prominent in numerous writings and advertisements. Particularly important was Walt Disney's Epcot Center, a sort of permanent world's fair White City where corporations treated millions of visitors to an image of progress, rational and irresistible, toward an orderly future run by corporate experts. There were also a few books showing technology as an integral part of a future where humanistic and ecological goals were paramount. But the most ubiquitous expression of social transformation was one-sided: the vision built in the sweeping highways, the amazing communications satellites, the urban spaces where shimmering glass towers rose above plazas lined with neat rows of trees.[20]

All these views of transmutation were valuable but incomplete. To look for the victory of spontaneity over authority, or of civilization over wilderness, or any other victory, is contrary to the teaching that transmutation is won through a marriage of opposites. It is not news that as individuals we need to reconcile reason with feelings. It is not news that individual efforts on this problem, and efforts to ameliorate the entire problem of human evil, can only succeed in a society that sustains such efforts at reconciliation—a society that avoids both the dehumanized obedience of the soldiers in *Level 7* and the brutal license of the boys in *Lord of the Flies.* All sides have agreed on this idea of balance. Yet it has seldom been fully achieved.

In our times, too many artists have weighted the balance heavily toward mystical communion with nature while rejecting rational organization; too many technologists, in their own works of metal, have weighted the balance heavily in the opposite direction. We do not yet possess an entirely convincing image of a society that will merge Arcadia with the White City, a society where the citizen will sing with both poets and engineers, and a society that will not encourage our cruel desire to prepare ever greater weapons. Yet work toward such a condition has been pursued by many, and not without partial success. We can still find our way.

Perhaps the best response has come from people I have not mentioned in this book, people who did not bother with nuclear imagery nor even transmutation imagery. They knew the lessons of fallout and missiles: that we must see everything as linked together, that the destiny of the citizens of Moscow is the destiny of the citizens of New York and of generations to come and even the fish in the sea. But they understood that to reach a union we do not need to tear open vast, angry secrets (except perhaps our own personal ones), that to live we do not need to destroy. Such people, whether artists or scientists or ordinary citizens, setting aside their fears of nuclear energy and their fantasies of magic transformation, work directly to help us understand the world, to cherish it, and to improve it.

Conclusion

Modern thinking about nuclear energy employs imagery that can be traced back to a time long before the discovery of radioactivity. That fact is disturbing, for it shows that such thinking has less to do with current physical reality than with old, autonomous features of our society, our culture, and our psychology. The way nuclear thinking emerged from these features has been the subject of this book, and to conclude I will now bring the main points together.

Most of the beliefs and symbols that gathered around nuclear energy were already associated with one another centuries earlier, in a highly structured cluster centered on the tremendous concept I have called transmutation—the passage through destruction to rebirth. By the nineteenth century the themes had become separated from one another, attaching themselves to the new wonders of electricity or to occultism or fantasy tales, surviving only as isolated cultural fragments. But in the early decades of the twentieth century the cluster drew together again around the discovery of nuclear energy. By the 1930s most people vaguely associated radioactivity with uncanny rays that brought hideous death or miraculous new life; with mad scientists and their ambiguous monsters; with cosmic secrets of death and life; with a future Golden Age, perhaps reached only through an apocalypse; and with weapons great enough to destroy the world, except perhaps for a few survivors. In sum, nuclear energy had become a symbolic representation for the magical transmutation of society and the individual.

The news from Hiroshima did little at first to change this bipolar structure of hopeful and fearful images. It only brought the whole structure into prominence, for what had seemed idle speculation turned into vivid expectations at the center of public attention. As the cliché expert Mr. Arbuthnot put it, the "fantastic prophecies" of Jules Verne and H. G. Wells seemed on their way to becoming "stern reality." Many complained that science was advancing so fast that human thinking could not adjust fast enough—but in fact there had been warnings about atomic bombs for decades, and another decade would pass before the threat became acute. The problem was that thinking remained stuck in old patterns, and would continue there except when overwhelming new facts forced it into new paths.[1]

One response to the widespread dismay and confusion was the building up of a national "security" apparatus, both to discover spies and to make more and greater weapons. Another response was the Atoms for Peace crusade, an attempt to counter dread by invoking hope from the same source. These responses brought some calm, but not for long. Thoughts about the future became thoughts about the present during the 1950s as hydrogen bombs and intercontinental missiles arrived. Meanwhile, fallout from bomb tests convinced people that bombs and radioactivity equaled blasphemous contamination.

By the early 1960s the symmetry was broken: for most people nuclear hope could no longer balance nuclear fear. An entire technology and indeed an entire part of physical reality, nuclear energy itself—something we all carry within the atoms of our bodies—was now regarded with irremediable distrust. Along with this attitude came new stereotypes of military officers and eventually other authorities as dangerous men not unlike mad scientists. The dreaded energy lurked not only within matter but within the human soul.

In the mid-1960s the public, exhausted by war scares and longing to deny the threat hanging over them, turned their attention away from bombs; a great silence descended. Distrust of nuclear authorities remained, however, breaking forth during the 1970s in the movement against reactors. Civilian nuclear power had genuine problems, but these would not have inspired such furious opposition had there not been increasing misgivings about all modern civilization, including its structure of authority and even its commitment to technology and rationality. This was a gravely significant trend, and it was furthered by the dread of bombs. That dread persisted even when not spoken aloud, contributing to a climate in which anxiety and despair for the human future were more visible than they had been for centuries.

Feelings about the bombs broke into the open once more in the early 1980s, then subsided back toward silence. At that point nuclear imagery was still usually expressed in much the same terms that had been common for generations. The waves of attention had come and gone without working any fundamental change in public thinking (see Figure 2, p. 387). Only in a tentative way, most visibly among intellectuals and artists, was a deeper understanding gradually taking hold.

In building and maintaining the public image of nuclear energy, the sequence of historical events pulled in tandem with contemporaneous social pressures. A number of groups actively promoted extreme imagery. From the turn of the century into the 1950s, thinking on nuclear energy was shaped chiefly by people who stressed that discoveries could bring a Golden Age or amazing weapons; these people meant to encourage public support for scientific research. Almost every well-known "fact" about the potential of nuclear energy was first offered to the public by one scientist or another. They were reinforced by admirers of science who drew their status and pay more indirectly from scientific advance, such as science journalists and science fiction writers, men who not only used nuclear images to impress their readers but were themselves impressed by the imagery's visceral power.

The teachings of these informed groups were reinforced from 1945 on by openly political groups. First came liberal atomic scientists, then pacifists and world federalists, civil defense officials, Communist propagandists, Air Force officers, and military industrialists, all with their own reasons and resources for focusing public fear upon nuclear weapons. Meanwhile, individuals ranging from grass-roots populists to national politicians, preoccupied with "control" and "security," particularly emphasized the idea of forbidden cosmic secrets. Beginning in the mid-1950s another miscellaneous class of people, remote from industrial hierarchies but skilled in communications, promoted additional fear of bombs, of radioactive contamination, and eventually of reactors. Their arguments drew upon an ideology of "natural" intuition as opposed to mechanical logic, an ideology that supported their goal of weakening the social authority of hierarchies.

All these different groups found symbols of transmutation useful, for the structure of imagery embraced opposites: mad scientists and victims, world doom and a Golden Age, death and life. There was imagery to serve Senator McMahon when he demanded that countless hydrogen bombs be built, and also Bertrand Russell when he demanded that every last bomb be destroyed; there was imagery to help Glenn Seaborg promise that nuclear towns would bring a utopia of clean technology, and to help Ernest Sternglass exclaim that reactors would befoul the entire planet. All parties joined to build up the theme at the center of the paradoxical duality—the myth of transmutational power from beyond the mortal sphere.

Nuclear images were used so widely because they could arouse emotions. In propagating the imagery, then, historical and social forces were reinforced by psychological pressures. I was especially struck by the role of personality when I looked into the background of people who were important in spreading ideas about world doom: I found that of those men and women whom I investigated, almost every one had a history of anguish relating to a mother who abandoned the child either emotionally or through death. (On the other hand, although the evidence is thinner here, it seems that unabashed prophets of a Golden Age like Seaborg and William Laurence had more nurturing mothers.) Similarly, when I looked at the individuals who were most important in creating mad-scientist tales, I usually found a history of uncommonly difficult relations with the father. I have emphasized these connections only because they clearly demonstrate that group beliefs about nuclear energy are burdened with projections of feelings that stem from personal problems.

The archaic symbols of nuclear legends resonate with universal anxieties and hopes. Peeking at forbidden secrets; punishment through abandonment or other victimization by an authority; a corresponding all-destructive rage; homicidal and suicidal urges and the accompanying guilt; struggle through chaos; heroic triumph over peril; miraculous life and regeneration of self; world rebirth through a marriage of survivors; entry to a joyful community: these are the stages of transmutation imagery in ancient tradition, repeated in a thousand modern works on nuclear energy. The process could go astray amid aggressive urges toward self-sufficient life power, or become stuck halfway in the stage of despair, or it could go ahead to loving union with others and the reconciliation of opposites. Nuclear energy, with its wealth of ambiguous associations, served well as a receptacle for projections of these hidden thoughts. Impossible hopes could seem almost plausible when attached to possession of bombs and reactors. And evils we denied within ourselves could be projected as a property of the men who controlled nuclear devices or of the devices themselves.

What was the effect of the images on history? In one sense imagery has no effect of its own. People do not take an action just because some symbol makes them feel like doing it; they act in response to a variety of pressures, as much imposed by their situation as generated within themselves. Yet in another sense nothing is so effective as imagery—when it harnesses a particular set of traditions, social tensions, and personal impulses, joining them into an alliance to work upon minds. In this sense of a focus of specific forces, it was imagery that in the matter of reactors concentrated the attention of everyone from schoolchildren to experts on the most global and awesome of possible outcomes. As a result, less attention was left for more prosaic but more serious problems of reactors and other technologies. Likewise, it was imagery that concentrated panicky attention on stockpiling thousands of apocalyptic weapons or on attempting to ban them altogether. Thus less attention was left for the ordinary weapons that were more likely to be used, and for nonmilitary approaches to the problems between nations and within them.

The effects of imagery were strongest where they joined with the laws of physics, for these laws too had a role in events. Devices using radioactivity and nuclear energy had genuine concentrated power and genuine effects on health, taking lives by tens of thousands and saving lives by the million. This was no accident. The physical laws were discovered and the weapons, reactors, and medical tools developed in a deliberate attempt, partly successful, to take control over the forces of life and death. Myth had acquired a new power because hardware had been purposefully brought into realms once inhabited only by myth.[2]

If it was not in us, it would not be in our creations. People have not only projected their feelings onto bombs and reactors but have built these devices purposely to be what they are, turning visionary fires into real ones. Our secret thoughts have come into the open at last, taking form in metal so that we can deny them no longer.

This development has had some solid benefits. Nuclear weapons have brought home to everyone the horrors of war, and they have even begun to force into the open the darker attractions of weaponry. Meanwhile, in an era when existing technology already suffices to make all the world prosperous, reactor debates properly focus attention on the question of just which technology should be used, and whose, and how. Nuclear images themselves are so bizarre that more and more people, facing the propaganda that exploits the images, have begun to ask whether the messages are realistic and relevant, or only an appeal to old tales and emotions, used by one or another group for their own purposes.

When imagery is used to manipulate others the process by its very nature is undemocratic, and thus is liable to promote undesirable ends. But imagery has another and more creative side, provided it is constantly tested against reality and human feelings. The beginning of a solution is to imagine ourselves, and our society, reformed in such a fashion that we apply our energy only in worthy directions. This work is well under way. The individual and collective mechanisms through which we approach technology are increasingly addressed in commonly understood symbols, from the nuclear Apocalypse to the futuristic flying-saucer civilization, as well as in more subtle works of art and literature. Most of this creative work has looked at the individual personality, explaining with growing power and honesty our tendencies to do evil and the painful path to redemption. A body of work is also gradually accumulating that speaks to the related social questions. Images of better ways, of a better world, persist and grow. This imagery too will make its mark on history.

A Personal Note

Sources and Methodology

Notes

Index

A Personal Note

Every work of history speaks implicitly to current problems, and I feel a responsibility to make plain my personal opinions about what we should do. The following conclusions are not "lessons" from the history of imagery, and indeed have little to do with that; they are largely based on facts outside the scope of this book.

In all that I have read I am most impressed by ideas that David Lilienthal offered in the 1960s and later. As head of the AEC and afterward he struggled with nuclear energy fantasies, finally realizing that he and others had abandoned the normal way of dealing with problems through piecemeal compromises: "Our obsession with the Atom drove us to seek a Grand Solution." Some people dreamed of doing away with every nuclear bomb under a system of world control, while others wanted to dismay their enemies by building more and more bombs without limit. As for reactors, some people said the solution was to tear them all down regardless of what the alternatives might be, while others said civilization could survive only through the use of fission. Lilienthal saw that these extraordinary ideas rested on apocalyptic myth, and he rejected them all. Real solutions would be more complex and modest.[1]

First, energy supply. Much more electricity will be needed before the entire world reaches minimal prosperity, and none of the ways to make this electricity is fully satisfactory. I think we should keep nuclear reactors at least as an option. As Lilienthal said, writing at the end of his life in moderate defense of reactors, we should especially continue the work now under way to develop new kinds of reactors, for prototypes already exist that are physically incapable of a massive meltdown no matter what mistakes the operators make. But even with current reactors, in terms of human and environmental health over both the short and long term, I would personally prefer to get my electricity from a nuclear reactor (provided it was built by a group with proven competence) rather than from a coal-fired plant. In any case the matter should not dominate public attention. Reactors are far from the worst health problem faced by the developed nations, let alone the less developed ones.

Second, war. Nuclear weapons too deserve less exclusive attention than they have received, for today a war between advanced nations could be intolerable even if they used only their non-nuclear weapons. Our main business is to understand and uproot the domestic and international causes of war itself. So this book has not been about our real problems at all; it has been about what distracts us from them. It is in non-military areas, as Lilienthal stressed, that we will build up true security, and it is also there that we may find, instead of hostile calculations, the pleasure of productive work.

Our work over the long term should begin with the fact, scarcely known or appreciated, that democratically elected governments never make war on one another.[2] We can be safe not just for a while, but forever, if the complete lack of military threats between the democracies can be gradually extended to the rest of the world. Thus the survival of civilization is tied up with the struggle against institutions whose authority rests on force rather than free consent. That struggle will not be won with our own aggressive force, and least of all with nuclear weapons. In fact the more we rely on such weapons the farther we ourselves move from democracy, as we grant unelected officials an authority to censor scientific knowledge and even to hold the literal keys to survival. The way to safety is to hold fast to an idea of the free and peaceful society we finally want, working toward it not with hostile demands but with a calm appeal.

To survive in the meantime we must do something soon about our tendency to build ever more weapons. My advice is that several billion dollars be cut from the annual budget for the development of new weapons (a budget that in the United States has more than doubled in recent years). This proposal may seem too modest, but I prefer it to a Grand Solution. The history of the twentieth century shows that research and development do much to determine what eventually comes to pass; anyone who tries to shift this little lever will learn what tremendous forces rest upon it.[3]

Finally, what of nuclear fear itself? And nuclear hope, which can also go to harmful extremes? I wish we could have a moratorium on the archaic images that incite such emotions. Exclaiming that reactors contain astonishing power terrifies the citizen; exclaiming that bombs can lay waste whole nations tempts the citizen to possess more. If any group has been like children playing with matches, it has been those who thought they could evoke fantastic images to good purpose.

The best way to affect imagery is to alter reality. For example, shifting an important fraction of scientific and engineering talent from weapons of mass destruction toward, say, energy supply would bring hope immediately as a gesture, and more hope over decades as the changed research effort changed our actual situation. We can also work with imagery directly if we approach the task with full respect for truth. Let us tell each new generation in new ways the fact that someday every town in the world might look like central Hiroshima in August 1945. More important, let us spread the word that every town can look like the finest quarters of Paris or better, merging into a healthy countryside, with nature and civilization, order and freedom, feelings and logic, all working in cooperation. Nuclear energy and its imagery have made familiar the dramatic possibilities for gain and loss that our own human energy can release. In the long run we may find that they have greatly aided us toward the true transmutations we need.

Sources and Methodology

This book approaches its subject by trying to take into account every force that has mattered, from the known laws of physics to the largely unknown influences of psychology: total history. The sources are correspondingly diverse. Many of the writings on nuclear energy listed in the notes represent types; for each citation the scholar can easily find a dozen similar examples. Even the scholarly works I consulted on such matters as, for example, anthropology, anxiety, and apocalypse are only a few of many possible entries to the subjects. The following list offers starting points for those wishing to dig deeper; I emphasize works with good bibliographies.

For a survey of the overt civilian and military history (which I have scarcely touched) one might begin with Bertrand Goldschmidt, *The Nuclear Complex: A Worldwide Political History of Nuclear Energy* (La Grange Park, Ill.: American Nuclear Society, 1982), balanced by the equally partisan, more complete, but less uniformly accurate Peter Pringle and James J. Spigelman, *The Nuclear Barons* (New York: Holt, Rinehart & Winston, 1981), with extensive bibliographical notes. For reactors and their history see Walter Patterson, *Nuclear Power,* 2nd ed. (Harmondsworth, England: Penguin, 1983); Steven L. Del Sesto, *Science, Politics, and Controversy: Civilian Nuclear Power in the United States, 1946–1974* (Boulder, Colo.: Westview, 1979); Peter deLeon, *Development and Diffusion of the Nuclear Power Reactor: A Comparative Analysis* (Cambridge, Mass.: Ballinger, 1979). On weapons, Michael Mandelbaum, *The Nuclear Question: The United States and Nuclear Weapons, 1946–1976* (Cambridge: Cambridge University Press, 1979), is good for diplomacy, and Lawrence Freedman, *The Evolution of Nuclear Strategy* (New York: St. Martin's, 1981), for strategy.

Among the many other historical works, three are noteworthy for their insights into ways of thinking: Norman Moss, *Men Who Play God: The Story of the H-Bomb and How the World Came to Live with It* (New York: Harper & Row, 1968); Fred Kaplan, *The Wizards of Armageddon* (New York: Simon & Schuster, 1983); and Sheldon Novick, *The Electric War: The Fight over Nuclear Power* (San Francisco: Sierra Club Books, 1976).

A useful bibliography on weapons is Dietrich Schroeer and John Dowling, "Resource Letter PNAR-1: Physics and the Nuclear Arms Race," *American Journal of Physics* 50 (1982): 786–795. Karl Hufbauer's 1986 "Bibliography of Historical Studies of the Nuclear Age Published Since 1/1/1975," on weapons, is available from the Institute of Global Conflict and Cooperation, Q-060, University of California, La Jolla. An introduction to the literature on civilian industry is Samuel Glasstone and Walter H. Jordan, *Nuclear Power and Its Environmental Effects* (LaGrange Park, Ill.: American Nuclear Society, 1981); see also George T. Mazuzan, "Nuclear Energy: A Subject in Need of Historical Research," *Technology and Culture* 27 (1986): 123–126.

Brief introductions to current debates are Michio Kaku and Jennifer Trainer, eds., *Nuclear Power: Both Sides. The Best Arguments For and Against the Most Controversial Technology* (New York: W. W. Norton, 1982) and Nigel Calder, *Nuclear Nightmares: An Investigation into Possible Wars* (New York: Viking, 1979). A particularly good appraisal of scientists and weapons is Solly Zuckerman, *Nuclear Illusion and Reality* (New York: Viking, 1982). Current bibliographical information on reactors is available from the Atomic Industrial Forum, Bethesda, Md. (pro) and from the Information and Resources Service, Washington, D.C. (anti); current books on weapons and sometimes energy are noted in *Foreign Affairs* magazine. Specialized works issued periodically by the U.S. Department of Energy, the London-based International Institute for Strategic Studies, and many others are useful, but skimming *The Bulletin of the Atomic Scientists* and *Science* will keep a general reader in touch with most issues. To be in touch in not only intellectual but practical ways, write to the Council for a Livable World, 11 Beacon St., Boston, Mass.

For apocalyptic and transmutational thinking one starting point in the huge literature is Eric S. Rabkin, Martin H. Greenberg, and Joseph D. Olander, eds., *The End of the World* (Carbondale: Southern Illinois University Press, 1983). The anthropological literature is also valuable; see, for example, Mircea Eliade, *The Forge and the Crucible: The Origins and Structure of Alchemy,* trans. Stephen Corrin (New York: Harper & Row, 1971). A psychological introduction is Ernest Becker, *The Denial of Death* (New York: Free Press, 1973).

Nuclear imagery itself is only beginning to be explored. Robert Jay Lifton, *The Broken Connection: On Death and the Continuity of Life* (New York: Simon & Schuster, 1979), is especially important; a more readable summmary is Robert Jay Lifton and Richard Falk, *Indefensible Weapons: The Political and Psychological Case Against Nuclearism* (New York: Basic Books, 1982). Peter M. Sandman and JoAnn M. Valenti, "Scared Stiff—or Scared into Action," *Bulletin of the Atomic Scientists* 42, no. 1 (Jan. 1986): 12–16, specifically addresses the use of nuclear fear. Philip D. Segal, "Imaginative Literature and the Atomic Bomb: An Analysis of Representative Novels, Plays, and Films from 1945 to 1972" (Ph.D. diss., Yeshiva University, 1973), has a large bibliography of nuclear war books and films to about 1970. The pioneering insights of Gerald G. Ringer, "The Bomb as a Living Symbol: An Interpretation" (Ph.D. diss., Florida State University, 1966), are still worth reading. For further provocative suggestions see Ira Chernus, *Dr. Strangegod: On the Symbolic Meaning of Nuclear Weapons* (Columbia: University of South Carolina Press, 1986).

I. F. Clarke, *The Pattern of Expectation, 1644–2001* (New York: Basic Books, 1979), gives an introduction to futuristic writings, supplemented by W. Warren Wagar, *Terminal Visions: The Literature of Last Things* (Bloomington: Indiana University Press, 1982). For recent views and a bibliography see "Nuclear War and Science Fiction," special issue of *Science-Fiction Studies* 13, pt. 2 (July 1986).

On films, one introduction with bibliography is Douglas Menville and R. Reginald, *Things to Come: An Illustrated History of the Science Fiction Film* (New York: New York Times Books, 1977); see also Jack G. Shaheen, ed., *Nuclear War Films* (Carbondale: Southern Illinois University Press, 1978). The most useful film collection is at the Library of Congress in Washington, D.C., but many may be found on late-night television or videotape, and scripts for many old films are available from the Cultural Education Center, New York State Archives, Albany. For performing arts information the Lincoln Center branch of the New York Public Library and the Museum of Broadcasting in New York City are justly famed.

A history of public images might in principle rely only on published sources, but I have also been curious about who decided to propagate certain images, and why. The principal archives include the collections of the Niels Bohr Library at the American Institute of Physics in New York City; the National Archives in Washington, D.C., and its presidential libraries elsewhere; the Library of Congress; the United States Department of Energy in Germantown, Maryland, and its National Laboratories elsewhere; and the Nuclear Regulatory Commission in Washington, D.C. Among many other useful archives are those in the libraries of the University of California at Berkeley, the University of Chicago, Columbia University, the Massachusetts Institute of Technology, and Princeton University. Materials at the Public Record Office, London, and the Radium Institute, Paris, are useful into the early 1950s. Much of this subject matter is alive in memory, and as the notes indicate, I have also interviewed a variety of people.

The methodology and vocabulary for this book had to be developed as it proceeded. For useful hints see Kenneth E. Boulding, *The Image: Knowledge and Life in Society* (Ann Arbor: University of Michigan Press, 1956); Jacques Le Goff, "Les Mentalités: une historie ambiguë," in *Faire de l'histoire: nouveaux objets*, ed. J. Le Goff and Pierre Nora (Paris: Gallimard, 1974), 76–94; Hans H. Toch and Clay Smith, eds., *Social Perception* (Princeton, N.J.: Van Nostrand, 1968), pt. 3. See also David E. Stennard, *Shrinking History: On Freud and the Failure of Psychiatry* (New York: Oxford University Press, 1980); Thomas A. Kohut, "Psychohistory as History," *American Historical Review* 91 (1986): 336–354; and Chapter 13, n. 71.

I do not usually use the term "projections" in the psychoanalytic sense of a disavowed emotion imagined in an outside person, although that can perhaps be subsumed in the more general process. Most other definitions important in this book are, like this one, implicit in the Preface. M. Fishbein and I. Ajzer, *Belief, Attitude, Intention and Behavior: An Introduction to Theory and Research* (Reading, Penn.: Addison-Wesley, 1975), 11–15, relate classical conditioning to "belief" as an association between things; their "attitude" is a consequent association of affect which can lead to action. I use "feelings" to include more generally all associated emotions. I use "image" in two related senses: first, a simple mental representation such as a picture or description (see Ned Block, ed., *Imagery*, Cambridge, Mass.: MIT Press, 1981); second, a "cluster" of such things combined with their associated beliefs and feelings; one form is the "common ideas" of Philippe Ariès, *The Hour of Our Death*, trans. Helen Weaver (New York: Knopf, 1981), 332. My more structured "patterns" are like the "bundles" of Claude Lèvi-Strauss in *Myth: A Symposium*, ed. Thomas A. Sebeok (Bloomington: Indiana University Press, 1965), 87. The most widespread and structured cases I call a "symbol" in narrow cases and "symbolic representation" or "myth" in broader cases. All these things are "public" when shared by many people. For myth see also Joseph Campbell, *The Masks of God,* vol. 1, *Primitive Mythology* (New York: Viking, 1969), pt. 1.

One recent review of media effectiveness is Aimée Dorr, "Contexts for Experience with Emotion, with Special Attention to Television," in *The Socialization of Emotion,* ed. Michael Lewis and Carolyn Saarni (New York: Plenum, 1985), 55–85. More generally, see the Preface in Jacques Ellul, *Propaganda: The Formation of Men's Attitudes,* trans. Konrad Kellen and Jean Lerner (New York: Knopf, 1968; reprint, Vintage, 1973). For my views on forces other than imagery see Spencer Weart, *Scientists in Power* (Cambridge, Mass.: Harvard University Press, 1979), Afterword.

Notes

I do not cite the majority of the sources I used, but only basic references for quotes, statistics, and obscure facts, and some starting points into the literature. Often I do not list important works that may be found through the bibliographies of later works I do cite. Books and movies may have various editions or distributors, and I cite whichever version I read or saw.

ABBREVIATIONS

AIP	Niels Bohr Library, American Institute of Physics, New York City
BAS	*Bulletin of the Atomic Scientists*
DDE	Dwight D. Eisenhower Library, Abilene, Kan.
JCAE	Records of the U.S. Congress Joint Committee on Atomic Energy, RG 128, National Archives, Washington, D.C.
JFK	John F. Kennedy Library, Boston, Massachusetts
MB	Museum of Broadcasting, New York City
NYPL/TC	New York Public Library, Theater Collection, Lincoln Center, New York City
NYT	*New York Times*

1. RADIOACTIVE HOPES

1. Muriel Howorth, *Pioneer Research on the Atom: . . . The Life Story of Frederick Soddy* (London: New World, 1958), 83–84. See also Thaddeus J. Trenn, *The Self-Splitting Atom: The History of the Rutherford-Soddy Collaboration* (London: Taylor & Francis, 1977); Alfred Romer, ed., *The Discovery of Radioactivity and Transmutation* (New York: Dover, 1964).

2. Jost Weyer, "The Image of Alchemy in 19th and 20th Century Histories of Chemistry," *Ambix* 23 (1976): 65–79. H. Carrington Bolton, "The Revival of Alchemy," *Science* 6 (10 Dec. 1897): 853–863. See Soddy to Rutherford, 7 Aug. 1903, Rutherford Papers, Cambridge, England, microfilm copies at repositories of the Archives for History of Quantum Physics, including AIP.

3. "Inexhaustible": Frederick Soddy, "Some Recent Advances in Radioactivity," *Contemporary Review* 83 (May 1903): 708–720. Steamship: *The Morning Herald* (Perth), 25 July 1904, p. 6; I am grateful to John G. Jenkin for sharing this. See his "Frederick Soddy's 1904 Visit to Australia . . . ," *Historical Records of Australian Science* 6 (1985): 153–169. "Dragonfly": Soddy, "The Energy of Radium," *Harper's Monthly* 120 (Dec. 1909): 52–59. "Race which could transmute": Soddy, *The Interpretation of Radium* (London: Murray, 3rd ed. 1912), 251. See also his *Matter and Energy* (London: Williams & Norgate, 1912).

4. See I. F. Clarke, *The Pattern of Expectation 1644–2001* (New York: Basic Books, 1979); Robert Nisbet, *History of the Idea of Progress* (New York: Basic Books, 1980). Camille Matignon, "Souvenirs sur Marcellin Berthelot," *Revue de Paris* 34 (1927): 362–380; M. Berthelot, "In the Year 2000" (1894), trans. A. Ettinghausen, *Chemistry in Britain* 15 (1979): 250–251.

5. Reid Badger, *The Great American Fair: The World's Columbian Exposition and American Culture* (Chicago: Nelson Hall, 1979). See Henry Adams, *The Education of Henry Adams* (1906; reprint, New York: Modern Library, 1931), chap. 22.

6. William Graves Hoyt, *Lowell and Mars* (Tucson: University of Arizona Press, 1976).

7. W. Stanley Jevons, *The Coal Question* (London: Macmillan, 1865). See Sir William Thomson, "Available Energy of Nature," *Popular Science Monthly* 20 (1881): 87–95; George Wise, "Predictions of the Future of Technology: 1890–1940," Ph.D. diss., Boston University, 1976; I am grateful to Wise for giving me a copy.

8. "Rivers are clean": Samuel P. Langley, *The New Astronomy* (Boston: Houghton Mifflin, 1896), 115.

9. "World's demand": Soddy, "The Energy of Radium," 58; "Coming struggle": Soddy, "Transmutation, the Vital Problem of the Future," *Scientia* 11 (1912): 199, as quoted in Thaddeus J. Trenn, "The Central Role of Energy in Soddy's Holistic and Critical Approach to Nuclear Science, Economics, and Social Responsibility," *British Journal for the History of Science* 12 (1979): 261–276.

10. "Rivals the sun": H. C. Bolton, "New Source of Light," *Popular Science Monthly* 57 (1900): 318–322, quoted in Lawrence Badash, "Radium, Radioactivity, and the Popularity of Scientific Discovery," *Proceedings of the American Philosophical Society* 122 (1978): 145–154; see also Badash, *Radioactivity in America: Growth and Decay of a Science* (Baltimore: Johns Hopkins University Press, 1979). Many clippings on radium are found in the William Hammer Collection, National Museum of American History, Smithsonian Institution, Washington, D.C. G. Le Bon, *L'Evolution de la Matière* (Paris:

Flammarion, 1909), 57; see Mary Jo Nye, "Gustave Le Bon's Black Light," *Historical Studies in the Physical Sciences* 4 (1975): 163–195.

11. Museum: Salem, Mass. *News*, 29 Sept. 1903, reprinted from N.Y. *Sun*, Hammer Collection, box 44.

12. Ronald C. Tobey, *The American Ideology of National Science, 1919–1930* (Pittsburgh: University of Pittsburgh Press, 1971). J. G. Crowther, *Fifty Years with Science* (London: Barrie & Jenkins, 1970), 47–51. W. Kaempffert, "Science Presses On Toward New Goals," *NYT Magazine* (28 Jan. 1934), pp. 6–7. See Kaempffert, *Science Today and Tomorrow* (New York: Viking, 1945); "Kaempffert," *NYT*, 28 Nov. 1956, p. 35.

13. See, for example, "When Fuel Gives Out," *Literary Digest* 61 (31 May 1919): 100–103; John K. Barnes, "The Crisis in the World's Oil Supply," *World's Work* 37 (1921): 69–70. Wise, "Predictions," chap. 6; Aaron Wildavsky and Ellen Tenenbaum, *The Politics of Mistrust: Estimating American Oil and Gas Resources* (Beverly Hills, Calif.: Sage, 1981).

14. Kaempffert, "Science Presses On."

15. "Famous problem": John A. Eldridge, *The Physical Basis of Things* (New York: McGraw-Hill, 1934), 330, 333. Ernest Rutherford, *The Newer Alchemy* (Cambridge: Cambridge University Press, 1937). On transmutation in 1930s physics see Spencer Weart, "The Discovery of Fission and a Nuclear Physics Paradigm," in *Otto Hahn and the Rise of Nuclear Physics*, ed. William R. Shea (Dordrecht: Reidel, 1983), 91–133.

16. The literature on alchemy is enormous. A clear introduction is included in Betty Jo Teeter Dobbs, *The Foundations of Newton's Alchemy, or "The Hunting of the Greene Lyon"* (Cambridge: Cambridge University Press, 1975). Of special interest is Carl G. Jung, *Collected Works*, vols. 12–14, trans. R. F. C. Hull (Princeton: Princeton University Press, 1968–1972).

17. Mircea Eliade, *The Forge and the Crucible: The Origins and Structure of Alchemy*, trans. Stephen Corrin (New York: Harper & Row, 1962), 169. "Divine furnace": Evelyn Underhill, *Mysticism: A Study in the Nature and Development of Man's Spiritual Consciousness* (12th ed., 1930; reprint, New York: Dutton, 1961), 140–148 and chap. 9; see p. 221. See Jacques Le Goff, *The Birth of Purgatory*, trans. Arthur Goldhammer (Chicago: University of Chicago Press, 1984), 7–11, 44, which also has comments on the history of images.

18. Mircea Eliade, *The Myth of the Eternal Return, or, Cosmos and History*, trans. Willard R. Trask (Princeton: Princeton University Press, 1965). Yi-Fu Tuan, *Landscapes of Fear* (New York: Pantheon, 1979), 56–61. Stith Thompson, *Motif Index of Folk-Literature* (Bloomington: Indiana University Press, rev. ed., 1955–1958), A1030 in vol. 1, p. 89; A1006.1–2 in vol. 1, p. 183; and A1038. A convenient anthology is William Griffith, ed., *Endtime: The Doomsday Catalog* (New York: Collier, 1979). One entry point to the historical literature is Robert E. Lerner, "The Black Death and Western European Eschatological Mentalities," *American Historical Review* 86 (1981): 533–552. For golden ages see the journal *Alternative Futures* and Frank E. Manuel and Fritzie P. Manuel, *Utopian Thought in the Western World* (Cambridge, Mass.: Harvard University Press, 1979).

19. "Something more noble": quoted in Edward N. da Costa Andrade, *Sir Isaac Newton* (London: Collins, 1934), 132–133. The ongoing debate on the nature of the scientific revolution may be tracked in the journal *Isis*.

2. RADIOACTIVE FEARS

1. "Storehouse": Frederick Soddy, "Some Recent Advances in Radioactivity," *Contemporary Review* 83 (May 1903): 708–720.

2. Le Bon quoted in New York *World*, 30 Aug. 1903, and other clippings, William Hammer Collection, National Museum of American History, Smithsonian Institution, Washington, D.C., box 42. "It is conceivable": W. C. D. Whetham, "Matter and Electricity," *Quarterly Review* no. 397 (Jan. 1904): 126.

3. "Some fool": Whetham to Rutherford, 26 July 1903, Rutherford Papers, Cambridge, England, microfilm copy at AIP.

4. Jean-Baptiste Cousin de Grainville, *Le Dernier Homme* (1805; Paris: Ferra et Deterville, 1811; Geneva: Slatkine, 1976). See I. F. Clarke, *The Pattern of Expectation 1644–2001* (New York: Basic Books, 1979), 45–47.

5. Mary Shelley, *The Last Man* (1826; Lincoln: University of Nebraska Press, 1965).

6. Camille Flammarion, *La Fin du monde* (Paris: Flammarion, 1894). Movies: Douglas Menville and R. Reginald, *Things to Come: An Illustrated History of the Science Fiction Film* (New York: New York Times Books, 1977).

7. S. Newcomb, "The End of the World, A Story," *McClure's Magazine* 21 (May 1903): 13. For further references see Yuri Rubinsky and Ian Wiseman, *A History of the End of the World* (New York: Quill, 1982).

8. Frederic C. Jaher, *Doubters and Dissenters: Cataclysmic Thought in America, 1885–1918* (New York: Free Press of Glencoe, 1964).

9. W. Warren Wagar, *Terminal Visions: The Literature of Last Things* (Bloomington: Indiana University Press, 1982).

10. E. M. Butler, *The Myth of the Magus* (Cambridge: Cambridge University Press, 1948).

11. E. M. Butler, *The Fortunes of Faust* (Cambridge: Cambridge University Press, 1952); J. W. Smeed, *Faust in Literature* (London: Oxford University Press, 1975). Robert Darnton, *Mesmerism and the End of the Enlightenment in France* (Cambridge, Mass.: Harvard University Press, 1968); Geoffrey Sutton, "Electric Medicine and Mesmerism," *Isis* 72 (1981): 375–392; Fred Kaplan, "'The Mesmeric Mania:' The Early Victorians and Animal Magnetism," *Journal of the History of Ideas* 35 (1974): 691–702. Taylor Stoehr, *Hawthorne's Mad Scientists: Pseudoscience and Social Science in Nineteenth-Century Life and Letters* (Hamden, Conn.: Archon, 1978).

12. Marquis de Sade, *The 120 Days of Sodom* (1785; New York: Grove, 1966), 364, as quoted in Wagar, *Terminal Visions*, 87. Shelley, *Frankenstein*, chap. 20.

13. Butler, *Myth of the Magus*, 3–6.

14. J. Verne, *Five Weeks in a Balloon* (1862; Westport, Conn.: Associated

Booksellers, 1958), 93. Verne, *For the Flag* (1896; Westport, Conn.: Associated Booksellers, 1961).

15. Matthew P. Shiel, *The Purple Cloud* (London: Chatto & Windus, 1901). See also my chap. 12.

16. Chemist: W. Nernst, *Das Weltgebaüde im Lichte der neueren Forschung* (Berlin: Springer, 1921), 24. Textbook: H. A. Kramers and Helge Holst, *The Atom and the Bohr Theory of Its Structure,* trans. R. B. Lindsay and R. T. Lindsay (New York: Knopf, 1926), 103. See also J. B. S. Haldane, *Daedalus, or Science and the Future* (New York: E. P. Dutton, 1924), 3–4. "Whole universe": T. J. C. Martyn, "In Science Lies the Challenge to War," *NYT Magazine* (30 June 1929): 4–5, 20.

17. "Shambles": Joseph Conrad, *The Secret Agent* (Garden City, N.Y.: Doubleday, 1958), 303. Anatole France, *Penguin Island,* trans. A.W. Evans (New York: Dodd, Mead, 1909), book 8.

18. John B. Priestley, *The Doomsday Men: An Adventure* (London: Heinemann, 1938), 277.

19. Robert Nichols and Maurice Browne, *Wings over Europe; a dramatic extravaganza on a pressing theme* (1928; New York: S. French, 1935).

20. George Perry and Alan Aldridge, *The Penguin Book of Comics* (Harmondsworth, England: Penguin, 1967), 61. Michael L. Fleisher, *The Encyclopedia of Comic Book Heroes,* vol. 1, *Batman* (New York: Collier, 1976), s.v. "Dr. Radium." See Thomas D. Clareson, ed., *SF: The Other Side of Realism* (Bowling Green, Ohio: Bowling Green University Popular Press, 1971), 18–20.

21. As reported in New York *Press,* 8 Feb. 1903, and other clippings, Hammer Collection, box 42. See E. E. Fournier d'Albe, *The Life of Sir William Crookes* (London: Unwin, 1923).

22. Soddy to Rutherford, 19 Feb. 1903, Rutherford Papers.

23. Wyn Wachhorst, *Thomas Alva Edison: An American Myth* (Cambridge, Mass.: MIT Press, 1981), 102–103. Soddy, "Radium," *Professional Papers of the Corps of Engineers* 29 (1903): 250–251, quoted in a draft paper kindly shown me by Richard E. Sclove, "Atomic PreMunitions: Frederick Soddy and the Social Responsibility of Scientists" (1980).

24. H. G. Wells, *The World Set Free* (1913; New York: Dutton, 1914), 222.

25. Mark R. Hillegas, *The Future as Nightmare: H. G. Wells and the Anti-Utopians* (New York: Oxford University Press, 1967). H. G. Wells, *Experiment in Autobiography* (New York: Macmillan, 1934), chap. 4. See Frank M. Turner, "Public Science in Britain, 1880–1919," *Isis* 71 (1980): 589–608.

26. Spencer R. Weart, *Scientists in Power* (Cambridge, Mass.: Harvard University Press, 1979), chap. 2. Story counts: Wagar, *Terminal Visions,* 27, 110.

27. W. Churchill, *Pall Mall,* 24 Sept. 1924, as quoted in Raymond B. Fosdick, *The Old Savage in the New Civilization* (Garden City, N.Y.: Doubleday, 1931), 24–25. Sigmund Freud, *Civilization and Its Discontents,* trans. J. Rivière (London: Hogarth, 1930), 144. Deserts: Robert A. Millikan, *Science and Life* (Boston: Pilgrim, 1924), 75.

28. S. Chase, "The Two Hour War," *New Republic* 58 (8 May 1929): 326–327. See I. F. Clarke, *Voices Prophesying War* (London: Oxford University

Press, 1966); John Newman and Kathleen J. Kruger, "Imaginary War Fiction in Colorado State University Libraries: A Bibliography," *Bulletin of Bibliography* 35 (1978): 157–171.

29. See Royal D. Sloan, "The Politics of Civil Defense: Great Britain and the United States," Ph.D. diss., University of Chicago, 1958.

30. Sloan, "Politics of Civil Defense"; George H. Quester, *Deterrence Before Hiroshima: The Airpower Background of Modern Strategy* (New York: John Wiley, 1966).

31. G. B. Shaw quoted in Sloan, "Politics of Civil Defense," 72. Chase, "Two Hour War," 327.

32. A. Nobel to Bertha von Suttner, in *Nobel: The Man and His Prizes,* ed. Nobel Foundation (Amsterdam: Elsevier, 1962), 528.

33. "Lay down their arms": F. W. Parsons, "Stupendous Possibilities of the Atom," *World's Work* 42 (May 1921): 35.

34. "Religion may preach": W. Kaempffert, *Science Today and Tomorrow,* 2nd ser. (New York: Viking, 1945), 266.

35. S. Newcomb, *His Wisdom, the Defender* (New York: Harper, 1900).

36. Juvenile fiction: Clareson, *SF,* 16. *Things to Come* (London Film, 1936); discussed in John Brosnan, *Future Tense: The Cinema of Science Fiction* (New York: St. Martin's, 1978). Cf. H. G. Wells, *The Shape of Things to Come* (New York: Macmillan, 1933).

37. Muriel Howorth, *Pioneer Research on the Atom: . . . The Life Story of Frederick Soddy* (London: New World, 1958), 274. Frederick Soddy, *Science and Life: Aberdeen Addresses* (London: John Murray, 1920), 36.

38. Frederick Soddy, Foreword to *The Frustration of Science* (1935), p. 24 in *Frederick Soddy,* ed. George B. Kauffman (Dordrecht: Reidel, 1986). Gary Werskey, *The Visible College* (London: Allen Lane, 1978). Atomic energy was promised, for example, by J. G. Crowther, *The Progress of Science* (London: Paul, Trench, Trubner, 1934), 45.

39. Kaempffert, *Science Today,* 90–91, see 73. A recent survey of urban utopias with a good bibliography is Thomas Bender, *Toward an Urban Vision: Ideas and Institutions in Nineteenth Century America* (Lexington: University Press of Kentucky, 1975). See Roderick Nash, *Wilderness and the American Mind* (New Haven: Yale University Press, rev. ed. 1973); Leo Marx, *The Machine in the Garden: Technology and the Pastoral Ideal in America* (New York: Oxford University Press, 1964); Frank E. Manuel and Fritzie P. Manuel, *Utopian Thought in the Western World* (Cambridge, Mass.: Harvard University Press, 1979).

40. Lowell Tozer, "A Century of Progress, 1833–1933: Technology's Triumph over Man," *American Quarterly* 4 (1952): 78–81; Robert W. Rydell, "The Fan Dance of Science: American World's Fairs in the Great Depression," *Isis* 76 (1985): 525–542.

41. Richard Olson, *Science Deified and Science Defied: The Historical Significance of Science in Western Culture,* vol. 1, *From the Bronze Age . . . to ca.* A.D. *1640* (Berkeley: University of California Press, 1982). Fritz Stern, *The Politics of Cultural Despair: A Study in the Rise of the German Ideology* (Berke-

ley: University of California Press, 1961). Daniel Kevles, *The Physicists: The History of a Scientific Community in Modern America* (New York: Knopf, 1978), 180–183. Marcel E. C. La Follette, "Authority, Promise, and Expectation: The Images of Science and Scientists in American Popular Magazines, 1910–1955," Ph.D. diss., Indiana University, 1979, 227–242.

42. For a review of the vast, uneven, and controversial literature see Daniel Lawrence O'Keefe, *Stolen Lightning: The Social Theory of Magic* (New York: Random House, 1982). Especially relevant are Mary Douglas, *Purity and Danger: An Analysis of Concepts of Pollution and Taboo* (Harmondsworth, England: Penguin, 1966); John Putnam Demos, *Entertaining Satan: Witchcraft and the Culture of Early New England* (Oxford: Oxford University Press, 1982); Norman Cohn, *Europe's Inner Demons: An Inquiry Inspired by the Great Witch-Hunt* (New York: Basic Books, 1975).

43. Kevles, *The Physicists,* 248–249. Fosdick, *The Old Savage,* 23–24.

44. "A new world": Robert A. Millikan, *Science* 54 (1921): 59–67; see Robert Kargon, *The Rise of Robert Millikan: Portrait of a Life in American Science* (Ithaca, N.Y.: Cornell University Press, 1982). "Bad small boy": Millikan, *Science and the New Civilization* (New York: Scribner's, 1930), 58–59, see 94–96, 111–113.

45. "Surrendering": "Intra-atomic Energy," *Outlook* 151 (1929): 420–421.

46. "Working on the subject": Bertrand Russell, *The ABC of Atoms* (New York: E. P. Dutton, 1923), 8. "Moonshine": *NYT,* 12 Sept. 1933, p. 1.

47. "Such a result": R. A. Millikan, H. G. Gale, and W. R. Pyle, *Elements of Physics* (Boston: Ginn, 1927), 479; repeated from Millikan and Gale, *A First Course in Physics* (Boston: Ginn, 1906), 482.

48. Weart, *Scientists in Power,* chap. 2. For another example see Leo Szilard, *Leo Szilard: His Version of the Facts,* ed. Spencer Weart and Gertrud Weiss Szilard (Cambridge, Mass.: MIT Press, 1978).

49. Raymond B. Fosdick, *Chronicle of a Generation: An Autobiography* (New York: Harper & Brothers, 1958), 271.

50. La Follette, "Authority, Promise, and Expectation," 283.

51. Jerome M. Gilison, *The Soviet Image of Utopia* (Baltimore: Johns Hopkins University Press, 1975). See A. N. Krivomazov, "The Reception of Soddy's Work in the U.S.S.R.," pp. 115–140 in Kauffman, ed., *Frederick Soddy.*

3. RADIUM: ELIXIR OR POISON?

1. News clippings, William Hammer Collection, National Museum of American History, Smithsonian Institution, Washington, D.C.; *Bibliographie der deutschen Zeitschriftenliteratur* (1903–1906), s.v. "Radium."

2. William J. Hammer, *Radium and Other Radio-Active Substances* (New York: Van Nostrand, 1904), 27, 29. Marie Curie, *Pierre Curie,* trans. C. Kellogg and V. Kellogg (New York: Macmillan, 1923), 117–118.

3. Frederick Soddy, *British Medical Journal* 2 (25 July 1903): 197–199, as cited in Michael I. Freedman, "Frederick Soddy and the Practical Significance of Radioactive Matter," *British Journal for the History of Science* 12 (1979):

257–260. The most feared disease was "consumption," according to an early survey: Thomas S. Lowden, "A Study in Personal Hygiene," *The Pedagogical Seminary* 13, no. 1 (1906): 4.

4. "Old Age May be Stayed": Salt Lake City *Telegraph*, 6 Nov. 1903. "Blind girl": Newark *Evening News*, 24 Aug. 1903. Raise dead: New London, Conn. *Day*, 27 Aug. 1903. All in Hammer Collection, boxes 42–44. Bouillon: *NYT*, 21 June 1905, p. 1; W. Ramsay, "Can Life be Produced by Radium?" *Independent* (New York) 59 (1905): 554.

5. G. Le Bon, *L'Evolution de la Matière* (Paris: Flammarion, 1909). "All Nature": *Review of Reviews* 28 (1903): 490. "Secret of Life": Los Angeles *Herald*, 6 Oct. 1903, Hammer Collection, box 42.

6. "Radium Christians": *Christian Endeavor World*, quoted in Boston *Star*, 4 Feb. 1904, and other items, Hammer Collection. Spiritualists: Atlanta *Journal*, 24 Jan. 1904, Hammer Collection, box 43.

7. Robert A. Millikan, *Science and Life* (Boston: Pilgrim, 1924), 27.

8. Frederick Soddy, *The Interpretation of Radium* (London: Murray, 3rd ed., 1912), 250.

9. See Carl G. Jung, *Mysterium Coniunctionis: An Inquiry into the Separation of Psychic Opposites in Alchemy*, trans. R. F. C. Hull (Princeton, N.J.: Princeton University Press, rev. ed., 1970), vol. 14 of Jung, *Collected Works*, 90–92; Mircea Eliade, *The Forge and the Crucible: The Origins and Structure of Alchemy*, trans. Stephen Corrin (New York: Harper & Row, 1962), 154–155. See also Joseph Needham, *Science and Civilisation in China*, vol. 5, *Chemistry and Chemical Technology*, especially part IV with Ho Ping-Yü, Lu Gwei-Djen, and Nathan Sivin, *Spagyrical Discovery and Invention: Apparatus, Theories and Gifts* (Cambridge: Cambridge University Press, 1980), 363–364; Melanie Klein, *The Psycho-Analysis of Children*, trans. Alix Strachey, rev. with H. A. Thorner (New York: Dell, 1975). On sexual symbolism in general see Paul Kline, *Fact and Fantasy in Freudian Theory* (London: Methuen, 1972).

10. See Chap. 1, n. 18.

11. Richard B. Onians, *The Origins of Euorpean Thought* (Cambridge: Cambridge University Press, 1951), 219–223, 289–291. Mircea Eliade, *Shamanism: Archaic Techniques of Ecstasy*, trans. Willard R. Trask (Princeton, N.J.: Princeton University Press, 1964), 470–477. Jean Piaget, *The Child's Conception of the World*, trans. Joan Tomlinson and Andrew Tomlinson (Totowa, N.J.: Littlefield, Adams, 1969), chaps. 5–6; Piaget, *The Child's Conception of Physical Causality*, trans. Marjorie Gabain (Totowa, N.J.: Littlefield, Adams, 1966), 215–218, 229, and chap. 5.

12. Onians, *Origins of European Thought*, 76–77, 155–158, 165–167. Needham, *Science and Civilization in China*, esp. vol. 5:2, 120 and vol. 5:4, 245, 292–297. A. C. Crombie, *Robert Grosseteste and the Origins of Experimental Science 1100–1700* (Oxford: Oxford University Press, 1953), 104–116. Eliade, *Forge and Crucible*, 79–80, 170–171. I am also indebted to the work of B. J. T. Dobbs, David Lindberg, and Nathan Sivin. Sun brings conception: Stith Thompson, *Motif Index of Folk-Literature* (Bloomington: Indiana University Press, rev. ed., 1955–1958), T521 in vol. 5, p. 393.

13. J. L. Heilbron, *Electricity in the 17th and 18th Centuries: A Study of Early Modern Physics* (Berkeley: University of California Press, 1979), 283–285, 435–437. Philip C. Ritterbush, *Overtures to Biology: The Speculations of Eighteenth-Century Naturalists* (New Haven, Conn.: Yale University Press, 1964).

14. National festivals: George L. Mosse, *The Nationalization of the Masses: Political Symbolism and Mass Movements in Germany . . .* (New York: New American Library, 1975), 41–42. Wyn Wachhorst, *Thomas Alva Edison: An American Myth* (Cambridge, Mass.: MIT Press, 1981), 23–26. Nancy Roth, *Electrotherapy in the United States* (Minneapolis: Medtronic, 1977). *Exploits of Elaine* (Pathe, 1914), as discussed in Douglas Menville and R. Reginald, *Things to Come: An Illustrated History of the Science Fiction Film* (New York: New York Times Books, 1977), 15.

15. David B. Wilson, "The Thought of Late Victorian Physicists: Oliver Lodge's Ethereal Body," *Victorian Studies* 15 (1971): 29–48.

16. Mary Jo Nye, "N-rays: An Episode in the History and Psychology of Science," *Historical Studies in the Physical Sciences* 11:1 (1980): 125–156.

17. "Secret of Sex": New York *Evening Journal,* 28 Jan. 1904, Hammer Collection, box 43. Autry: *The Phantom Empire,* episode 5 (Mascot, 1935).

18. For witches, see Chap. 2, n. 42. Maya: Anthony F. Aveni, *Skywatchers of Ancient Mexico* (Austin: University of Texas Press, 1980), 186.

19. See Chap. 2, n. 11 and Charles C. Gillispie, *Science and Polity in France at the End of the Old Regime* (Princeton, N.J.: Princeton University Press, 1980), chap. 4, sect. 2.

20. Impotence: Emmanuel Le Roy Ladurie, "The Aiguillette: Castration by Magic," in Ladurie, *The Mind and Method of the Historian,* trans. Sian Reynolds and Ben Reynolds (Chicago: University of Chicago Press, 1981), 84–96. Australian tribe: Géza Róheim, *Magic and Schizophrenia* (Bloomington: Indiana University Press, 1970), 37.

21. Maria M. Tatar, *Spellbound: Studies in Mesmerism and Literature* (Princeton: Princeton University Press, 1978), chap. 2. L. Edward Purcell, "Trilby and Trilby-Mania, the Beginning of the Bestseller System," *Journal of Popular Culture* 11 (1977): 62–76.

22. For an analysis of the evil eye see Sigmund Freud, "The 'Uncanny,'" in Freud, *Collected Papers,* ed. Joan Riviere and J. Strachey, vol. 4, trans. Joan Riviere (London: Hogarth, 1949), 368–407. For bibliography see Brian Spooner, "The Evil Eye in the Middle East," in *Witchcraft: Confessions and Accusations,* ed. Mary Douglas (London: Tavistock, 1970), 311–319. "Penetrating": quoted in Edmund Morris, *The Rise of Theodore Roosevelt* (New York: Ballantine, 1979), 547.

23. Thompson, *Motif Index,* N271.1 in vol. 5, p. 91. Otto Glasser, *Dr. W. C. Röntgen* (Springfield, Ill.: C. Thomas, 1945), 82. Lawrence S. Kubie, "Unsolved Problems of the Scientific Career," in *The Sociology of Science,* ed. Bernard Barber and Walter Hirsch (New York: Free Press of Glencoe, 1962), 208. Many stories on rays and eyes can be found in the psychological literature. See Karl Abraham, "Restrictions and Transformations of Scopophobia," in Abraham,

Selected Papers on Psycho-analysis, trans. D. Bryan and A. Strachey (New York: Basic Books, 1953).

24. See, for example, Harold Searles, *Collected Papers on Schizophrenia and Related Subjects* (London: Hogarth, 1965), 476. Radium tube: Róheim, *Magic and Schizophrenia*, 95–96, see 110–113. Black sun: R. D. Laing, *The Divided Self: An Existential Study in Sanity and Madness* (Harmondsworth, England: Penguin, 1965), chap. 11.

25. E. Bulwer-Lytton, *The Coming Race* (London: Routledge, 1874).

26. New York *Evening Journal*, 28 Jan. 1904, Hammer Collection, box 43. "Rays That Can Blow Up Battleships," *Current Opinion* 56 (June 1914): 36–37. Jack London, "Goliah" (1908), in *The Science Fiction of Jack London*, ed. Richard Gid Powers (Boston: Gregg, 1975), 101–144.

27. Wilbur Knorr, "The Geometry of Burning-Mirrors in Antiquity," *Isis* 74 (1983): 53–73. W. Irving, "The Men of the Moon" (1809), in *Future Perfect: American Science Fiction of the Nineteenth Century*, ed. H. Bruce Franklin (London: Oxford University Press, rev. ed. 1978), 314–318. Glasser, *Röntgen*, 60. George Wise, "Predictions of the Future of Technology: 1890–1940," Ph.D. diss., Boston University, 1976, 228–229, 268. E. E. Slosson, "Death-Dealing and Labor-Saving Rays," *Scientific Monthly* 20 (Jan. 1925): 108–111.

28. Denis Gifford, *The Science Fiction Film* (New York: E. P. Dutton, 1971), 67–70; Menville and Reginald, *Things to Come*, 15, 22, 27, 29. "Buck Rogers in the 25th Century" (CBS radio, 1931–1939). "Adventures of Captain Marvel" (Republic, 1941); from *Whiz Comics* (1940–1941). The Russian film was based on Alexei Tolstoi, *The Death Box*, trans. Bernard G. Guerney (London: Methuen, 1936).

29. Garry Wills, *Reagan's America: Innocents at Home* (Garden City, N.Y.: Doubleday, 1987), 361, 447. Radar: Ronald W. Clark, *Tizard* (Cambridge, Mass.: MIT Press, 1965), 108–117.

30. Daniel Paul Serwer, "The Rise of Radiation Protection: Science, Medicine and Technology in Society, 1896–1935" (Upton, N.Y.: Brookhaven National Laboratory Informal Report BNL 22279, 1976). I am grateful to Barton C. Hacker (Reynolds Electrical Co., Las Vegas), for drafts of "Elements of Controversy: A History of Radiation Safety in the Nuclear Test Program," chaps. 1, 2, 1979–1980. Rudyard Kipling, "As Easy as A.B.C." (1912), in Kipling, *Writings*, vol. 26, *A Diversity of Creatures* (New York: Scribner's, 1917), 21–22.

31. Lawrence Badash, *Radioactivity in America: Growth and Decay of a Science* (Baltimore: Johns Hopkins University Press, 1979), chap. 9. My figures are derived from Ruth and Edward Brecher, *The Rays: A History of Radiology in the United States and Canada* (Baltimore: Williams and Wilkins, 1969), 211–213, 279. J. B. S. Haldane, *Possible Worlds and Other Essays* (London: Chatto & Windus, 1927), 147. Gynecology: U. V. Portmann, "Roentgen Therapy," in *The Science of Radiology*, ed. Otto H. Glasser (Springfield, Ill.: C. C. Thomas, 1933), 210–241; see also H. H. Bowing and R. E. Fricke in Glasser, ed., *Science of Radiology*, 284–286.

32. Elof Axel Carlson, *Genes, Radiation, and Society: The Life and Work of H. J. Muller* (Ithaca, N.Y.: Cornell University Press, 1981).

33. "Exactly as physicists": John Langdon-Davies, "Radiation and Evolution," *Forum* 90 (Oct. 1933): 214–217. H. J. Muller, *Science* 66 (1927): 84–87.

34. See, for example, Edmond Hamilton, "The Man Who Evolved" (1931), in *Before the Golden Age: A Science Fiction Anthology of the 1930s,* ed. Isaac Asimov (Garden City, N.Y.: Doubleday, 1974), 25–40.

35. "Radioactivity": William Seifriz, "The Gurwitsch Rays," in Glasser, ed., *Science of Radiology,* 412–427. J. G. Crowther, *The Progress of Science* (London: Paul, Trench, Trubner, 1934), 77–80. See Alexander Hollaender, "The Problem of Mitogenetic Rays," in *Biological Effects of Radiation,* ed. Benjamin M. Duggar (New York: McGraw-Hill, 1936) 2: 919–959.

36. "Many will be surprised": "Deadly Radium Gas," *Literary Digest* 101 (15 June 1929): 19.

37. *Gehes Codex der Bezeichnungen von Arzneimitteln* . . . (Dresden: Schwarzeck, 5th ed., 1929), s.v. "Radi-".

38. Oral history interview of Robley D. Evans by Charles Weiner, 1972, AIP, p. 48. Alexander O. Gettler and Charles Norris, "Poisoning from Drinking Radium Water," *Journal of the American Medical Association* 100 (1933): 400–402. Needham, *Science and Civilization in China,* vol. 5:2, 282, 291–292.

39. "Radium Poisoning," *Literary Digest* 113 (16 April 1932): 13; "Radioactive Waters," *Literary Digest* 114 (2 July 1932): 22.

40. William D. Sharpe, "The New Jersey Radium Dial Painters: A Classic in Occupational Carcinogenesis," *Bulletin of the History of Medicine* 52 (1979): 560–570. Roger J. Cloutier, "Florence Kelly and the Radium Dial Painters," *Health Physics* 39 (1980): 711–716.

41. "Insanity Gas," *Literary Digest* 83 (22 Nov. 1924): 18.

42. Jack Schubert and Ralph E. Lapp, *Radiation: What It Is and How It Affects You* (New York: Viking, 1957), 100–103. Dean B. Cowie and Leonard A. Scheele, "A Survey of Radiation Protection in Hospitals," *Journal of the National Cancer Institute* 1 (1941): 767–787, reprinted in *Health Physics* 38 (1980): 929–947.

43. For cyclotrons see the oral history interviews by Charles Weiner, AIP, and the E. O. Lawrence papers, Bancroft Library, University of California, Berkeley. John H. Lawrence, "Early Experiences in Nuclear Medicine," *Journal of Nuclear Medicine* 20 (1979): 561–564.

4. THE SECRET, THE MASTER, AND THE MONSTER

1. "Deepest secrets": Ernest Rutherford, "The Transmutation of the Atom," 13th BBC National Lecture (London: British Broadcasting Corp., 1933), 25. "Unveiling": Radio addresses, 1929–1930, printed as Hans Reichenbach, *Atom and Cosmos: The World of Modern Physics,* trans. and rev. Edward S. Allen (New York: Macmillan, 1933), 221. Waldemar Kaempffert, "Ultimate Truths Sought in the Atom," *NYT Magazine* (24 May 1936): pp. 6ff.

2. Here much of Freud is relevant; for an introduction see Abraham H. Maslow, *Toward a Psychology of Being* (New York: Van Nostrand Reinhold, 2nd ed., 1968), 60–63. Cultural transmission: John Bowlby, *Attachment and*

Loss, vol. 2, *Separation: Anxiety and Anger* (London: Hogarth, 1973), 131–132, 159–160.

3. Melanie Klein, "On the Theory of Anxiety and Guilt," in Klein et al., *Developments in Psychoanalysis*, ed. Joan Riviere (London: Hogarth, 1952), 271–291. Melanie Klein, *The Psycho-Analysis of Children*, trans. Alix Strachey, rev. with H. A. Thorner (New York: Dell, 1975), 173–175.

4. Carolyn Merchant, *The Death of Nature: Women, Ecology, and the Scientific Revolution* (San Francisco: Harper & Row, 1980). See also Brian Easlea, *Science and Sexual Oppression: Patriarchy's Confrontation with Women and Nature* (London: Weidenfeld and Nicolson, 1981), esp. chap. 3. Evelyn Fox Keller, *Reflections on Gender and Science* (New Haven: Yale University Press, 1985). "Interrogated nature:" Jean-Baptiste Cousin de Grainville, *Le Dernier Homme* (1805; Paris: Ferra et Deterville, 1811; Geneva: Slatkine, 1976), 141.

5. Bohr Institute (with neutron and neutrino confused): George Gamow, *Thirty Years That Shook Physics: The Story of Quantum Theory* (Garden City, N.Y.: Doubleday, 1966), 165–218. See Johann Wolfgang von Goethe, *Aus meiner Leben: Dichtung und Wahrheit* (1811–1814), book 8.

6. "Les propriétés les plus intimes de la matière:" Poincaré et al. to Nobel Prize for Physics Committee, Jan. 1903, Protokoll vol. 3, Swedish Academy of Sciences, Stockholm, p. 134; I thank John Heilbron for this quote and for much else. "Cold-chisel": said of J. J. Thomson in Henry Bumstead to Rutherford, 30 Sept. 1903, Rutherford Papers, Cambridge, microfilm copy at AIP; my thanks to Lawrence Badash. "Satisfaction": Robert Millikan, *Science and the New Civilization* (New York: Scribner's, 1930), 60. Anyone reading the literature of physics can find many other examples.

7. Waldemar Kaempffert, "Science Launches New Attack on the Atom's Citadel," *NYT*, 15 Nov. 1931, sec. 9, p. 4. Kaempffert, "Atomic Energy—Is It Nearer?" *Scientific American* 147 (Aug. 1932): 79–81.

8. See Karl Čapek, *The Absolute at Large* (New York: Macmillan, 1927); Colette Guedeney and Gérard Mendel, *L'Angoisse atomique et les centrales nucléaires: Contribution psychoanalytique et sociopsychanalytique à l'étude d'un phénomène collectif* (Paris: Payot, 1973), 208–213.

9. Autry: *The Phantom Empire* (Mascot, 1935). Corrigan: *The Undersea Kingdom* (Republic, 1936), episode 1. Atom furnace: *Spaceship to the Unknown* (Universal, 1936). See Denis Gifford, *The Science Fiction Film* (New York: E. P. Dutton, 1971), 3, 40; Douglas Menville and R. Reginald, *Things to Come: An Illustrated History of the Science Fiction Film* (New York: New York Times Books, 1977), 68–72.

10. Mark R. Hillegas, *The Future as Nightmare: H. G. Wells and the Anti-Utopians* (New York: Oxford University Press, 1967), 37–38 and passim.

11. Bowlby, *Attachment and Loss*, vol. 2, *Separation*. Ernest Becker, *The Denial of Death* (New York: Free Press, 1973). See also Sigmund Freud, *Inhibitions, Symptoms and Anxiety*, trans. and ed. James Strachey, in *The Standard Edition of the Complete Psychological Works of Sigmund Freud* (London: Hogarth, 1959), vol. 20, chaps. 8, 10, pp. 77–175. Klein, *Psycho-Analysis of Children*, 99, 173–175, 243–244. Joseph C. Reingold, *The Mother, Anxiety,*

and Death: The Catastrophic Death Complex (London: Churchill, 1967). Of special value in this particular context are Robert Jay Lifton, *The Broken Connection: On Death and the Continuity of Life* (New York: Simon & Schuster, 1979), 53–57; Franco Fornari, *Psicanalisa della situazione atomica* (Milan: Rizzoli, 1970), 199–200; Dario De Martis, "Note sui deliri di negazione," *Rivista sperimentale di Freniatria* 91 (1967): 1119–43.

12. Muriel Howorth, *Pioneer Research on the Atom: . . . The Life Story of Frederick Soddy* (London: New World, 1958), 29. Bowlby, *Attachment and Loss*, vol. 3, *Loss: Sadness and Depression* (New York: Basic Books, 1980), 225, chap. 4, and passim; also vol. 2, *Separation*, xii and passim.

13. Matthew P. Shiel, *The Purple Cloud* (London: Chatto & Windus, 1901). Another example is Nathaniel Hawthorne, "The Birthmark" (1843), in *Future Perfect: American Science Fiction of the Nineteenth Century*, ed. H. Bruce Franklin (London: Oxford University Press, rev. ed., 1978), 24–40.

14. Daniel Paul Schreber, *Denkwürdigkeiten eines Nervenkranken* (Leipzig: Oswald Mutze, 1903), 18–19, 282–294; discussed in Lifton, *Broken Connection*, 226–233; see especially the Niederland papers cited by Lifton.

15. Harold Searles, *Collected Papers on Schizophrenia and Related Subjects* (London: Hogarth, 1965), chap. 17. Becker, *Denial of Death*. Karl Jaspers, *General Psychopathology*, trans. J. Hoenig and Marian W. Hamilton (Manchester: University of Manchester Press, 1963), 294–296. Walter Bromberg and Paul Schilder, "The Attitude of Psychoneurotics towards Death," *Psychoanalytic Review* 23 (1936): 1–25. Ralph J. Hallman, *Psychology of Literature: A Study of Alienation and Tragedy* (New York: Philosophical Library, 1961); David Ketterer, *New Worlds for Old: The Apocalyptic Imagination, Science Fiction, and American Literature* (Garden City, N.Y.: Doubleday, 1974). For hallucinogens and mysticism see, for example, Timothy Leary, Ralph Metzner, and Richard Alpert, *The Psychedelic Experience: A Manual Based on the Tibetan Book of the Dead* (New Hyde Park, N.Y.: University Books, 1964). For the first medieval chiliast see Robert E. Lerner, "The Black Death and Western European Eschatological Mentalities," *American Historical Review* 86 (1981): 533–552.

16. "Penetrate": Mary Shelley, *Frankenstein* (1818; New York: Bantam, 1981), 33. This edition has a good introduction and bibliography by Diane Johnson. Another entry-point to the literature is Aija Ozolins, "Recent Work on Mary Shelley and *Frankenstein*," *Science-Fiction Studies* 8 (1976): 187–200. See especially Martin Tropp, *Mary Shelley's Monster* (Boston: Houghton Mifflin, 1976). Movies: *Frankenstein* (Universal, 1931); *The Bride of Frankenstein* (Universal, 1935). On the "Oedipal Project" of mastering life see Norman O. Brown, *Life Against Death: The Psychoanalytical Meaning of History* (New York: Random House, 1959).

17. *Dr. Cyclops* (Paramount, 1940). Cf. Karl Abraham, *Selected Papers on Psycho-Analysis*, trans. D. Bryan and A. Strachey (New York: Basic Books, 1953), 179–182.

18. "Golem," *Encyclopaedia Judaica* 7 (1971); Gershom Scholem, *On the Kabbalah and Its Symbolism*, trans. Ralph Manheim (New York: Schocken, 1969), chap. 5. Carlos Clarens, *An Illustrated History of the Horror Film* (New

York: Capricorn, 1967), 11–13, 20–21. For witches' beasts see, for example, John Putnam Demos, *Entertaining Satan: Witchcraft and the Culture of Early New England* (Oxford: Oxford University Press, 1982).

19. "Docile servant": Michael Mok, "Radium: Life-giving Element Deals Death in Hands of Quacks," *Popular Science Monthly* 121, no. 1 (July 1932): 9ff. "Golem": Philip H. Leib, *NYT,* 7 June 1931, sec. 3, p. 2, referring to A. H. Compton. Raymond B. Fosdick, *The Old Savage in the New Civilization* (Garden City, N.Y.: Doubleday, 1931), 23–24.

20. See references in n. 16 above. Film robot: *The Master Mystery* (Octagon, 1918), as described in Menville, *Things to Come.*

21. The best-known example is Freud's 1923 theory of the Id-Ego split, developed while he was working on the death instinct and mob behavior. Ronald W. Clark, *Freud: The Man and the Cause* (New York: Random House, 1980). For mobs see, for example, Gareth Stedman-Jones, *Outcast London: A Study in the Relationship between Classes in Victorian Society* (Oxford: Oxford University Press, 1971).

22. Leo Marx, "American Literary Culture and the Fatalistic View of Technology," *Alternative Futures* 3, no. 2 (Spring 1980): 45–70. Herbert L. Sussman, *Victorians and the Machine: The Literary Response to Technology* (Cambridge, Mass.: Harvard University Press, 1968).

23. Karel Čapek, *R.U.R. (Rossum's Universal Robots),* trans. Paul Selver (Garden City, N.Y.: Doubleday, Page, 1923). William E. Harkins, *Karel Čapek* (New York: Columbia University Press, 1962), 1–13, 94.

24. I am indebted here to the works of Claude Lévi-Strauss and other "structuralists." See my Chapters 9 and 18.

25. *The Invisible Ray* (Universal, 1936). I am grateful to J. Timko for loaning me his videotape of this hard-to-find movie. I also used a script from the New York State Archives, Albany. *Batman* (Dec. 1941–Jan. 1942) as described in Michael L. Fleisher, *The Encyclopedia of Comic Book Heroes,* vol. 1, *Batman* (New York: Collier, 1976), s.v. "Dr. Radium."

26. See Rollo May, *The Meaning of Anxiety* (New York: Washington Square Press, rev. ed., 1970), 35–38.

27. Klein, "On the Theory of Anxiety and Guilt." Searles, *Collected Papers,* 500.

28. De Martis, "Note sui deliri," 1124. Cf. Tropp, *Mary Shelley's Monster,* 31.

29. "He sought to destroy": Universal Studios, *Advance Publicity* (Pressbook), 1936, NYPL/TC, n.c. 240. Autry: *The Phantom Empire* (Mascot, 1935).

30. Karel Čapek, *An Atomic Phantasy: Krakatit,* trans. Lawrence Hyde (London: Allen & Unwin, 1948). See Harkins, *Karel Čapek,* 13, 106–108.

31. Becker, *Denial of Death.* Victor E. Frankl, *Man's Search for Himself: An Introduction to Logotherapy* (New York: Washington Square Press, 1963). Lifton, *Broken Connection.*

32. Kenneth MacDonald Jones, "Science, Scientists and Americans: Images of Science and the Formation of Federal Science Policy, 1945–1950," Ph.D. diss., Cornell University, 1975, 37–42. Walter Hirsch, "The Image of the Scientist in Science Fiction: A Content Analysis," *American Journal of Sociology*

63 (1958): 506–512, reprinted in Bernard Barber and Walter Hirsch, eds., *The Sociology of Science* (New York: Free Press of Glencoe, 1962), 259–268. Marcel E. C. La Follette, "Authority, Promise, and Expectation: The Images of Science and Scientists in American Popular Magazines, 1910–1955," Ph.D. diss., Indiana University, 1979. For scientists' views see oral history interviews, AIP.

33. *Madame Curie* (Metro-Goldwyn-Mayer, 1944). Eve Curie, *Madame Curie,* trans. Vincent Sheean (Garden City, N.Y.: Doubleday, Doran, 1938), 247–248. Georges Sagnac to Jakob Bjerknes, 5 May 1906, with letter of 23 March 1906, Bjerknes Papers, Oslo. My thanks to Robert Friedman for use of his microfilm.

34. La Follette, "Authority, Promise, and Expectation," 93–99. See Wyn Wachhorst, *Thomas Alva Edison: An American Myth* (Cambridge, Mass.: MIT Press, 1981).

35. Spencer R. Weart, *Scientists in Power* (Cambridge, Mass.: Harvard University Press, 1979), chaps. 2, 3; cartoon in clippings file, Curie Papers, Radium Institute, Paris.

36. For an introduction to the history of nuclear physics and a bibliography see Daniel Kevles, *The Physicists: The History of a Scientific Community in Modern America* (New York: Knopf, 1978). For fission see William R. Shea, ed., *Otto Hahn and the Rise of Nuclear Physics* (Dordrecht: Reidel, 1983).

37. See Jacques Ellul, *Propaganda: The Formation of Men's Attitudes,* trans. Konrad Kellen and Jean Learner (New York: Knopf, 1968; reprint, Vintage, 1973), chap. 1.3; David E. Nye, *Image World: Corporate Identities at General Electric, 1890–1930* (Cambridge, Mass.: MIT Press, 1985).

5. WHERE EARTH AND HEAVEN MEET

1. From this point on the literature is enormous. There is a good bibliography in Martin Sherwin, *A World Destroyed: The Atomic Bomb and the Grand Alliance* (New York: Knopf, 1975). The one indispensable book on the Manhattan Project is Richard G. Hewlett and Oscar E. Anderson, Jr., *A History of the United States Atomic Energy Commission,* vol. 1, *The New World 1939/1946,* WASH 1214 (1962; Springfield, Va.: National Technical Information Service, 1972). A good popular history is Richard Rhodes, *The Making of the Atomic Bomb* (New York: Simon and Schuster, 1986). For 1939–1940 see Spencer R. Weart, *Scientists in Power* (Cambridge, Mass.: Harvard University Press, 1979).

2. *NYT,* 3 Feb. 1939, p. 14; see 5 March 1939, sec. 2, p. 9, and 30 April 1939, p. 35. *Newsweek* (27 March 1939): 32; also (13 May 1940): 41. *Scientific American* 161 (1939): 2, 214–216; also 162 (1940): 268.

3. Leo Szilard, *Leo Szilard: His Version of the Facts,* ed. Spencer Weart and Gertrud Weiss Szilard (Cambridge, Mass.: MIT Press, 1978).

4. A. Troller, "Les Transmutations de l'uranium," *La Nature* 67, pt. 2 (15 Dec. 1939): 197–201. Since I have already speculated that early preoccupations with death and abandonment could lay a foundation for later thoughts of atomic apocalypse, it is worth mentioning that Halban's mother died when he

was born. Halban's co-worker Kowarski also had an absent mother in his childhood, and Fermi's mother put him out to a wet-nurse for his first two and a half years. I have anecdotal private evidence for similar, less acute but still unusual problems of maternal relations for three other key pioneers of fission, more than one would expect on the basis of chance.

5. Jean Thibaud, *Vie et transmutations des atomes* (Paris: A. Michel, rev. ed. 1942), 98–99. Bruce Bliven, "The World-Shaking Promise of Atomic Research," *Reader's Digest* (July 1941): 103–106, from *New Republic* (16 June 1941). *NYT,* 5 March 1939, sec. 2, p. 9.

6. See, for example, William L. Laurence, *NYT,* 5 May 1940, p. 1. See also Lawrence Badash, Elizabeth Hodes, and Adolph Tiddens, "Nuclear Fission: Reaction to the Discovery in 1939," *Proceedings of the American Philosophical Society* 130 (1986): 196–231.

7. John W. Campbell, Jr., "The Brain Stealers of Mars" (1936), in *Before the Golden Age: A Science Fiction Anthology of the 1930s,* ed. Isaac Asimov (Garden City, N.Y.: Doubleday, 1974), 837–856. See Campbell editorials in *Astounding Science-Fiction* 21, no. 4 (June 1938): 21; 23, no. 2 (April 1939): 6; 26, no. 2 (Oct. 1940): 6.

8. H. Bruce Franklin, *Robert A. Heinlein: America as Science Fiction* (New York: Oxford University Press, 1980), chaps. 1, 2.

9. Robert Heinlein, "Blowups Happen," *Astounding Science-Fiction* 26, no. 1 (Sept. 1940): 51–85; see Campbell on pp. 5–6.

10. "Angus MacDonald" [Robert Heinlein], "Solution Unsatisfactory," *Astounding Science-Fiction* 27, no. 3 (May 1941): 56–86. See Paul A. Carter, *The Creation of Tomorrow: Fifty Years of Magazine Science Fiction* (New York: Columbia University Press, 1977), 24–25.

11. "V-3?" *Time* 44 (27 Nov. 1944): 88.

12. Bohr announcement: *NYT,* 30 April 1939, sec. 1, p. 35.

13. Margaret M. Gowing, *Britain and Atomic Energy, 1939–1945* (New York: St. Martin's, 1964).

14. George H. Quester, *Deterrence before Hiroshima: The Airpower Background of Modern Strategy* (New York: John Wiley, 1966). Harold Nicolson, *Public Faces* (Boston: Houghton Mifflin, 1933), 18.

15. Gowing, *Britain and Atomic Energy;* Churchill quote, p. 106.

16. "More valuable Christian": Roger H. Stuewer, *The Compton Effect: Turning Point in Physics* (New York: Science History Publications, 1975), 92.

17. Arthur Holly Compton, *Atomic Quest: A Personal Narrative* (New York: Oxford University Press, 1956), 136–139. See also, n. 24 below.

18. Albert Wattenberg, "December 2, 1942," *BAS* 38, no. 10 (Dec. 1982): 22–32.

19. Compton, *Atomic Quest,* 144.

20. "Military risks": A. H. Compton to Robert Stone, MUC-AC-1257, 4 Jan. 1945, "Health and Safety" file, University of Chicago Library, also MUC files, Argonne National Laboratory, Argonne, Ill. I thank Carol Gruber for a copy.

21. Here and below I use Robert S. Stone, ed., *Industrial Medicine on the*

Plutonium Project (New York: McGraw-Hill, 1951); Stone, "Health Protection Activities of the Plutonium Project," *Proceedings of the American Philosophical Society* 90 (1946): 11–19; Barton C. Hacker, "Elements of Controversy: A History of Radiation Safety in the Nuclear Test Program" (draft), chaps. 2, 3, 1980–1981, published as *The Dragon's Tail: Radiation Safety in the Manhattan Project, 1942–1946* (Berkeley: University of California Press, 1987).

22. Heinlein, "Solution Unsatisfactory," 62.

23. Plutonium toxicity: *Health Physics* 41 (1981): 706.

24. A. H. Compton, E. Fermi, R. S. Stone, MUC-AC-81, 6 Feb. 1943, MUC files. Compton to Gen. Groves, "Chain Reacting Unit on University of Chicago Campus," MUC-AC-72, 2 Feb. 1943, MUC files.

25. A. H. Compton to Roger Williams, MUC-AC-89, 16 Feb. 1943; see also MUC-AC-88, 25 Feb. 1943, MUC files.

26. Szilard, *Leo Szilard: His Version*, 170–171.

27. David Irving, *The Virus House: Germany's Atomic Research and Allied Counter-Measures* (London: William Kimber, 1967). See Mark Walker, "Uranium Machines, Nuclear Explosives, and National Socialism: The German Quest for Nuclear Power, 1939–1949," Ph.D. diss., Princeton University, 1987.

28. A. H. Compton, Preliminary draft of report to President of the National Academy of Sciences, Oct. 1941, File 1015:1/1, E. O. Lawrence Papers, Bancroft Library, University of California, Berkeley. Compton to James B. Conant, MUC-AC-2235, 15 Aug. 1944, MUC files.

29. See oral history interviews of F. Bloch, E. U. Condon, and M. White, AIP; George Kistiakowsky in Lawrence Badash, Joseph O. Hirschfelder, and Herbert P. Broida, eds., *Reminiscences of Los Alamos 1943–1945* (Dordrecht and Boston: D. Reidel, 1980), 49. S. R. Weart, "The Road to Los Alamos," *Journal de Physique* Colloque C8, suppl. to vol. 43, no. 12 (Dec. 1982): C8/301–321.

30. Archibald C. Bell, *A History of the Blockade of Germany . . . 1914–1918* (1937; London: H.M.S.O., 1961), chap. 34.

31. For an introduction and a bibliography of the extensive literature see R. J. Overy, *The Air War 1939–1945* (New York: Stein & Day, 1980). I use especially Quester, *Deterrence*, and Sir Charles Webster and Noble Frankland, *The Strategic Air Offensive against Germany* (London: H.M.S.O., 1961). "Raise the stakes": Winston Churchill, *The Second World War*, vol. 2, *Their Finest Hour* (Boston: Houghton Mifflin, 1949), 342. For morality see Robert C. Batchelder, *The Irreversible Decision* (Boston: Houghton Mifflin, 1962), chap. 15.

32. See John W. Dower, *War Without Mercy: Race and Power in the Pacific War* (New York: Pantheon, 1986). Harrison F. Salisbury, *The 900 Days: The Siege of Leningrad* (New York: Harper & Row, 1969).

33. Ronald Schaffer, *Wings of Judgment: American Bombing in World War II* (New York: Oxford University Press, 1985), esp. chap. 9.

34. Michael J. Yavendetti, "American Reactions to the Use of Atomic Bombs on Japan, 1945–1947," Ph.D. diss., University of California, Berkeley, 1970, 8–18; *Public Opinion Quarterly* 9 (1945): 94, 533.

35. "Out of all proportion": Overy, *Air War,* 206–208. *Victory through Air Power* (Disney-United Artists, 1943). Robert A. Divine, *Second Chance: The Triumph of Internationalism During World War II* (New York: Atheneum, 1971), 80, 83, 122, 311–312.

36. James P. Baxter, 3rd, *Scientists Against Time* (Boston: Little, Brown, 1946), 291–293; Schaffer, *Wings of Judgment,* chap. 6.

37. Cost: H. H. Arnold in Dexter Masters and Katharine Way, eds., *One World or None* (New York: McGraw-Hill, 1946), 26.

38. Sherwin, *World Destroyed.*

39. Alice Kimball Smith and Charles Weiner, eds., *Robert Oppenheimer: Letters and Recollections* (Cambridge, Mass.: Harvard University Press, 1980). J. Robert Oppenheimer, *Science and the Common Understanding* (New York: Simon & Schuster, 1954), 37, 98; Oppenheimer, *The Open Mind* (New York: Simon & Schuster, 1955), 92, 94, 129.

40. Quotes: Smith and Weiner, *Oppenheimer,* 250, 317. See Oppenheimer, *Open Mind,* 61. A. K. Smith, personal communication.

41. Sherwin, *A World Destroyed,* is again a good guide to the literature, although all of it is too narrowly focused. See my comments at end of Chapter 10 and note 41 there; see also Dower, *War Without Mercy.* "Vital war plant": Interim Committee Minutes, 31 May 1945, reprinted with other useful documents in Robert C. Williams and Philip L. Cantelon, eds., *The American Atom: A Documentary History of Nuclear Policies from the Discovery of Fission to the Present* (Philadelphia: University of Pennsylvania Press, 1984), 62. For a summary of the scientists' and Air Force views see Schaffer, *Wings of Judgment,* 149–162.

42. E. Rabinowitch, "Five Years After," in *The Atomic Age: Scientists in National and World Affairs. Articles from the Bulletin of the Atomic Scientists 1945–1962,* ed. Morton Grodzins and Eugene Rabinowitch (New York: Basic Books, 1963), 156–162.

43. A.H. Compton to James B. Conant, 15 Aug. 1944; for this and what follows see Alice Kimball Smith, *A Peril and a Hope: The Scientists' Movement in America, 1945–47* (Cambridge, Mass.: MIT Press, rev. ed. 1970).

44. Calculations: R. Cornoy, NDN-A42942, 10 April 1944, Misc. Chron. file, Argonne National Laboratory, Argonne, Ill. See entire "Nucleonics" folder, Misc. Declassified Documents file, Argonne National Lab.

45. K. K. Darrow to R. S. Mullikan, 4 June 1945, "Nucleonics" folder, Argonne National Lab.

46. For this and the following see William L. Laurence, oral history interviews by Louis M. Starr, 1956–1957, and by Scott Bruns, 1964, Columbia University Library; "Science of Reporting: Career of New York *Times* Science Editor," *Time* 82 (27 Dec. 1963): 32.

47. W. Laurence, *Dawn over Zero: The Story of the Atomic Bomb* (New York: Knopf, 2nd ed., 1946; Westport, Conn.: Greenwood, 1972), vii-x. Compton, *Atomic Quest,* xviii.

48. W. Laurence, *Dawn over Zero,* 116; see Laurence, *Men and Atoms: The Discovery, the Uses, and the Future of Atomic Energy* (New York: Simon &

Schuster, 1959), 112.

49. Laurence, *Men and Atoms,* 97–98.

50. The latest word on the question is Thomas A. Weaver and Lowell Wood, *Physical Review* A20 (1979): 316.

51. Lansing Lamont, *Day of Trinity* (New York: Atheneum, 1965), 235–236; Farrell's report is in Sherwin, *A World Destroyed,* App. P, 312–313. Other reactions are described in many sources; see Robert Jay Lifton, *The Broken Connection: On Death and the Continuity of Life* (New York: Simon & Schuster, 1979), 370–375. "Reverence": "O. E. Lawrence Thoughts" (*sic*), 16 July 1945, File 4, "Top Secret" Correspondence of the Manhattan Engineer District (RG 77), microfilm 1108, National Archives, Washington, D.C.

52. Laurence, *Men and Atoms,* 117–120; *Bhagavad-Gita,* XI.

53. W. Laurence, oral history interview, 319. See W. Laurence, "Things Were Never the Same Again," *NYT Book Review,* 18 July 1965, p. 1.

54. Truman Journal, 25 July 1945, as reported in *NYT,* 2 June 1980, p. A14.

55. Messages: Laurence, *Dawn over Zero,* 206.

6. THE NEWS FROM HIROSHIMA

1. Leslie R. Groves, *Now It Can Be Told: The Story of the Manhattan Project* (New York: Harper & Row, 1962), 327–328. William L. Laurence, "Tentative draft of radio address by President Truman," 17 May 1945, File 4, and Groves to Chief of Staff, 6 Aug. 1945, File 5B, "Top Secret" Correspondence of the Manhattan Engineer District (RG 77), microfilm 1108, National Archives, Washington, D.C.

2. *NYT,* 8 Aug. 1945, p. 1; 9 Aug. 1945, p. 1; 10 Aug. 1945, p. 1. These have the by-line W. H. Lawrence, another reporter who was fed his stories by W. L. Laurence; William L. Laurence, oral history interviews by Louis M. Starr, 1956–1957, and by Scott Bruns, 1964, Columbia University Library, p. 355.

3. "Atom Bomb" (Paramount newsreel, Aug. 1945), text courtesy Public Information Office, Argonne National Laboratory, Argonne, Ill.

4. "Serve": for example, Harry S. Truman, *Memoirs,* vol. 1, *Year of Decision* (Garden City, N.Y: Doubleday, 1955), 523–524.

5. "Darkening heavens": War Dept. release in Henry D. Smyth, *Atomic Energy for Military Purposes: The Official Report . . .* (Princeton, N.J.: Princeton University Press, 1946), App. 6. Cf. Lewis Wood, *NYT,* 7 Aug. 1945, p. 1.

6. Cartoons: for example, reproduced in *NYT,* 18 Nov. 1945, sec. 4, p. 3; 30 Nov. 1945, sec. 4, p. 9; 30 June 1946, sec. 4, p. 3. H. V. Kaltenborn, NBC radio, 6 Aug. 1945, R78:0345, MB.

7. For early polls see Cornell University (Richard Crutchfield et al.), *Public Reaction to the Atomic Bomb and World Affairs: A Nation-wide Survey* (Ithaca, N.Y: Cornell University, 1947), summarized in Leonard S. Cottrell, Jr., and Sylvia Eberhart, *American Opinion on World Affairs in the Atomic Age* (1948; New York: Greenwood, 1969). See also Michael J. Yavendetti, "American Reactions to the Use of Atomic Bombs on Japan, 1945–1947," Ph.D. diss., University of California, Berkeley, 1970. For British reactions and other informa-

tion see Harley J. Stucky, *August 6, 1945: The Impact of Atomic Energy* (New York: American Press, 1964), 101–110. American influence on opinion is surveyed in Bill S. Caldwell, "The French Socialists' Attitude Toward the Use of Nuclear Weapons, 1945–1978," Ph.D. diss., University of Georgia, 1980, 46–55.

8. Frank Sullivan, "The Cliché Expert Testifies on the Atom," in Sullivan, *A Rock in Every Snowball* (Boston: Little, Brown, 1946), 28–36. An actual example of "harnessed . . . unleashed" is Lyle Van, NBC radio, 6 Aug. 1945, R78:0344, MB.

9. William L. Laurence in "The Quick and the Dead" (NBC radio, 1950), on RCA Victor records, copy at AIP. Pea: Sullivan, "Cliché Expert," 31.

10. Cottrell and Eberhart, *American Opinion*, 13–14. "Atoms . . . broken loose," Reuters, 4 Feb. 1946, in Baltimore *Evening Sun*, 5 Feb. 1946; see *NYT*, 4 Feb. 1946, p. 15.

11. Edward A. Shils, *The Torment of Secrecy: The Background and Consequences of American Security Policies* (Glencoe, Ill.: Free Press, 1956), 71.

12. See the popular poem by Hermann Hagedorn, *The Bomb That Fell on America* (New York: Association Press, 1946; rev. ed., 1950). On these points I am grateful for discussions with Paul Boyer. See Boyer, *By the Bomb's Early Light: American Thought and Culture at the Dawn of the Atomic Age* (New York: Pantheon, 1985). Norman Cousins, "Modern Man Is Obsolete," *Saturday Review* 28 (18 Aug. 1945): 1; a book version was published (New York: Viking, 1945).

13. Here and below I use Robert Jay Lifton, *Death in Life: Survivors of Hiroshima* (New York: Simon & Schuster, 1967). For comparable experiences elsewhere see Irving L. Janis, *Air War and Emotional Stress: Psychological Studies of Bombing and Civilian Defense* (New York: McGraw-Hill, 1951), 8–23, 53–66.

14. "Coming together": Lifton, *Death in Life*, 486.

15. Approval of bomb: 1945 Roper poll cited in Boyer, *By the Bomb's Early Light*, 183. John Hersey, *Hiroshima* (New York: Knopf, 1946). ABC radio script is S77:001, MB. See Yavendetti, "American Reactions," 356–370.

16. "John Hersey," *Current Biography* (1944). "Obsession": Hersey, *Here to Stay* (New York: Knopf, 1964), vii–viii.

17. Joseph Luft and W. M. Wheeler, "Reaction to John Hersey's 'Hiroshima,'" *Journal of Social Psychology* 18 (1948): 138–140.

18. Rumors and other local effects: United States Strategic Bombing Survey, *The Effects of Atomic Bombs on Hiroshima and Nagasaki* (Washington, D.C.: Government Printing Office, 1946), 22. For the chemist Harold Jacobson's announcement see Yavendetti, "American Reactions," 297; New York *Journal American*, 8 Aug. 1945, p. 1; H. G. Nicholas, ed., with Isaiah Berlin, *Washington Despatches, 1941–1945: Weekly Political Reports from the British Embassy* (Chicago: University of Chicago Press, 1981), 598–599.

19. Yagi's dream: quoted in Harry S. Hall, *Congressional Attitudes towards Atomic Energy* (New York: Arno, 1979), 116.

20. "Atomic plague": Peter Burchett, London *Daily Express*, 5 Sept. 1945;

see Wilfrid Burchett, *Passport: An Autobiography* (Melbourne: T. Nelson, 1969), 120, 162–176.

21. William A. Shurcliff, *Bombs at Bikini: The Official Report of Operation Crossroads* (New York: W. H. Wise, 1947). *NYT,* 25 Jan. 1946, p. 1; 23 May 1946, p. 3; 2 July 1946, p. 19; 4 July 1946, p. 4. Stephen B. Withey, *4th Survey of Public Knowledge and Attitudes Concerning Civil Defense* (Ann Arbor: Survey Research Center, University of Michigan, 1954), 44–45. John McLean, "Atom Bomb" (CBS radio, 21 June 1946), script S76:024, MB. French poll: Institut Français d'Opinion Publique, *Sondages,* February 1947, as cited in J. M. Fourgous, J. F. Picard, and C. Raguenel, "Les Français et l'énergie. Recueil d'enquêtes et de sondages d'opinion . . . de 1945 à nos jours" (Paris: CNRS Centre de Documentation Sciences Humaines; Electricité de France, Direction des Etudes et Recherches, 1980), 7; see this compilation also for nuclear war attitudes in general. Sign: Paramount newsreel, 9 May 1947, C-9172, Sherman Grinberg Film Library, New York City.

22. "Atomic Bomb!" (Paramount newsreel, 6 Aug. 1946), C-7686, Grinberg Library. *NYT,* 8 Aug. 1946, p. 18.

23. David Bradley, *No Place to Hide* (Boston: Little, Brown, 1948); see Boyer, *By the Bomb's Early Light,* 91–92.

24. Hersey, *Hiroshima* (1946; New York: Bantam, 1959), p. 89. Aldous Huxley, *Ape and Essence* (New York: Harper, 1948).

25. Bernard S. Wolf, "Medical Aspects of Radiation Safety," *Nucleonics* (Oct. 1946): 25–26. W. Laurence, *NYT,* 28 Nov. 1948, sec. 4, p. 9. Physicians died: H. M. Parker, "Health Physics Instrumentation and Radiation Protection" (1947), reprinted in *Health Physics* 38 (1980): 957–999, p. 980. Oak Ridge worker: "Radioactivity Scare," *Time* (1 Sept. 1947): 56–57.

26. Fear and persuasion: M. Fishbein and I. Ajzer, *Belief, Attitude, Intention and Behavior: An Introduction to Theory and Research* (Reading, Pa.: Addison-Wesley, 1975), 497–509; Seymour Feshbach, Ronald W. Rogers, and C. Ronald Mewborn, pp. 513–531 in *Psychology and the Prevention of Nuclear War: A Book of Readings,* ed. Ralph K. White (New York: New York University Press, 1986); and especially Peter M. Sandman and JoAnn M. Valenti, "Scared Stiff—Or Scared into Action," *BAS* 42, no. 1 (Jan. 1986): 12–16.

27. Alice Kimball Smith, *A Peril and a Hope: The Scientists' Movement in America, 1945–47* (Cambridge, Mass.: MIT Press, rev. ed. 1970). "Supernatural": Harry S. Hall, "Scientists and Politicians," *BAS* (Feb. 1956), reprinted in *The Sociology of Science,* ed. Bernard Barber and Walter Hirsch (New York: Free Press of Glencoe, 1962), 269–287. See Kenneth MacDonald Jones, "Science, Scientists, and Americans: Images of Science and the Formation of Federal Science Policy, 1945–1950," Ph.D. diss., Cornell University, 1975.

28. *Time* 48 (1 July 1946), cover and p. 52. Note that the first scientific papers on fission sometimes calculated the energy not with $E = mc^2$ but with simple electrical repulsion of the two fragments. By "molecular mass" I mean the (tiny) mass-energy associated with the electrical forces involved in oxidation.

29. "Guilty men": *Time* (5 Nov. 1945): 27, quoted in Yavendetti, "American

Reactions," see 174–175, 272–274. "Touched very deeply": Oppenheimer, "Atomic physics in civilization," ms., box 29, Bulletin of Atomic Scientists Collection, University of Chicago Library, quoted in Yavendetti, "American Reactions," p. 182. E. Teller, "Atomic Scientists Have Two Responsibilities," *BAS* 3, no. 12 (Dec. 1947): 356. "Known sin": J. Robert Oppenheimer, *The Open Mind* (New York: Simon & Schuster, 1955), 88. See Philip M. Stern with Harold P. Green, *The Oppenheimer Case: Security on Trial* (New York: Harper & Row, 1969), 90. Alan J. Friedman and Carol C. Donley, *Einstein as Myth and Muse* (Cambridge: Cambridge University Press, 1985), 178–180, 194.

30. John Bowlby, *Attachment and Loss*, vol. 3, *Loss: Sadness and Depression* (New York: Basic Books, 1980). Cancer: personal communication from Albert Rothenberg.

31. Stucky, *August 6, 1945*, 58–60. Freeman Dyson, *Disturbing the Universe* (New York: Harper & Row, 1979), 60–61.

32. Oppenheimer, *Open Mind*, 65. Harold Urey with Michael Amrine, "I'm a Frightened Man," *Collier's* 117 (5 Jan. 1946): 18ff. Chicago *Tribune*, 26 Oct. 1945, quoted in Stucky, *August 6, 1945*, 69. Los Alamos sand: Smith, *A Peril and a Hope*, 288.

33. Daniel Lang, *Early Tales of the Atomic Age* (Garden City, N.Y.: Doubleday, 1948), 75, 90. "Action-goading fear": *NYT*, 26 May 1946, p. 7.

34. David L. Hill, Eugene Rabinowitch, and John A. Simpson, Jr., "The Atomic Scientists Speak Up," *Life* (19 Oct. 1945), reprinted in *BAS* 37, no. 1 (Jan. 1981): 23–25.

35. Raymond B. Fosdick, "The Challenge: One World or *None*," *NYT Magazine* (2 Sept. 1945), p. 8. Bertrand Russell in British Broadcasting Corp., *Atomic Challenge: A Symposium* (London: Winchester, 1947), 155. Sullivan, "Cliché Expert," 32.

36. "Too revolutionary": Truman to Congress, 3 Oct. 1945, quoted in Richard G. Hewlett and Oscar E. Anderson, Jr., *A History of the United States Atomic Energy Commission*, vol. 1, *The New World 1939/1946*, WASH 1214 (1962; Springfield, Va.: National Technical Information Service, 1972), 426.

37. "Oppie's plan": Herbert Marks in Lang, *Early Tales*, 102. See notes, Oppenheimer to Lilienthal, 2 Feb. 1946, in "State Dept. Board of Consultants," Govt. File, box 191, Oppenheimer Papers, Library of Congress, Washington, D.C.

38. Smith, *A Peril and a Hope*. "One World—or None" (National Committee on Atomic Information, 1947). Copy in Library of Congress film collection.

39. Arm: "Unhappy Birthday" (ABC radio, 6 Aug. 1946), R77:0236, MB. See Lawrence S. Wittner, *Rebels Against War: The American Peace Movement, 1941–1960* (New York: Columbia University Press, 1969), 134–143, 170–174. Fonda: SANE Education Fund, "Shadows of the Nuclear Age: American Culture and the Bomb" (WGBH-FM broadcast and cassettes, 1980), cassette 1.

40. Joseph Alsop and Stewart Alsop, "Your Flesh *Should* Creep," *Saturday Evening Post* 219 (13 July 1946), 9. Herbert Block, *The Herblock Book* (Boston: Beacon Press, 1952), 35. See Janet Besse and Harold Lasswell, "Our Columnists on the A-Bomb," *World Politics* 3 (Oct. 1950): 72–87.

41. David Holloway, *The Soviet Union and the Arms Race* (New Haven,

Conn.: Yale University Press, 1983). Arnold Kramish, *Atomic Energy in the Soviet Union* (Stanford, Ca.: Stanford University Press, 1959), chap. 7. See *NYT,* 14 Dec. 1945, p. 9; 25 Sept. 1946, p. 3; 23 Dec. 1950, p. 5. Soviet statements: *Current Digest of the Soviet Press* 1, no. 27 (1949): 3–7; 2, no. 4 (1950): 22. Alex Inkeles, *Public Opinion in Soviet Russia: A Study in Mass Persuasion* (Cambridge, Mass.: Harvard University Press, 1951), chap. 16.

42. Richard F. Rosser, *An Introduction to Soviet Foreign Policy* (Englewood Cliffs, N.J.: Prentice-Hall, 1969), 262–266. Marshall D. Shulman, *Stalin's Foreign Policy Reappraised* (Cambridge, Mass.: Harvard University Press, 1963), chap. 4.

43. Harriman quoted in Alva Myrdal, *The Game of Disarmament: How the United States and Russia Run the Arms Race* (New York: Pantheon, 1976), 333, see 379n13.

44. Soviet quotes: *Current Digest of the Soviet Press* 2, no. 30 (1950): 27; 2, no. 29 (1950): 39; 2, no. 28 (1950): 46; see 3, no. 39 (1951): 14–15.

45. Alsop and Alsop, "Your Flesh *Should* Creep." Boyer, *By the Bomb's Early Light,* chap. 13.

46. See James R. Newman and Byron S. Miller, *The Control of Atomic Energy: A Study of Its Social, Economic, and Political Implications* (New York: McGraw-Hill, 1948).

47. Groves, *Now It Can Be Told,* 415, 438–439; Stephane Groueff, *Manhattan Project: The Untold Story of the Making of the Atomic Bomb* (Boston: Little, Brown, 1967), 3, 31–32.

48. "Best-kept secret": *NYT,* 9 Aug. 1945, p. 8, often repeated elsewhere. "Vital secrets": Groves, *Now It Can Be Told,* 140.

49. *NYT* index: my counts, Oct. 1945. Radio: my counts from Executive Office of the President, Division of Press Intelligence, "Atomic Energy," 22 April–25 July, 1947, folder "Radio & Press References," box 7, JCAE. Cabinet: Truman, *Memoirs,* 1: 525–526.

50. News article: David E. Lilienthal, *The Journals of David Lilienthal,* vol. 2, *The Atomic Energy Years 1945–1950* (New York: Harper & Row, 1964), 288. Jones, "Science, Scientists, and Americans," 393, see 115–118, 229–262.

51. For summary and bibliography see Montgomery Hyde, *The Atom Bomb Spies* (New York: Atheneum, 1980). For Groves's role see Gregg Herken, "'A Most Deadly Illusion:' The Atomic Secret and American Nuclear Weapons Policy, 1945–1950," *Pacific History Review* 49 (Feb. 1980): 51–76.

52. David Caute, *The Great Fear: The Anti-Communist Purge under Truman and Eisenhower* (New York: Simon & Schuster, 1978), chap. 5 and p. 541. Edward A. Shils, *Torment of Secrecy.* For general ideas see Sissela Bok, *Secrets: On the Ethics of Concealment and Revelation* (New York: Pantheon, 1982). Harold P. Green, "Q-Clearance: the Development of a Personnel Security Program," *BAS* 20, no. 5 (May 1964): 9–15.

53. Spencer R. Weart, *Scientists in Power* (Cambridge, Mass.: Harvard University Press, 1979), chap. 18. Margaret M. Gowing with Lorna Arnold, *Independence and Deterrence: Britain and Atomic Energy, 1945–1952,* 2 vols. (New York: St. Martin's, 1974), 1: 28–29, 51–55, 179–184, 331.

54. Tadeusz Wittlin, *Commissar: The Life and Death of Lavrenty Pavlov-*

ich Beria (New York: Macmillan, 1972). On Stalin's sadism see Erich Fromm, *The Anatomy of Human Destructiveness* (New York: Holt, Rinehart & Winston, 1973), 280–299. Cf. Paul Kline, *Fact and Fantasy in Freudian Theory* (London: Methuen, 1972), 339.

55. Zhores A. Medvedev, *Soviet Science* (New York: Norton, 1978), chap. 5 and pp. 148–149.

56. Khrushchev: Holloway, *Soviet Union and the Arms Race*, 127, see chap. 2. See also David Yergin, *Shattered Peace: The Origins of the Cold War and the National Security State* (Boston: Houghton Mifflin, 1977), 37, 168–169, 322–324.

57. Andrew Dowdy, *The Films of the Fifties: The American State of Mind* (New York: William Morrow, 1973), 30. Gary H. Grossman, *Saturday Morning TV* (New York: Dell, 1981), 331–334. Annie: Oct.–Dec. 1945, in Harold Gray, *The Life and Hard Times of Little Orphan Annie 1935–1945* (New Rochelle, N.Y.: Arlington, 1970).

58. *Kiss Me Deadly* (United Artists, 1955). *The Atomic Kid* (Republic, 1954). *Atomic Captive* in series *The Adventures of Superman* (Superman Inc., 1950); I used summary in Library of Congress film collection, Lp10527. *The Atomic Man* (Todon-Allied Artists, 1955); I used production sheet in Library of Congress film collection, Lp5991. Signs: *Atomic City* (Paramount, 1952).

59. Quotes: Jones, "Science, Scientists, and Americans," 391, 141. "Little boys": Sullivan, "Cliché Expert," 36.

60. *Seven Days to Noon* (British Lion, 1950).

61. Stewart Alsop, introduction to Ralph Lapp, *The New Force: The Story of Atoms and People* (New York: Harper, 1953), ix. Parliament: Gowing, *Independence and Deterrence*, 1:52. See Robert R. Wilson, "Atomic Taboos," *BAS* 9, no. 1 (Feb. 1953): 7, 12.

62. "Torn from nature": "The Story of Five Bombs" (U.S. Dept. of Defense, 1946?), text in Public Information Office, Argonne National Lab. "Probe": *Congressional Record* 83:1, vol. 1C (1953): 239, as quoted in Gerald J. Ringer, "The Bomb as a Living Symbol: An Interpretation," Ph.D. diss., Florida State University, 1966, 116–117.

63. Edith Sitwell, "Dirge for the New Sunrise," in *The Collected Poems of Edith Sitwell* (New York: Vanguard, 1954), 364–365.

64. See Robert Jay Lifton, *The Broken Connection: On Death and the Continuity of Life* (New York: Simon & Schuster, 1979), 354–357.

65. Boyer, *By the Bomb's Early Light*, 106 and chap. 9 passim.

7. NATIONAL DEFENSES

1. Cyclotrode X: Denis Gifford, *Science Fiction Film* (London: Studio Vista; New York: E. P. Dutton, 1971), 144. Harold Gray, *The Life and Hard Times of Little Orphan Annie 1935–1945* (New Rochelle, N.Y: Arlington, 1970), 30 Oct. 1945.

2. The most complete survey of polls for ca. 1950–1963 is Jiri Nehnevajsa, "Civil Defense and Society," prepared by the Dept. of Sociology, University of

Pittsburgh, for the Office of Civil Defense, U.S. Army, Report AD–445285 (1964), available from Defense Technical Information Center, Alexandria, Va. See also Leonard S. Cottrell, Jr., and Sylvia Eberhart, *American Opinion on World Affairs in the Atomic Age* (1948; New York: Greenwood, 1969); Stephen B. Withey, *4th Survey of Public Knowledge and Attitudes Concerning Civil Defense* (Ann Arbor: Survey Research Center, University of Michigan, 1954).

3. See, for example, Donald Monson, "City Planning in Project East River," *BAS* 9, no. 7 (Sept. 1953): 265–267, 268; Associated Universities, Inc., *Report of Project East River*, part V, *Urban Vulnerability* (New York: Brookhaven National Laboratory, 1952). "Drive cities": Frank Sullivan, "The Cliché Expert Testifies on the Atom," in Sullivan, *A Rock in Every Snowball* (Boston: Little, Brown, 1946), 34.

4. For this and the following see Thomas J. Kerr, "The Civil Defense Shelter Program: A Case Study of the Politics of National Security Policy Making," Ph.D. diss., Syracuse University, 1969; Royal D. Sloan, "The Politics of Civil Defense: Great Britain and the United States," Ph.D. diss., University of Chicago, 1958; Allan M. Winkler, "A 40-Year History of Civil Defense," *BAS* 40, no. 6 (June 1984): 16–22. The rapid rise of interest is plain in the *NYT* index and the *Readers' Guide to Periodical Literature.* United States Executive Office of the President, National Security Resources Board, Civil Defense Office, *Survival under Atomic Attack: The Official U.S. Government Booklet* (Washington, D.C: Government Printing Office, 1950). For this and other 1949–1951 matters I have benefited from talks with Paul Boyer and from Ira Robert Bashkow, "Bandaids for Armageddon: Public Information on the Early Civil Defense Program," B.A. thesis, Harvard University, 1984.

5. For inoculation see Irving L. Janis, *Air War and Emotional Stress: Psychological Studies of Bombing and Civilian Defense* (New York: McGraw-Hill, 1951), 221, 224; Val Peterson, "Panic: The Ultimate Weapon?" *Collier's* 132, no. 6 (21 Aug. 1953): 100–109. The third reason was first suggested to me by McGeorge Bundy, personal communication; note that much of the propaganda begged for "support" for civil defense. See Harry B. Yoshpe, "Our Missing Shield: The U.S. Civil Defense Program in Historical Perspective" (Washington, D.C.: Federal Emergency Management Agency, 1981). "Criminally stupid": Ralph Lapp, "An Interview with Governor Val Peterson," *BAS* 9, no. 7 (Sept. 1953): 241.

6. "See It Now" (CBS-TV, 29 June 1952), T79:0489, MB. "Bomb Target—USA" (CBS radio, 20 March 1953), transcript in "Ground Observer Corps Misc.," box 13, James M. Lambie, Jr., Papers, DDE. In addition to the Lambie Papers see oral history interview of Lambie by Ed Edwin, 1968, Columbia University Library, New York City. "Dangerous reductions": Lambie to Sherman Adams, 9 July 1953, folder "Candor (1)," box 12, White House Central Files, Confidential File, DDE.

7. "War games": for example, Minutes of Cabinet Meeting, 10 June 1955, box 5, Ann Whitman Cabinet Files, DDE. See the cabinet minutes for spring and summer of 1956; Wilson quote in 13 July 1956, box 7. See also "Operation Alert" files, box 49, White House Central Files, Confidential File, DDE.

8. See Wylie to Sen. Brien McMahon, 1951, in folder "Wylie, Philip," box

719, JCAE; Truman F. Keefer, *Philip Wylie* (Boston: Twayne, 1977), 109, 124–128. Philip Wylie, *Tomorrow!* (New York: Rinehart, 1954). I read this first in the late 1950s and attest that some of the scenes are unforgettable. Another realistic book was Judith Merril [Judith Merril Pohl], *Shadow on the Hearth* (Garden City, N.Y: Doubleday, 1950). "We have taught": Wylie in Reginald Bretnor, ed., *Modern Science Fiction: Its Meaning and Its Future* (New York: Coward-McCann, 1953), 240.

9. "Atomic Explosion: Yucca Flats, Nevada" (CBS-TV, 17 March 1953), T77:0328, MB. Withey, *4th Survey of Public Knowledge,* 73–76. Wylie, "Agenda for a Bull Session," to Brien McMahon, 28 Jan. 1952, in folder "Wylie," box 719, JCAE.

10. On civilians witnessing tests see Daniel Lang, *The Man in the Thick Lead Suit* (New York: Oxford University Press, 1954), chap. 4; I also use various personal communications. Michael Uhl and Tod Ensign, *GI Guinea Pigs* (n.p.: Playboy Press, 1980), chaps. 4–5, is useful if its strong bias is allowed for.

11. I draw here on many personal communications. See *School Life,* special issue, vol. 35 suppl. (Sept. 1953); Albert Furtwangler, "Growing Up Nuclear," *BAS* 37, no. 1 (Jan. 1981): 44–48. "Clothing can be destroyed": William M. Lamers, "Identification for School Children," *National Education Association Journal* 41 (1952): 99.

12. *Survival Under Atomic Attack* (Castle Films, 1951). Boy in rubble: *Atomic Alert* (Encyclopedia Brittanica Films, 1951). For other films see *Film News* 11, no. 8 (Sept. 1951) and collections in Library of Congress. A good survey is "No Place to Hide" (Media Study-Public Broadcasting System, 1982).

13. "Please don't let them": SANE Education Fund, "Shadows of the Nuclear Age: American Culture and the Bomb" (WGBH-FM broadcast and cassettes, 1980), cassette 4. "Most terrifying" and a shelter nightmare: Edwin S. Shneidman, *Deaths of Man* (New York: Quadrangle, 1973), 179–185. Other anecdotes may be gathered from any group of adult Americans.

14. Michael J. Carey, "Psychological Fallout," *BAS* 38, no. 1 (Jan. 1982): 20–24; cf. Robert Jay Lifton, *The Broken Connection: On Death and the Continuity of Life* (New York: Simon & Schuster, 1979), 363–365.

15. Eddie McCloskey to D. D. Eisenhower, 25 March 1954, folder "133B Civil Defense," box 656, White House Central Files, Official File, DDE.

16. Cornell University (Richard Crutchfield et al.), *Public Reaction to the Atomic Bomb and World Affairs: A Nation-wide Survey* (Ithaca, N.Y: Cornell University, 1947) (summarized in Cottrell and Eberhart, *American Opinion*); see part II, p. 83. *Public Opinion Quarterly* 9 (Winter 1945–46): 530. For foreign polls, not only French, see France, Institute of Public Opinion, *Sondages* 7 (1945): 183; 9 (1947): 23–24; 10 (1948): 91; 11 (1949): 27; 17 (1955): 23; 19 (1957): 3–51.

17. "Calf crop": "Atomic Bomb—Operation Crossroads" (CBS radio, 28 May 1946), transcript S76:0502, MB, p. 12.

18. Report of the Special Committee of the American Society of Newspaper Editors on Atomic Information Problems, 16 April 1948, in folder "AEC Office of Information," box 59, JCAE. See Elizabeth Douvan and Stephen Withey,

"Some Attitudinal Consequences of Atomic Energy," *Annals of the American Academy of Political and Social Science* 290 (1953): 108–117. "Supernatural . . . heads in the sand": Rensus Lickert on "You and the Atom" (CBS radio, 30–31 July 1946), R76:0223–0224, MB.

19. Janis, *Air War and Emotional Stress,* 239.

20. Wylie, "Panic, Psychology, and the Bomb," *BAS* 10, no. 2 (Feb. 1954): 37–39. See also Withey, *4th Survey of Public Knowledge,* 72. For all these matters see Nehnevajsa, "Civil Defense and Society." For fear and persuasion see Chap. 6, n. 26. Wylie referred to hysterical blindness, which is related to denial; see Chap. 13, n. 67.

21. A. J. Goodpaster, Memo. of Conference (19 Aug. 1956), 30 Aug. 1956, folder "Aug. '56 Diary—Staff Memos," box 17, Ann Whitman DDE Diaries, DDE. Frank Fremont-Smith et al., "The Human Effects of Nuclear Weapons Development," Report to the President, 21 Nov. 1956, folder "AEC 1955–56 (1)," box 4, Ann Whitman Administration Files, DDE.

22. For Britain see Sloan, *Politics of Civil Defense.* For other countries see U.S. Congress, House 86:1, Committee on Government Operations, *Civil Defense in Western Europe and the Soviet Union. 5th Report . . .* (Washington, D.C.: Government Printing Office, 1959). Foster Hailey, *NYT,* 2 Oct. 1949, sec. 4, p. 5.

23. Leon Gouré, *Civil Defense in the Soviet Union* (Berkeley: University of California Press, 1962); U.S. Congress, *Civil Defense in Western Europe and the Soviet Union.* See *Current Digest of the Soviet Press* 6 (1954), no. 19: 22; no. 31: 3–5; *NYT,* 27 March 1954, p. 8; 9 Jan. 1955, p. 9.

24. Bernard Brodie, ed., *The Absolute Weapon* (New York: Harcourt, Brace, 1946), 74. On development of strategy see Fred Kaplan, *The Wizards of Armageddon* (New York: Simon & Schuster, 1983); Gregg Herken, *The Winning Weapon: The Atomic Bomb in the Cold War, 1945–1950* (New York: Knopf, 1980); and my chap. 12.

25. "As other munitions": National Security Council NSC 20/4, Nov. 1948, repeated verbatim in NSC 162/2, 30 Oct. 1953, folder "National Security Council (1)," box 45, White House Central Files, Confidential File, DDE. First published in *The Pentagon Papers: The Defense Department History of United States Decisionmaking on Vietnam,* ed. Mike Gravel (Boston: Beacon Press, 1971), 1: 428. Curtis E. LeMay with MacKinlay Kantor, *Mission with LeMay: My Story* (Garden City, N.Y.: Doubleday, 1965), 481–482, 561.

26. "Brand-new": Wylie, *Tomorrow!,* 367. See Keefer, *Philip Wylie;* Lifton, *Broken Connection,* 332, 350–351.

27. "Immediate destruction": Memo, Lauris Norstad to L. R. Groves, 15 Sept. 1945, File 3, "Top Secret" Correspondence of the Manhattan Engineer District (RG 77), microfilm 1108, National Archives, Washington, D.C. See Herken, *Winning Weapon.* LeMay told designers: Sam Cohen, *The Truth about the Neutron Bomb: The Inventor of the Bomb Speaks Out* (New York: William Morrow, 1983), 30.

28. P. M. S. Blackett, *Fear, War, and the Bomb* (New York: McGraw-Hill, 1949). "Blow a hole"; "not a military weapon": David E. Lilienthal, *The Jour-*

nals of David Lilienthal, vol. 2, *The Atomic Energy Years 1945–1950* (New York: Harper & Row, 1964), 473–474, 391.

29. Drew Pearson, ABC radio, 3 Dec. 1950, in folder "Broadcasts—Pearson," box 106, JCAE. See *NYT,* 1 Dec. 1950, pp. 1, 4, 7–10; 2 Dec. 1950, pp. 3, 6, 12, 23, 46; also 1 Sept. 1950, p. 50. Robert J. Donovan, *Tumultuous Years: The Presidency of Harry S. Truman, 1949–1953* (New York: W. W. Norton, 1982), 307–310.

30. My conclusions come from a wide range of evidence, and this book is not the place for a discussion. See *Pentagon Papers;* Herken, *Winning Weapon;* Edward Friedman, "Nuclear Blackmail and the End of the Korean War," *Modern China* 1 (1975): 75–91.

31. "Frightened": William S. White, "The Atomic Senator," *NYT Magazine* (12 Feb. 1950): 19ff. See folder "Committee Members—McMahon," box 182, and materials in boxes 516–17, JCAE. I am grateful to McMahon's widow, the Baroness Silvercruys, for an interview.

32. "World revolves": Lilienthal, *Journals,* 2: 585.

33. McMahon, "No Miracle in the Marshall Plan," *Collier's* 121, no. 14 (3 April 1948): 22.

34. "No matter how many": Brien McMahon, "Survival—The Real Issue of Our Times," *BAS* 8, no. 6 (Aug. 1952): 173–175.

35. Stanley A. Blumberg and Gwinn Owens, *Energy and Conflict: The Life and Times of Edward Teller* (New York: Putnam's, 1976). Cf. Steve J. Heims, *John von Neumann and Norbert Wiener: From Mathematics to the Technologies of Life and Death* (Cambridge, Mass.: MIT Press, 1980), 37–38 and discussion of von Neumann, passim. Lifton, *Broken Connection,* 421–431 discusses Teller's motivation as "nuclearism;" I question whether it is useful to invent pejoratives for traits we all share.

36. Oral history interviews of Martin Schwarzschild, 1977, pp. 70–71, and of S. Chandrasekhar, 1977, p. 98, by Spencer Weart, AIP.

37. "Really delightful": Blumberg and Owens, *Energy and Conflict,* 119.

38. "It's a boy": Richard G. Hewlett and Francis Duncan, *Atomic Shield, 1947/1952* (University Park: Pennsylvania State University Press, 1969), 542. See Hans Bethe, "Comments on the History of the H-Bomb" (1954), *Los Alamos Science* (Fall 1982): 43–53.

39. Herbert F. York, *The Advisors: Oppenheimer, Teller and the Superbomb* (San Francisco: W. H. Freeman, 1976); David Rosenberg, "American Atomic Strategy and the Hydrogen Bomb Decision," *Journal of American History* 66 (1979): 62–87. "Evil thing": Enrico Fermi and I. I. Rabi, in York, *Advisors,* App.

40. *Life* 28, no. 9 (27 Feb. 1950): 100. True figures: David Alan Rosenberg, "U.S. Nuclear Stockpile, 1945 to 1950," *BAS* 39, no. 5 (May 1982): 25–30.

41. "Frankenstein": Lilienthal, *Journals,* 2: 581. See Donovan, *Tumultuous Years,* chap. 14.

42. Hewlett and Duncan, *Atomic Shield,* 522–587.

43. For these major historical developments see Bertrand Goldschmidt, *The Atomic Complex: A Worldwide Political History of Nuclear Energy* (La Grange Park, Ill.: American Nuclear Society, 1982); Peter Pringle and James J.

Spigelman, *The Nuclear Barons* (New York: Holt, Rinehart & Winston, 1981), with good bibliography.

44. For military public relations see Adam Yarmolinsky, *The Military Establishment: Its Impacts on American Society* (New York: Harper & Row, 1971). On SAC attitudes see Richard G. Hubler, *SAC: The Strategic Air Command* (New York: Duell, Sloan & Pearce, 1958). Godfrey: Thomas M. Coffey, *Iron Eagle: The Turbulent Life of General Curtis LeMay* (New York: Crown, 1986), 326, 337, 348.

45. John Prados, *The Soviet Estimate: U.S. Intelligence Analysis & Russian Military Strength* (New York: Dial, 1982), 42–43, 77.

46. Coffey, *Iron Eagle,* 56–57. "More machine than man": W. B. Huie, "A-Bomb General of Our Air Force," *Coronet* 28 (Oct. 1950): 89. "Irrefutable logic": Ernest Havemann, "Toughest Cop of the Western World," *Life* 36 (14 June 1954): p. 136. LeMay, *Mission,* 495. For the purposes of this book it is irrelevant whether it was LeMay or his biographer who actually wrote these words.

47. Note that I am not relying on Klein's theories but on her clinical findings, which have been repeatedly confirmed. Melanie Klein, *The Psycho-Analysis of Children,* trans. Alix Strachey, rev. with H. A. Thorner (New York: Dell, 1975).

48. For information I thank Jay Spenser. See Ian Logan and Henry Nield, *Classy Chassy* (New York: A & W Visual Library, 1977). For other sexy bombers see Peter Biskind, *Seeing Is Believing: How Hollywood Taught Us to Stop Worrying and Love the Fifties* (New York: Pantheon, 1983), 64–70.

49. "Like a woman": "Self Preservation against Atomic Attack" (Cascade–U.S. Army, 1951). The original "Blonde Bombshell" was a 1933 movie. On dangerous females see Joseph C. Rheingold, *The Mother, Anxiety, and Death: The Catastrophic Death Complex* (London: Churchill, 1967), 66, and many nursery tales.

50. H. G. Wells, *The World Set Free* (1913; New York: Dutton, 1914), 110. *The Beginning or the End* (Metro-Goldwyn-Mayer, 1946). On sexual/military/nuclear ideas see Charles Spretnak, "Naming the Cultural Forces That Push Us toward War," *Journal of Humanistic Psychology* 23, no. 3 (Summer 1983): 104–114.

51. Ronald Schaffer, *Wings of Judgment: American Bombing in World War II* (New York: Oxford University Press, 1985), 17–18. "Mechanically": Harold H. Martin, "Are Our Big Bombers Ready to Go?" *Saturday Evening Post* 223 (30 Dec. 1950): p. 65.

52. "Harder than the metal": Beirne Lay, Jr., and Sy Bartlett, *Twelve O'Clock High!* (1948; New York: Ballantine, 1965), 164. The film (Twentieth Century-Fox, 1950), made the same points but less fully. Cf. *Command Decision* (Metro-Goldwyn-Mayer, 1949). "You have to turn": LeMay, *Mission,* 425, see 383.

53. Morris Janowitz, *The Professional Soldier: A Social and Political Portrait* (Glencoe, Ill.: Free Press, 1960), 189–194; David A. Anderton, *Strategic Air Command: Two-Thirds of the Triad* (London: Allan, 1975), 24, 27.

54. *Above and Beyond* (Metro-Goldwyn-Mayer, 1953). For this and other movies I used reviews in NYPL/TC, especially from the *NYT* and the *New York Herald Tribune.* See also folder "Clippings—Atomic Bomb," NYPL/TC. Paul W. Tibbets, Jr., with Clair Stebbins and Harry Franken, *The Tibbets Story* (New York: Stein & Day, 1978), 167, 262–266. Coffey, *Iron Eagle,* 293.

55. *Strategic Air Command* (Paramount, 1955). *Bombers B-52* (Warner Bros., 1957). *A Gathering of Eagles* (Universal, 1963). See Julian Smith, *Looking Away: Hollywood and Vietnam* (New York: Scribner's, 1975), chap. 6; Lawrence H. Suid, *Guts & Glory: Great American War Movies* (Reading, Mass.: Addison-Wesley, 1978), 187–215; Biskind, *Seeing Is Believing,* 64–70.

56. Wayne: see, for example, *Fort Apache* (RKO, 1948); *Flying Leathernecks* (RKO, 1951). "Divided man": "Wagon Train: The Fort Pierce Story" (ABC-TV, 23 Sept. 1963), T78:0610–0611, MB. This view seems to have been especially common in the 1950s, and probably relates to the poorly-understood connections between the Cold War and changing sexual roles.

57. Press release: Smith, *Looking Away,* 192–193.

58. "Forerunner": Hubler, *SAC,* 262, see 280 and passim. For Wellsian spirit see also Janowitz, *Professional Soldier,* 307; Havemann, "Toughest Cop," 147.

59. P. M. S. Blackett, *Studies of War* (New York: Hill & Wang, 1962), 93.

60. "Robot-soldier": *Current Digest of the Soviet Press* 1, no. 33 (1949).

8. ATOMS FOR PEACE

1. Public knowledge: "The 'Hydrogen Bomb' Story," *BAS* 8, no. 9 (Dec. 1952): 297–300. Pearson, ABC radio, 1 Jan. 1950, in folder "Broadcasts—Pearson," box 106, JCAE. See also William L. Laurence, *The Hell Bomb* (New York: Knopf, 1951). Speculations: for example, *Time* (13 Feb. 1950): 49. Special atom-smasher: Oak Ridge National Laboratory, *ORNL Review* 9, no. 4 (Fall 1976): 73; George A. Kolstad, ms. autobiography, 1982, AIP.

2. C. D. Jackson to Walter B. Smith, 10 Nov. 1953, folder "OCB—Misc. Memos (2)," box 1, Jackson Records, DDE. See also Jackson Papers, DDE, and materials in boxes 24 and 32, Ann Whitman Administrative Files, DDE.

3. Robert J. Donovan, *Eisenhower: The Inside Story* (New York: Harper & Row, 1956), chap. 13. J. R. Oppenheimer, "Atomic Weapons and Foreign Policy," *Foreign Affairs* 31 (1953): 525–535.

4. Lambie to Sherman Adams, 9 July 1953, folder "Candor (1)," box 12, White House Central Files, Confidential File, DDE. See also Lambie Papers, DDE.

5. "Scare the country": probably a paraphrase of Jackson's recollection, John Lear, "Ike and the Peaceful Atom," *The Reporter* 14, no. 1 (12 Jan. 1956): 11. "Bang-bang": Jackson to Lewis Strauss, folder "Atoms for Peace," box 5, Ann Whitman Administration Files, DDE; see Jackson to Horace Craig, 13 Aug. 1953, folder "Candor (12)," box 12, White House Central Files, Confidential File, DDE.

6. IVY showing: Robert Cutler, memorandum, 29 May 1953, folder "108A Atomic Weapons (1)," box 525, White House Central Files, Official Files, DDE.

7. Soviets behind U.S.: Herbert F. York, *The Advisors: Oppenheimer, Teller and the Superbomb* (San Francisco: W. H. Freeman, 1976); note that this book draws on knowledge of classified material.

8. "Peaceful uses": p. 12 of draft in Jackson to Horace Craig, 13 Aug. 1953, folder "Candor (12)," box 12, White House Central Files, Confidential File, DDE. W. Sterling Cole to D. D. Eisenhower, 21 Aug. 1953, and reply, 25 Aug., folder "President's Papers 1953 (3)," box 1, Records of the Office of the Special Assistant for National Security Affairs, Special Asst. Series, Presidential Subseries, DDE.

9. Robert Cutler to Lewis Strauss and C. D. Jackson, 10 Sept. 1953, copy in folder "Atoms for Peace," box 5, Ann Whitman Administration File, DDE. See Thomas F. Soapes, "A Cold Warrior Seeks Peace: Eisenhower's Strategy for Nuclear Disarmament," *Diplomatic History* 4 (1980): 62.

10. William Laurence, "Paradise or Doomsday?" *Woman's Home Companion* 75 (May 1948): 33.

11. I counted items or measured lengths of columns in the *Readers' Guide to Periodical Literature,* s.v. "Atom . . . ," "Hydrogen Bomb," "Nuclear . . . ," "Radiation," "Radioact . . . ," "Radium," and cross-references from those rubrics in particular volumes (for example, "Milk contamination," "Reactors"). Assignment into categories was subjective, and proved stable within 10 percent when I did recounts four years later. I normalized all figures to the total number of articles in the volume, estimating from the number of pages. Newsreels: survey, s.v. "Atom" of card catalog for Paramount, Pathe, and some others, Sherman Grinberg Film Library, New York City. *France-Soir:* J. P. Delzani and C. Carde, "Approche psychologique de l'évolution du thème nucléaire à partir d'un quotidien populaire de 1950 à 1976," in *Colloque sur les implications psycho-sociologique du développement de l'industrie nucléaire,* ed. M. Tubiana (Paris: Société Française de Radioprotection, 1977), 354–374.

12. Kaempffert, *NYT,* 28 Jan. 1947, p. 16. See Zalmay Khalilzad and Cheryl Benard, "Energy: No Quick Fix for a Permanent Crisis," *BAS* 36, no. 10 (Dec. 1980): 15–20. Soviet: A. Vishinsky, quoted in *NYT,* 11 Nov. 1949, p. 1; see 13 Nov. 1949, p. 32.

13. David E. Lilienthal, *TVA: Democracy on the March* (New York: Harper, 1944); Lilienthal, *The Journals of David Lilienthal* (New York: Harper & Row, 1964–1983).

14. Lilienthal, *Journals,* 2: 16–17.

15. Lilienthal, *Change, Hope, and the Bomb* (Princeton, N.J.: Princeton University Press, 1963), 23.

16. "Atoms for Peace": Lilienthal, *Journals,* 2: 635, see 474–475. The phrase was used earlier, for example, "Atoms for Peace," *Newsweek* 28 (5 Aug. 1946): 33.

17. "Fascination": Thomas E. Murray, *Nuclear Policy for War and Peace* (Cleveland: World, 1960), 8. Lilienthal, *Change, Hope,* 109.

18. Richard G. Hewlett and Francis Duncan, *Atomic Shield, 1947/1952* (University Park: Pennsylvania State University Press, 1969), 435–438. See United States Congress, 82:2, Joint Committee on Atomic Energy, *Atomic Power and Private Enterprise* (Washington, D.C.: Government Printing Office, 1952).

19. On the British see Margaret M. Gowing with Lorna Arnold, *Independence and Deterrence: Britain and Atomic Energy, 1945–1952*, 2 vols. (New York: St. Martin's, 1974), 2: 290–301; on the Soviets see Arnold Kramish, *Atomic Energy in the Soviet Union* (Stanford, Ca.: Stanford University Press, 1959); George A. Modelski, *Atomic Energy in the Communist Bloc* (Melbourne: University of Melbourne Press, 1959). "Goal of national importance": Report by Atomic Energy Commission, 6 March 1953, folder "NSC 145," box 4, White House Office of the Special Assistant for National Security Affairs, NSC Series, Policy Papers, DDE. U.S. Congress, 83:1, Joint Committee on Atomic Energy, *Atomic Power Development and Private Industry: Hearings* (Washington, D.C.: Government Printing Office, 1953).

20. "Wonderful": Burton R. Fisher, C. A. Metzner, and B. J. Darsky, *Peacetime Uses of Atomic Energy*, 2 vols. (Ann Arbor: Survey Research Center, University of Michigan, 1951), 2: 56–57. "To the general public": Leonard S. Cottrell, Jr., and Sylvia Eberhart, *American Opinion on World Affairs in the Atomic Age* (Princeton, N.J.: Princeton University Press, 1948), 36. "Restricted to the upper": Elizabeth Douvan and Stephen Withey, "Public Reaction to Nonmilitary Aspects of Atomic Energy," *Science* 119 (1954): 1–3. Opinion leaders: Central Surveys, Inc., report to Electric Companies Public Information Program, 12 Aug. 1953, folder "108 Atomic Energy 1954," box 523, White House Central Files, Official Files, DDE. See also L. W. Kay and I. J. Gitlin, "Atomic Energy or the Atomic Bomb: A Problem in the Development of Morale and Opinion," *Journal of Social Psychology* 29 (1949): 57–84.

21. "Direct challenge": Jackson to OCB, 9 Dec. 1953, folder "OCB—Misc. memos (2)," box 1, Jackson Records, DDE. See Peter Pringle and James J. Spigelman, *The Nuclear Barons* (New York: Holt, Rinehart & Winston, 1981), 125–126. "Detracted popular attention": Stefan Possony, "The Atoms for Peace Program," in F. L. Anderson Panel, "Psychological Aspects of United States Strategy: Source Book . . . ," Nov. 1955, folder "Rockefeller (5)," box 61, White House Central Files, Confidential Files, DDE, p. 203.

22. I studied systematically the card catalogs of the New York Public Library, New York City, the Bibliothèque Nationale, Paris, and the Deutsches Museum, Munich, under appropriate rubrics. The Paramount-Pathe newsreel index, s.v. "Atom . . ." and "Aviation—Air Raids—Protection" lists 15 peaceful and 49 military items in 1949–53, but 42 peaceful and 27 military items in 1954–1956.

23. International Atomic Energy Agency, *Films on the Peaceful Uses of Atomic Energy* (Vienna: IAEA, 1962).

24. Laura Fermi, *Atoms for the World: United States Participation in the Conference on the Peaceful Uses of Atomic Energy* (Chicago: University of Chicago Press, 1957), 64, see 189–193.

25. "Aladdins Wunderlampe": Karl Winnacker, *Nie den Mut verlieren. Erinnerungen . . .* (Düsseldorf: Econ, 1971), 311–312; see Pringle and Spigelman, *Nuclear Barons,* 172–175.

26. Jack M. Holl, "Eisenhower's Peaceful Atomic Diplomacy: Atoms for Peace in the Public Interest," in *Papers Presented at the Conference on Energy in American History* (Blacksburg, Va.: Virginia Tech Center for the Study of Science in Society, 1982). See Possony, "Atoms for Peace Program."

27. "Condense decades": M. Ahmed quoted in Khalilzad and Benard, "Energy: No Quick Fix," 18. See James Everett Katz and Oukar S. Marwah, eds., *Nuclear Power in Developing Countries* (Lexington, Mass.: D.C. Heath, 1982), 162 and passim. Numbers of agreements: Lewis Strauss to D. D. Eisenhower, 28 Feb. 1958, folder "AEC 1958 (4)," box 5, Ann Whitman Administration File, DDE. Lilienthal, *Change, Hope,* 111–112.

28. *Izvestia: Current Digest of the Soviet Press* 2, no. 13 (1950): 31. See Modelski, *Atomic Energy,* 103–107. Paul R. Josephson, "The Historical Roots of the Chernobyl Disaster," preprint, 1987.

29. Harrison Salisbury, *NYT,* 2 Oct. 1949, sec. 4, p. 5; Modelski, *Atomic Energy,* 13.

30. "Additional exhibits": Memo of conference, 14 Jan. 1955, folder "AEC 1955–56 (8)," box 5, Ann Whitman Administration File, DDE. Poll: *NYT,* 20 Dec. 1948, p. 27. AEC Public Information Service, list of stock film footage, 1952, folder "Civilian Defense," Library of Congress film collection, Washington, D.C. "Atoms for Peace" film: folder "AEC—Motion Pictures 1955," JCAE. On all this see United States Congress, 84:2, Joint Committee on Atomic Energy, *Peaceful Uses of Atomic Energy: Background Material for the Report of the Panel on the Impact of the Peaceful Uses . . .* , vol. 2 of McKinney Panel report (Washington, D.C.: 1956), 565–574.

31. Lewis L. Strauss, *Men and Decisions* (Garden City, N.Y.: Doubleday, 1952), 4, 429. "Beneficent use": Strauss, remarks at Rockhurst College, Mo., 24 May 1955, fiche SPCH–1, AEC speeches, Public Document Room, Nuclear Regulatory Commission, Washington, D.C., p. 30.

32. Strauss, remarks for National Association of Science Writers, New York City, 16 Sept. 1954; my thanks to George Mazuzan for a copy of this AEC press release. William Laurence, *Men and Atoms: The Discovery, the Uses, and the Future of Atomic Energy* (New York: Simon & Schuster, 1959), 245.

33. "I realized": Victor Cohn in Sharon Friedman, *Science in the Newspaper,* no. 1 (Washington, D.C.: American Association for the Advancement of Science, 1974), 21. See Hillier Krieghbaum, *Science and the Mass Media* (New York: New York University Press, 1967), 72–73, 92–94.

34. Arnold Kramish and Eugene M. Zuckert, *Atomic Energy for Your Business: Today's Key to Tomorrow's Profits* (New York: McKay, 1956), 56. "Contest is on": Francis K. McCune, "Atomic Power—A Challenge to U.S. Leadership," *General Electric Review* (Nov. 1955): 10.

35. *Fortune* (Mar. 1952): 112.

36. Folders "Atomic Industrial Forum," Carroll L. Wilson Papers, Massachusetts Institute of Technology Archives, Cambridge, Mass.; I thank Helen

Slotkin for assistance. Members in 1956 are listed in Kramish and Zuckert, *Atomic Energy for Your Business,* App. E, 248–256.

37. L. W. Cronkhite in Atomic Industrial Forum, *Atomic Energy, a Realistic Appraisal. Proceedings of a Meeting . . .* (New York: AIF, 1955), 1; Cisler, ibid., 99–105.

38. "Cold-war weapon": Lambie to Sherman Adams, 6 Oct. 1954, folder "Atomic Industrial Forum," box 11, Lambie Papers, DDE. For GE see Harold Beaudoin in Atomic Industrial Forum, *Public Relations for the Atomic Industry. Proceedings of a Meeting . . .* (New York: AIF, 1956), 107–108.

39. For churches' responses and other matters see Erwin N. Hiebert, *The Impact of Atomic Energy: A History of Responses by Governments, Scientists and Religious Groups* (Newton, Kan.: Faith and Life Press, 1961).

40. Johan Galtung, *Atoms for Peace: A Comparative Study of Students' Attitudes* (Oslo: Institut for samfunnsforskning Report 5–1, 1960), in *Peaceful Uses of Atomic Energy,* ed. Otto Klineberg (Paris: UNESCO, 1964), 37–39. See also France, Institute of Public Opinion, *Sondages* 19, no. 3 (1957): 16–17; 20, no. 1–2 (1958): 113; 26, no. 4 (1964): 40.

41. See special issues of *School Life:* 31 suppl. (Mar. 1949); 35 suppl. (Sept. 1953). Beaudoin in Atomic Industrial Forum, *Public Relations for the Atomic Industry. A Is for Atom* (General Electric, 1952).

42. *Our Friend the Atom* (Walt Disney, 1956). Heinz Haber, *The Walt Disney Story of Our Friend the Atom* (New York: Simon & Schuster, 1956).

43. "Good Atoms": George L. Glasheen, "What Schools Are Doing in Atomic Energy Education," *School Life* 35 suppl. (Sept. 1953): p. 153.

9. GOOD AND BAD ATOMS

1. I use "structure" in the sense developed by Claude Lévi-Strauss. Books that I analyzed with structure specifically in mind included *Anwendung der Atomenergie für friedliche Zwecke,* trans. from *Primenenie atomnoj energii v. mirnych celjach* (Moscow, 1956) by Victor Bredel (Leipzig and Jena: Urania, 1956); Albert Bouzat, *L'Energie atomique,* Que sais-je? no. 317 (1949; Paris: Presses universitaires de France, 1957); David Dietz, *Atomic Science, Bombs, and Power* (New York: Mead, 1954; New York: Collier, rev. ed., 1962); Jean-Jacques Libert, *L'Energie Atomique* (Paris: Hachette, 1959); Gerhard Löwenthal and Josef Hansen, *Wir werden durch Atome leben* (West Berlin: Lothar Blanvalet, 1956); Martin Mann, *Peacetime Uses of Atomic Energy* (New York: Crowell, 1957; rev. ed., 1961); James Stokley, *The New World of the Atom* (New York: Washburn, 1957; rev. ed., 1970); and David O. Woodbury, *Atoms for Peace* (New York: Dodd, Mead, 1955).

Films analyzed included *A Is for Atom* (General Electric, 1952); *The Atom Comes to Town* (U.S. Chamber of Commerce, 1957); "Atoms for Peace" (Warner News–U.S. Information Agency, 6 parts, 1955–1957); French newsreel (April 1950), Paramount D–3519, Sherman Grinberg Film Library, New York City; and "Atoms for Peace" (Moscow Popular Science Studios, 1956) as reviewed in *NYT* and *New York Herald Tribune,* 25 June 1956, and *Izvestiia* in *Current Digest of the Soviet Press* 8, no. 14 (1956): 40.

Also John M. McCullough, "Atomic Energy: Utopia or Oblivion?" reprinted articles from the Philadelphia *Inquirer* (Philadelphia: The Inquirer, 1947); Soviet 1958 Geneva exhibit as described in *L'Energie atomique à des fins pacifiques* (Geneva, 1958); and a comic book by Joe Musial, "Dagwood Splits the Atom!" (King Features, 1949).

2. William Laurence, *Dawn over Zero: The Story of the Atomic Bomb* (New York: Knopf, 2nd ed., 1946; Westport, Conn.: Greenwood, 1972), 254.

3. Jack Schubert and Ralph E. Lapp, *Radiation: What It Is and How It Affects You* (New York: Viking, 1957), chaps. 5, 6, 8. Dade W. Moeller, James G. Terrill, Jr., and Samuel C. Ingraham II, "Radiation Exposure in the United States," *Public Health Reports* 68 (1953): 57–65.

4. Ruth Ashton, "The Sunny Side of the Atom" (CBS radio, 30 June 1947), transcript in box 22, Federation of Atomic Scientists Collection, University of Chicago Library; see Paul Boyer, *By the Bomb's Early Light* (New York: Pantheon, 1985), 299–301.

5. "Mystery of photosynthesis": A. Kursanov, *Current Digest of the Soviet Press* 5, no. 48 (1953): 11–12, 35. Peanuts: *The Atom Comes to Town.*

6. For AEC budget see Harold Orlans, *Contracting for Atoms: A Study of Public Policy Issues . . .* (Washington, D.C.: Brookings Institute, 1967), 112.

7. Piston rings: AEC Press Release no. 153, 28 Jan. 1949, Records of the AEC, Germantown, Md. *The Magic of the Atom* (Lee Handel, 1955–1956), summaries in Library of Congress film collection, MP7121.

8. See Kenneth MacDonald Jones, "Science, Scientists, and Americans: Images of Science and the Formation of Federal Science Policy, 1945–1950," Ph.D. diss., Cornell University, 1975, 69–73. My thanks to Paul Josephson for information on Russian toys. Victor Appleton II, *Tom Swift and His Triphibian Aircar* (New York: Grosset and Dunlap, 1962). Ford: *NYT,* 2 Oct. 1951, p. 38. David Sarnoff, "The Fabulous Future," *Fortune* 51 (Jan. 1955): 83.

9. "The Fight for an Ultimate Weapon," *Newsweek* 47 (4 June 1956): 55–60. See John Tierney, "Take the A-Plane," *Science 82* 3, no. 1 (Jan.–Feb. 1982): 47–55.

10. Richard G. Hewlett and Francis Duncan, *Nuclear Navy, 1946–1962* (Chicago: University of Chicago Press, 1974); Norman Polmar and Thomas B. Allen, *Rickover* (New York: Simon & Schuster, 1982).

11. For safe reactors see Freeman Dyson, *Disturbing the Universe* (New York: Harper & Row, 1979), 95–104. Irving C. Bupp and Jean-Claude Derian, *Light Water: How the Nuclear Dream Dissolved* (New York: Basic Books, 1978).

12. Richard Rhodes, "A Demonstration at Shippingport," *American Heritage* 32, no. 4 (June-July 1981): 66–73.

13. For reactor development see Peter deLeon, *Development and Diffusion of the Nuclear Power Reactor: A Comparative Analysis* (Cambridge, Mass.: Ballinger, 1979); Steven L. Del Sesto, *Science, Politics, and Controversy: Civilian Nuclear Power in the United States, 1946–1974* (Boulder, Colo.: Westview, 1979).

14. Bruce C. Netschert, *The Future Supply of Oil and Gas . . . Through 1975* (Baltimore, Md.: Johns Hopkins Press, 1958); Gerald D. Nash, *United States*

Oil Policy, 1890–1964 (Pittsburgh: University of Pittsburgh Press, 1968). M. King Hubbert, "Nuclear Energy and the Fossil Fuels," in *Drilling and Production Practice—1956* (New York: American Petroleum Institute, 1957), 7–25.

15. One brief introduction is Claude Lévi-Strauss, "The Structural Study of Myth," pp. 81–106 in Thomas A. Seboek, ed., *Myth: A Symposium* (Bloomington: Indiana University Press, 1958). Ira Chernus, *Dr. Strangegod: On the Symbolic Meaning of Nuclear Weapons* (Columbia: University of South Carolina Press, 1986), 6–7 and passim. "Good & Bad Atoms," *Time* 49 (31 Mar. 1947): 81.

16. "Hard core": Lebaron Foster, "Public Thinking on the Peacetime Atom," in Atomic Industrial Forum, *Public Relations for the Atomic Industry. Proceedings of a Meeting . . .* (New York: AIF, 1956), 85. For similar results see Burton R. Fisher, C. A. Metzner, and B. J. Darsky, *Peacetime Uses of Atomic Energy*, 2 vols. (Ann Arbor: Survey Research Center, University of Michigan, 1951), 56–57; F. Fagnani, ed., *Le Débat nucléaire en France: Acteurs sociaux et communication de Masse* (Grenoble: Université des Sciences Sociales, Institut de Recherche Economique, 1977), 4–5, 13.

17. Frank Pittman in Atomic Industrial Forum, *Public Relations*, 94–95.

18. "Humanité de plus en plus mécanisé": Charles-Noël Martin, *Promesses et menaces de l'énergie nucléaire* (Paris: Presses Universitaires de France, 1960), 250.

19. "Obedient and tireless servant": Eisenhower's message to 1955 Geneva Atoms for Peace Conference, repeated by Richard Nixon for the 1972 Conference, in International Conference on the Peaceful Uses of Atomic Energy, Fourth, *Peaceful Uses of Atomic Energy* (New York: United Nations, 1972), 1: 86. Productions confusing reactors with bombs included Edward R. Murrow, "See It Now: Atomic Timetable" (CBS-TV, 24 Nov. 1957; McGraw-Hill Films, 1958), T77:0150, MB; *Our Friend the Atom* (Walt Disney, 1956).

20. L. W. Kay and I. J. Gitlin, "Atomic Energy or the Atomic Bomb: A Problem in the Development of Morale and Opinion," *Journal of Social Psychology* 29 (1949): 57–84; their sample (N = 33) is small, but I confirmed the findings by research in Brookhaven records and publications. I am grateful to Allan Needell and the BNL staff for invaluable assistance.

21. "A lot of thought": Michael Amrine to Morse Salisbury, 3 Feb. 1949, folder "Public Relations 1948–50," temp. box 37, Haworth Papers, Brookhaven National Laboratory, Upton, N.Y., microfilm at AIP. Compare study of Marcoule center by Isac Chiva, "Imagination Collective et Inconnu," in *Echanges et Communications: Mélanges offerts à Claude Lévi-Strauss . . .* (Paris: Mouton, 1970): 162–168. Gases etc.: Daniel Lang, *Early Tales of the Atomic Age* (Garden City, N.Y.: Doubleday, 1948), 191–192.

22. "Would not be hampered": Eldon Sharp and Leland Haworth to Wilbur F. Kelley, 28 May 1949, folder "Public Relations 1948–50," temp. box 37, Haworth Papers, Brookhaven National Laboratory, Upton, N.Y., microfilm at AIP. "Actually safer": Laurence Swart, "Brookhaven Laboratory and Long Island" (WHLI radio, 8 June 1948), transcript in folder "Speeches–1948," box 23, Records of Associated Universities, Inc., Brookhaven National Laboratory. I am grateful to J. B. Horner Kuper for a tape-recorded interview, 1980.

23. "Things dangerous to touch": Fisher, *Peacetime Uses* 2: 12–16, see 25–28. Workers: Joseph Blank, "Atomic Tragedy in Texas," *Look* 21 (3 Sept. 1957): 25–29; George T. Mazuzan and J. Samuel Walker, *Controlling the Atom: The Beginnings of Nuclear Regulation 1946–1962* (Berkeley: University of California Press, 1984), 327–332. Mazuzan and Walker generously shared the drafts of this book with me. Glowing man: for example, *The Atomic Kid* (Mickey Rooney Productions, 1954).

24. Complaints: for example, Ritchie Calder, *Living with the Atom* (Chicago: University of Chicago Press, 1962), 41. *Scientific American* case: Hillier Krieghbaum, *Science and the Mass Media* (New York: New York University Press, 1967), 194–195.

25. See John Major, *The Oppenheimer Hearing* (New York: Stein & Day, 1971); Barton J. Bernstein, "'In the Matter of J. Robert Oppenheimer,'" *Historical Studies in the Physical Sciences* 12 (1982): 195–252, and the references they cite. For Alsop friendship see folder "Alsop," box 15, General Case File, Oppenheimer Papers, Library of Congress, Washington, D.C.

26. Heinar Kipphardt, *In the Matter of J. Robert Oppenheimer,* trans. Ruth Speirs (1964; New York: Hill and Wang, 1968), 127; *The Day after Trinity* (Pyramid, 1981).

27. Aaron Wildavsky, *Dixon-Yates: A Study in Power Politics* (New Haven, Conn.: Yale University Press, 1962). For background see Sheldon Novick, *The Electric War: The Fight over Nuclear Power* (San Francisco: Sierra Club, 1976). "Virginity": Harold P. Green and Alan Rosenthal, *Government of the Atom: The Integration of Powers* (New York: Prentice-Hall, 1963), 64–65. For public reaction see box 328, White House Central Files, General File, DDE.

10. THE NEW BLASPHEMY

1. Here and in the following see Robert A. Divine, *Blowing on the Wind: The Nuclear Test Ban Debate 1954–1960* (New York: Oxford University Press, 1978); IVY film: *NYT,* 1 April 1954, p. 1; 2 April 1954, pp. 5, 35; 8 April 1954, p. 8.

2. Release of film: "Film on Operation Ivy," 8 Dec. 1953, AEC 483/47, folder "MRA7 Ivy vol. 3," box 1262, AEC Secretariat Records, Dept. of Energy, Germantown, Md. I thank Roger Anders for help with these materials. See also folder "108D Operation Ivy Film," box 526, White House Central Files, Official File, DDE; *Fortune* (Nov. 1953): 121.

3. Great Britain, Commons, *Debates* 315 (1 March 1955): 1895.

4. My counts of various indexes and library book catalogs. World press: World Health Organization, "Mental Health Aspects of the Peaceful Uses of Atomic Energy: Report of a Study Group," Technical Report Series no. 151 (Geneva: W.H.O., 1958), 19, 50–53.

5. My counts from lists in Philip D. Segal, "Imaginative Literature and the Atomic Bombs: An Analysis of Representative Novels, Plays, and Films from 1945 to 1972," Ph.D. diss., Yeshiva University, New York, 1973. See also John Newman and Kathleen J. Kruger, "Imaginary War Fiction in Colorado State University Libraries: A Bibliography," *Bulletin of Bibliography* 35 (1978): 157–

171; Denis Gifford, *The Science Fiction Film* (New York: E. P. Dutton, 1971), 144–148.

6. David B. Parker, "Mist of Death over New York," *Reader's Digest* (April 1947): 7–10.

7. "AEC sources": Roger Sprague to William Knowland, 10 May 1952, folder "AEC motion pictures 1952–54," JCAE. Tom Lehrer, "The Wild West" (1953) in *Too Many Songs from Tom Lehrer* (New York: Pantheon, 1981). I have benefited from Barton C. Hacker, "Elements of Controversy: A History of Radiation Safety in the Nuclear Test Program," draft chapters, published as *The Dragon's Tail: Radiation Safety in the Manhattan Project, 1942–1946* (Berkeley, Calif.: University of California Press, 1987). Also useful is Richard G. Hewlett, "Nuclear Weapon Testing and Studies Related to Health Effects: An Historical Summary," in U.S. Dept. of Health and Human Services, Interagency Radiation Research Committee, *NIH Publication No. 81–507* (Washington, D.C.: National Institutes of Health, 1980), 24–101, see 25–31.

8. Howard Ball, *Justice Downwind: America's Testing Program in the 1950's* (New York: Oxford University Press, 1986). Cronkite: "The U.N. in Action" (CBS-TV, 17 March 1953), T77:0329, MB.

9. *Asahi*, 17 March 1954, trans. in folder "Weapons Tests 1954," box 712, JCAE. See Ralph Lapp, *The Voyage of the Lucky Dragon* (New York: Harper, 1958). Coral: Press Release, 31 March 1954, in folder "Weapons Tests (Eniwetok)," box 712, JCAE.

10. See Eugene J. Rosi, "Elite Political Communication: Five Washington Columnists on Nuclear Weapons Testing," *Social Research* 34 (1967): 713; Rosi, "How 50 Periodicals and the *Times* Interpreted the Test Ban Controversy," *Journalism Quarterly* 41 (1964): 545–556.

11. Lapp, *Voyage of the Lucky Dragon*; Herbert Passin, "Japan and the H-Bomb," *BAS* 11, no. 8 (Oct. 1955): 289–292.

12. Nahun Z. Medalia and Otto N. Larsen, "Diffusion and Belief in a Collective Delusion: The Seattle Windshield Pitting Epidemic," *American Sociological Review* (1958): 180–186.

13. Kefauver, *NYT*, 17 Oct. 1956, p. 1. *The Day the Earth Caught Fire* (scripted 1954, produced 1961), discussed in John Brosnan, *Future Tense: The Cinema of Science Fiction* (New York: St. Martin's, 1978), 144–147. Khrushchev, *NYT*, 31 May 1957, p. 8. Earthquakes discussed in minutes of AEC meetings 1274, 27 March 1957, and 1277, 17 April 1957, Records of the AEC, Dept. of Energy.

14. U.S. Congress, 84:1, Joint Committee on Atomic Energy, *Hearings . . . on Health and Safety Problems and Weather Effects Associated with Atomic Explosions* (Washington, D.C.: Government Printing Office, 1955). Many letters are found in the records of the Joint Committee, the AEC, and Eisenhower. France, Institute of Public Opinion, *Sondages* 26, no. 4 (1964): 40.

15. See Mary Douglas and Aaron Wildavsky, *Risk and Culture: An Essay on the Selection of Technical and Environmental Dangers* (Berkeley: University of California Press, 1982); Claude Lévi-Strauss, *Introduction to a Science of Mythology*, vol. 1, *The Raw and the Cooked*, trans. John Weightman and

Doreen Weightman (New York: Harper & Row, 1969). Sigmund Freud, *Totem and Taboo,* trans. James Strachey (New York: Norton, 1950) has useful ideas as well as fantasy. See also references in my Chap. 2, n. 43. H. G. Wells, *Tono-Bungay* (1908; reprint, Garden City, N.Y.: Garden City, 1927), 386–387.

16. "Unkindly copulations": Giambattista della Porta, *Natural Magick* (1558; trans. 1658; New York: Basic Books, 1957). See Herbert Maisch, *Incest,* trans. Colin Bearne (London: Deutsch, 1973), 39–42; George Devereux, *A Study of Abortion in Primitive Societies* (London: Yoseloff, 1960), 8–9, 141–142; Norman Cohn, *Europe's Inner Demons: An Inquiry Inspired by the Great Witch-Hunt* (New York: Basic Books, 1975), 16, 40, 46, 53. Maya: Bernardino de Sahagùn, *Florentine Codex: General History of the Things of New Spain,* trans. Arthur Anderson and Charles Dibble (Ogden: University of Utah Press, 1953), part 8: 8–9. The matter is unexplored and beset with taboos; why, for example, were witches accused of attacking childbirth and infants but almost never of causing birth defects?

17. Susan Sontag, *Illness as Metaphor* (New York: Vintage, 1979).

18. Joseph Alsop and Stewart Alsop, 18 Jan. 1950, quoted in Norman Moss, *Men Who Play God: The Story of the H-Bomb and How the World Came to Live with It* (New York: Harper & Row, 1968), 33. William Randolph Hearst, Los Angeles *Herald Express,* 16 March 1954. Pius XII in *NYT,* 19 April 1954, p. 12; 11 April 1955, pp. 1, 27.

19. Boise pastors, 16 April 1954, and similar letters in folder "Weapons Tests (Eniwetok) 1954," box 712, JCAE.

20. *Gojira* (Toho, 1954; in English with additions, 1955).

21. *Them!* (Warner Brothers, 1954). Moviegoers: *NYT* and *New York Herald Tribune* reviews, 17 June 1954. See Dennis Saleh, *Science Fiction Gold: Film Classics of the 50s* (New York: McGraw-Hill, 1979), 99–121; Frederick Pohl and Frederick Pohl IV, *Science Fiction: Studies in Film* (New York: Grosset & Dunlap, 1981).

22. *It Came from Beneath the Sea* (Columbia, 1955); *Attack of the Giant Leeches* (American International, 1959); *The Black Scorpion* (Warner Bros., 1957); *Tarantula* (Universal, 1955); *Attack of the Crab Monsters* (Allied, 1957). Grasshoppers: *The Beginning of the End* (ABPT, 1957).

23. The comic book literature must be studied by haphazard sampling. I thank Spike Barkin and the Museum of Cartoon Art, Port Chester, N.Y. See especially *Weird Science* (EC, 1950; reprinted in 4 vols., West Plains, Mo.: Russ Chochran, n.d.). Michael L. Fleisher, *The Encyclopedia of Comic Books,* vol. 3, *The Great Superman Book* (New York: Warner, 1978), 369 and s.v. "Krypton." Readership: S. E. Finer, "A Profile of Science Fiction," *Sociological Review* n.s. 2 (1954): 239–255; John W. Campbell, Jr., in Regnald Bretnor, ed., *Modern Science Fiction: Its Meaning and Its Future* (New York: Coward-McCann, 1953), 21.

24. *The Beast from 20,000 Fathoms* (Warner Brothers, 1953).

25. One cheap 1950s film showed atomic explosions but little more: *Invasion U.S.A.* (Columbia, 1953). Faraway wars: *Rocketship X-M* (Lippert, 1950); *This Island Earth* (Universal, 1955); *The Mysterians* (Toho, 1957); *Teenage*

Cave Man (American International, 1955); *World Without End* (Allied, 1955); *The Time Travelers* (American International, 1964). *The Deadly Mantis* (Universal, 1957); *Rodan* (Toho, 1957).

26. Susan Sontag, "The Imagination of Disaster," in Sontag, *Against Interpretation and Other Essays* (New York: Delta, 1966), 208–225.

27. *X the Unknown* (Hammer, 1958); cf. *The Magnetic Monster* (United Artists, 1953).

28. In the literature and in conversations with several psychologists I have found only one case of a man convinced he was bombarded with nuclear rays. The gap was briefly noticed in World Health Organization, "Mental Health Aspects," 17, and a few subsequent studies.

29. *The Quatermass Experiment* (Hammer, 1955), U.S. release *The Creeping Unknown* (United Artists, 1957); see Brosnan, *Future Tense*, 108–112. Similarly, *The Day the World Ended* (American-Golden State, 1955). *The Thing* (RKO, 1951).

30. Paul A. Carter, *The Creation of Tomorrow: Fifty Years of Magazine Science Fiction* (New York: Columbia University Press, 1977), 160–161, 230, 250–253; see Groff Conklin, ed., *Science Fiction Adventures in Mutation* (New York: Vanguard, 1955). Brosnan, *Future Tense*, 95.

31. J. Fred MacDonald, "The Cold War as Entertainment in 'Fifties Television," *Journal of Popular Film and Television* 7 (1978): 3–31; Pohl and Pohl, *Science Fiction: Studies in Film*, 105–106.

32. Sontag, "Imagination of Disaster," 223; Andrew Dowdy, *The Films of the Fifties: The American State of Mind* (New York: William Morrow, 1973), chap. 9. See *The Blob* (Paramount, 1958); *Invaders from Mars* (Twentieth Century-Fox, 1953). Giant baby: *The Amazing Colossal Man* (American International, 1957); *War of the Colossal Beast* (American International, 1958).

33. "Prone to terror": review of *The Magnetic Monster, New York Herald Tribune*, 14 May 1953; like other reviews I found this in NYPL/TC. "It's radioactive!": for example, *The Crawling Eye*, alt. title *The Trollenberg Terror* (Eros, 1958); *War of the Worlds* (Paramount, 1953).

34. See George Basalla, "Pop science: The Depiction of Science in Popular Culture," in *Science and Its Public: The Changing Relationship*, ed. Gerald Holton and William A. Blanpied (Boston: Reidel, 1976), 261–278.

35. Robert Jay Lifton, *Death in Life: Survivors of Hiroshima* (New York: Simon & Schuster, 1967), 110 and passim. See Committee for the Compilation of Materials on Damage Caused by the Atomic Bombs, *Hiroshima and Nagasaki: The Physical, Medical, and Social Effects of the Atomic Bombings*, trans. Eisei Ishikawa and David L. Swan (New York: Basic Books, 1981), 420–427, 481ff.

36. In addition to the works cited in n. 35 see Donald Richie, "Mono no aware: Hiroshima in Film," in *Film: Book Two. Films of Peace and War*, ed. Robert Hughes (New York: Grove, 1962), 67–86; A. M. Halpern, "Changing Japanese Attitudes Toward Atomic Weapons," Project Rand Research Memorandum RM-1331 (Santa Monica, Ca.: Rand Corp., 1 Sept. 1954); Yasumasa Tanaka, "Japanese Attitudes Toward Nuclear Arms," *Public Opinion Quar-*

terly 34 (1970): 26–42. "Fürchterlichen Experiment": Ernst H. Krause, *Atom am Horizont* (Leipzig: Urania, 1960), 175.

37. Guinea pigs: Lifton, *Death in Life,* 512. Film by A. Kurosawa, *Ikimono no Kiroku* (1955), English title *I Live in Fear* or *Record of a Living Being.* J. Victor Koschmann, "Postwar Democracy and the Japanese Ban-the-Bomb Movement," talk presented at American Historical Association meeting, New York City, 29 Dec. 1985.

38. The *NYT* Index and the *Saturday Review* are good guides. Michael Yavendetti, "The Hiroshima Maidens and American Benevolence in the 1950s," *Mid-America* 64 (April-July 1982): 21–39; Yavendetti, "American Reactions to the Use of Atomic Bombs on Japan, 1945–1947," Ph.D. diss., University of California, Berkeley, 1970; Bernard M. Kramer, S. Michael Kalick, and Michael A. Milburn, "Attitudes Toward Nuclear Weapons and Nuclear War: 1945–1982," *Journal of Social Issues* 39 (1983): 7–24.

39. I heard these rumors ca. 1953. Paul W. Tibbets, Jr., with Clair Stebbins and Harry Franken, *The Tibbets Story* (New York: Stein & Day, 1978), 7; Joseph L. Marx, *Seven Hours to Zero* (New York: Putnam's, 1967), 9, 205–209, 245.

40. Claude R. Eatherly and Gunther Anders, *Burning Conscience: The Case of the Hiroshima Pilot* (New York: Monthly Review Press, 1962). Cf. William B. Huie, *Hiroshima Pilot: The Case of Major Claude Eatherly* (New York: Putnam's, 1964).

41. A handy compendium is Paul R. Baker, ed., *The Atomic Bomb: The Great Decision* (Hinsdale, Ill.: Dryden, rev. ed., 1976). Martin Sherwin, *A World Destroyed: The Atomic Bomb and the Grand Alliance* (New York: Knopf, 1975) is the best to date. See Martin J. Sherwin, "Old Issues in New Editions," *BAS* 41, no. 11 (Dec. 1985): 40–44; John Lewis Gaddis, "A Time of Confrontation and Confusion," *Times Literary Supplement* (8 May 1987): 479–480. Five films about the decision are listed in John Dowling, *War Peace Film Guide* (Chicago: World Without War, rev. ed., 1980). The scholars' and propagandists' views should be compared with those like "Hiroshima: A Soldier's View," *New Republic* 185 (22 Aug. 1981): 26–30; Laurens van der Post, *The Prisoner and the Bomb* (New York: William Morrow, 1971); Hisako Matsubara, *Cranes at Dusk,* trans. Leila Vennewitz (Garden City, N.Y.: Doubleday-Dial, 1985).

11. DEATH DUST

1. Thomas Murray to Eisenhower, folder "1953–54 AEC (5)," box 4, Ann Whitman Administration File, DDE. Here and below see Robert A. Divine, *Blowing on the Wind: The Nuclear Test Ban Debate 1954–1960* (New York: Oxford University Press, 1978) and Robert Gilpin, *American Scientists and Nuclear Weapons Policy* (Princeton, N.J.: Princeton University Press, 1962).

2. Nehru quoted in Divine, *Blowing on the Wind,* 20.

3. Elof Axel Carlson, *Genes, Radiation, and Society: The Life and Work of H. J. Muller* (Ithaca, N.Y.: Cornell University Press, 1981), 337–339 and chap. 25. "Custodians": Muller in *The Science of Radiology,* ed. Otto H. Glas-

ser (Springield, Ill.: C. C. Thomas, 1933), 314. Muller in *NYT,* 26 April 1955, p. 17.

4. *NYT,* 13 June 1956, p. 1. Here and below see George T. Mazuzan and J. Samuel Walker, *Controlling the Atom: The Beginnings of Nuclear Regulation 1946–1962* (Berkeley: University of California Press, 1984), chap. 2.

5. See Eugene J. Rosi, "How 50 Periodicals and the *Times* Interpreted the Test Ban Controversy," *Journalism Quarterly* 41 (1964): 545–556; Frederick M. O'Hara, Jr., "Attitudes of American Magazines toward Atmospheric Nuclear Testing, 1945–1965," Ph.D. diss., University of Illinois at Urbana, 1974. *U.S. News & World Report* 38 (25 March 1955): 21–26.

6. "Face the Nation" (CBS-TV and radio, 19 June 1955), transcript in folder "Broadcasts—general," box 106, JCAE.

7. Survey: Charles E. Osgood, George J. Suci, and Percy H. Tannenbaum, *The Measurement of Meaning* (Urbana: University of Illinois Press, 1957), 105–109.

8. "Most dreadful poison": John B. Martin, *Adlai Stevenson and the World* (Garden City, N.Y.: Doubleday, 1977), 373; Adlai E. Stevenson, "Why I Raised the H-bomb Question," *Look* 21 (5 Feb. 1957): 23–25.

9. Letters in folders "155-B Sept. 1956 (1,2)," box 1215, White House Central Files, General File, DDE.

10. Norman Cousins, *Dr. Schweitzer of Lambarèné* (New York: Harper, 1960).

11. Linus Pauling, *No More War!* (New York: Dodd, Mead & Co., 1958); see Carolyn Kopp, "The Origins of the American Scientific Debate over Fallout Hazards," *Social Studies of Science* 9 (1979): 403–422.

12. Linus Pauling, *Science* 128 (14 Nov. 1958): 1183–86.

13. Journals: O'Hara, "Attitudes of American Magazines." For Joint Committee see Clinton P. Anderson with Milton Viorst, *Outsider in the Senate* (New York: World, 1970): 198–203.

14. Edward Teller and Albert Latter, "The Compelling Need for Nuclear Tests," *Life* 44 (19 Feb. 1958): 64–72.

15. Letters: box 1215, White House Central Files, General File, DDE. For novels see Philip D. Segal, "Imaginative Literature and the Atomic Bombs: An Analysis of Representative Novels, Plays, and Films from 1945 to 1972," Ph.D. diss., Yeshiva University, New York, 1973, 58–60. Hazel Gaudet Erskine, "The Polls: Defense, Peace, and Space," *Public Opinion Quarterly* 25 (1961): 478–489; Eugene J. Rosi, "Mass and Tentative Opinion on Nuclear Weapons Tests and Fallout, 1954–1963," *Public Opinion Quarterly* 29 (1965): 280–297; France, Institute of Public Opinion, *Sondages* 19, no. 3 (1957): 19. David Lilienthal to Carroll L. Wilson, 26 May 1958, folder "Lilienthal," Wilson Papers, Massachusetts Institute of Technology Archives.

16. Herbert F. York, "Sakharov and the Nuclear Test Ban," *BAS* 31, no. 9 (Nov. 1981): 33–37. Resolution of USSR Supreme Soviet, 31 March 1958, in *The Soviet Stand on Disarmament* (New York: Crosscurrents, 1962), 90. See Andrei V. Lebedinsky, ed., *What Russian Scientists Say about Fallout* (Moscow, 1959), trans. George Yankovsky (New York: Collier, 1962), also published as *Soviet Scientists Concerning the Dangers of Nuclear Weapons Tests,* OTS

publication 59–11, 547 (Washington, D.C.: Office of Technical Services, U.S. Dept. of Commerce, 1959).

17. *NYT,* 16 April 1958, p. 9; 31 Oct. 1961, p. 14. Soviet statement of 31 Aug. 1961, *Soviet Stand on Disarmament,* 101.

18. *NYT,* 18 May 1967, p. 2; 3 Feb. 1958, p. 1, 13; 26 Oct. 1961, p. 12; 30 April 1962, p. 3.

19. See letters from servicemen in box 310, JCAE. Much is currently being written on this matter. "See It Now: Atomic Timetable Part 2, Fallout" (CBS-TV, 30 March 1958), T77:0151, MB.

20. I am following the classic definition of anxiety in Sigmund Freud, *Inhibitions, Symptoms and Anxiety,* trans. and ed. James Strachey, in *The Standard Edition of the Complete Psychological Works of Sigmund Freud* (London: Hogarth, 1959), vol. 20, addendum B.

21. Mitsuo Taketani in *Time* 69 (3 June 1957): 65.

22. For this and the following see especially Committee for the Compilation of Materials on Damage Caused by the Atomic Bombs, *Hiroshima and Nagasaki: The Physical, Medical, and Social Effects of the Atomic Bombings,* trans. Eisei Ishikawa and David L. Swan (New York: Basic Books, 1981); Itsuzo Shigematsu and Abraham Kagan, eds., *Cancer in Atomic Bomb Survivors* (New York: Plenum, 1986); Gary D. Fullerton et al., eds., *Biological Risks of Medical Irradiations,* American Association of Physicists in Medicine, Medical Physics Monograph no. 5 (New York: American Institute of Physics, 1980); National Academy of Sciences, National Research Council, Committee on the Biological Effects of Ionizing Radiation, *The Effects on Populations of Exposure to Low Levels of Ionizing Radiation: 1980,* BEIR III (Washington, D.C.: National Academy Press, 1980); John D. Boice, Jr., and Joseph J. Fraumeni, Jr., eds., *Radiation Carcenogenesis: Epidemiology and Biological Significance* (New York: Raven, 1984).

23. T. D. Luckey, "Physiological Benefits from Low Levels of Ionizing Radiation," *Health Physics* 43 (1982): 771–789; Richard J. Hickey et al., "Low-Level Ionizing Radiation and Human Mortality: Multi-Regional Epidemiological Studies. A Preliminary Report," *Health Physics* 40 (1981): 625–641; "Radiation Hormesis," special issue of *Health Physics* 52, no. 5 (May 1987).

24. See, for example, "Radiation Hazards from Fallout and X-Rays," *Consumer Reports* 23, no. 9 (Sept. 1958): 484–488. Critics' views: Pauling in "See It Now . . . Fallout;" Cousins, *Dr. Schweitzer,* 173, 175.

25. "Not to scare": Minutes of Cabinet Meeting, 12 May 1960, box 16, Ann Whitman Cabinet Files, DDE. "Security": for example, announcement at bomb test on "Atomic Bomb Blast" (CBS-TV, 5 May 1955), T78:0333, MB.

26. Adm. Arthur Radford, 15 June 1957, folder "Radford (1)," box 32, Ann Whitman Administration File, DDE, p. 9.

27. "Death ray": Sen. Thomas Dodd quoted in Divine, *Blowing on the Wind,* 305–306. *NYT,* 28 June 1957, p. 22.

28. Edward Teller with Allen Brown, *The Legacy of Hiroshima* (Garden City, N.Y.: Doubleday, 1962), 81–91; Teller, "We're Going to Work Miracles," *Popular Mechanics* 113 (Mar. 1960): 97ff.

29. John McPhee, *The Curve of Binding Energy* (New York: Farrar, Straus

and Giroux, 1974), 157–159. Victor Appleton II, *Tom Swift and His Atomic Earth Blaster* (New York: Grosset & Dunlap, 1954).

30. Teller and Brown, *Legacy of Hiroshima,* 56. *Crack in the World* (Security-Paramount, 1965). "Digging too deep": C. M. Kornbluth, "Gomez," in Anthony Boucher, ed., *A Treasury of Great Science Fiction* (Garden City, N.Y.: Doubleday, 1959), 1: 305.

31. See Anderson, *Outsider in the Senate,* and folder "Anderson vs. Adm. Strauss," box 806, Clinton P. Anderson Papers, Library of Congress, Washington, D.C.

32. "Falling Out" (Lincoln Diamant, 1961). See letters in box 1215, White House Central Files, General File, DDE. Walter R. Guild in John M. Fowler, ed., *Fallout: A Study of Superbombs, Strontium-90, and Survival* (New York: Basic Books, 1960), 91.

33. Divine, *Blowing on the Wind,* 323.

34. John Bowlby, *Attachment and Loss,* vol. 3, *Loss: Sadness and Depression* (New York: Basic Books, 1980), chap. 4; David E. Stennard, *Shrinking History: On Freud and the Failure of Psychiatry* (New York: Oxford University Press, 1980), 93–94, 107; Paul Kline, *Fact and Fantasy in Freudian Theory* (London: Methuen, 1972), 181–182, 355.

35. See n. 24, above.

36. "The Contaminators," *Playboy* 6, no. 10 (Oct. 1959): 38. See advertisements in *NYT,* 5 July 1962, p. 54; 18 April 1962, p. 26.

37. *NYT,* 24 Jan. 1962, p. 1; 7 Oct. 1959, p. 39; 30 July 1960, p. 25; 28 Aug. 1962, p. 56. See letters in folder "Fallout vol. 8," JCAE.

38. Cartoon: *Washington Post,* 24 Oct. 1961, p. A14. Snow: Benjamin Spock, "Do Your Children Worry About War?" *Ladies' Home Journal* 79, no. 8 (Sept. 1962): 48.

12. THE IMAGINATON OF SURVIVAL

1. *Public Papers of the Presidents of the United States. John F. Kennedy, 1961* (Washington, D.C.: Government Printing Office, 1962), 625. This whole series is a rich source of clichés. See Folder "UN speech—Memoranda," box 64, Sorenson Papers, JFK. *Good Will to Men* (Metro-Goldwyn-Mayer, 1955), following *Peace on Earth* (Metro-Goldwyn-Mayer, 1939), scripts in Library of Congress motion picture division, Washington, D.C.

2. Murray in *NYT,* 18 Nov. 1955, pp. 1, 13. Vienna appeal quoted in Committee for the Compilation of Materials on Damage Caused by the Atomic Bombs, *Hiroshima and Nagasaki: The Physical, Medical, and Social Effects of the Atomic Bombings,* trans. Eisei Ishikawa and David L. Swan (New York: Basic Books, 1981), 577.

3. For U.S. polls see Chap. 7 n. 2. "Germany . . . a desert": Hans Speier, *German Rearmament and Atomic War: The Views of German Military and Political Leaders* (Evanston, Ill.: Row, Peterson, 1957), 249; see Elisabeth Noelle and Erich P. Neumann, eds., *The Germans: Public Opinion Polls 1947–1966,* trans. Gerard Finan (Allensbach: Verlag für Demoskopie, 1967), 594.

4. Reversal of views documented in Jiri Nehnevajsa, "Civil Defense and Society," prepared by the Dept. of Sociology, University of Pittsburgh, for the Office of Civil Defense, U.S. Army, Report AD-445285 (1964), pp. 34–35, available from Defense Technical Information Center, Alexandria, Va. On Sputnik see Walter A. McDougall, "Technocracy and Statecraft in the Space Age—Toward the History of a Saltation," *American Historical Review* 87 (1982): 1010–40.

5. "Ten tons": White House Press Release, 21 Jan. 1964.

6. Nevil Shute [Nevil S. Norway], *Slide Rule: The Autobiography of an Engineer* (New York: Morrow, 1954), 140 and passim; Julian Smith, *Nevil Shute (Nevil Shute Norway)* (Boston: Twayne, 1976), 124–128.

7. Nevil Shute [Nevil Shute Norway], *Whatever Happened to the Corbetts* (1938), published in U.S. as *Ordeal* (New York: William Morrow, 1939). Shute, *On the Beach* (New York: William Morrow, 1957). Sales figures from *Publishers' Weekly* 117 (25 Jan. 1960), 218, and jacket blurb of 1982 Signet printing.

8. Stanley Kramer in *New York Herald Tribune,* 13 Dec. 1959, sec. 4, p. 1. Sales calculated allowing for inflation from listings in *Variety,* 9 Jan. 1980. "Too stark": *New York Herald Tribune,* 18 Dec. 1959. See "On the Beach" clippings file, NYPL/TC.

9. Cabinet meeting, 11 Dec. 1956, box 15, Ann Whitman Cabinet Series, DDE.

10. David Hawkins, Edith C. Truslow, and Ralph Carlisle Smith, *Project Y: The Los Alamos Story,* LAMS-2532 (1947; Los Angeles: Tomash, 1983), 187. Edward Teller with Allen Brown, *The Legacy of Hiroshima* (Garden City, N.Y.: Doubleday, 1962), 239.

11. Teller and Brown, *Legacy of Hiroshima,* 241.

12. Smith, *Nevil Shute,* 133. See also Joseph Keyerleber in *Nuclear War Films,* ed. Jack G. Shaheen (Carbondale: Southern Illinois University Press, 1978), 31–38; Robert Hughes, ed., *Film: Book Two. Films of Peace and War* (New York: Grove, 1962), 92–96. Another discouraging after-the-bombs book written on a deathbed is Robert C. O'Brien, *Z for Zachariah* (New York: Atheneum, 1975).

13. "Who survives?" John M. McCullough, "Atomic Energy: Utopia or Oblivion?" reprinted articles from the Philadelphia *Inquirer* (Philadelphia: The Inquirer, 1947), 27. Upton Sinclair, *A Giant's Strength: Drama in Three Acts* (Monrovia, Calif.: By the Author, 1948). Aldous Huxley, *Ape and Essence* (New York: Harper, 1948). *Rocketship X-M* (Lippert, 1950); *Captive Women* (RKO, 1952); *World Without End* (Allied Artists, 1955); *Teenage Cave Man* (American International, 1958); *The Time Machine* (Metro-Goldwyn-Mayer, 1960); *The Time Travelers* (American International, 1964); *Planet of the Apes* (Twentieth Century-Fox, 1968); *Beneath the Planet of the Apes* (1970); *Escape from the Planet of the Apes* (1971); *Conquest of the Planet of the Apes* (1972); *Battle of the Planet of the Apes* (1973); *Back to the Planet of the Apes* (1974). Short stories are collected in the useful volumes by H. Bruce Franklin, ed., *Countdown to Midnight: Twelve Great Stories about Nuclear War* (New York: Daw, 1984), and Walter M. Miller, Jr., and Martin H. Greenberg, eds., *Beyond Ar-*

mageddon (New York: Donald Fine, 1985). For a bibliography of books see John Newman and Kathleen J. Kruger, "Imaginary War Fiction in Colorado State University Libraries: A Bibliography," *Bulletin of Bibliography* 35 (1978): 157–171. See Paul Brians, "Resources for the Study of Nuclear War in Fiction," *Science-Fiction Studies* 13 (1986): 193–197.

14. "Good habits": quoted in Mikiso Hane, *Peasants, Rebels, and Outcastes: The Underside of Modern Japan* (New York: Pantheon, 1982), 72, see p. 36. Antimodernism in general is discussed in many works; for specific stories see Paul A. Carter, *The Creation of Tomorrow: Fifty Years of Magazine Science Fiction* (New York: Columbia University Press, 1977), 231–233, 241–244. Richard Jefferies, *After London; or Wild England* (1885; London: Duckworth, 1911).

15. Jack London, *The Scarlet Plague* (1915) in *The Science Fiction of Jack London: An Anthology*, ed. Richard G. Powers (Boston: Gregg, 1975), 285–455, see pp. xvi-xvii. Richard Rafael in *Astounding Science-Fiction* 27 (May 1941), quoted in Carter, *Creation of Tomorrow*, 242. Ray Bradbury, *Fahrenheit 451* (New York: Random House, 1953); similarly Bradbury, *The Martian Chronicles* (Garden City, N.Y.: Doubleday, 1958).

16. *Five* (Columbia, 1951); *The World, the Flesh, and the Devil* (Metro-Goldwyn-Mayer, 1959). See Shaheen, *Nuclear War Films*. Similarly *Panic in Year Zero* (United, 1962) and Ray Bradbury, "The Million-Year Picnic," in Bradbury, *Martian Chronicles*. Teen fantasies: Robert Jay Lifton, *The Broken Connection: On Death and the Continuity of Life* (New York: Simon & Schuster, 1979), 364, and private communications.

17. Milne's autobiography and other material cited in Robert Plank, "The Lone Survivor," in *The End of the World*, ed. Eric S. Rabkin, Martin H. Greenberg, and Joseph D. Olander (Carbondale: Southern Illinois University Press, 1983), 20–52. Anna Freud, *The Ego and the Mechanisms of Defense*, rev. ed., in *The Writings of Anna Freud*, vol. 2 (New York: International Universities Press, 1966), 170–171.

18. Television: see Marc Scott Zicree, *The Twilight Zone Companion* (New York: Bantam, 1982), s.v. "Two" (Sept. 1961) and "Probe 7—Over and Out" (Nov. 1963).

19. For origins of films see clippings files and pressbooks, NYPL/TC. "Bomb would kill": Benjamin Spock, "Do Your Children Worry About War?" *Ladies' Home Journal* 79, no. 8 (Sept. 1962): 48.

20. *You Can Beat the A Bomb* (RKO, 1950). See *Panic in Year Zero; The Atomic Cafe* (Archives Project, 1982); *No Place to Hide* (Media Study, 1982).

21. Pat Frank, *Alas, Babylon* (New York: Lippincott, 1959; New York, Bantam, 1980); Frank, "Hiroshima: Point of No Return," *Saturday Review* (24 Dec. 1960): 25.

22. Walter M. Miller, Jr., *A Canticle for Liebowitz* (Philadelphia: Lippincott, 1959). See David Ketterer, *New Worlds for Old: The Apocalyptic Imagination, Science Fiction, and American Literature* (Garden City, N.Y.: Doubleday Anchor, 1974), 139–148; David N. Samuelson, "The Last Canticles of Walter M. Miller, Jr.," *Science-Fiction Studies* 8 (1976): 3–24; Dominic Man-

ganiello, "History as Judgment and Promise in *A Canticle for Leibowitz*," *Science-Fiction Studies* 13 (1986): 159–168.

23. The Twilight Zone, "Where Is Everybody?" (CBS-TV, 2 Oct. 1959), T77:0039, MB; "Time Enough at Last" (20 Nov. 1959), T82:0044, MB.

24. John Wyndham, *Re-Birth,* also titled *The Chrysalids* (1955), in *The John Wyndham Omnibus* (New York: Simon & Schuster, 1964). Cf. the frequently reprinted André Norton [Alice Mary Norton], *Star Man's Son* (New York: Harcourt, Brace, Jovanovich, 1952).

25. See especially *The Day the World Ended* (Golden State, 1955).

26. The mad general of *Dr. Strangelove* came in 1963 and I discuss him at that point, in Chapter 14 (the 1958 novel where he first appeared was little read). Another exception, much later, is in Philip K. Dick and Roger Zelazny, *Deus Irae* (New York: Dell, 1976).

27. *The Forbidden Planet* (Metro-Goldwyn-Mayer, 1956).

28. *20,000 Leagues Under the Sea* (Buena Vista, 1954).

29. Destroy all humanity: Karl Menninger, *Man Against Himself* (1938; New York: Harcourt, Brace, Jovanovich, 1966), 180, see also pt. 2, passim; Walter M. Miller, Jr., in Miller and Greenberg, *Beyond Armageddon,* 46.

30. Hitler: Albert Speer, *Inside the Third Reich,* trans. Richard Winston and Clara Winston (New York: Macmillan-Collier, 1970), 438–440. For suicide and references see Lifton, *Broken Connection,* chap. 17; Ira Chernus, *Dr. Strangegod: On the Symbolic Meaning of Nuclear Weapons* (Columbia: University of South Carolina Press, 1986), chap. 9, see also chap. 8 on sacrifices and martyrdom; Daniel Lawrence O'Keefe, *Stolen Lightning: The Social Theory of Magic* (New York: Random House, 1982), 306–307, 318–319. "Samson": Donald N. Michael, "The Psychopathology of Nuclear War," *BAS* 18, no. 5 (May 1962): 28–29.

31. The best-described case is Frederick C. Crews, *The Sins of the Fathers: Hawthorne's Psychological Themes* (New York: Oxford University Press, 1966), 241–242 and passim. See Jean Jules Verne, *Jules Verne: A Biography,* trans. Roger Greaves (New York: Taplinger, 1976), 7; William E. Harkins, *Karel Čapek* (New York: Columbia University Press, 1972). Wells was a particularly complex case, for his relation with both parents was ambiguous, as was his work taken as a whole. See Anthony West, *H. G. Wells: Aspects of a Life* (New York: Random House, 1984).

32. Jack London, "A Thousand Deaths" (1899), in *Future Perfect: American Science Fiction of the Nineteenth Century,* ed. H. Bruce Franklin (London: Oxford University Press, rev. ed., 1978), 222–239. Joan London, *Jack London and His Times: An Unconventional Biography* (1939; Seattle: University of Washington Press, 1968); Robert Barltop, *Jack London: The Man, the Writer, the Rebel* (London: Pluto, 1976).

33. Edwin Balmer and Philip Wylie, *When Worlds Collide* (Philadelphia: Lippincott, 1933); Wylie, *Tomorrow!* (New York: Rinehart, 1954); Wylie, *Triumph* (Garden City, N.Y.: Doubleday, 1963).

34. Truman F. Keefer, *Philip Wylie* (Boston: Twayne, 1977), 19, 127, and passim.

35. Philip Wylie, "Blunder: A Story of the End of the World," *Collier's* 117 (12 Jan. 1946): 11ff.

36. For example, Ross A. Lewis in *Milwaukee Journal,* 16 May 1951, in *The Cartoons of R. A. Lewis,* ed. George Lockwood (Milwaukee: The Journal, 1968), 51. See Gerald J. Ringer, "The Bomb as a Living Symbol: An Interpretation," Ph.D. diss., Florida State University, 1966, 249–251. Precursor cartoons are H. Daumier, "Equilibre Européen," Dec. 1866, April 1867. Raymond L. Garthoff, "Cuba: Even Dicier Than We Knew," *Newsweek* 110, no. 17 (26 Oct. 1987): 34.

37. For overview and references to the original works (which I have used but do not list here) see Michael Mandelbaum, *The Nuclear Question: The United States and Nuclear Weapons, 1946–1976* (Cambridge: Cambridge University Press, 1979); Fred Kaplan, *Wizards of Armageddon* (New York: Simon & Schuster, 1983); Lawrence Freedman, *The Evolution of Nuclear Strategy* (New York: St. Martin's, 1981); Laurence Martin, ed., *Strategic Thought in the Nuclear Age* (Baltimore: Johns Hopkins University Press, 1979); Joseph D. Douglass, Jr., *Soviet Military Strategy in Europe* (New York: Pergamon, 1980); Joseph D. Douglass, Jr., and Amoretta M. Hoeber, *Soviet Strategy for Nuclear War* (Stanford, Calif.: Hoover Institution, 1979); David Holloway, *The Soviet Union and the Arms Race* (New Haven, Conn.: Yale University Press, 1983); Michael MccGwire, *Military Objectives in Soviet Foreign Policy* (Washington, D.C.: Brookings Institution, 1987).

38. Churchill in Great Britain, Commons, *Debates* 537 (1 March 1955), 1899, see 1902.

39. A good summary is Thomas Powers, "Choosing a Strategy for World War III," *Atlantic* 250 (Nov. 1982): 82–110.

40. Herman Kahn, *On Thermonuclear War* (Princeton, N.J.: Princeton University Press, 2nd ed. 1961), see 145ff.

41. Herman Kahn, *Thinking about the Unthinkable* (1962; New York: Avon, rev. ed., 1966), 198, quoting Bertrand Russell, *Common Sense and Nuclear Warfare* (New York: Simon & Schuster, 1959), 30. See Freedman, *Evolution of Nuclear Strategy,* 188–189, 219–222.

42. Nixon: Jonathon Schell, *The Time of Illusion* (New York: Knopf, 1976), chap. 6; Seymour M. Hersh, *The Price of Power: Kissinger in the Nixon White House* (New York: Simon & Schuster, 1983), 51–53, 124–125.

43. For example, George F. Kennan, *The Nuclear Delusion: Soviet-American Relations in the Atomic Age* (New York: Pantheon, 1982), 142–143. The contradictions of strategists' thinking will be explored in a forthcoming work by Steven Kull, whom I thank for discussions.

44. Quoted in Thomas Powers, *Thinking about the Next War* (New York: Knopf, 1983), 51.

45. U.S. Congress, *Congressional Record* 114 (2 Oct. 1968): 29175.

46. SANE Education Fund, "Shadows of the Nuclear Age: American Culture and the Bomb" (WGBH-FM broadcast and cassettes), 1980, cassette VIII.

47. Bonestell paintings in the New-York Historical Society Museum. See, for example, *Collier's* 126, no. 6 (5 Aug. 1950): cover and 12–13. Similarly, Los Angeles *Times,* 12 March 1961, p. C1.

48. Stephen B. Withey, *4th Survey of Public Knowledge and Attitudes Concerning Civil Defense* (Ann Arbor: Survey Research Center, University of Michigan, 1954), 72.

49. *The War Game* (BBC-TV, 1966); see Shaheen, *Nuclear War Films*, 109–115.

50. Nehnevajsa, "Civil Defense and Society."

51. National Academy of Sciences, National Research Council, "Emergency Planning and Behavioral Research" (Washington, D.C: National Academy of Sciences, 1962), copy in folder "ND2 Civil Defense 6/62–3/63," box 596, Central Subject Files, JFK, p. 5.

52. Postwar conditions: Val Peterson quoted in Kahn, *On Thermonuclear War*, 438, see 71–95. Arthur M. Katz, *Life after Thermonuclear War: The Economic and Social Impacts of Nuclear Attacks on the United States* (Cambridge, Mass.: Ballinger, 1981); cf. review by Joseph Romm, *BAS* 38, no. 6 (June 1982): 48–49.

53. Khrushchev in *NYT*, 17 Jan. 1963, pp. 1–2. Suslov in *Current Digest of the Soviet Press* 16, no. 13 (1964): 10. See *Current Digest* 16, no. 38 (1964): 23; 14, no. 4 (1962): 19–20.

54. Leon Gouré, *Civil Defense in the Soviet Union* (Berkeley: University of California Press, 1962), 119; Ray Garthoff, *The Soviet Image of Future War* (Washington, D.C.: Public Affairs Press, 1959), chap. 4.

55. See the sources in n. 37 above.

56. I am grateful to Mark Azbel, Leon Gouré, Loren Graham, Gregory Guroff, Krysztof Szymborski, and a dozen anonymous émigrés for discussions. See Gregory Guroff and Steven Grant, "Soviet Elites: World View and Perceptions of the U.S," United States International Communication Agency report R-19-81 (Sept. 1981).

13. THE POLITICS OF SURVIVAL

1. Photographs: Folders "Atom Protest," Wide World agency, New York City.

2. For Britain see Christopher Driver, *The Disarmers: A Study in Protest* (London: Hodder & Stoughton, 1964); Frank Parkin, *Middle Class Radicalism: The Social Bases of the British Campaign for Nuclear Disarmament* (Manchester: Manchester University Press, 1968); William P. Snyder, *The Politics of British Defense Policy, 1945–1962* (Bowling Green: Ohio State University Press, 1964). For the U.S. see Milton Steven Katz, "Peace, Politics, and Protest: SANE and the American Peace Movement, 1957–1972," Ph.D. diss., St. Louis University, 1973; Lawrence S. Wittner, *Rebels Against War: The American Peace Movement, 1941–1960* (New York: Columbia University Press, 1969). For Japan see Committee for the Compilation of Materials on Damage Caused by the Atomic Bombs, *Hiroshima and Nagasaki: The Physical, Medical, and Social Effects of the Atomic Bombings*, trans. Eisei Ishikawa and David L. Swan (New York: Basic Books, 1981).

3. Campaign for Nuclear Disarmament, "The Bomb and You," pamphlet,

n.d. "Act Now or Perish!" quoted in Norman Moss, *Men Who Play God: The Story of the H-Bomb and How the World Came to Live with It* (New York: Harper & Row, 1968), 182. See Charles Wyatt Lomas and Michael Taylor, eds., *The Rhetoric of the British Peace Movement* (New York: Random House, 1971); David Boulton, ed., *Voices from the Crowd: Against the H-Bomb* (Philadelphia: Dufour, 1964).

4. Bertrand Russell, *The Autobiography of Bertrand Russell: 1872–1914* (Boston: Little, Brown, 1967), 3–4, 220–221 and passim; Russell, *The Final Years 1944–1969* (Boston: Little, Brown, 1967). Ronald W. Clark, *The Life of Bertrand Russell* (New York: Knopf, 1976), 84–86, 264.

5. Denis Healey, Great Britain, Commons, *Debates* 537 (1 March 1955): 1936–37. "Göttingen declaration," *BAS* 13, no. 6 (June 1957): 228.

6. "No basis": Catherine McArdle Kelleher, *Germany and the Politics of Nuclear Weapons* (New York: Columbia University Press, 1975), 113. See Karl A. Otto, *Vom Ostermarsch zur APO: Geschichte der ausserparlamentarische Opposition in der Bundesrepublik, 1960–1970* (Frankfurt: Campus, 1977), 56–64; Hans Karl Rupp, *Ausserparlamentarische Opposition in der Ara Adenauer: Der Kampf gegen die Atombewaffnung in den fünfziger Jahren* (Cologne: Pahl-Rugenstein, 1970), 127–135.

7. Bill S. Caldwell, "The French Socialists' Attitude Toward the Use of Nuclear Weapons, 1945–1978," Ph.D. diss., University of Georgia, 1980; S. Gervis, "France's Ban-the-Bombers," *The Nation* 197 (24 Aug. 1963): 91–93. I have discussed nuclear politics with a number of French people, but the subject needs further research.

8. Gerald J. Ringer, "The Dawn of Terror in the American Novel: The Bomb, 1946–1960," M.A. thesis, Florida State University, 1961. Parkin, *Middle Class Radicalism;* see Richard Taylor and Colin Pritchard, *The Protest Makers: The British Nuclear Disarmament Movement of 1958–1965, Twenty Years On* (Oxford: Pergamon, 1980). Shunsuke Tsurumi, *An Intellectual History of Wartime Japan* (London: KPI; Routledge & Kegan Paul, 1986), 101. Rupp, *Ausserparlamentarische Opposition,* Anhang IV, p. 288.

9. Midge Decter, "The Peace Ladies," *Harper's* 226 (March 1963): 48–53. Whether women are inherently more fearful than men is moot, but their greater social tendency to display fearfulness is well documented.

10. Parkin, *Middle Class Radicalism,* 58–59.

11. J. B. Priestley, *Instead of the Trees* (London: Heinemann, 1977), 85–87. See Priestley, *Margin Released: A Writer's Reminiscences and Reflections* (London: Heinemann, 1962); Priestley, *The Writer in a Changing Society* (Aldington, Kent: Hand and Flower Press, 1956); Susan Cooper, *J. B. Priestley: Portrait of an Author* (London: Heinemann, 1970); John Braine, *J. B. Priestley* (London: Weidenfeld and Nicholson, 1977). I learned of Priestley's maternal deprivation only after I selected him for analysis as representative of CND attitudes.

12. Rev. Kenneth Rawlings, quoted in Parkin, *Middle Class Radicalism,* 67.

13. Campaign for Nuclear Disarmament, "The Bomb and You." Heinrich Vogel, "Atomwaffen sind Sünde," in *Kampf dem Atomtod* (Bonn: Arbeitsaus-

schuss "Kampf dem Atomtod," 1958), 31. See Erwin N. Hiebert, *The Impact of Atomic Energy: A History of Responses by Governments, Scientists and Religious Groups* (Newton, Kans.: Faith and Life Press, 1961).

14. George Clark, quoted in Driver, *Disarmers,* 126, see 128. Cf. Adalbert Bärwolf, *Da hilft nur beten* (Düsseldorf: Muth, 1956).

15. One entry to this large and important topic is Norman Cohn, *The Pursuit of the Millennium* (New York: Oxford University Press, 3rd ed., 1970). For summary and bibliography see Yuri Rubinsky and Ian Wiseman, *A History of the End of the World* (New York: Quill, 1982).

16. Cartoons by Mauldin (14 Sept. 1961) and Alexander (28 July 1963), and the general point, are from Gerald J. Ringer, "The Bomb as a Living Symbol: An Interpretation," Ph.D. diss., Florida State University, 1966, 249.

17. Other armed conflicts between Communists were the Vietnamese invasion of Kampuchea and border skirmishes between the Soviet Union and China; also relevant, and with no equivalent in democratic nations, are the internal upheavals that killed millions in the Soviet Union, China, and Kampuchea. By "democratically elected" I mean that a majority of adult males have an effective franchise including freedom of speech. Conflicts such as the War of 1812, the American Civil War, and the Boer War were not between such governments; indeed, in each case the lack of full representation played a role in the origins of the war. The limits of the definition are set by the French occupation of the Ruhr in 1923, not quite a war, and the Indo-Pakistani war of 1965, with Pakistan not quite a democracy. The definition remains broad enough to embrace scores of democracies interacting peacefully over nearly two centuries. This important fact was approached but not discovered by David Wilkinson, *Deadly Quarrels: Lewis F. Richardson and the Statistical Study of War* (Berkeley: University of California Press, 1980), 69–70, 102.

18. Marghanita Laski, *The Offshore Island* (1954; London: Cresset, 1959). On nonnuclear deterrence see Gene Sharp, *Making Europe Unconquerable: The Potential of Civilian-based Deterrence and Defence* (Cambridge, Mass.: Ballinger, 1985).

19. Moss, *Men Who Play God,* 198. Susan T. Fiske, Felicia Pratto, and Mark A. Pavelchak, "Citizens' Images of Nuclear War: Content and Consequences," *Journal of Social Issues* 19 (1983): 41–65.

20. Archibald MacLeish, "The Poet and the Press," *Atlantic* 203 (March 1959): 40–46.

21. Herman Kahn, *Thinking about the Unthinkable* (1962; New York: Avon, rev. ed., 1966), chap. 1.

22. For criticism of strategy see P. M. S. Blackett, "Critique of Some Contemporary Defense Thinking," in Blackett, *Studies of War* (New York: Hill & Wang, 1962), 129–146; James Fallows, *National Defense* (New York: Random House, 1981), 140.

23. John Newman, *Scientific American* 204 (March 1961): 197–200. Norman Thomas, *Saturday Review* 44 (4 Feb. 1961): 17, 33.

24. J. B. Priestley, "Sir Nuclear Fission," *BAS* 11, no. 8 (Oct. 1955): 293–294. Pristley, *The Doomsday Men: An Adventure* (London: Heinemann, 1938).

25. Robert Wallace, "A Deluge of Honors for an Exasperating Admiral," *Life* 45 (8 Sept. 1958), 104ff.

26. "Talk of sex": J. Robert Moskin, "Polaris," *Look* 25 (29 Aug. 1961): 17–31. Norman Polmar and Thomas B. Allen, *Rickover* (New York: Simon & Schuster, 1982), chaps. 13–15 and p. 550; Hanson W. Baldwin, *NYT,* 16 Feb. 1964, p. 1; A. Satloff, "Psychiatry and the Nuclear Submarine," *American Journal of Psychiatry* 124 (1967): 547–549.

27. Richard B. Stolley, "How It Feels to Hold the Nuclear Trigger," *Life* 57 (6 Nov. 1964): 34–41.

28. Eugene Rabinowitch, "The First Year of Deterrence," *BAS* 13, no. 1 (Jan. 1957): 2–8.

29. Maurice Friedberg, *A Decade of Euphoria: Western Literature in Post-Stalin Russia, 1954–64* (Bloomington: Indiana University Press, 1977), 146–154, 308–314; Friedberg, "Reading for the Masses: Popular Soviet Fiction, 1976–80," United States International Communication Agency report R-13-81 (June 1981). Malinovsky in *Current Digest of the Soviet Press* 16, no. 5 (1964): 3–7; see also 15, no. 15 (1963): 19–20; 17, no. 46 (1965): 26–27; 21, no. 47 (1969): 6–7.

30. Max Born, "What Is Left to Hope For?" *BAS* 20 (April 1964): 4. See Paul Fussell, *The Great War and Modern Memory* (London: Oxford University Press, 1975).

31. Half of all U.S. physicists ca. 1978 were doing work of direct military value according to E. L. Woollett, "Physics and Modern Warfare: The Awkward Silence," *American Journal of Physics* 48 (1980): 104–111. M. Grigoryev in *Current Digest of the Soviet Press* 21, no. 47 (1969): 7. Wayne C. Miller, *An Armed America: Its Face in Fiction. A History of the American Military Novel* (New York: New York University Press, 1970), chap. 7.

32. John Hersey, *The War Lover* (New York: Knopf, 1959); *The War Lover* (film; Columbia, 1962). See Lawrence H. Suid, *Guts & Glory: Great American War Movies* (Reading, Mass.: Addison-Wesley, 1978), chaps. 9–10.

33. John Prados, *The Soviet Estimate: U.S. Intelligence Analysis & Russian Military Strength* (New York: Dial, 1982), chaps. 5–8.

34. Thomas J. Kerr, "The Civil Defense Shelter Program: A Case Study of the Politics of National Security Policy Making," Ph.D. diss., Syracuse University, 1969, chaps. 3–4. Philip Wylie in *BAS* 13, no. 4 (April 1957): 146.

35. Charles Haskins to McGeorge Bundy, 21 Feb. 1961, folder "Civil Defense," box 295, National Security Files, JFK. See folders "Office of Emergency Planning," box 85, President's Office Files, JFK.

36. "Stiffen": McGeorge Bundy, memorandum, folder "Civil Defense 4/61–6/61," box 295, National Security Files, JFK. See "ND2 Civil Defense 1/61–5/61," Central Subject Files, JFK. I am grateful to McGeorge Bundy for comments.

37. Drafts, 24–25 July 1961, folder "Berlin Speech," box 60, Theodore Sorenson Papers, JFK.

38. Volume of mail: Thomas Hagen to Pierre Salinger, 11 Aug. 1961, folder "ND2 Civil Defense 8/61–9/61," box 595, Central Subject Files, JFK. Canvass:

Arthur Schlesinger, Jr., "Reflections on Civil Defense," folder "Civil Defense 12/61," box 295, National Security Files, JFK. Chancellor quoted in Arthur I. Waskow and Stanley L. Newman, *America in Hiding* (New York: Ballantine, 1962), 33; see this for the entire shelter debate.

39. Copy of bank advertisement in folder "ND 2–3, 9–10/61," box 598, Central Subject Files, JFK.

40. Waskow and Newman, *America in Hiding*, 101–124 and chap. 10; Kerr, "Civil Defense Shelter Program," chap. 5.

41. John Galbraith to J. F. Kennedy, folder "Civil Defense 10–11/61," box 295, National Security Files, JFK. Twilight Zone: "The Shelter" (CBS-TV, 29 Sept. 1961). See *Ladybug, Ladybug* (United Artists, 1963). Shelter owners: Jiri Nehnevajsa, "Civil Defense and Society," prepared by the Dept. of Sociology, University of Pittsburgh, for the Office of Civil Defense, U.S. Army, Report AD-445285 (1964), available from Defense Technical Information Center, Alexandria, Va., 387, 390. Cf. "Gun Thy Neighbor?" *Time* 78 (18 Aug. 1961), 58. Every man for himself ("*Sauve-qui-peut*"): Schlesinger, "Reflections on Civil Defense."

42. *Time* 78 (20 Oct. 1961): 25. *Life* 51 (15 Sept. 1961): 95–108.

43. I have heard shelter recollections from a number of people, mostly women. See Nehnevajsa, "Civil Defense and Society," 218–222, 285–291. "Coffin": Malinovski as quoted by Malcolm Mackintosh, *BAS* 18, no. 6 (June 1962): 38–39; see *Current Digest of the Soviet Press* 14, no. 4 (1962): 19–20.

44. Michigan State University Survey in Nehnevajsa, "Civil Defensc and Society," 153, see 157; Ralph L. Garrett, "Civil Defense and the Public: An Overview of Public Attitude Studies" (Washington, D.C.: Federal Emergency Management Agency, 1979).

45. "Fallout Protection: What to Know and Do About Nuclear Attack" (Washington, D.C.: Government Printing Office, 1962).

46. Robert F. Kennedy, *Thirteen Days* (New York: W. W. Norton, 1969), 79, 87–90, 98, 180. See also Elie Abel, *The Missile Crisis* (Philadelphia: Lippincott, 1966). For other references and a good analysis see Michael Mandelbaum, *The Nuclear Question: The United States and Nuclear Weapons, 1946–1976* (Cambridge: Cambridge University Press, 1979), chap. 6.

47. Driver, *Disarmers*, 146–147. Mark Chesler and Richard Schmuck, "Student Reactions to the Cuban Crisis and Public Dissent," *Public Opinion Quarterly* 28 (1964): 467–482. George W. Ball, *The Past Has Another Pattern: Memories* (New York: W. W. Norton, 1982), 304–305. Moss, *Men Who Play God*, 307–308.

48. Joseph B. Perry, Jr., "Fear in Response to the Threat of Nuclear War," *Kansas Journal of Sociology* 2 (1966): 100–106. G. N. Levine and J. Modell, "American Public Opinion and the Fallout-Shelter Issue," *Public Opinion Quarterly* 29 (1965): 270–279. My counts from *NYT* index. *Reader's Guide*, s.v. "Civil(ian) Defense" and "Air Raid (Atomic Bomb) Shelters," shows a six-fold decline from the mid-1950s to the late 1960s.

49. Oral history interview of Frederick Osborn by Richard McKinzie, 1974, Harry S Truman Library, Independence, Mo., 59–60.

50. Hazel Gaudet Erskine, "The Polls: Defense, Peace, and Space," *Public Opinion Quarterly* 29 (1961): 487–488; France, Institute of Public Opinion, *Sondages* 25, no. 3 (1963), 91; *Sondages*, 26, no. 3 (1964); Elisabeth Noelle and Erich P. Neumann, eds., *The Germans: Public Opinion Polls 1947–1966*, trans. Gerard Finan (Allensbach: Verlag für Demoskopie, 1967), 601. Philip D. Segal, "Imaginative Literature and the Atomic Bombs: An Analysis of Representative Novels, Plays, and Films from 1945 to 1972," Ph.D. diss., Yeshiva University, New York, 1973, 43–52.

51. David Lilienthal, *Change, Hope, and the Bomb* (Princeton, N.J.: Princeton University Press, 1963), 50, see 36, 43–51.

52. My counts of *NYT* index columns, s.v. "Armament-Control," "Arms Control," "Atomic Energy—International Control." The nadir was 0.06 percent of all items in 1952, the zenith 0.9 percent in 1963.

53. See sources in nn. 2 and 6 above; Paul Boyer, "From Activism to Apathy: America and the Nuclear Issue, 1963–1980," *BAS* 40, no. 7 (Aug. 1984): 14–23.

54. See Chap. 15 n. 16. After doing these counts I found similar if less complete work in Mary P. Lowther, "The Decline of Public Concern over the Atom Bomb," *Kansas Journal of Sociology* 9 (1973): 77–88; Rob Paarlberg, "Forgetting about the Unthinkable," *Foreign Policy* 10 (1973): 132–140. In John Newman and Kathleen Kruger, "Imaginary War Fiction in Colorado State University Libraries: A Bibliography," *Bulletin of Bibliography* 35 (1978): 157–171, I count eight significant nuclear war novels in the 9 years 1945–53, seventeen in the 11 years 1954–64, and only four in the 13 years 1965–77. See Louis Timbal-Duclaux, "La Représentation de l'atome au cinéma . . . de 1945 à 1975," in *Colloque sur les implications psycho-sociologiques du développement de l'industrie nucléaire*, ed. M. Tubiana (Paris: Société Française de Radioprotection, 1977), 419–428. Robert M. Overstreet, *The Comic Book Price Guide* (New York: Crown, 1982).

55. Levin, "American Public Opinion"; Nehnevajsa, "Civil Defense and Society," 89.

56. National Academy of Sciences "Emergency" report, folder "ND2 Civil Defense 6/62–3/63," box 596, Central Subject Files, JFK. For a review of the theory see M. Fishbein and I. Ajzer, *Belief, Attitude, Intention and Behavior: An Introduction to Theory and Research* (Reading, Pa.: Addison-Wesley, 1975), chap. 10.

57. Nehnevajsa, "Civil Defense and Society," 87.

58. Ibid., 31, 38, 527–531; F. Kenneth Berrier, "Shelter Owners, Dissonance and the Arms Race," *Social Problems* 11 (1963): 87–91; Erskine, "The Polls: Defense, Peace, and Space," 488–489; Noelle and Neumann, *The Germans*, 604–605; France, Institute for Public Opinion, *Sondages* 19, no. 3 (1957): 19; 25, no. 3 (1963): 89; 26, no. 3 (1964): 57; 29, no. 4 (1967): 73.

59. Jiri Nehnevajsa, "Issues of Civil Defense: Vintage 1978. Summary Results of the 1978 National Survey" (Pittsburgh: University of Pittsburgh Center for Social and Urban Research, 1979). Albert H. Cantril and Charles W. Roll, Jr., *Hopes and Fears of the American People* (New York: Universe, 1971), 21–23.

60. Sidney Kraus, Reuben Mehling, and Elaine El-Assad, "Mass Media and the Fallout Controversy," *Public Opinion Quarterly* 27 (1963): 191–205; Eugene J. Rosi, "Mass and Tentative Opinion on Nuclear Weapons Tests and Fallout, 1954–1963," *Public Opinion Quarterly* 29 (1965): 280–297.

61. Levine and Modell, "American Public Opinion."

62. Sibylle K. Escalona, "Children and the Threat of Nuclear War," in *Behavioral Science and Human Survival,* ed. Milton Schwebel (Palo Alto, Calif.: Science and Behavior Books, 1965), 201–209. R. Serge Denisoff and Mark H. Levine, "The Popular Protest Song: The Case of 'Eve of Destruction,'" *Public Opinion Quarterly* 35 (1971): 117–122.

63. British Council of Churches, *The Era of Atomic Power* (1946), 17, as quoted by Driver, *Disarmers,* 206. Sibylle K. Escalona, "Growing Up with the Threat of Nuclear War: Some Indirect Effects of Personality Development," *American Journal of Orthopsychiatry* 52 (1982): 600–607; see John Mack, pp. 590–599 of the same issue. See also George P. Elliott, *David Knudsen* (New York: Random House, 1962).

64. Robert Liebert, *Radical and Militant Youth: A Psychoanalytic Inquiry* (New York: Praeger, 1971), 234.

65. Ernest Becker, *The Denial of Death* (New York: Free Press, 1973); Paul Kline, *Fact and Fantasy in Freudian Theory* (London: Methuen, 1972), 170–181.

66. "The Face of the Future," *Look* 29, no. 1 (12 Jan. 1965): 73. I am skipping over the distinction between conscious denial or suppression of knowledge, and unconscious repression; for introduction and references see Sissela Bok, *Secrets: On the Ethics of Concealment and Revelation* (New York: Pantheon, 1982), chap. 5.

67. William Abbott Scott, "The Avoidance of Threatening Material in Imaginative Behavior," in *Motives in Fantasy, Action and Society,* ed. John W. Atkinson (Princeton, N.J.: Van Nostrand, 1958), 572–585; I thank John Popplestone for bringing this to my attention. See Robert Jay Lifton and Richard Falk, *Indefensible Weapons: The Political and Psychological Case against Nuclearism* (New York: Basic Books, 1982). See also n. 70 below.

68. I surveyed *Social Studies and Humanities Index, Psychological Abstracts,* and the *Comprehensive Dissertations Index.* "Great anxiety": Helen Swick Perry, "Selective Inattention as an Explanatory Concept for U.S. Public Attitudes Toward the Atomic Bomb," *Psychiatry* 17 (1954): 225–242; cf. 18 (1955): 196–203. "Overwhelmed": Lifton and Falk, *Indefensible Weapons,* 109–110. I have felt this too, nightmares and all.

69. Kahn, *Thinking about the Unthinkable,* 21–31. John P. Rose, *The Evolution of U.S. Army Nuclear Doctrine, 1945–1980* (Boulder, Colo.: Westview, 1980), 56–57, 92–93.

70. B. Ashem, "The Treatment of a Disaster Phobia by Systematic Desensitization," *Behavior Research and Therapy* 1 (1963): 81–84.

71. A good review of these mechanisms is Calvin Springer Hall and Gardner Lindzey, *Theories of Personality* (New York: Wiley, 3rd ed., 1978), 613–616. Note that I am avoiding the distinctions between Freud and Pavlov, between psychoanalysis and learning theory. For example, denial is probably akin to

learned helplessness; it is when people feel they are helpless that they minimize threats (see Chap. 6, n. 26). Note also that it is basic to my argument here and elsewhere that when stimuli affect many people together, as news events or movies do, they forge associations that are similar and widespread, that is, public images. This is an assumption, a frontier for research. See Fishbein and Ajzer, *Belief.* Bertrand Russell, "My View of the Cold War," reprinted in Boulton, *Voices from the Crowd,* 142–145.

72. Ellul, *Propaganda,* 271–272 and passim.

14. FAIL/SAFE

1. Leonard Wibberley, *The Mouse That Roared* (Boston: Little, Brown, 1955); *The Mouse That Roared* (film; Columbia, 1959).

2. *NYT,* 12 March 1958, p. 1; 13 March 1958, p. 3. On all these matters see Joel Larus, *Nuclear Weapons Safety and the Common Defense* (Columbus: Ohio State University Press, 1967); Paul Bracken, *The Command and Control of Nuclear Forces* (New Haven, Conn.: Yale University Press, 1983).

3. *NYT,* 19 April 1958, pp. 1, 4; see Herman Kahn, *On Thermonuclear War* (Princeton, N.J.: Princeton University Press, 2nd ed., 1961), 190–210. Charles Yulish to Arthur Sylvester, 4 Feb. 1968, folder "Airplane Crash in Spain (Correspondence)," JCAE.

4. Surveys: Jiri Nehnevajsa, "Civil Defense and Society," prepared by the Dept. of Sociology, University of Pittsburgh, for the Office of Civil Defense, U.S. Army, Report AD-445285 (1964), pp. 69–70, available from Defense Technical Information Center, Alexandria, Va.; H. Ornauer et al., eds., *Images of the World in the Year 2000: A Comparative Ten Nation Study* (Atlantic Highlands, N.J.: Humanities, 1976), 661. See Geoffrey Blainey, *The Causes of War* (New York: Free Press, 1973), chap. 9.

5. Eugene Burdick and Harvey Wheeler, *Fail-Safe* (New York: McGraw-Hill, 1962). Alice Payne Hackett, *80 Years of Best Sellers* (New York: Bowker, 1977).

6. Sidney Hook, *The Fail-Safe Fallacy* (New York: Stein & Day, 1963), 15.

7. "Vocabulary": Harvey Wheeler in *Saturday Review* 45, no. 42 (20 Oct. 1962): 22, 38. *Fail Safe* (Columbia, 1964). See Lawrence H. Suid, *Guts & Glory: Great American War Movies* (Reading, Mass.: Addison-Wesley, 1978), 195–201.

8. Peter Bryant [Peter Bryan George], *Two Hours to Doom* (London: Clark Boardman, 1957), published in the United States as *Red Alert* (New York: Ace, 1958). George, *Commander–1* (London: Heinemann, 1965). See Julian Smith, *Looking Away: Hollywood and Vietnam* (New York: Scribner's, 1975), chap. 6.

9. *Dr. Strangelove: Or, How I Stopped Worrying and Learned to Love the Bomb* (Columbia, 1964). This film is discussed in many works, for example, Charles Maland, "Dr. Strangelove (1964): Nightmare Comedy and the Ideology of Liberal Consensus," *American Quarterly* 31 (1979): 697–717.

10. Robert Brustein, *New York Review of Books,* 6 Feb. 1964, pp. 3–4. *Metropolis* (UFA, 1926).

11. *Time* 84 (25 Sept. 1964): 15–19. Theodore H. White, *The Making of the President: 1964* (New York: Atheneum, 1965), 300, see 296–300, 321–328, 373–374.

12. *The Bedford Incident* (Columbia, 1965). Fletcher Knebel and Charles W. Bailey, *Seven Days in May* (New York: Harper & Row, 1962); *Seven Days in May* (film; Paramount, 1964).

13. For audience and critical reactions to *Strangelove* see clippings in NYPL/TC.

14. For transition see, for example, C. N. Barclay: "A Very Real Risk—War by Accident," *NYT Magazine* (5 May 1963), pp. 17, 114.

15. *Dr. No* (United Artists, 1962); see also Ian Fleming, *Moonraker* (New York: Macmillan, 1955), and most other Bond books and films. *Goldfinger* (United Artists, 1964) and *Thunderball* (United Artists, 1965) are the only nuclear-related films in the top 100 rental grosses listed in *Variety,* 4 May 1983, pp. 15, 206. "Toy": John Brosnan, *Future Tense: The Cinema of Science Fiction* (New York: St. Martin's, 1978), 163. See Louis Timbal-Duclaux, "La Représentation de l'atome au cinéma . . . de 1945 à 1975," in *Colloque sur les implications psycho-sociologiques du développement de l'industrie nucléaire,* ed. M. Tubiana (Paris: Société Française de Radioprotection, 1977), 419–428.

16. Vannevar Bush, *Endless Horizons* (Washington, D.C.: Public Affairs Press, 1946), 105; see Bush, draft memorandum, 10 June 1944, file 2, Harrison-Bundy Files (RG77), microfilm M1108, National Archives, Washington, D.C.

17. Edward Teller, "Reactor Hazards Predictable," *Nucleonics* 11, no. 11 (Nov. 1953): 80; *NYT,* 13 Sept. 1953, p. 34.

18. C. Rogers McCullough, Mark M. Mills, and Edward Teller, "The Safety of Nuclear Reactors," International Conference on the Peaceful Uses of Nuclear Energy A/CONF.8/P/853, July 1955; *NYT,* 11 Aug. 1955, p. 11.

19. A. P. Armagnac, "How Safe Are Our A-Power Plants?" *Popular Science* 169 (Nov. 1956): 134ff.

20. France: *NYT,* 9 March 1947, p. 13. Belgium: Lewis Strauss to D. D. Eisenhower, 14 June 1956, folder "AEC 1955–56 (4)," box 4, Ann Whitman Administration Files, DDE. Stories: for example, Pat Frank, *Mr. Adam* (Philadelphia: Lippincott, 1946); Lester Del Rey, *Nerves* (New York: Ballantine, 1956).

21. Teller, "Reactor Hazards Predictable," 80. *Atoms for Peace* 7, no. 23 (28 March 1957).

22. The American study was privately conducted, and I have been asked not to cite the source. P. Sivadon and A. Fernandez, "Etude des attitudes psychologiques des travailleurs nucléaires" (1968), summarized in Colette Guedeney and Gérard Mendel, *L'Angoisse atomique et les centrales nucléaires* (Paris: Payot, 1973), 43–45, 105, and Guedeney's own study, 75–76, 91–92, 101.

23. Richard G. Hewlett and Francis Duncan, *Atomic Shield, 1947/1952* (University Park: Pennsylvania State University Press, 1969), 186–187. See Minutes of meetings of AEC, Department of Energy archives, Germantown, Md.

24. Hewlett and Duncan, *Atomic Shield,* 200–203, 208, 217–219; Edward Teller with Allen Brown, *The Legacy of Hiroshima* (Garden City, N.Y.: Doubleday, 1962), 102–104.

25. Paul Schmitz and Roger Griffin, "Power Reactor Containment," *Nucleonics* 23 (Oct. 1965): 50–55. Interview with Henry Hurwitz, 1980.

26. See the AEC Minutes and George T. Mazuzan and J. Samuel Walker, *Controlling the Atom: The Beginnings of Nuclear Regulation 1946–1962* (Berkeley: University of California Press, 1984). I have also used interviews with Richard Adams, Karl Cohen, Robert Creagan, Milton Edlund, Paul Fields, Henry Hurwitz, John Kelly, J. B. Horner (Desmond) Kuper, Rollin Taecker, Alvin Weinberg, Edward Wimunc, and others, 1980–1982, and oral history interviews of Milt Levenson and Walter Zinn by D. F. Wood, 1980, at Argonne National Laboratory, Argonne, Ill. I have further benefited from reading the minutes and some correspondence of the Argonne Reactor Safety Review Committtee.

27. Margaret Gowing with Lorna Arnold, *Independence and Deterrence: Britain and Atomic Energy, 1945–1952,* 2 vols. (New York: St. Martin's, 1974), 2:110–212, 352–353. Horst Reichmann, "Soziologische Aspekte der Verwendung von Kernkraft," *Kölner Zeitschrift für Soziologie und Sozial-Psychologie* 13 (1961): 217–226.

28. The only published Soviet reference to a Soviet accident that I know is oblique: I. N. Golovin, *I. V. Kurchatov: A Socialist-Realist Biography of the Soviet Nuclear Scientist,* trans. William H. Dougherty (Bloomington, Ind.: Selbstverlag Press, 1968), 61. Unpublished rumor tells of additional problems. I thank Paul Jacobson and Arnold Kramish for discussion of these issues.

29. R. F. Pocock, *Nuclear Power: Its Development in the United Kingdom* (Old Woking, Surrey: Unwin, 1977), 66–71; Ritchie Calder, *Living with the Atom* (Chicago: University of Chicago Press, 1962), 24–25; *NYT,* 31 Oct. 1957, p. 3.

30. Leo Szilard, *Leo Szilard: His Version of the Facts,* ed. Spencer Weart and Gertrud Weiss Szilard (Cambridge, Mass.: MIT Press, 1978), 199–200; Westinghouse Company, "Infinite Energy" pamphlet, quoted in Stephen Hilgartner, Richard C. Bell, and Rory O'Connor, *Nukespeak: Nuclear Language, Visions, and Mindset* (San Francisco: Sierra Club, 1982), 190.

31. For a balanced view and references see Mazuzan and Walker, *Controlling the Atom,* chaps. 5–6. C. Rogers McCullough to K. E. Fields, 6 June 1956, and other materials in folder "IR&A6, Reg. Power Development Co.," box 8331, AEC Secretariat File, archives of the Nuclear Regulatory Commission, Washington, D.C.

32. In addition to the NRC records see Minutes of AEC meetings, Department of Energy archives, Germantown, Md. Clinton P. Anderson with Milton Viorst, *Outsider in the Senate* (New York: World, 1970), chaps. 6–7.

33. Lewis Strauss, Memo of Conference with the President, 6 Aug. 1956, folder "Aug. '56 Diary—Staff Memos," box 17, Ann Whitman DDE Diaries, DDE. Strauss to Atomic Industrial Forum, 29 Oct. 1957, in "Reply Brief of Intervenors" in *PRDC Hearing Proceedings,* vol. 3, 1957, Docket 50–16, Pub-

lic Document Room, Nuclear Regulatory Commission, Washington, D.C. Press release by Rep. George Meader, 1 Sept. 1956, folder 1, box 8331, AEC Secretariat File, NRC archives. Sen. Pat McNamara to Walker Cisler, 12 Sept. 1956, same folder. For Gore-Holifield Bill politics see Frank G. Dawson, *Nuclear Power: Development and Management of a Technology* (Seattle: University of Washington Press, 1976), 101–108.

34. "Atomic bomb": John G. Fuller, *We Almost Lost Detroit* (New York: Random House, 1975), 54. "A-plant Safe": E. Pauline Alexanderson, ed., *Fermi-I: New Age for Nuclear Power* (LaGrange Park, Ill.: American Nuclear Society, 1979), 119–120.

35. Steven L. Del Sesto, *Science, Politics, and Controversy: Civilian Nuclear Power in the United States, 1946–1974* (Boulder, Colo.: Westview, 1979), 122–135.

36. Atomic Industrial Forum statement, 14 Nov. 1955, in United States Congress, 84:2, Joint Committee on Atomic Energy, *Peaceful Uses of Atomic Energy: Background Material for the Report of the Panel on the Impact of the Peaceful Uses . . .*, vol. 2 of McKinney Panel report (Washington, D.C.: 1956), 599.

37. "Resemble studies": Louis Roddis, Jr., to Leland Haworth, 8 Aug. 1956, folder "Reactor Safety Project—AEC," Leland Haworth Papers, Brookhaven National Laboratory, Upton, N.Y., microfilm at AIP. Interview with J. B. Horner Kuper, 1980.

38. Edward Teller to Clifford Beck, 9 Jan. 1957, folder "Reactor Safety Project—AEC," Leland Haworth Papers, Brookhaven National Laboratory, Upton, N.Y., microfilm at AIP.

39. "Did not anticipate": Herbert Kouts to Editor, *Natural History*, 23 July 1969, copy in folder "Public Relations 1964–1972," Maurice Goldhaber Files, Brookhaven National Laboratory, Upton, N.Y. See Fuller, *We Almost Lost Detroit*; *NYT*, 10 Nov. 1974, p. 1.

40. David Lilienthal, *Change, Hope, and the Bomb* (Princeton, N.J.: Princeton University Press, 1963), 101–103, 106–107. Lilienthal, *The Journals of David E. Lilienthal*, vol. 3, *Venturesome Years 1950–1955* (New York: Harper & Row, 1966), 21; vol. 4, *The Road to Change 1955–1959* (New York: Harper & Row, 1969), 165, 491.

41. *NYT*, 15 June 1963, pp. 1, 17; 26 Aug. 1963, p. 26. Folder "IR&A6 Reg. Con-Edison—Ravenswood," box 8339, AEC Seretariat File, NRC archives. David Okrent, *Nuclear Reactor Safety: On the History of the Regulatory Process* (Madison: University of Wisconsin Press, 1981), 75.

42. David E. Lilienthal, *Atomic Energy: A New Start* (New York: Harper & Row, 1980), 32–39. See Lilienthal, *Journals* 4: 521–522.

15. REACTOR POISONS AND PROMISES

1. Rodney Southwick to Joseph Fouchard, 10 March 1962, folder "IR&A6, Reg. PG&E—Bodega Bay," box 8330, AEC Secretariat files, Nuclear Regulatory Commission, Washington, D.C. For the Bodega battle see John Holdren and

Phillip Herrera, *Energy: A Crisis in Power* (San Francisco: Sierra Club, 1971), 195–203; David Okrent, *Nuclear Reactor Safety: On the History of the Regulatory Process* (Madison: University of Wisconsin Press, 1981), 263–273.

2. R. F. Pocock, *Nuclear Power: Its Development in the United Kingdom* (Old Woking, Surrey: Unwin, 1977), 94–96. Glenn Anderson to Glenn Seaborg, 13 Aug. 1963, folder "IR&A6, Reg. PG&E—Bodega Bay," box 8330, AEC Secretariat files, archives of The Nuclear Regulatory Commission, Washington, D.C.

3. Jessie V. Coles to the Dairymen of Sonoma and Marin Counties, 25 March 1963, folder "IR&A6, Reg. PG&E—Bodega Bay," box 8330, AEC Secretariat files, NRC archives.

4. The *NYT* index lists ten items on wastes in 1955, three or four in the preceding and following years. Brian Balogh, "Experts Everywhere: Nuclear Reactor Safety, 1947–1973," talk presented at History of Science meeting, Raleigh, N.C., 30 Oct. 1987; Abel Wollman, oral history interview by Walter Hollander, Jr., Columbia University Library, New York City; George T. Mazuzan and J. Samuel Walker, *Controlling the Atom: The Beginnings of Nuclear Regulation 1946–1962* (Berkeley: University of California Press, 1984), chap. 12.

5. Pare Lorentz, "The Fight for Survival," *McCall's* 84, no. 4 (Jan. 1957): 29, 73–74. Charles-Noel Martin, *L'Atome, maître du monde* (Paris: Centurion, 1956), 173–194. Robert Rienow and Leona Train Rienow, *Our New Life with the Atom* (New York: Crowell, 1958), 35–36, 100–102, 142, 160.

6. Letter in folder "JCAE Radioactive Waste Disposal," box 818, Clinton P. Anderson Papers, Library of Congress, Washington, D.C. *The Giant Behemoth* (Allied Artists, 1959). *Providence Journal,* 19 July 1959, as quoted in Mazuzan and Walker, *Controlling the Atom,* 358.

7. Riviera: *NYT,* 12 Oct. 1960, p. 11.

8. "Crapped up": Paul Loeb, *Nuclear Culture: Living and Working in the World's Largest Atomic Complex* (New York: Coward, McCann & Geoghegan, 1982), 12. Association test on *déchets* ("wastes"): Christine Blanchet in *Colloque sur les implications psycho-sociologiques du développement de l'industrie nucléaire,* ed. M. Tubiana (Paris: Société Française de Radioprotection, 1977), 225.

9. Melanie Klein, *The Psycho-Analysis of Children,* trans. Alix Strachey, rev. with H. A. Thorner (New York: Dell, 1975), 129, 145, 239. For anality and aggression see Seymour Fisher and Roger P. Greenberg, *The Scientific Credibility of Freud's Theories and Therapy* (New York: Basic Books, 1977), 154–158. Edward Jablonski, *Flying Fortress: The Illustrated Biography of the B-17s and the Men Who Flew Them* (Garden City, N.Y.: Doubleday, 1965); my thanks to Jay Spenser for help. Leo Szilard, *Leo Szilard: His Version of the Facts,* ed. Spencer Weart and Gertrud Weiss Szilard (Cambridge, Mass.: MIT Press, 1978), 185.

10. *NYT,* 14 April 1958, p. 1; 24 April 1958, p. 14. Zhores A. Medvedev, *Nuclear Disaster in the Urals* (New York: Norton, 1979). *Science* 216 (1982): 274.

11. Film counts from U.S. Library of Congress, *National Union Catalog: Motion Pictures & Filmstrips* (New York: Rowman & Littlefield). Audience statistics: Glenn Seaborg, address, 25 May 1970, box 78, JCAE.

12. "Atomic Power Today: Service with Safety" (Atomic Industrial Forum, 1966). Laura Fermi, *The Story of Atomic Energy* (New York: Random House, 1961, 1966).

13. Harold P. Green and Alan Rosenthal, *Government of the Atom: The Integration of Powers* (New York: Prentice-Hall, 1963), chap. 1 and 267–273. David Lilienthal, *Change, Hope, and the Bomb* (Princeton, N.J.: Princeton University Press, 1963), 126–128.

14. Ritchie Calder, "As the Hare Said to the Tortoise," *BAS* 16, no. 9 (Nov. 1960): 352–354.

15. My counts, combining the *Bibliographie der deutschen Zeitschriften-literatur* and the *Bibliographie der fremdsprachigen Zslit.* and normalizing them at 1962–1965 to the succeeding *Internationale Bibl. d. Zslit.* J. P. Delzani and C. Carde, "Approche psychologique de l'évolution du thème nucléaire à partir d'un quotidien populaire de 1950 à 1976," in *Colloque sur les implications psycho-sociologiques du développement de l'industrie nucléaire,* ed. M. Tubiana (Paris: Société Française de Radioprotection, 1977), 354–374.

16. My counts, normalized to total number of pages. I use all "atomic," "nuclear," and "radioactive" categories. I have confirmed this with counts from the *NYT* index; see especially s.v. "Atomic . . . Radiation Hazards." For cruder but similar results done independently see Allan Mazur, "Opposition to Technological Innovation," *Minerva* 13 (1975): 58–81; Mazur, *The Dynamics of Technical Controversy* (Washington, D.C.: Communications Press, 1981); C. Hohenemser, R. Kasperson, and R. Kates, "The Distrust of Nuclear Power," *Science* 196 (1977): 25–34; Roger E. Kasperson et al., "Public Opposition to Nuclear Energy: Retrospect and Prospect," *Science, Technology & Human Values* 5, no. 31 (Spring 1980): 11–23.

17. E. Pauline Alexanderson, ed., *Fermi-I: New Age for Nuclear Power* (LaGrange Park, Ill.: American Nuclear Society, 1979), chap. 11; John Fuller, *We Almost Lost Detroit* (New York: Random House, 1975), chaps. 13–14; cf. Earl M. Page, *We Did* Not *Almost Lose Detroit* (Detroit: Detroit Edison, 2nd ed., 1976).

18. World Health Organization, "Mental Health Aspects of the Peaceful Uses of Atomic Energy: Report of a Study Group," Technical Report Series no. 151 (Geneva: W.H.O., 1958), 14.

19. Rep. Craig Hosmer, 17 Feb. 1969, in David Okrent, "On the History of the Evolution of Light Water Reactor Safety in the United States" (ca. 1979), copy at AIP, p. 2-411.

20. "Clots": Boris Pregel, "The Impact of the Nuclear Age," in *America Faces the Nuclear Age.* ed. Johnson E. Fairchild and David Landman (New York: Sheridan House, 1961), 37. Glenn T. Seaborg and William R. Corliss, *Man and Atom: Building a New World through Nuclear Technology* (New York: E. P. Dutton, 1971), chap. 5. Alvin M. Weinberg, "Nuclear Energy: A Prelude to H. G. Wells' Dream," *Foreign Affairs* 49 (1971): 407–418.

21. David E. Lilienthal, *Atomic Energy: A New Start* (New York: Harper & Row, 1980), 73. Alvin M. Weinberg, "Social Institutions and Nuclear Energy," *Science* 177 (1972): 27–34.

22. Glenn T. Seaborg to President Kennedy, 20 Nov. 1962, in AEC Press Release E-425, 22 Nov. 1962. Glenn T. Seaborg, "Large-Scale Alchemy: Twenty-fifth Anniversary at Hanford-Richland" (1968), in Seaborg, *Nuclear Milestones: A Collection of Speeches* (San Francisco: W. H. Freeman, 1972), 166.

23. Peter deLeon, *Development and Diffusion of the Nuclear Power Reactor: A Comparative Analysis* (Cambridge, Mass.: Ballinger, 1979), chaps. 6–7.

24. Okrent, *Nuclear Reactor Safety*; Elizabeth S. Rolph, *Nuclear Power and the Public Safety: A Study in Regulation* (Lexington, Mass.: D. C. Heath, 1979), 80–86; folder "ACRS vol. 3, 5/65–12/68," JCAE. A key document is Report of the AEC Advisory Task Force on Power Reactor Emergency Cooling (William K. Ergen, chair), Oct. 1967, copy in library of Argonne National Laboratory, Argonne, Ill.

25. This criticism is now widely understood among the nuclear engineers I have talked with.

26. Airplane: Minutes of ACRS-AEC-Industry meeting, 27–28 Feb. 1968, quoted in Okrent, "On the History," p. 6-126.

27. In addition to the sources cited in the preceding notes and several interviews, I am indebted to Robert Gillette, "Nuclear Safety," *Science* 177 (1972): 771–772, 867–871, 970–975, 1080–82. "Discipline": for example, Milton Shaw to American Nuclear Society, "AEC Views on Quality Assurance," 2 Nov. 1966, folder "Shaw," box 78, JCAE.

28. Interviews with André Gauvenet and Bertrand Goldschmidt, 1980–1981.

29. Interviews with Richard Adams, Karl Cohen, Robert Creagan, Henry Hurwitz, John Kelly, Rollin Taecker, Alvin Weinberg, Ed Wimunc, and others, 1980–1981.

16. THE DEBATE EXPLODES

1. Two less radical technologies, pesticides and the supersonic transport airplane, ran into trouble slightly earlier.

2. See references in Chap. 12, n. 37. See also Ralph E. Lapp, *The Weapons Culture* (New York: W.W. Norton, 1968); Ernest J. Yanarella, *The Missile Defense Controversy: Strategy, Technology, and Politics, 1955–1972* (Lexington: University Press of Kentucky, 1977); Anne Hessing Cahn, *Eggheads and Warheads: Scientists and the ABM* (Cambridge, Mass.: MIT Science and Public Policy Program, 1971).

3. David Inglis, "H-Bombs in the Back Yard," *Saturday Review* 51 (21 Dec. 1968): 11. "Anywhere except": quoted in Joel Primack and Frank von Hippel, *Advice and Dissent: Scientists in the Political Arena* (New York: Basic Books, 1974), 194n26.

4. Student quoted by Edwin S. Shneidman, *Deaths of Man* (New York: Quadrangle, 1973), 185. *NYT,* 24 March 1969, p. 12.

5. McNamara address, 18 Sept. 1967, in Lapp, *Weapons Culture,* App. 12. "Colored": Lapp, *Weapons Culture,* 29. For further comments and references see Lawrence Freedman, *The Evolution of Nuclear Strategy* (New York: St. Martin's, 1981); Volker R. Berghahn, *Militarism: The History of an International Debate 1861–1979* (Cambridge: Cambridge University Press, 1984), chap. 5.

6. Leslie J. Freeman, *Nuclear Witnesses: Insiders Speak Out* (New York: W. W. Norton, 1981), chap. 3. Philip M. Boffey, "Ernest J. Sternglass: Controversial Prophet of Doom," *Science* 166 (1969): 195–200. *NYT,* 17 March 1968, sec. 4, p. 13.

7. Transcript of debate, WMAL-TV, 29 July 1969, and other materials, folder "Radio and TV," box 7551, AEC Secretariat Files, Department of Energy, Germantown, Md.

8. Steven L. Del Sesto, *Science, Politics, and Controversy: Civilian Nuclear Power in the United States, 1946–1974* (Cambridge, Mass.: Ballinger, 1979), chap. 6 and 155–156; Ralph E. Lapp, *The Radiation Controversy* (Greenwich, Conn.: Reddy, 1979).

9. Freeman, *Nuclear Witnesses,* 76; Boffey, "Sternglass."

10. Freeman, *Nuclear Witnesses,* 114. See John Gofman, *"Irrevy": An Irreverent, Illustrated View of Nuclear Power* (San Francisco: Committee for Nuclear Responsibility, 1979), chap. 8.

11. John Gofman and Arthur Tamplin, *"Population Control" through Nuclear Pollution* (Chicago: Nelson-Hall, 1970). See Allan Mazur, *The Dynamics of Technical Controversy* (Washington, D.C.: Communications Press, 1981), chap. 2.

12. "Powers That Be" (KNBC-TV, 18 May 1971), transcript and other materials in folder "TV," box 7809, AEC Secretariat Files, archives of The Nuclear Regulatory Commission, Washington, D.C.

13. George T. Mazuzan and J. Samuel Walker, *Controlling the Atom* (Berkeley: University of California Press, 1984), chap. 11; H. Peter Metzger, *The Atomic Establishment* (New York: Simon & Schuster, 1972), chap. 4. See *Health Physics* 38 (1981): 1061–74.

14. Metzger, *Atomic Establishment,* chap. 5.

15. Robert Gillette, "Radiation Spill at Hanford: The Anatomy of an Accident," *Science* 181 (1973): 728–730.

16. For review with bibliography see Ronnie D. Lipschutz, *Radioactive Waste: Politics, Technology, and Risk* (Cambridge, Mass.: Ballinger, 1980), chap. 4; for more recent debate see Luther J. Carter, "Nuclear Imperatives and Public Trust: Dealing with Radioactive Waste," *Issues in Science and Technology* 3, no. 2 (Winter 1987): 46–61. The German case was similar.

17. My *Readers' Guide* counts, 1945–1980: about half the negative titles were general or addressed problems only temporarily salient; reactor accidents and fuel cycle problems accounted for another quarter each. Polls: Barbara D. Melber et al., "Nuclear Power and the Public: Analysis of Collected Survey

Research," Report PNL-2430 (Seattle, Wash.: Battelle Human Affairs Research Centers, 1977), 129–133, 170–177. See Stanley M. Nealey, Barbara D. Melber, and William L. Rankin, *Public Opinion and Nuclear Energy* (Lexington, Mass.: D. C. Heath, 1983), 95, 103; F. Fagnani, ed., *Le Débat nucléaire en France: Acteurs sociaux et communication de Masse* (Grenoble: Université des Sciences Sociales, Institut de Recherche Economique, 1977).

18. House without toilets: John W. Powell, "Nuclear Power in Japan," *BAS* 39, no. 5 (May 1983): 33–39, and personal communications. Heap of excrement ("Tas de merde"): quoted in Colette Guedeney and Gérard Mendel, *L'Angoisse atomique et les centrales nucléaires* (Paris: Payot, 1973), 234. Dieter Rucht, *Von Wyhl nach Gorleben: Bürger gegen Atomprogramme und nukleare Entsorgung* (Munich: Beck, 1977), 127.

19. "Infection": William S. Maynard et al., "Public Values Associated with Nuclear Waste Disposal," Report BNWL-1997 (Seattle, Wash.: Battelle Human Affairs Research Centers, 1976), 173. Peter Faulkner, ed., *The Silent Bomb: A Guide to the Nuclear Energy Controversy* (New York: Random House, 1977), see chap. 9 and ix, 118–119.

20. *Doomwatch* (British Film Productions, 1972), part of an environmentalist series. *Empire of the Ants* (AIP, 1977).

21. One introduction to this period is Richard S. Lewis, *The Nuclear Power Rebellion: Citizens vs. the Atomic Industrial Establishment* (New York: Viking, 1972). See also Jim Falk, *Global Fission: The Battle over Nuclear Power* (Melbourne: Oxford University Press, 1982).

22. Primack and von Hippel, *Advice and Dissent*, 208–235. Gary F. Downey, "Looking for Anomalies: The Union of Concerned Scientists as Technological Revolutionaries," talk presented at Society for History of Technology meeting, Pittsburgh, 25 October 1986. See Daniel Ford, *The Cult of the Atom: The Secret Papers of the Atomic Energy Commission* (New York: Simon & Schuster, 1982), pt. 2.

23. Metzger, *Atomic Establishment.*

24. *Red Alert* (Paramount; CBS-TV, 18 May 1977). Peter Bryant [Peter Bryan George], *Two Hours to Doom* (London: Clark Boardman, 1957), published in U.S. as *Red Alert* (New York: Ace, 1958).

25. Helen Caldicott, *Nuclear Madness: What You Can Do!* (New York: Bantam, 1980), 1–7; "Pediatrician Helen Caldicott," *People* 16, no. 22 (30 Nov. 1981): 90–93. Sheldon Novick, *The Careless Atom* (Boston: Houghton Mifflin, 1969). See Dorothy Nelkin, "Anti-nuclear Connections: Power and Weapons," *BAS* 37, no. 4 (April 1981): 36–40. Bulletin boards: my own observations in several U.S. cities.

26. Nealey, Melber, and Rankin, *Public Opinion and Nuclear Energy*, 79–83; Commission of the European Community, Euro-baromètre survey no. 16 (April 1982), quoted in *Health Physics Society Newsletter* 12, no. 9 (Sept. 1984): 19.

27. "The Plutonium Connection" (PBS, 9 March 1975). See *NYT*, 27 Feb. 1975, p. 12.

28. For discussion and bibliography see Gene I. Rochlin, *Plutonium, Power,*

and Politics: International Arrangements for the Disposition of Spent Nuclear Fuel (Berkeley: University of California Press, 1979). Experts surveyed by John P. Holdren, "Environmental Liabilities of Nuclear Power," *BAS* 36, no. 1 (Jan. 1980): 31–32. Bertrand Goldschmidt, *The Atomic Complex: A Worldwide Political History of Nuclear Energy* (La Grange Park, Ill.: American Nuclear Society, 1982), gives a pro-reactor historical survey; Amory B. Lovins and L. Hunter Lovins, *Energy/War: Breaking the Nuclear Link* (San Francisco: Friends of the Earth, 1980), give the opposite viewpoint.

29. William L. Rankin, Stanley M. Nealey, and Daniel E. Montano, "Analysis of Print Media Coverage of Nuclear Power Issues" (Seattle: Battelle Memorial Institute, Human Affairs Research Centers, 1978); overall this parallels my *Readers' Guide* study, suggesting a good correlation between the titles and the contents of the articles. Melber et al., "Nuclear Power and the Public," 125, 185–187; D. Denoncin and C. Blanchet in *Colloque sur les implications psycho-sociologiques du développement de l'industrie nucléaire,* ed. M. Tubiana (Paris: Société Française de Radioprotection, 1977), 251–252.

30. My counts and J. P. Delzani and C. Carde, "Approche psychologique de l'évolution du thème nucléaire à partir d'un quotidien populaire de 1950 à 1976," in *Colloque sur les implications,* 354–374.

31. David E. Lilienthal, *Atomic Energy: A New Start* (New York: Harper & Row, 1980), 22–23.

32. See Chap. 11, n. 34 for displacement in general. For reactors see Alain Touraine et al., *La Prophétie anti-nucléaire* (Paris: Seuil, 1980), 59, 88, 96–99.

33. Equally likely: Jiri Nehnevajsa, "Issues of Civil Defense: Vintage 1978. Summary Results of the 1978 National Survey" (Pittsburgh: University of Pittsburgh Center for Social and Urban Research, 1979), 16, 23. Other polls: for example, H. Ornauer et al., eds., *Images of the World in the Year 2000: A Comparative Ten Nation Study* (Atlantic Highlands, N.J.: Humanities, 1976), 321, 347, 350, 366.

34. 1966–1973 polls: Amitai Etzioni and Clyde Nunn, "The Public Appreciation of Science in Contemporary America," *Daedalus* 130, no. 3 (Summer 1974): 191–205; see Gerald Holton and William A. Blanpied, eds., *Science and Its Public: The Changing Relationship* (Boston: Reidel, 1976). Ornauer et al., *Images of the World,* 57–63. "Unspoken fear": Hillier Krieghbaum, *Science, the News and the Public* (New York: New York University Press, 1958), 30. John L. Martin, "Science and the Successful Society," *Public Opinion* 4, no. 3 (June/July 1981): 16. Todd R. La Porte and Daniel Metlay, "Public Attitudes toward Present and Future Technologies: Satisfaction and Apprehension," *Social Studies of Science* 5 (1975): 373–398. Seymour Martin Lipset and William Schneider, *The Confidence Gap: Business, Labor, and Government in the Public Mind* (New York: Macmillan–Free Press, 1983). Cora Bagley Marrett, "Public Concerns about Nuclear Power and Science," in *Public Reactions to Nuclear Power: Are There Critical Masses?,* ed. William R. Freudenburg and Eugene A. Rosa, AAAS Selected Symposium 93 (Boulder, Colo.: Westview, 1984), 307–328.

35. Walter A. Rosenbaum, *The Politics of Environmental Concern* (New

York: Praeger, 1973); Cynthia W. Enloe, *The Politics of Pollution in Comparative Perspective: Ecology and Power in Four Nations* (New York: McKay, 1975); J. Clarence Davies, III, and Barbara S. Davies, *The Politics of Pollution*, 2nd ed. (Indianapolis: Pegasus, 1975). Maurice Friedberg, "Reading for the Masses: Popular Soviet Fiction, 1976–80," United States International Communication Agency report R-13-81 (June 1981), 16–19; Boris Komarov [pseud.], *The Destruction of Nature in the Soviet Union*, trans. Michael Vale and Joe Hollander (White Plains, N.Y.: M. E. Sharpe, 1980), 6, 16–17, 137–138.

36. Rosenbaum, *Politics of Environmental Concern*, 63–71. Gordon Rattray Taylor, *The Doomsday Book: Can the World Survive?* (Greenwich, Conn.: Fawcett, 1971). I thank Evan Vlachos for sharing his unpublished drafts, "Catastrophic Dread and Collective Mobilization," 1980, and "Apocalyptic Strains and the Potential for Utopian Movements in the United States," 1975.

37. Barry Commoner, *The Closing Circle: Nature, Man, and Technology* (New York: Knopf, 1971), 52, see 65. "Nature was forever": Frank Graham, Jr., *Since Silent Spring* (Boston: Houghton Mifflin, 1970), 13–14. Rachel Carson, *Silent Spring* (New York: Fawcett, 1962), 14, 18, see 187–190. H. M. Hoover, *Children of Morrow* (New York: Four Winds, 1973); John Wyndham, *Re-Birth*, also titled *The Chrysalids* (1955), in *The John Wyndham Omnibus* (New York: Simon & Schuster, 1964).

38. Harris Poll cited in Del Sesto, *Science, Politics, and Controversy*, 193. Allan Mazur, "Opposition to Technological Innovation," *Minerva* 13 (1975) 58–81.

39. See *NYT*, for example 5 Nov. 1971, pp. 20, 23.

40. Sheldon Novick, *The Electric War: The Fight over Nuclear Power* (San Francisco: Sierra Club, 1976), 241.

41. "Grim flavor": McKinley C. Olson, "Reacting to the Reactors," *The Nation* 220 (11 Jan. 1975): 15–17.

42. Bronk Evans, "Sierra Club Involvement in Nuclear Power: An Evolution of Awareness," *Oregon Law Review* 54 (1975): 607–622; oral history interview of William E. Siri by Ann Lage, 1976, Bancroft Library, University of California, Berkeley. The Sierra Club Records in the Bancroft have only a few relevant materials.

43. I use my *NYT* index and *Readers' Guide* surveys. See Nealey, Melber, and Rankin, *Public Opinion and Nuclear Energy*, 129–135; D. L. Sills, "The Environmental Movement and Its Critics," *Human Ecology* 3 (1975): 1–41, polls at p. 7; David J. Webber, "Is Nuclear Power Just Another Environmental Issue? An Analysis of California Voters," *Environment & Behavior* 14 (1982): 72–83.

17. ENERGY CHOICES

1. "L'âge d'or atomique" ("atomic golden age"): Charles-Noel Martin, *L'Atome, maître du monde* (Paris: Centurion, 1956), 194.

2. Géza Róheim, *Magic and Schizophrenia* (Bloomington: Indiana University Press, 1970), 175–180. For "City of the Sun" see Frances A. Yates, *Gior-*

dano Bruno and the Hermetic Tradition (New York: Random House, 1964), 367–373.

3. Robert Heinlein, "Let There Be Light," in Heinlein, *The Man Who Sold the Moon* (New York: New American Library, 1951).

4. Stanley M. Nealey, Barbara D. Melber, and William L. Rankin, *Public Opinion and Nuclear Energy* (Lexington, Mass.: D. C. Heath, 1983), 133–137. "Ideally suited": Amory Lovins, "Energy Strategy: The Road Not Taken?" *Foreign Affairs* 55 (Oct. 1976): 89. Zalmay Khalilzad and Cheryl Benard, "Energy: No Quick Fix for a Permanent Crisis," *BAS* 36, no. 10 (Dec. 1980): 15–20.

5. One bibliographic guide is Laurent Hodges, "Resource Letter SE-2: Solar Energy," *American Journal of Physics* 50 (1982): 876–881.

6. W. Ostwald's commandment quoted in George Basalla, "Energy and Civilization," in *Science, Technology, and the Human Prospect,* ed. Chauncey Starr and Philip C. Ritterbush (New York: Pergamon, 1980), 45.

7. One review is Lon C. Ruedisili and Morris W. Firebaugh, eds., *Perspectives on Energy: Issues, Ideas, and Environmental Dilemmas* (New York: Oxford University Press, 1982), pt. 2.

8. See Carroll L. Wilson, *Coal—Bridge to the Future: Report of the World Coal Study* (Cambridge, Mass.: Ballinger, 1980).

9. Dam collapse: *NYT,* 3 Dec. 1959, p. 1; 20 Dec. 1959, p. 7.

10. Herbert Kouts to editor, *Natural History,* 23 July 1969, copy in folder "Public Relations 1964–72," Maurice Goldhaber files, Brookhaven National Laboratory, Upton, N.Y.

11. Ernest Siddall, "Statistical Analysis of Reactor Safety Standards," *Nucleonics* 17, no. 2 (Feb. 1959): 64–69.

12. The pioneer was Chauncey Starr, "Social Benefit vs. Technological Risk," *Science* 185 (1969): 1232–38. See Edmund A. Crouch and Richard Wilson, *Risk/Benefit Analysis* (Cambridge, Mass.: Ballinger, 1982). The first attempt at popularization was Petr Beckmann, *The Health Hazards of NOT Going Nuclear* (Boulder, Colo.: Golem, 1976). See Linda Gaines, Stephen Berry, and Thomas Veach Long II, *TOSCA: The Total Social Cost of Coal and Nuclear Power* (Cambridge, Mass.: Ballinger, 1978). There is now a large literature on risk and social justice; see special issue of *Science* 236 (17 April 1987).

13. The equivalence is very rough and still under debate; see *Health Physics* 1981– . A summary is Richard Wilson et al., *Health Effects of Fossil Fuel Burning: Assessment and Mitigation* (Cambridge, Mass.: Ballinger, 1980).

14. For summaries of main points see Fred A. Donath and Robert O. Pohl, articles in *Nuclear Power: Both Sides. The Best Arguments For and Against the Most Controversial Technology,* ed. Michio Kaku and Jennifer Trainer (New York: W. W. Norton, 1982).

15. Norman Glass, ed., "Environmental Effects of Increased Coal Utilization: Ecological Effects . . . ," Report EPA-600/7-78-108, NTIS PB-285-440 (Corvallis, Ore.: Corvallis Environmental Research Laboratory, 1978). Many subsequent studies have reached similar conclusions. For a review, and for carcinogen fears in general, see Edith Efron, *The Apocalyptics: Cancer and the Big Lie* (New York: Simon & Schuster, 1984), 98–101 and passim.

16. A partisan but accurate review is Bernard L. Cohen, *Before It's Too Late: A Scientist's Case for Nuclear Energy* (New York: Plenum, 1983).

17. U.S. Nuclear Regulatory Commission, "Reactor Safety Study: An Assessment of Accident Risks in U.S. Commercial Nuclear Power Plants" [Rasmussen Report], WASH-1400, NUREG 75/014 (Washington, D.C.: NRC, 1975), available from National Technical Information Service, Springfield, Va. For a summary see Samuel Glasstone and Walter H. Jordan, *Nuclear Power and Its Environmental Effects* (LaGrange Park, Ill.: American Nuclear Society, 1981), 98–108.

18. Thomas H. Moss and David L. Sills, eds., *The Three Mile Island Nuclear Accident: Lessons and Implications,* Annals of the New York Academy of Sciences, vol. 365 (New York: New York Academy of Sciences, 1981).

19. See Richard Wilson, "A Visit to Chernobyl," *Science* 236 (1987): 1636–1640, and references therein. On radiation see Chapter 11, especially n. 23.

20. For references and further comparisons see Fred Hoyle and Geoffrey Hoyle, *Commonsense in Nuclear Energy* (San Francisco: Freeman, 1980).

21. "Far cleaner": *Energy War: Reports from the Front* (Westport, Conn.: Lawrence Hill, 1979), 225. Ignorant assumptions: for cxample, Fred C. Shapiro, *Radwaste* (New York: Random House, 1981), 21.

22. "Il est politique" (It's political): Alain Touraine et al., *La Prophétie antinucléaire* (Paris: Seuil, 1980), 71, see 70–79, 153.

23. "Elders . . . unreliable": Robert K. Musil, "Growing Up Nuclear," *BAS* 38, no. 1 (Jan. 1982): 19. Survey: Sybille K. Escalona, "Children and the Threat of Nuclear War," in *Behavioral Science and Human Survival,* ed. Milton Schwebel (Palo Alto, Calif.: Science and Behavior Books, 1965), 201–209. See Edwin S. Shneidman, *Deaths of Man* (New York: Quadrangle, 1973), 183–196; Michael Mandelbaum, *The Nuclear Revolution: International Politics Before and After Hiroshima* (Cambridge: Cambridge University Press, 1981), 212–218; Kenneth Keniston, *Young Radicals: Notes on Committed Youth* (New York: Harcourt, Brace & World, 1968), 247–252; Robert Liebert, *Radical and Militant Youth: A Psychoanalytic Inquiry* (New York: Praeger, 1971), 234–240.

24. Hans Karl Rupp, *Ausserparlamentarische Opposition in der Ara Adenauer: Der Kampf gegen die Atombewaffnung in den fünfziger Jahren* (Cologne: Pahl-Rugenstein, 1970), 236–239, 260–261, 275–282; Karl A. Otto, *Vom Ostermarsch zur APO: Geschichte der ausserparlamentarische Opposition in der Bundesrepublik, 1960–1970* (Frankfurt: Campus, 1977).

25. Glenn Seaborg in *NYT,* 10 June 1969, p. 63. Changed composition of opponents: for example, Jean-François Picard, "Les Français et l'énergie," *Revue générale nucléaire* no. 2 (March-April 1981): 134–140. "Hippy characters": L. D. Smith, "Evolution of Opposition to the Peaceful uses of Nuclear Energy," *Nuclear Engineering International* 17 (1972): 461–468. See Steven L. Del Sesto, *Science, Politics, and Controversy: Civilian Nuclear Power in the United States, 1964–1974* (Boulder, Colo.: Westview, 1979), 182–190; Dorothy Nelkin and Michael Pollak, *The Atom Besieged: Extraparliamentary Dissent in France and Germany* (Cambridge, Mass.: MIT Press, 1981), 18, 21–24, 173–

175, 193. In addition to these and many other references, I am of course drawing on my own experience.

26. Jerry Ackerman, "[Nuclear] Energy's 'Public Enemy No. 1,'" Boston *Globe*, 8 Oct. 1978. Lovins, "Energy Strategy." E. E. Schumacher, *Small Is Beautiful: Economics As If People Mattered* (London: Blond & Briggs, 1973).

27. Lovins, "Energy Strategy," 93.

28. Robert Jungk, *Strahlen aus der Asche: Geschichte einem Wiedergeburt* (Bern: Scherz, 1959), 318. Jungk, *Brighter Than a Thousand Suns: A Personal History of the Atomic Scientists*, trans. James Cleugh (New York: Harcourt Brace Jovanovich, 1958); my negative evaluation of this book is shared by all knowledgeable historians. Jungk, *The New Tyranny: How Nuclear Power Enslaves Us*, trans. Christopher Trump (New York: Grosset and Dunlap, 1979).

29. "Choice": Wasserman, *Energy War*, xii. On all this see Nelkin and Pollak, *Atom Besieged.*

30. In addition to Nelkin and Pollak, *Atom Besieged*, and Touraine et al., *La Prophétie anti-nucléaire*, for here and below some useful works are Michael T. Hatch, *Politics and Nuclear Power: Energy Policy in Western Europe* (Lexington: University Press of Kentucky, 1986); International Conference on Nuclear Power and Its Fuel Cycle (1977), *Nuclear Power and Its Fuel Cycle*, vol. 7, *Nuclear Power and Public Opinion, and Safeguards* (Vienna: International Atomic Energy Agency, 1977); Irving C. Bupp and Jean-Claude Derian, *Light Water: How the Nuclear Dream Dissolved* (New York: Basic Books, 1978), esp. 105–117, 138–144; R. F. Pocock, *Nuclear Power: Its Development in the United Kingdom* (Old Woking, Surrey: Unwin, 1977), 221–281; Smith, "Evolution of Opposition;" Martine Chaudron and Yves LePape, "Le Mouvement écologique dans la lutte anti-nucléaire," in *Nucléopolis: matériaux pour l'analyse d'une société nucléaire*, ed. Francis Fagnani and Alexandre Nicolon (Grenoble: Presses universitaires de Grenoble, 1979), 52–78.

31. Hans H. Wüstenhagen, *Bürger gegen Kernkraftwerke: Wyhl—der Anfang?* (Reinbek: Rowohlt, 1975). Dieter Rucht, *Von Wyhl nach Gorleben: Bürger gegen Atomprogramme und nukleare Entsorgung* (Munich: Beck, 1977).

32. Nelkin and Pollak, *Atom Besieged*, chap. 5; Touraine et al., *La Prophétie anti-nucléaire*, 49–50, 144–145.

33. Nelkin and Pollak, *Atom Besieged;* Michael W. Golay, Isi I. Saragossi, and Jean-Marc Willefert, "Comparative Analysis of United States and French Nuclear Power Plant . . . Regulatory Policies . . . ," Massachusetts Institute of Technology Energy Laboratory Report MIT-EL-77-044-WP (Cambridge, Mass.: MIT Energy Laboratory, 1977); Hatch, *Politics and Nuclear Power.*

34. Hatch, *Politics and Nuclear Power*, 104–106.

35. Brian Dumaine, "Nuclear Scandal Shakes the TVA," *Fortune* 114 (27 Oct. 1986): 40–48. The most widely read discussion of reactor economics was in Amory B. Lovins and L. Hunter Lovins, *Energy/War: Breaking the Nuclear Link* (New York: Harper and Row, 1981). Studies favoring reactors, by utilities, were mostly unpublished.

18. CIVILIZATION OR LIBERATION?

1. See Stephen Cotgrove, *Catastrophe or Cornucopia: The Environment, Politics and the Future* (New York: John Wiley, 1982), 33, 88–89, 98, 117.

2. Todd R. La Porte and Daniel Metlay, "Technology Observed: Attitudes of a Wary Public," *Science* 188 (1975): 121–127; Mark P. Lovington and Robert G. Horne, "Project on Public Images of Nuclear Power and Technical Advance," Thesis no. 78C0051 (Worcester, Mass.: Worcester Polytechnic Institute, 1978); John A. Meyer III, "1986 Nuclear Opinion Study," Thesis no. 122JMW0321 (Worcester, Mass.: Worcester Polytechnic Institute, 1986); and many other polls.

3. See, for example, Donald Stever, *Seabrook and the Nuclear Regulatory Commission* (Hanover, N.H.: University Presses of New England, 1980); Brian Wynne, *Rationality and Ritual: The Windscale Inquiry and Nuclear Decisions in Britain* (Calfont St. Giles: British Society for the History of Science, 1982), 97–111.

4. "Revolutionary": Joe Shapiro, "The Anti-nuclear Movement," *Science for the People* 12, no. 4 (July-Aug. 1980): 16–21.

5. "Earth raped": Suzanne Gordon, "From Earth Mother to Expert," *Nuclear Times* 1, no. 7 (May 1983): 13–16. "Patriarchy": Mary Daly, *Gyn/Ecology* (Boston: Beacon Press, 1978), as quoted in Dorothy Nelkin, "Nuclear Power as a Feminist Issue," draft, 1980; I thank Nelkin for sharing this and other drafts.

6. See, for example, Howard Kohn, *Who Killed Karen Silkwood?* (New York: Summit, 1981).

7. "The Plutonium Incident" (CBS-TV, 1980).

8. *Silkwood* (ABC—Twentieth-Century Fox, 1983). "Contamination": Stephen Schiff in *Vanity Fair* 47, no. 2 (Feb. 1984): 28. *The China Syndrome* (Columbia, 1979).

9. One review is J. R. Ravetz, "Criticism of Science," in Ina Spiegel-Rösing and Derek de Solla Price, *Science, Technology, and Society: A Cross-Disciplinary Perspective* (London: Sage, 1977), 71–89. See Richard Olson, *Science Deified and Science Defied: The Historical Significance of Science in Western Culture,* vol. 1, *From the Bronze Age . . . to ca.* A.D. *1640* (Berkeley: University of California Press, 1982). Roger Sperry, "Some Effects of Disconnecting the Cerebral Hemispheres," *Science* 217 (1982): 1223–26.

10. Alain Touraine et al., *La Prophétie anti-nucléaire* (Paris: Seuil, 1980), 40–41, 66–67, 73, 93–95, 201, 206–207.

11. Sarah Lichtenstein et al., "Judged Frequency of Lethal Events," *Journal of Experimental Psychology: Human Learning and Memory* 4 (1978): 551–578. Robert C. Mitchell, "Rationality and Irrationality in the Public's Perception of Nuclear Power," in *Public Reactions to Nuclear Power: Are There Critical Masses?,* ed. William R. Freudenburg and Eugene A. Rosa, AAAS Selected Symposium 93 (Boulder, Colo.: Westview, 1984), 137–179. Baruch Fischhoff, Paul Slovic, and Sarah Lichtenstein, "'The Public' vs. 'The Experts': Perceived vs. Actual Disagreements About Risks of Nuclear Power," in *The Analysis of Actual versus Perceived Risks,* ed. Vincent T. Covello et al. (New York:

Plenum, 1983), 235–249. See also Paul Slovic, "Images of Disaster: Perception and Acceptance of Risks from Nuclear Power," in *Perceptions of Risk: Proceedings of the 15th Annual Meeting, 1979* (Washington, D.C.: National Council on Radiation Protection and Measurements, 1980), 34–55; C. Starr and Chris Whipple, "Risks of Risk Decisions," *Science* 208 (1980): 1114–20; Paul Slovik, "Perception of Risk," *Science* 236 (1987): 280–285.

12. "Best hope": quoted by Philip M. Boffey, "Plutonium: Its Morality Questioned by National Council of Churches," *Science* 192 (1976): 356–359. Hans Matthoefer, *Interviews und Gespräche zur Kernenergie* (Karlsruhe: C. F. Müller, 1977), 15, as cited in Michael T. Hatch, *Politics and Nuclear Power: Energy Policy in Western Europe* (Lexington: University Press of Kentucky, 1986), 78. See Michael Barkum, *Disaster and the Millennium* (New Haven, Conn.: Yale University Press, 1974), 145–152; Mary Douglas and Aaron Wildavsky, *Risk and Culture: An Essay on the Selection of Technical and Environmental Dangers* (Berkeley: University of California Press, 1982), 123–124.

13. Some starting points are C. Glacken, *Traces on the Rhodian Shore: Nature and Culture in Western Thought . . .* (Berkeley: University of California Press, 1967); Roderick Nash, *Wilderness and the American Mind* (New Haven, Conn.: Yale University Press, rev. ed., 1973); Carol P. MacCormack and Marilyn Strathern, *Nature, Culture and Gender* (Cambridge: Cambridge University Press, 1980); Ronald Inglehart, *The Silent Revolution: Changing Values and Political Styles Among Western Publics* (Princeton, N.J.: Princeton University Press, 1977). Nature : culture should not be confused with other polarities that are uncorrelated (for example, individual : society, where individualism or submersion in society can each be either urban or rural) or partly correlated (for example, Max Weber's *gemeinschaft : gesellschaft*).

14. Gerard Duménil, "Energie nucléaire et opinion publique," in *Le Débat nucléaire en France: Acteurs sociaux et communication de Masse*, ed. F. Fagnani (Grenoble: Université des Sciences Sociales, Institut de Recherche Economique, 1977), 362. Stanley M. Nealey, Barbara D. Melber, and William L. Rankin, *Public Opinion and Nuclear Energy* (Lexington, Mass.: D. C. Heath, 1983), 153–154. Among many other polls are D. Agrafiotis, G. Morlat, and J. P. Pages, "Le Public et le Nucléaire," in International Conference on Nuclear Power and Its Fuel Cycle, *Nuclear Power and Its Fuel Cycle*, vol. 7, *Nuclear Power and Public Opinion, and Safeguards* (Vienna: International Atomic Energy Agency, 1977), 309–324; H. J. Otway, "Review of Research on Identification of Factors Influencing Social Response to Technological Risks," in *Nuclear Power and Its Fuel Cycle*, vol. 7, 95–117.

15. William L. Rankin and Stanley M. Nealey, "The Relationship of Human Values and Energy Beliefs to Nuclear Power Attitudes" (Seattle: Battelle Memorial Institute Human Affairs Research Centers, 1978), 18.

16. Introductions are Alvin W. Gouldner, *The Future of Intellectuals and the Rise of the New Class* (New York: Seabury, 1979); B. Bruce-Briggs, ed., *The New Class?* (New Brunswick, N.J.: Transaction Books, 1979). S. Robert Lichter, Stanley Rothman, and Linda S. Lichter, *The Media Elite* (Bethesda, Md.: Adler & Adler, 1986), 37–38; *NYT*, 27 Nov. 1984, p. C18.

17. Dorothy Nelkin and Michael Pollak, *The Atom Besieged: Extraparliamentary Dissent in France and Germany* (Cambridge, Mass.: MIT Press, 1981), 105–108; Touraine et al., *La Prophétie anti-nucléaire,* 61 and passim. A more general argument is William Tucker, *Progress and Privilege: America in the Age of Environmentalism* (Garden City, N.Y.: Doubleday, 1982); cf. Frederick H. Buttel and William L. Flinn, "Social Class and Mass Environmental Beliefs: A Reconsideration," *Environment and Behavior* 10 (1978): 433–448.

18. Cotgrove, *Catastrophe or Cornucopia.* Jon D. Miller and Kenneth Prewitt, "A National Survey of the Non-Governmental Leadership of American Science and Technology" (DeKalb: Northern Illinois University Public Opinion Laboratory, 1982), 61–64.

19. Richard Curtis and Elizabeth Hogan, *Perils of the Peaceful Atom: The Myth of Safe Nuclear Power Plants* (Garden City, N.Y.: Doubleday, 1969), xiii. Robert Jungk, *The New Tyranny: How Nuclear Power Enslaves Us,* trans. Christopher Trump (New York: Grosset and Dunlap, 1979), 10.

20. Curtis and Hogan, *Perils,* 159, 257. Jungk, *New Tyranny,* 10, 91. On Faust see Chap. 2, n. 11; Alvin M. Weinberg, "Social Institutions and Nuclear Energy," *Science* 177 (1972): 27–34; Günther Schwab, *Morgen holt dich der Teufel . . .* (Salzburg: Bergland, 1968). *The Chosen* (Embassy-Aston-American International, 1978), later on CBS-TV as "Holocaust 2000." Nelkin and Pollak, *Atom Besieged,* 136–142.

21. Sharon M. Friedman, "A Case of Benign Neglect: Coverage of Three Mile Island before the Accident," in Friedman, Sharon Dunwoody, and Carol L. Rogers, *Scientists and Journalists: Reporting Science as News* (New York: Macmillan Free Press, 1986), 182–201. "Energy Independence": for example, *NYT Magazine,* 5 July 1981, pp. 2–3. See United States Atomic Energy Commission, *Combined Film Catalog* (Washington, D.C.: AEC). Tradition: David E. Nye, *Image World: Corporate Identities at General Electric, 1890–1930* (Cambridge, Mass.: MIT Press, 1985), 129. "Shivering in the dark": Craig Hosmer in John O'Neill, "Hit Hard at Enemy," *Nuclear Industry* (Feb. 1978): 12–17.

22. Colette Guedeney and Gérard Mendel, *L'Angoisse atomique et les centrales nucléaires* (Paris: Payot, 1973), 99–100. See Lichter et al., *Media Elite;* also Stanley Rothman and S. Robert Lichter, "The Nuclear Energy Debate: Scientists, the Media, and the Public," *Public Opinion* 5, no. 4 (Aug.–Sept. 1982): 47–52. Thinking/feeling study: Raymond Johnson and W. Larry Petcovic, "The 10/90 Rule for Credibility in Risk Communication," *Health Physics Society Newsletter* (July 1986): 14–15.

23. Steven L. Del Sesto, "Uses of Knowledge and Values in Technical Controversies: The Case of Nuclear Reactor Safety in the United States," *Social Studies of Science* 13 (1983): 395–416. Cf. Douglas and Wildavsky, *Risk and Culture.*

24. In Joann Davis and Wendy Smith, "A Checklist of Nuclear Books," *Publishers Weekly* (26 March 1982): 45–51, roughly four overtly antinuclear paperbacks or hardcover books are listed for each pronuclear one. In the American and European bookstores and paperback racks I have seen, the ratio was even higher.

25. John G. Fuller, *We Almost Lost Detroit* (1975; New York: Ballantine, 1978); cf. Petr Beckmann, *The Health Hazards of NOT Going Nuclear* (Boulder, Colo.: Golem, 1976), 75–76. Fuller, *Arigo: Surgeon of the Rusty Knife* (New York: Crowell, 1974); cf. review by Martin Gardner, *New York Review of Books* (16 May 1974): 18–19. Fuller, *The Ghost of Flight 401* (New York: Berkley, 1978). Fuller, *The Interrupted Journey: Two Lost Hours "Aboard a Flying Saucer"* (New York: Dial, 1966); cf. Robert Sheaffer, *The UFO Verdict: Examining the Evidence* (Buffalo, N.Y.: Prometheus, 1981), 34–44, 111–119.

26. E. Pauline Alexanderson, ed., *Fermi-I: New Age for Nuclear Power* (La Grange Park, Ill.: American Nuclear Society, 1979).

27. Hans Heinrich Ziemann, *The Accident* (New York: St. Martin's, 1979). Ron Kytle, *Meltdown* (New York: McKay, 1976). Bett L. Pohnka and Barbara C. Griffin, *The Nuclear Catastrophe* (Port Washington, N.Y.: Ashley, 1977). In a class by itself is Alexander Sidar III, *The Dorset Disaster* (New York: Grosset & Dunlap, 1980). Burton Wohl, *The China Syndrome* (New York: Bantam, 1979), novelized the film.

28. Survey: Lichter et al., *Media Elite,* 188–197. "60 Minutes" (CBS-TV, 8 Feb. 1976). "Dramatic": Stephen Lesher, *Media Unbound: The Impact of Television Journalism on the Public* (Boston: Houghton Mifflin, 1982), 35, see 31–35, 169–181. "Interesting": John J. O'Connor, *NYT,* 30 May 1976, sec. 2, p. 19. See clippings s.v. Drama—Subject—Atomic Energy, NYPL/TC.

29. Lauriston S. Taylor, "Some Nonscientific Influences on Radiation Protection Standards and Practice," *Health Physics* 39 (1980): 868. Similarly, Ralph E. Lapp, *The Radiation Controversy* (Greenwich, Conn.: Reddy, 1979), 138–140. Bernard L. Cohen, "A Poll of Radiation Health Scientists," *Health Physics* 50 (1986): 639–644.

30. Among other surveys are my own analysis of *Readers' Guide* titles; J. P. Delzani and C. Carde, "Approche psychologique de l'évolution du thème nucléaire à partir d'un quotidien populaire de 1950 à 1976," in *Colloque sur les implications psycho-sociologiques du développement de l'industrie nucléaire,* ed. M. Tubiana (Paris: Société Française de Radioprotection, 1977), 354–374; William L. Rankin, Stanley M. Nealey, and Daniel E. Montano, "Analysis of Print Media Coverage of Nuclear Power Issues" (Seattle: Battelle Memorial Institute, Human Affairs Research Centers, 1978); Lichter et al., *Media Elite,* chap. 6; Robert L. DuPont, "Nuclear Phobia—Phobic Thinking about Nuclear Power" (Washington, D.C.: The Media Institute, 1980).

31. Lichter et al., *Media Elite,* 166–167, 178–184, 216–217.

32. My typology borrows from Daniel Joseph, "Sondage SOFRES: Le public et le nucléaire," in *Colloque sur les implictions,* 113–138. Barbara D. Melber et al., "Nuclear Power and the Public: Analysis of Collected Survey Research," Report PNL-2430 (Seattle: Battelle Human Affairs Research Centers, 1977); Nealey, Melber, and Rankin, *Public Opinion and Nuclear Energy;* Connie DeBoer, "The Polls: Nuclear Energy," *Public Opinion Quarterly* 41 (1977): 402–411; William R. Freudenburg and Eugene A. Rosa, eds., *Public Reactions to Nuclear Power: Are There Critical Masses?* AAAS Selected Symposium 93 (Boulder, Colo.: Westview, 1984).

33. Some European polls are cited in Nelkin and Pollak, *Atom Besieged,*

108–112; Allan Mazur, *The Dynamics of Technical Controversy* (Washington, D.C.: Communications Press, 1981), 108–109. Citizens' anxieties: for example, Girogii Tsistsishvili, Tbilisi *Komunisti*, 18 Feb. 1981, p. 3, in *Daily Report: Soviet Union*, 1 May 1981, S4-S5; my thanks to L. Gouré for this item and other information. Shanghai: Shirley Aronson, private communication, 1985. Émigrés: eight physicists.

34. Melber et al., "Nuclear Power and the Public," 103–106, 116–119; Fagnani, *Débat nucléaire en France*, 10; Donald A. Clelland and Michael D. Bremseth, "Explanations of Public Response to a Nuclear Energy Alternative," talk presented at meeting of the American Association for the Advancement of Science, San Francisco, 6 Jan. 1980.

35. Roger E. Kasperson et al., "Public Opposition to Nuclear Energy: Retrospect and Prospect," in *Sociopolitical Effects of Energy Use and Policy*, ed. Charles T. Unseld et al. (Washington, D.C.: National Academy of Sciences—National Research Council, 1979), 261–292; see also polls cited previously. Charles J. Brody, "Nuclear Power: Sex Differences in Public Opinion," Ph.D. diss., University of Arizona, 1981. Mitchell, "Rationality and Irrationality." It is not gender associations but one's sex itself that matters, according to Scott M. Favreau and John F. Curry, "Attitudes toward Nuclear Power: An Examination of the Sex Differential on Four College Campuses," thesis 1-28-JMW-9559, Worcester Polytechnic Institute, Worcester, Mass.

36. Arkady Strugatsky and Boris Strugatsky, *Roadside Picnic*, trans. Antonina W. Bouis (New York: Macmillan, 1977). *Stalker* (Mosfilm, 1979).

37. Stan Lee, *Origins of Marvel Comics* (New York: Simon & Schuster, 1974). *Dr. Solar* no. 29 (Whitman Comics, 1981); see also *The Fury of Firestorm the Nuclear Man* (DC Comics, 1981-).

38. *Superman and Spiderman* (New York: Warner, 1981); "Marvel Family" (NBC-TV, 12 Dec. 1981).

39. Christine Blanchart in *Colloque sur les implications*, 226–227.

40. Isaac Chiva, "Imagination collective et inconnu," in *Echanges et Communications: Mélanges offerts à Claude Lévi-Strauss . . .* (Paris: Mouton, 1970): 162–168. See also William Beardslee and John Mack, "The Impact on Children and Adolescents of Nuclear Developments," in *Psychosocial Aspects of Nuclear Developments*, American Psychiatric Association Task Force Report 20 (Washington, D.C.: American Psychiatric Association, 1982), 64–93; M. Schwebel and B. Schwebel, "Children's Reactions to the Threat of Nuclear Plant Accidents," *American Journal of Orthopsychiatry* 51 (1981): 260–270.

41. My account of reactions to Chernobyl is drawn from many press reports and personal communications. One thoughtful account is Mary J. Salter, "Living with Fallout," *Atlantic* 259, no. 1 (Jan. 1987): 30–35. *Newsweek* 107, no. 19 (12 May 1986): 40.

42. "No longer control": Harrison Salisbury, "Gorbachev's Dilemma," *NYT Magazine* (27 July 1986), p. 18. *New Yorker* 62, no. 12 (12 May 1986), 29. *NYT*, 15 May 1986, p. A10; 7 July 1986, p. 1 (quote from Revelation 8:11); 22 Feb. 1987, p. E23.

43. Philip L. Cantelon and Robert C. Williams, *Crisis Contained: The De-*

partment of Energy at Three Mile Island (Carbondale: Southern Illinois University Press, 1982), chaps. 3–5, Cronkite quoted on p. 58. See Daniel Martin, *Three Mile Island: Prologue or Epilogue?* (Cambridge, Mass.: Ballinger, 1980); David M. Rubin in *The Three Mile Island Nuclear Accident: Lessons and Implications,* ed. Thomas H. Moss and David L. Sills, Annals of the New York Academy of Sciences, vol. 365 (New York: New York Academy of Sciences, 1981), 95–106; Peter M. Sandman and Mary Paden, "At Three Mile Island," *Columbia Journalism Review* 18 (July-Aug. 1979): 43–58; articles by Ann Marie Cunningham and Sharon M. Friedman in Friedman, *Scientists and Journalists,* 182–211; J-F. Picard and J-M. Fourgous, "Analyse des déformations de l'information donnée par les média en France à propos de l'accident de Three Mile Island" (Le Chesnay: Electricité de France, Comité de Radioprotection, 1979). David E. Lilienthal, *The Journals of David E. Lilienthal,* vol. 7, *Unfinished Business 1968–1981* (New York: Harper & Row, 1983), 753–765.

44. Michael Mufson, "Three Mile Island: Psychological Effects of a Nuclear Accident and Mass Media Coverage," in *Psychosocial Aspects* (American Psychiatric Association), 42–54. Allan Mazur, "Media Influence on Public Attitudes Toward Nuclear Power," in *Public Reactions to Nuclear Power: Are There Critical Masses?* ed. William R. Freudenburg and Eugene A. Rosa, AAAS Selected Symposium 93 (Boulder, Colo.: Westview, 1984), 97–114; Paul Slovic, Baruch Fischoff, and Sarah Lichtenstein, "Perception and Acceptability of Risk from Energy Systems," in *Public Reactions to Nuclear Power,* 115–135.

45. *NYT,* 23 April 1980, p. 1; 19 April 1981, p. 6E.

19. THE WAR FEAR REVIVAL

1. "Hiroshima/Nagasaki 1945"; see Jack Gould, *NYT,* 4 Aug. 1970. Two separate fainting cases were reported to me.

2. Dorothy Nelkin, "Anti-nuclear Connections: Power and Weapons," *BAS* 37, no. 4 (April 1981): 36–40. Helen Caldicott, *Nuclear Madness: What You Can Do!* (New York: Bantam, 1980), 61; Caldicott, "PSR History," *BAS* 42, no. 3 (March 1986): 57.

3. Alistair MacLean, *Goodbye California* (Garden City, N.Y.: Doubleday, 1978). Larry Collins and Dominique Lapierre, *The Fifth Horseman* (New York: Simon & Schuster, 1980). Robert Ludlum, *The Parsifal Mosaic* (New York: Random House, 1982), 616; best-seller according to *NYT Book Review,* 2 Jan. 1983, p. 22.

4. Joann Davis and Wendy Smith, "A Checklist of Nuclear Books," *Publishers' Weekly* (26 March 1982): 45–51. Jonathan Schell, *The Fate of the Earth* (New York: Knopf, 1982); see "The Fate of the Book," *BAS* 38, no. 8 (Oct. 1982): 63.

5. Stockholm International Peace Research Institute, *World Armaments and Disarmament: SIPRI Yearbook* (Cambridge, Mass.: Oelgeschlager, Gunn & Hain; later Oxford: Oxford University Press, 1968-); Tom Gervasi, *The Myth of Soviet Military Supremacy* (New York: Harper & Row, 1986), App. F.

6. Sen. Daniel P. Moynihan, "The Nuclear Challenge," address presented

at Daemen College, 15 May 1983, courtesy Sen. Moynihan's office. On rise in fears see Connie DeBoer, "The Polls: Our Commitment to World War III," *Public Opinion Quarterly* 45 (1981): 126–134. Jeremy W. Sanders, *Peddlers of Crisis: The Committee on the Present Danger and the Politics of Containment* (Boston: South End Press, 1983); Gervasi, *Myth of Soviet Supremacy,* 75 and chap. 31.

7. James E. Dougherty and Robert L. Pfaltzgraff, Jr., eds., *Shattering Europe's Defense Consensus: The Antinuclear Protest Movement and the Future of NATO* (Washington, D.C.: Pergamon-Brassey's, 1985). For U.S. origins see Fox Butterfield, "Anatomy of the Nuclear Protest," *NYT Magazine* (11 July 1982), p. 14.

8. Jiri Nehnevajsa, "Issues of Civil Defense: Vintage 1978. Summary Results of the 1978 National Survey" (Pittsburgh: University of Pittsburgh Center for Social and Urban Research, 1979); Nehnevajsa, "1982: Some Public Views of Civil Defense" (Pittsburgh: University of Pittsburgh Center for Social and Urban Research, 1983). H. Ornauer et al., *Images of the World in the Year 2000: A Comparative Ten Nation Study* (Atlantic Highlands, N.J.: Humanities Press, 1976), 86–90, 641–642, 660. Bernard M. Kramer, S. Michael Kalick, and Michael A. Milburn, "Attitudes Toward Nuclear Weapons and Nuclear War: 1945–1982," *Journal of Social Issues* 39 (1983): 7–24.

9. Rita R. Rogers, "On Emotional Responses to Nuclear Issues and Terrorism," in *Psychosocial Aspects of Nuclear Developments,* American Psychiatric Association Task Force Report 20 (Washington, D.C.: American Psychiatric Association, 1982), 14. Janet Morris, ed., *Afterwar* (New York: Baen, 1985), 8, 12. Gallup poll, *Newsweek* (5 Oct. 1981): 35.

10. William Beardslee and John Mack, "The Impact on Children and Adolescents of Nuclear Developments," in *Psychosocial Aspects of Nuclear Developments,* American Psychiatric Association, 64–93. Student ignorance reported to me by teachers from three different regions.

11. Michael MccGwire, *Military Objectives in Soviet Foreign Policy* (Washington, D.C.: Brookings Institution, 1987); Leon Gouré, *War Survival in Soviet Strategy: USSR Civil Defense* (Miami: University of Miami Center for Advanced International Studies, 1976), chaps. 5, 9–11; cf. Fred M. Kaplan, "Soviet Civil Defence: Some Myths in the Western Debate," *Survival* 20 (1978): 113–120. "Blowing up": Brezhnev to Moscow World Peace Congress, 1973, quoted by Robert L. Arnett, "Soviet Attitudes towards Nuclear War: Do They Really Think They Can Win?" *Journal of Strategic Studies* (Sept. 1979): 178. Stephen Shenfield, "Soviet Thinking about the Unthinkable," *BAS* 41, no. 2 (Feb. 1985): 23–25. In addition to these and many other publications I have used talks with émigrés and travelers.

12. Eric Chivian, John Mack, and Jeremy Waletzky, "Soviet Children and the Threat of Nuclear War: A Preliminary Study," *American Journal of Orthopsychiatry* 55 (1985): 484–502; see report in *Psychology Today* 18, no. 4 (April 1984): 20. Film: *NYT,* 8 Feb. 1987, p. H32. P. T. Yegorov, I. A. Shlyakhov, and N. I. Albin, *Civil Defense* (Moscow, 2nd ed. 1970), trans. Oak Ridge National Laboratory (Washington, D.C.: Government Printing Office, 1977), 217.

13. "The Defense of the United States" (CBS-TV, 14 June 1981). "The Day After" (ABC-TV, 20 Nov. 1983). Also "A Guide to Armageddon" (BBC-TV, 1982) and "Threads" (BBC-TV, 1984), both shown in the United States.

14. John Hackett et al., *The Third World War: August 1985* (New York: Macmillan, 1979). "World War III" (NBC-TV, 1982). Roger Zelazny, "Damnation Alley" (1967), reprinted in Zelazny, *The Last Defender of Camelot* (New York: Pocket Books, 1980), see pp. 2–3. *Damnation Alley,* also titled *Survival Run* (20th Century-Fox, 1977). The magazine is *Survival Weapons and Tactics* (1981-). See also, for example, John Nahmlos, *Survivors* (New York: Kensington, 1982); the series by David Robbins, *Endworld* (New York: Leisure Books), reached its 4th volume by 1987 and that by Ryder Stacy, *Doomsday Warrior* (New York: Kensington) its 10th; Barry Popkess, *The Nuclear Survival Handbook* (London: Arrow, 1980). Children's book: Louise Lawrence, *Children of the Dust* (New York: Harper & Row, 1985).

15. For example, Arthur M. Katz, *Life after Nuclear War: The Economic and Social Impacts of Nuclear Attacks on the United States* (Cambridge, Mass.: Ballinger, 1981). A fine exception is Thomas Powers, *Thinking About the Next War* (New York: Knopf, 1983), chaps. 6–7. See Jack H. Nunn, "Termination: The Myth of the Short, Decisive Nuclear War," *Parameters* 10, no. 4 (Dec. 1980): 36–41. For expert civil defense advocates and opponents see Jennifer Leaning and Langley Keyes, *The Counterfeit Ark: Crisis Relocation for Nuclear War* (Cambridge, Mass.: Ballinger, 1983); John Dowling and Evans Harnell, eds., *Civil Defense: A Choice of Disasters* (New York: American Institute of Physics, 1987).

16. My calculations, using Samuel Glasstone, ed., *The Effects of Nuclear Weapons* (Washington, D.C.: Government Printing Office, rev. ed., 1964), with 9 cal/cm^2 for 10 mi. visibility, ignoring both cloud cover and combined effects of simultaneous bursts, and excluding Alaska. I calculated the area for all sizes of Soviet warheads weighted by number, as given in *Physics Today* 36, no. 1 (Jan. 1983): 38; the area is about the same as if all warheads were one megaton.

17. Division of Biology and Medicine, "Summary Discussion of Effects on Humans, Agricultural Products, and Weather of a Projected Nuclear War," 9 Oct. 1956, Atomic Energy Commission, quoted in Jack Holl and Richard Hewlett, vol. 3 of history of the Commission, forthcoming. Ralph Lapp, *Kill and Overkill: The Strategy of Annihilation* (New York: Basic Books, 1962), 102–103. Barry Commoner, *Science and Survival* (New York: Viking, 1963), 122. *NYT,* 6 Sept. 1974, p. 1; 12 Nov. 1974, p. 38.

18. P. J. Crutzen and J. W. Binks, *Ambio* 11 (1982): 114. See *Physics Today* 37, no. 2 (Feb. 1984): 17–20; Lester Grinspoon, ed., *The Long Darkness: Psychological and Moral Perspectives on Nuclear Winter* (New Haven, Conn.: Yale University Press, 1986).

19. Stanley Feldman and Lee Sigelman, "The Political Impact of Prime-Time Television: 'The Day After,'" *Journal of Politics* 47 (1985): 556–578.

20. See Steven Kull, "Nuclear Nonsense," *Foreign Policy,* no. 58 (Spring 1985): 28–52.

21. Carl Sagan, "Nuclear War and Climatic Catastrophe: Policy Implica-

tions," pp. 7–62 in Grinspoon, *The Long Darkness,* discusses the intellectual impact, while Robert Jay Lifton, "Imagining the Real: Beyond the Nuclear End," pp. 79–99 in *The Long Darkness,* suggests how it might eventually affect imagery.

22. "PR splash": Alexander Haig as quoted in E. P. Thomson, ed., *Star Wars* (New York: Pantheon, 1985), 12; see chap. 1. George Keyworth as quoted in William Hartung, "Star Wars Pork Barrel," *BAS* 42, no. 1 (Jan. 1986): 20. One summary of the debate is H. Guyford Stever and Heinz R. Pagels, eds., *The High Technologies and Reducing the Risk of War* (New York: New York Academy of Sciences, 1986). Others of the many books (although not his own) are reviewed by Lord [Solly] Zuckerman, "Reagan's Highest Folly," *New York Review of Books* 34, no. 6 (9 April 1987): 35–41. Here and below I also rely on private communications.

23. Poll: *NYT,* 31 Oct. 1986, sec. 1, p. 36. Harrison Brown, "Draw the Line at Star Wars," *BAS* 43, no. 1 (Jan.-Feb. 1987): 3. For the American Physical Society study see *Physics Today* 40 (1987).

24. This is not hindsight; I wrote Chapter 3 (and the opening of Chapter 6, q.v.) almost entirely before 1983. Director George Lucas acknowledged that old comic books, films, and myths were main sources of *Star Wars,* later titled *Star Wars: A New Hope* (Twentieth Century-Fox, 1977). Magic flight: Mircea Eliade, *Shamanism: Archaic Techniques of Ecstasy,* trans. Willard R. Trask (Princeton, N.J.: Princeton University Press, 1964), 403–412, 477–482. Lawrence Suid, "Hollywood and Space," address presented to Society for History of Technology, Pittsburgh, October 1986. William J. Broad, *Star Warriors* (New York: Simon & Schuster, 1985), 118–120 and passim. On faith in technology see Jeff Smith, "Reagan, Star Wars, and American Culture," *BAS* 43, no. 1 (Jan.-Feb. 1987): 19–25.

25. Eymert den Oudsten, "Public Opinion on Peace and War," in Stockholm International Peace Research Institute, *SIPRI Yearbook 1986,* 17–35. *Wargames* (Metro-Goldwyn-Mayer/United Artists, 1983). The first explicitly post-nuclear Mad Max film was *Mad Max: Beyond Thunderdome* (Warner Brothers, 1985), especially notable for its scientist symbol. Others in this genre: *After the Fall of New York* (Nuova Dania-Medusa-Griffon, 1983); *Radioactive Dreams* (ITM, 1986). Another example, *Testament* (Paramount, 1983), was frequently compared with *On the Beach* but in fact avoided the latter's shallow, romantic resignation. I discuss artistic productions in Chapters 20 and 21. For more on the early 1980s see Paul Boyer, *By the Bomb's Early Light: American Thought and Culture at the Dawn of the Atomic Age* (New York: Pantheon, 1985), 361–367.

20. THE MODERN ARCANUM

1. W. H. Auden, *The Age of Anxiety: A Baroque Eclogue* (New York: Random House, 1947). Tests: *Science* 233 (1986): 1250. See Rollo May, *The Meaning of Anxiety* (New York: Washington Square Press, rev. ed., 1970), 3–12, 208–

211; Gerald J. Ringer, "The Bomb As a Living Symbol: An Interpretation," Ph.D. diss., Florida State University, 1966, 127–128, 143.

2. J. B. Priestley, *Thoughts in the Wilderness* (London: Heinemann, 1957), 57. Charles Glicksberg, *Literature and Religion: A Study in Conflict* (Dallas: Southern Methodist University Press, 1960), 4–7, 23–25. See Ringer, "Bomb as Living Symbol," chap. 3.

3. Ignace Lapp, *Death and Its Mysteries,* trans. Bernard Muchland (New York: Macmillan, 1968), as quoted in William Griffin, ed., *Endtime: The Doomsday Catalog* (New York: Collier, 1979), 58–59. See Robert Jay Lifton, *The Broken Connection: On Death and the Continuity of Life* (New York: Simon & Schuster, 1979). On these and other points a good collection is Jim Schley, ed., *Writing in a Nuclear Age* (Hanover, N.H.: University Press of New England, 1984).

4. Kurt Vonnegut, Jr., *Palm Sunday: An Autobiographical Collage* (New York: Laurel, 1984), 69. One introduction to the extensive critical literature is Jerome Klinkowitz and Donald L. Lawler, eds., *Vonnegut in America* (New York: Delta, 1977).

5. John Braine, "People Kill People," in *Voices from the Crowd: Against the H-Bomb,* ed. David Boulton (Philadelphia: Dufour, 1964), 181. Directors: *Show* (June 1962): 78–81.

6. Pat Frank, "Hiroshima: Point of No Return," *Saturday Review* (24 Dec. 1960): 25. Mordecai Roshwald, *Level 7* (New York: McGraw-Hill, 1960). See Philip D. Segal, "Imaginative Literature and the Atomic Bombs: An Analysis of Representative Novels, Plays, and Films from 1945 to 1972," Ph.D. diss., Yeshiva University, New York, 1973, 48–52.

7. Johns: for example, *Target with Four Faces* (1955); see Max Kozloff, *Jasper Johns* (New York: Meridien, 1974), 12. On Tinguely see Harold Rosenberg, *The De-Definition of Art: Action Art to Pop to Earthworks* (New York: Horizon, 1972), 156–166. Cf. Serge Guilbaut, *How New York Stole the Idea of Modern Art: Abstract Expressionism, Freedom, and the Cold War,* trans. Arthur Goldhammer (Chicago: University of Chicago Press, 1983), 96–97, 107–108.

8. On Evergood see Greta Berman and Jeffrey Wechsler, *Realism and Realities: The Other Side of American Painting 1940–1960* (New Brunswick, N.J.: Rutgers University Press, 1981), 57. One example of the 1980s movement was the ninety-artist segment "Preparing for War" in the "Terminal New York" show, Brooklyn, N.Y., October 1983. Ruined cities: Vitaly Komar and Alexander Melamid's series, *Scenes from the Future* (ca. 1983); cf. Max Ernst, *Europe after the Rains* (1940–42). Alex Grey, *Nuclear Crucifixion* (1980). "Elaborate despair": Alex Grey, "Victim Nightmares, Sacred Mirrors," *Co-Evolution Quarterly* no. 38 (Summer 1983): 35–39.

9. Paul Fussell, *The Great War and Modern Memory* (London: Oxford University Press, 1975), 8, 34–35, 203; cf. Dadaism. Kurt Vonnegut, *Cat's Cradle* (New York: Holt, Reinhart & Winston, 1965). Thomas Pynchon, *Gravity's Rainbow* (New York: Viking, 1973); see esp. the last few pages. Joseph Heller, *Catch-22* (New York: Simon & Schuster, 1961). "I wrote it" quoted in Lawrence H. Suid, *Guts & Glory: Great American War Movies* (Reading, Mass.:

Addison-Wesley, 1978), 271. James Rosenquist, "What Is the F-111?" in John Russell and Suzi Gablik, *Pop Art Redefined* (New York: Praeger, 1969).

10. Kurt Vonnegut, *Slaughterhouse-5: or the Children's Crusade* (New York: Delta, 1969), 17. See also Paul Boyer, *By the Bomb's Early Light: American Thought and Culture at the Dawn of the Atomic Age* (New York: Pantheon, 1985), chap. 20.

11. Fred Kirby quoted by Charles Wolfe, notes to the recording *Atomic Café* (Archives Project, 1981). See Michael J. Yavendetti, "American Reactions to the Use of Atomic Bombs on Japan, 1945–1947," Ph.D. diss., University of California, Berkeley, 1970, 255–256. Johnson: *NYT,* 2 Dec. 1967, p. 78.

12. "Basic power": for example, John Cockcroft in British Broadcasting Corp., *Atomic Challenge: A Symposium* (London: Winchester, 1947), 2. "Cure the sick": Spirit of Memphis Quartet, "Atomic Telephone" (1951), on recording *Atomic Café.* See Lifton, *Broken Connection,* 369–372.

13. Ringer, "Bomb as Living Symbol," 187–202.

14. On pathology of hope see Abraham H. Maslow, *Toward a Psychology of Being* (New York: Van Nostrand Reinhold, 2nd ed., 1968), chap. 8.

15. Lowell Blanchard, "Jesus Hits Like an Atom Bomb" (1949–50), on recording *Atomic Café.* Revelation 6–16; the Final Battle can be traced to Zoroaster. See "Sect Anticipates Atomic Armageddon," *Life* 37 (22 Nov. 1954), 176–177; Jane Allyn Hardyck and Marcia Braden, "Prophecy Fails Again: A Report of a Failure to Replicate," *Journal of Abnormal Social Psychology* 65 (1962): 136–141.

16. Hal Lindsey with C. C. Carlson, *The Late Great Planet Earth* (Grand Rapids, Mich.: Zondervan, 1970). See William Martin, "Waiting for the End," *Atlantic* 249, no. 6 (June 1982): 31–37; *NYT,* 21 Oct. 1984, p. 32. I also draw on personal communications. In his second 1984 debate with Walter Mondale the President declined to repudiate the view.

17. Jacques Le Goff, *The Birth of Purgatory,* trans. Arthur Goldhammer (Chicago: University of Chicago Press, 1984), 7–11, 44. Donald B. Kraybill, *Facing Nuclear War: A Plea for Christian Witness* (Scottdale, Pa.: Herald, 1982), 264. See Griffin, *Endtime,* chaps. 3, 4. Rev. Royce Elms, 1982, quoted in A. G. Motjabai, *Blessèd Assurance: At Home with the Bomb in Amarillo, Texas* (Boston: Houghton Mifflin, 1986), 166; see also passim.

18. Among others, George S. Albee, "The Next Voice You Hear," *Cosmopolitan* 125 (Aug. 1948): 34–35. *The Next Voice You Hear* (film; Metro-Goldwyn-Mayer, 1950). Philip Wylie, *The Answer* (New York: Rinehart, 1955), first published in *Saturday Evening Post,* 7 May 1955. Morris West, *The Clowns of God* (New York: William Morrow, 1981).

19. The Pope's homily, 8 Jan. 1979, as quoted in Griffin, *Endtime,* 322.

20. Arthur C. Clarke, "If I Forget Thee, O Earth . . . ," reprinted in Clarke, *Across the Sea of Stars* (New York: Harcourt, Brace & World, 1959), 63–67. See, for example, Stewart Brand, ed., *Space Colonies* (Sausalito, Calif.: Co-Evolution Quarterly, 1977); Louis J. Halle, "A Hopeful Future for Mankind," *Foreign Affairs* 58 (1980): 1129–36.

21. Walter Hirsch, "The Image of the Scientist in Science Fiction: A Content Analysis," *American Journal of Sociology* 63 (1958): 506–512, reprinted in *The Sociology of Science,* ed. Bernard Barber and Walter Hirsch (New York: Free Press of Glencoe, 1962), 263. Arthur C. Clarke, *Childhood's End* (New York: Ballantine, 1953); see also Clarke, *2001: A Space Odyssey* (New York: New American Library, 1968), which also sold millions of copies.

22. Madeleine L'Engle, *A Swiftly Tilting Planet* (New York: Farrar, Straus & Giroux, 1978). *The Day the Earth Stood Still* (Twentieth Century-Fox, 1951).

23. For this and the following see David Michael Jacobs, *The UFO Controversy in America* (Bloomington: Indiana University Press, 1975); Edward J. Ruppert, *The Report on Unidentified Flying Objects* (London: Victor Gollancz, 1956); Robert Sheaffer, *The UFO Verdict: Examining the Evidence* (Buffalo, N.Y.: Prometheus, 1981), 4–12, 139–140.

24. Ruppert, *Report,* chap. 15. Daniel Lang, *The Man in the Thick Lead Suit* (New York: Oxford University Press, 1954), chap. 2; *NYT,* 7 July 1947, p. 5. Note "UFOs Buzz American Weapons Base," *National Enquirer,* 6 March 1984, p. 2.

25. C. G. Jung, *Flying Saucers: A Modern Myth of Things Seen in the Skies,* trans. R. F. C. Hull (1964; Princeton, N.J.: Princeton University Press, 1978), see 14, 22–33, 55–62.

26. For example, George Adamski, *Inside the Space Ships* (New York: Abelard-Schuman, 1955); *The UFO Incident* (Universal; NBC-TV, 1975). See Jacobs, *UFO Controversy,* chap. 5. Schizophrenic man is in the family of an acquaintance. Disney movie: *Return to Witch Mountain* (Buena Vista, 1978).

27. Sheaffer, *UFO Verdict,* 237 (entire sentence italicized in original); see his chap. 17.

28. Michael J. Carey, "Psychological Fallout," *BAS* 38, no. 1 (Jan. 1982): 23. *Pravda,* 7 Aug. 1964, p. 3, in *Current Digest of the Soviet Press* 16, no. 32, p. 17.

29. O. R. Frisch quoted in Margaret M. Gowing, *Britain and Atomic Energy, 1939–1945* (New York: St. Martin's, 1964), app. 5; William Laurence, *Men and Atoms: The Discovery, the Uses, and the Future of Atomic Energy* (New York: Simon & Schuster, 1959); War Department release in Henry D. Smyth, *Atomic Energy for Military Purposes* (Princeton, N.J.: Princeton University Press, 1946), app.; Leslie Groves in Martin Sherwin, *A World Destroyed: The Atomic Bomb and the Grand Alliance* (New York: Knopf, 1975), app. P; E. Fermi quoted in Stephane Groueff, *Manhattan Project: The Untold Story of the Making of the Atomic Bomb* (Boston: Little, Brown, 1967), 356; "O. E. Lawrence Thoughts," 16 July 1945, File 4, "Top Secret" Correspondence of the Manhattan Engineer District (RG 77), microfilm 1108, National Archives, Washington, D.C. Lansing Lamont, *Day of Trinity* (New York: Atheneum, 1965), 238–239; J. Kennedy quoted in Glenn T. Seaborg, "History of Met Lab Sections C-I, vol. 4, May 1945 to May 1946," Lawrence Berkeley Laboratory Publication 112 (Berkeley, Calif.: Lawrence Berkeley Laboratory, 1980), 111. Japanese correspondent quoted in Wilfrid Burchett, *Passport: An Autobiography* (Mel-

bourne: T. Nelson, 1969), 168. William A. Shurcliff, *Bombs at Bikini: The Official Report of Operation Crossroads* (New York: W. H. Wise, 1947), 155–158, 168.

30. Robert G. Wasson and V. P. Wasson, *Mushrooms, Russia and History*, 2 vols. (New York: Pantheon, 1957). Trinity test: works cited in n. 29 above.

31. Robert G. Wasson, *Soma, Divine Mushroom of Immortality* (New York: Harcourt Brace Jovanovitch, 1971). Joseph Needham, *Science and Civilization in China*, vol. 5, *Chemistry and Chemical Technology*, pt. 2, *Spagyrical Discovery and Invention: Magisteries of Gold and Immortality* (Cambridge: Cambridge University Press, 1974), 115, 121–123. Richard Evans Schultes and Albert Hofmann, *Plants of the Gods: Origins of Hallucinogenic Use* (New York: McGraw-Hill, 1979). Jean de Brunhoff, *The Story of Babar*, trans. Merle Haas (New York: Random House, 1933), 34; on dying kings see of course James G. Frazer, *The Golden Bough*.

32. R. G. Wasson quoted in Ralph Metzner, "Mushrooms and the Mind," in *Psychedelics: the Uses and Implications of Hallucinogenic Drugs*, ed. Bernard Aaronson and Humphry Osmond (Garden City, N.Y.: Doubleday, 1970), 107.

33. "Atomic Tests in Nevada" (Atomic Energy Commission, 1955). Albert Q. Maisel, "Medical Dividend," *Collier's* 119 (3 May 1947): 14.

34. The Moore sculpture is at the University of Chicago; I thank S. Chandrasekhar for an introduction. Moore quoted in *NYT*, 27 March 1966, sec. 2, p. 30.

35. Anatole France, *Penguin Island*, trans. A. W. Evans (New York: Dodd, Mead, 1909), 336. News conference quoted in Glenn T. Seaborg with Benjamin S. Loeb, *Kennedy, Khrushchev, and the Test Ban* (Berkeley: University of California Press, 1981), 199.

36. Harvey E. White, *Physical Review* 38 (1931): 513. A typical early Bohr atom is in Arnold Sommerfeld, *Atombau und Spektrallinien* (Braunschweig: F. Vieweg, 2nd ed., 1921), 272.

37. Americo Favale to Roland Anderson, 1 Aug. 1956. My thanks to Richard Hewlett and Roger Anders for digging this letter out of the AEC Patent Branch records, Department of Energy, Germantown, Md.

38. Carl G. Jung, *Mandala Symbolism*, trans. R. F. C. Hull (Princeton, N.J.: Princeton University Press, 1972). Heinrich Zimmer, *Myths and Symbols in Indian Art and Civilization* (New York: Harper & Brothers, 1962), 139–148.

39. Carl G. Jung, *Mysterium Coniunctionis: An Inquiry into the Separation of Psychic Opposites in Alchemy*, trans. R. F. C. Hull (Princeton, N.J.: Princeton University Press, 2nd ed., 1970), 463.

40. For example, the American Atheists Society of Austin, Texas, uses atom rings on its letterhead, presumably to symbolize rational science. "The Atom" (Western Publishing, 1962–1968); the character originated in "All-Star Comics" no. 3 (1940) and has remained active. "The Fury of Firestorm the Nuclear Man" (DC Comics, 1981-). William J. Reedy, *The Story of Salvation*, Life and Light Series, book 1, 2nd ed. (New York: W. H. Sadlier, 1969).

41. See also Ira Chernus, *Dr. Strangegod: On the Symbolic Meaning of Nuclear Weapons* (Columbia: University of South Carolina Press, 1986).

21. ARTISTIC TRANSMUTATIONS

1. For Brecht's *Galileo* see Gerhard Szczesny, *Das Leben des Galilei und der Fall Bertolt Brecht* (Frankfurt: Ullstein, 1966); Eric Bentley, *Theatre of War* (New York: Viking, 1972); Alan J. Friedman and Carol C. Donley, *Einstein as Myth and Muse* (Cambridge: Cambridge University Press, 1985), 176–177. Heinar Kipphardt, *In the Matter of J. Robert Oppenheimer,* trans. Ruth Speirs (New York: Hill and Wang, 1968), 146–164. Cf. Robert O. Butler, *Countrymen of Bones* (New York: Horizon, 1983); Thomas McMahon, *Principles of American Nuclear Chemistry: A Novel* (Boston: Little, Brown, 1981).

2. Philip Wylie, "The Answer," *Saturday Evening Post,* 7 May 1955. *Good Will to Men* (William Hanna and Joseph Barbera, Metro-Goldwyn-Mayer, 1955). William Golding, *Lord of the Flies* (London: Faber & Faber, 1954). See Bernard S. Oldsey and Stanley Weintraub, *The Art of William Golding* (New York: Harcourt, Brace & World, 1965), chap. 2.

3. Carl Jung, "On the Nature of the Psyche" (1947), in Jung, *Collected Works,* vol. 8, trans. R. F. C. Hull (New York: Pantheon, 1960), 159–234, see 218, 222.

4. Anne Frank, *The Diary of a Young Girl,* trans. B. M. Mooyaart (Garden City, N.Y.: Doubleday, 1952); Francis Goodrich and Albert Hackett, "The Diary of Anne Frank" (1955), in *Best American Plays: Supplementary Volume 1918–1958,* ed. John Gassner (New York: Crown, 1961). See Sidra DeKoven Ezrahi, *By Words Alone: The Holocaust in Literature* (Chicago: University of Chicago Press, 1980), 34–35, 200–204 and passim; Alvin H. Rosenfeld, *A Double Dying: Reflections on Holocaust Literature* (Bloomington: Indiana University Press, 1980), esp. chap. 8. Bruno Bettelheim, "The Ignored Lesson of Anne Frank," in Bettelheim, *Surviving and Other Essays* (New York: Random House Vintage, 1980), 246–257.

5. See Robert J. Lifton, *Death in Life: Survivors of Hiroshima* (New York: Simon & Schuster, 1967), chaps. 10, 11, and app. Lifton is an exception: Robert Jay Lifton, *The Nazi Doctors: Medical Killing and the Psychology of Genocide* (New York: Basic Books, 1986). For another important work on Hiroshima, later extended to include Auschwitz see John W. Dower and John Junkerman, eds., *The Hiroshima Murals: The Art of Iri Maruki and Toshi Maruki* (Tokyo: Kodansha; New York: Harper & Row, 1985). Yasunari Kawabata, *Thousand Cranes,* trans. Edward G. Seidensticken (New York: Knopf, 1959).

6. Melanie Klein, *The Psycho-Analysis of Children,* trans. Alix Strachey, rev. with H. A. Thorner (New York: Dell, 1975), 127n4. Bruno Bettelheim, *Truants from Life: The Rehabilitation of Emotionally Disturbed Children* (Glencoe, Ill.: Free Press, 1955), 68, 215–216, 260. R. D. Laing and A. Esterson, *Sanity, Madness, and the Family: Families of Schizophrenics* (Harmondsworth, Engl.: Penguin, 1970), 75; cf. R. D. Laing, *The Divided Self: An Existential Study in Sanity and Madness* (Harmondsworth, Engl.: Penguin, 1965), 12, 93.

7. Frank Sullivan, "The Cliché Expert Testifies on the Atom," in Sullivan,

A Rock in Every Snowball (Boston: Little, Brown, 1946), 31. Karel Čapek, *An Atomic Phantasy: Krakatit,* trans. Lawrence Hyde (1924; London: Allen & Unwin, 1948).

8. Aldous Huxley, *Ape and Essence* (New York: Harper, 1948). Russell Hoban, *Riddley Walker* (New York: Summit, 1980). Denis Johnson, *Fiskadoro* (New York: Knopf, 1985). See also Carol Sternhell, *Golden Days* (New York: McGraw-Hill, 1986).

9. *The Seventh Seal* (Svensk filmindustrie, 1956). Max von Sydow, interview in *Omni* 3, no. 5 (Feb. 1981), 120. Stig Björkmann, Torsten Manns, and Jonas Sima, *Bergman on Bergman: Interviews with Ingmar Bergman,* trans. P. B. Austin (New York: Simon & Schuster, 1973), 117.

10. *Hiroshima mon Amour* (Argos-Daiei-Como-Pathé, 1960); Marguerite Duras, *Hiroshima mon Amour,* trans. Richard Seaves (New York: Grove, 1961). Cf. *La Jetée* (Argos, 1963).

11. M.-L. von Franz in Carl G. Jung et al., *Man and His Symbols* (Garden City, N.Y.: Doubleday, 1964), 222.

12. *I Live in Fear* (*Ikimono no Kiroku*), also titled *Record of a Living Being* (Toho, 1955). Tim O'Brien, *The Nuclear Age* (New York: Knopf, 1985). Butler, *Countrymen of Bones.* See also, for example, Hans Koning, *Acts of Faith* (London: Gollancz, 1986) and the explicitly antitransmutational Susan B. Weston, *Children of the Light* (New York: St. Martin's, 1985) and Bernard Malamud, *God's Grace* (New York: Farrar, Straus, Giroux, 1982). Kurt Vonnegut, *Deadeye Dick* (New York: Delacorte, 1985); Vonnegut, *Galápagos* (New York, Delacorte, 1985), 8.

13. Denise Levertov, *The Poet in the World* (New York: New Directions, 1973), 121–122. Levertov, *Candles in Babylon* (New York: New Directions, 1982), 73. Cf. Allen Ginsberg, "Plutonian Ode" (1978) in Ginsberg, *Plutonian Ode: Poems 1977–80* (San Francisco: City Lights, 1982); Gary Snyder, "LMFBR," in *The Postmoderns: The New American Poetry Revised,* ed. Donald Allen and George F. Butterick (New York: Grove, 1982), 281.

14. Doris Lessing, "My Father," in Lessing, *A Small Personal Voice: Essays, Reviews, Interviews,* ed. Paul Schleuter (New York: Knopf, 1974), 93.

15. Doris Lessing, *The Four-Gated City* (New York: Knopf, 1969). See also Lessing, *The Golden Notebook* (New York: Simon & Schuster, 1962). Lessing, *A Small Personal Voice,* 65–66, 70. Useful are Roberta Rubenstein, *The Novelistic Vision of Doris Lessing: Breaking the Forms of Consciousness* (Urbana: University of Illinois Press, 1979); Mary Ann Singleton, *The City and the Veld: The Fiction of Doris Lessing* (London: Associated University Presses, 1977).

16. Lessing, *A Small Personal Voice,* 9–10, see 7.

17. Lessing, *A Small Personal Voice,* 59–60. Bruno Bettelheim, *Truants from Life: The Rehabilitation of Emotionally Disturbed Children* (Glencoe, Ill.: Free Press, 1955), 251–253. See Ira Chernus, *Dr. Strangegod: On the Symbolic Meaning of Nuclear Weapons* (Columbia: University of South Carolina Press, 1986), 22.

18. Alchemy: Singleton, *City and Veld,* 120–126. Lessing, *Four-Gated City,* 605–608, see pt. 3, chap. 4 and 396. On mutation symbolism cf. Paul Zindel,

The Effect of Gamma Rays on Man-in-the-Moon Marigolds: A Drama in Two Acts (New York: Harper & Row, 1971).

19. Especially Doris Lessing, *Shikasta (Canopus in Argos: Archives: 1. Re: Colonised Planet 5: Shikasta)* (New York: Knopf, 1979). See also the remarkable work by Ursula LeGuin, *Always Coming Home* (New York: Harper & Row, 1985).

20. An outstanding work of technological hope is Brian Stableford and David Lankford, *The Third Millennium: A History of the World: AD 2000–3000* (New York: Knopf, 1985); of integration, Ernest Callenbach, *Ecotopia: The Notebooks and Reports of William Weston* (Berkeley, Calif.: Banyan Tree, 1975).

CONCLUSION

1. Frank Sullivan, "The Cliché Expert Testifies on the Atom," in Sullivan, *A Rock in Every Snowball* (Boston: Little, Brown, 1946), 33–34.

2. See also Spencer R. Weart, *Scientists in Power* (Cambridge, Mass.: Harvard University Press, 1979).

A PERSONAL NOTE

1. David Lilienthal, *Change, Hope, and the Bomb* (Princeton, N.J.: Princeton University Press, 1963), 20; Lilienthal, *Atomic Energy: A New Start* (New York: Harper & Row, 1980).

2. See Chap. 13, n. 17.

3. Reasons for cutting this budget are explained by Daniel S. Greenberg, "A Hidden Cost of Military Research: Less National Security," *Discover* 8, no. 1 (Jan. 1987): 94–101.

Index

Index